PRAISE FOR *The Icarus Syndrome*

"A highly readable and useful hundred-year account of American ventures abroad that can serve as a path to understanding the past failures and uncovering why policy renewal is now proving so elusive. . . . Beinart usefully grapples with the practical impediments to making good policy."
—Leslie H. Gelb, *New York Times Book Review*

"A rollicking history. . . . Beinart is a wonderful storyteller. There's not much that he leaves out of *The Icarus Syndrome*. (It's exactly the book I wish I'd had when I was teaching American foreign policy.) Beinart has the writing chops to infuse the story with the dramatic tension and flair it deserves. . . . He sincerely believes that ideas matter in world politics."
—*Newsweek*

"*The Icarus Syndrome* is a confident and contentious history of more than a century of American foreign policy and its recurring tragic flaws. Agree or not with all of Peter Beinart's specific interpretations, one can only admire his effort to understand the cycles of modern American diplomacy and statecraft, and his timely warnings about the temptations of pride."
—Sean Wilentz, author of *The Age of Reagan*

"Impressive. . . . Mr. Beinart has produced an original and ambitious study."
—*The Economist*

"*The Icarus Syndrome* is a readable survey of 'America in the world' over the past hundred years. Nothing is more chilling than Beinart's catalog of the continuous, wrongheaded invocation of 'Munich' and 'appeasement.' "
—Geoffrey Wheatcroft, *New York Review of Books*

"A brilliant new book about the pendulum swings of U.S. foreign policy between excessive ambition and excessive retrenchment."
—*Los Angeles Times*

"Beinart's *The Icarus Syndrome* is not of course the first study of America's tendency toward imperial temptations, but it is one of the best, and certainly one of the best-researched and best-written. And since the cyclical pattern he traces seems to be as pertinent today as it was in the eras of Woodrow Wilson and Lyndon Johnson, this is very much a book with a message: a cautionary message to avoid hubris and to recognize the messy reality of world politics."
—Paul Kennedy, author of *The Rise and Fall of the Great Powers*

"Beinart is at his most illuminating when he lingers on forgotten episodes that reveal how difficult it is to understand the implications of any event at any given moment—the extent to which everyone is a prisoner of past failure or past success. . . . *The Icarus Syndrome* is history with an argumentative purpose."
—George Packer, *The New Yorker*

"With this perceptive and provocative book, Peter Beinart has given us a compelling argument about our times. *The Icarus Syndrome* does what works of history and journalism do at their very best: use the past to illuminate, in often stark and surprising ways, the challenges of the present. This is an important book."
—Jon Meacham, author of *American Lion*

"Beinart possesses the analytical skills of a seasoned historian. . . . He's a smart, reasoned political analyst who doesn't resort to hyperbole and hysteria when making a point. . . . The result is a book that's generally enlightening."
—*San Francisco Chronicle*

"Riveting. . . . Insightful. . . . Beinart is a gifted writer who really knows how to tell a story, and in this case the story itself happens to be endlessly fascinating. . . . What the book perhaps does best is to link foreign-policy behavior not only to the personalities of presidents and cabinet officials, but also to the embodied philosophical currents of the day and to the social and economic circumstances in which they reside."
—Adam Garfinkle, *National Review*

"Informative and engaging. . . . Beinart's book tackles a great deal of material in an approachable, yet never simplistic, way. . . . *The Icarus Syndrome* is a valuable addition to the public debate about the United States's ever-evolving role in the world."
—*Boston Globe*

"A century of unwise American military adventures is probed in this perceptive study of foreign policy overreach. . . . Beinart's analyses are consistently lucid and provocative. . . . The book amounts to a brief for moderation, good sense, humility, and looking before leaping—virtues that merit Beinart's spirited, cogent defense."
—*Publishers Weekly*

"Peter Beinart has written a vivid, empathetic, and convincing history of the men and ideas that have shaped the ambitions of American foreign policy during the last century—a story in which human fallibility and idealism flow together. The story continues, of course, and so his book is not only timely; it is indispensible."
—Steve Coll, author of *Ghost Wars*

"Why do we succumb to hubris? Peter Beinart has written a highly intelligent and wonderfully readable book that answers the question by looking at a century of American foreign policy. As with everything Beinart writes, it is lucid, thoughtful, and strikingly honest." —Fareed Zakaria, author of *The Post-American World*

"Energetically researched and entertainingly written, Peter Beinart's *The Icarus Syndrome* is both a fascinating intellectual history and an important coming-of-age parable about his generation's hard-learned lesson in the limits of American power."
—Jane Mayer, author of *The Dark Side*

· THE ·
ICARUS
SYNDROME

· THE ·
ICARUS
SYNDROME

A HISTORY
OF
AMERICAN
HUBRIS

PETER BEINART

HARPER PERENNIAL

NEW YORK · LONDON · TORONTO · SYDNEY · NEW DELHI · AUCKLAND

A COUNCIL ON FOREIGN RELATIONS BOOK

HARPER PERENNIAL

A hardcover edition of this book was published in 2010 by HarperCollins Publishers.

Published in cooperation with the Council on Foreign Relations.

The Council on Foreign Relations (CFR) is an independent, nonpartisan membership organization, think tank, and publisher dedicated to being a resource for its members, government officials, business executives, journalists, educators and students, civic and religious leaders, and other interested citizens in order to help them better understand the world and the foreign policy choices facing the United States and other countries. Founded in 1921, CFR carries out its mission by maintaining a diverse membership, with special programs to promote interest and develop expertise in the next generation of foreign policy leaders; convening meetings at its headquarters in New York and in Washington, DC, and other cities where senior government officials, members of Congress, global leaders, and prominent thinkers come together with CFR members to discuss and debate major international issues; supporting a Studies Program that fosters independent research, enabling CFR scholars to produce articles, reports, and books and hold roundtables that analyze foreign policy issues and make concrete policy recommendations; publishing *Foreign Affairs,* the preeminent journal on international affairs and U.S. foreign policy; sponsoring Independent Task Forces that produce reports with both findings and policy prescriptions on the most important foreign policy topics; and providing up-to-date information and analysis about world events and American foreign policy on its Web site, www.cfr.org.

The Council on Foreign Relations takes no institutional position on policy issues and has no affiliation with the U.S. government. All statements of fact and expressions of opinion contained in its publications are the sole responsibility of the author or authors.

FIRST HARPER PERENNIAL EDITION PUBLISHED 2011.

Designed by William Ruoto

Library of Congress Cataloging-in-Publication Data has been applied for.

ISBN 978-0-06-145647-3 (pbk.)

11 12 13 14 15 OV/RRD 10 9 8 7 6 5 4 3 2 1

To Ezra and Naomi

CONTENTS

PART III: THE HUBRIS OF DOMINANCE

· THE ·
ICARUS
SYNDROME

INTRODUCTION

"If you want war, nourish a doctrine. Doctrines are the most frightful tyrants to which men are ever subject."

—WILLIAM GRAHAM SUMNER

A story: the reason I wrote this book. I was sitting at a restaurant on New York's Upper East Side in early 2006, squeezed into a corner table next to an elderly, elfin man sipping a martini. He was Arthur Schlesinger, Jr.; I was petrified. I'd hero-worshipped the man since high school, when I developed a bout of Kennedy mania not uncommon among young liberals in Reagan-era Boston. I'd consumed his memoir of the Kennedy White House, and his cold war liberal manifesto, *The Vital Center*, and stacks of his op-eds. I didn't really understand why he was having lunch with me, and as far as I could tell, none of my halting efforts at flattery and political insight were validating his decision. He mostly wanted to gossip, but even there, I was out of my depth.

Then his tone changed: "Why did your generation support this war?" I began to stammer, something along the lines of "Well, of course, not everyone did; they weren't all as dumb as me." No laughter. "But yes, perhaps there was something in our experience that, on average . . ." Then he spilled his martini, and for the first time looked frail. Waiters swarmed; I was off the hook. We never spoke again. Less than a year later, he died.

But as one often does after flubbing a pregnant moment, I kept going over the scene in my mind, trying to formulate an answer worthy of the question and the questioner, something that convinced him I wasn't a fool. His assumption was right: Young people had supported the Iraq invasion at higher rates than their elders. According to an October 2002 Pew Research Center poll, Democrats under the age of thirty were almost as pro-war as *Republicans* over the age of sixty-five. A survey of bloggers, pundits, and op-ed writers revealed the same generational skew. And the reason, the culprit . . . suddenly it hit me: One of the culprits was him!

It was Schlesinger, after all, who had written in January 1991 that the Gulf War "will most likely be bloody and protracted," that it "will be the most unnecessary war in American history, and it may well cause the gravest damage to the vital interests of the republic." He may not have remembered that op-ed, but I did. I was a sophomore in college then, a dove, a worshipper of Arthur Schlesinger. On the eve of the first major war of my adult life, I was looking for guidance, and everywhere I looked, from my professors on up, the guidance was: Don't do it. Remember Vietnam. Tremble when you contemplate war. And so I trembled, and war came, and it was neither protracted nor bloody, at least for our side. In fact, the Kuwaitis greeted us as liberators, and it turned out that Saddam Hussein had been hiding a crash nuclear program, which might have reached fruition had he not been thrashed—all of which made me tremble, in retrospect, about what might have happened had we *not* gone to war.

Then came Bosnia. In 1993, with Bill Clinton trying halfheartedly to convince America's European allies to support air strikes against the Serb army that was raping and murdering Bosnia's Muslims, Schlesinger warned that "the arguments used today for intervention in Bosnia have disquieting echoes of the arguments used 30 years ago for intervention in Vietnam." *Vietnam!* The word made me shiver. But two years later, Clinton did launch air strikes, and then he sent peacekeepers, and there was no quagmire. On the contrary, the rapes and murders stopped. The barbed wire from the concentration camps came down.

Finally, in November 2001, early in the Afghan War, my hero put pen to paper yet again, this time insisting that the Bush administration's effort to depose the Taliban was proving futile. "Perhaps," he wrote, "they should have reflected on Vietnam." But by now I was sick of reflecting on it. It had never been my war, after all. It was a bogeyman, conjured by well-meaning but patronizing elders—in the way a grandparent who had witnessed the Depression might warn against investing too heavily in stocks. By 2003, when America invaded Iraq, my generation had witnessed its own history, and it was not a history that made one tremble at the prospect of war. I was tired of hearing people cry wolf. *Vietnam*—just hearing it invoked in a foreign policy debate made me roll my eyes.

So I had my answer to Schlesinger's question. And I imagined—in that puffed-up way you imagine when you're on a pedestal speaking to yourself—that he would have been unsettled. And then I had another thought, which unsettled *me*: My answer wouldn't have surprised him a bit. He had gone through the same process himself. As a college student in the late 1930s, he too had heard his elders invoke a traumatic war—World War I—as a reason not to fight again. At first he believed them, clinging to isolationism until Nazi troops reached French soil. But then he saw his elders proved wrong: For America, World War II was not a disaster; it was a triumph. And after the war, the triumphs continued. If 1989 through 2003 were our golden years, 1945 through 1965 were his. He too had been a child of the age of victory. And then Vietnam came, and he tasted tragedy, and politically he was never the same.

Now history's wheel had turned again. Another generation—mine—had seen so much go right that we had difficulty imagining anything going wrong, and so many of us grew more and more emboldened until a war did go hideously wrong. Maybe there was an argument here about cycles—Schlesinger loved cycles—cycles of success leading to hubris leading to tragedy, leading, perhaps, to wisdom. *Hubris.* The word stuck in my brain.

It comes from the ancient Greeks, who defined it as insolence toward the gods. In Aeschylus's play *Agamemnon*, the great warrior, fresh from his triumph at Troy, strides upon a lush purple robe, even though he knows that purple symbolizes divine power. Angered, the gods withdraw their protection, and Agamemnon's vengeful wife hacks him to death in the bath. In another Aeschylus drama, *The Persians*, a young king, Xerxes, inherits a mighty empire from his father, Darius. From the start, Xerxes is tormented by feelings of inferiority, constantly reminded that Darius's empire "was won at spearpoint / while he [Xerxes] not half the man / secretly played toy spears at home / and added nothing / to inherited prosperity." Determined to surpass his father's accomplishments, and emboldened by Persia's military might, he traverses the holy Bosporus, a body of water humans are not supposed to cross. Once again the gods retract their favor, and Persia's enemies massacre Xerxes' forces. Upon learning the news, Darius's spirit rises from the grave to bemoan "my son in his ignorance / his reckless youth. . . . Mere man that he is / he thought, but not on good advice / he'd overrule all gods."

In Greek literature, people sometimes create their own hubris: Like Agamemnon, they win epic triumphs and thus decide they are more than human. And sometimes, like Xerxes, they inherit their hubris from the triumphs of generations past. Take the legend of Icarus. Father and son are

trapped on the island of Crete. But Daedalus, a resourceful man, builds
wings of feathers and wax. Don't fly too low, he cautions his son, a light-
hearted youth named Icarus, or too high: "Keep to a middle range if you
can, and don't try to show off." At first, Icarus—heeding his father's warning—flies cautiously. But in his exhilaration, he gradually forgets Daedalus's words. As the Roman poet Ovid recounts the story, "the youngster's
initial fears have been mostly calmed. His confidence now has developed.
He wonders what he can do with this splendid toy, what limits there are
to his father's invention. He flaps his wings and rises higher—but nothing bad happens. He figures he still has plenty of margin and rises higher
still." Watching from earth, observers assume that this winged creature
must be a god. "It's exciting, wonderful fun, as he soars and wheels, but he
doesn't notice the wax of his wings is melting and feathers are falling out."
He has flown too near the sun. The wings crumble, revealing him to be
mortal after all. And he plunges to his death, into the sea.

There's nothing intrinsically American about hubris. As Aeschylus and
Ovid testify, it's a vice that long predates us. But since it's an affliction
born from success, we've been especially prone. Over the last one hundred years, no world power has come close to matching our run of good
fortune. When your cities are bombed and your lands are plundered and
your government is toppled and your empire dissolves—which is what
happened, in varying degrees, to America's major competitors during
the twentieth century—you have lots of problems, but hubris is no longer one of them. We, on the other hand, who did much of the bombing
and toppling and dissolving, have as a result sometimes been tempted to
believe—to paraphrase the campaign slogan of a certain Texas governor
turned president—that whatever Americans can dream, Americans can
do. A rousing sentiment, but dangerous, the kind of thing that can get you
into trouble with the gods.

   This book is about American hubris, American tragedy, and the search
for American wisdom. It's about three moments in the last century when
a group of leaders and thinkers found themselves in possession of wings.
In each case, the Icarus generations flapped gently at first, unsure how
much weight the contraptions could bear. But the contraptions worked
marvelously, and so people gradually forgot that they were mere human
creations, finite and fragile. Politicians and intellectuals took ideas that
had proved successful in certain, limited circumstances and expanded
them into grand doctrines, applicable always and everywhere. They took

military, economic, and ideological resources that had proved remarkably potent, and imagined that they made America omnipotent.

In critical ways, each generation was different. Woodrow Wilson and the pro-war progressives were more like Agamemnon; they mostly generated rather than inherited their hubris. Lyndon Johnson and the Camelot intellectuals of the 1960s, by contrast, and George W. Bush and the post–cold war conservatives, were more like Xerxes and Icarus: eager to surpass the accomplishments of the Daedalus generations that came before. The ideas were different, too. Wilson succumbed to the hubris of reason: the belief that America was wise and powerful enough to turn the jungle of international affairs into a garden. Johnson and the Camelot intellectuals (including, for a time, Arthur Schlesinger) fell in love with toughness: the idea that through unyielding force America could halt communism's march anywhere in the world. George W. Bush and his conservative supporters (with an assist from liberal hawks like me) grew entranced by the idea of dominance: the belief that America could make itself master of every important region on earth.

In each case, politicians and intellectuals built on what came before, pushing ideas further and further, until, like a swelled balloon, they burst. The problem wasn't the ideas themselves; in limited form, they had often served the nation well. In fact, ironically enough, they had emerged partly as correctives to the hubris of a prior age, before inflating themselves. It is tempting, in the wake of foreign policy disaster, to damn an entire intellectual tradition for what it ultimately wrought. Thus, since Iraq, some liberals have projected back into history a unified conservative foreign policy lineage, always pregnant with the sins that the Bush administration would bring forth into the world. But I've grown distrustful of this way of thinking, which imagines that there are stained bloodlines and pure ones, and grown more partial to Reinhold Niebuhr's suggestion that in foreign policy, as in life, evil often grows from unchecked good. Harry Truman's efforts to contain communism in Europe were wise, even though they helped plow the ideological soil for Vietnam. Bush the father was right to fight the Gulf War, even though it helped breed the confidence that led his son to overthrow Saddam. Hubris is not the possession of any one party or intellectual tradition; it is any intellectual tradition taken too far. And foreign policy wisdom sometimes consists of understanding that the very conceptual seedlings you must plant now can, if allowed to grow wild, ravage the garden.

But how do you know when the former sapling, which once bore tasty fruit, is starting to strangle the other plants? In Greek literature, the gods don't tell you. Humans are simply expected to know. And to make matters even trickier, excessive ambition is not the only human pitfall; insufficient ambition can be deadly, too. Daedalus didn't just warn Icarus not to fly too high; he also warned him not to fly too *low*. In fact, Daedalus was, in his way, quite a risk-taker himself. He believed that he and his son could fly, which in itself might have been considered a transgression against the divine. The Greeks could have spun a legend that culminated with his death as well, thus implying that he should have remained earthbound and imprisoned. But they didn't, even though they had no evidence that humans could fly. They imagined it, and because of human ambition, now we can.

So where does ambition end and hubris begin? There's no formula for answering that. In fact, the belief that you've discovered a formula that works in all situations is itself a sign that you've crossed the line. To some degree, foreign policy is all about deciding in which direction you'd rather be wrong. Are you so intent on making sure America doesn't fly too high that you oppose not only invading Iraq, but saving Kuwait? Are you so determined to avoid flying too low that you support not merely World War II but Vietnam? Barely anyone will be right every time, because the gods don't speak to us. Or, as Warren Buffett has said about investing in a bull market, it's like Cinderella at the ball. She knows that if she stays too late her chariot will turn into a pumpkin and her gown will turn to rags. But she doesn't want to leave too early and miss meeting Prince Charming. The problem is that there are no clocks on the wall.

No clocks, but there are warning signs, the kind that someone with lots of experience at balls might notice. One warning sign is overconfidence: a political climate in which influential people assume that the war's outcome is preordained; that because of America's military prowess or economic resources or ideological appeal, we cannot possibly lose. When influential Americans talk that way, it's usually because America has not lost in a long time. We're all prisoners of analogy, and people tend to imagine that the future will be like the immediate past: that Vietnam will be like Korea or World War II; that Iraq will be like the Gulf War or the initial phase of Afghanistan. But that reliance on analogy often blinds us to the ways in which, rather than replicating the successes of the past, we are reaching beyond them, taking on more risk. We think we are running on a treadmill when we are actually ascending a ladder.

If overconfidence is one danger sign, unilateralism is another. France, Canada, and Britain, which had fought alongside the United States in World War II and Korea, refused to join us in Vietnam. France, Canada, and Germany, which had supported the United States in the Gulf War, the Balkans, and Afghanistan, opposed invading Iraq. The problem with failing to convince other countries—and particularly the Western democracies that broadly share our values and interests—to back us in war is not necessarily that we need their help to win. (These days, given the vast gap between our military capacity and theirs, they may be as much a burden as a help.) The problem is that their unwillingness to back us may be evidence that winning is impossible. What we need, in other words, is not our allies' tanks but their judgment. It's like a patient contemplating a high-risk medical procedure: You may not need several doctors to perform the operation, but you want several doctors to confirm that the operation can be successfully performed at all.

The sober judgment of allies is especially important for a nation intoxicated with success. (As the old saying goes, when three friends say you're drunk, lie down.) Americans sometimes dismiss our European allies as defeatist, so burdened by their tragic histories and so enfeebled by their military weakness that they instinctively choose retreat over confrontation. But it is precisely because they have been battered by history that they may be able to spot hubris when we, because of our more triumphant experience, cannot. Before Vietnam, and again before Iraq, French leaders urged the United States to learn from France's imperial misfortunes in Southeast Asia and the Arab world. But Lyndon Johnson and George W. Bush scoffed at the idea that we had anything to learn. The French, after all, were history's losers. We were its winners. They were mere mortals; we were America.

A third flashing light is excessive fear. Even when America's leaders fly nearest the sun, they generally insist that they are merely taking defensive measures against grave threats. Hubris rarely speaks its name. And to some extent, those leaders are right: Foreign threats usually have something to do with America's decision to launch a war. Woodrow Wilson would not have taken America into World War I just because he believed he could reorder world politics; German subs were sinking our ships. The United States would not have fought the Vietnam War had Ho Chi Minh not been a communist, no matter how confident we were of our military might. But the problem with explaining America's wars solely as a response to threats is that our threat perceptions vary wildly over time. Things we take in stride at one moment terrify us at another. In 1939, few American politicians believed that a Nazi takeover of Warsaw constituted a grave danger to the United States. By 1965, many believed we couldn't live with a North Vietnamese takeover of Saigon. In the 1980s, Americans lived peacefully, albeit anxiously, with thousands of Soviet nuclear warheads pointed our way. By 2003, many Washington commentators claimed that even Iraqi biological or chemical weapons put us in mortal peril. How threatened American policymakers feel is often a function of how much power they have. The more confident our leaders and thinkers become about the hammer of American force, the more likely they are to find nails.

That's the problem with explaining Iraq merely as a response to the terrorist attacks of September 11, 2001, as the Bush administration did. Obviously, 9/11 mattered: Without it, we would not have sent more than one hundred thousand U.S. troops into Iraq. But had 9/11 occurred in 1979 or 1985 or 1993, moments when America was less confident that its military was unstoppable, its economic resources plentiful and its ideals universal, it is unlikely that we would have responded by launching two distant wars in short succession, something the United States had never done in its entire history. In fact, even in 2002, had the war in Afghanistan not initially gone well, thus further buoying the Bush administration's self-confidence, America probably would not have invaded Iraq. And had Bush officials doubted that they could successfully invade and remake Iraq, they probably would not have declared Saddam an urgent and intolerable threat. America's leaders tend not to tell us we are in grave danger unless they think they can do something about it.

This is not to say it's always a mistake for the United States to wage war in places it once considered irrelevant. Over the course of the last century, as America has risen to global power, we have naturally come to worry about regions of the world to which we previously gave little thought. But when we allow our fears to swell so dramatically that quelling them would require virtually unlimited quantities of money and blood, something has gone wrong. Once Lyndon Johnson declared a communist takeover in backward and remote Vietnam to be a grave threat to the United States, then a communist takeover in virtually any country constituted a grave threat. Once the Bush administration said America was in mortal danger because Saddam was supporting terrorism (not even terrorism against the United States, just terrorism) and seeking weapons of mass destruction, America was suddenly in mortal danger from Iran, North Korea, Syria, and perhaps Pakistan, too.

A wise foreign policy starts with the recognition that since America's power is limited, we must limit our enemies. That's why Franklin Roosevelt hugged Stalin close until the Soviet Union had helped us defeat Nazi Germany, and why Richard Nixon opened relations with China, so the United States wasn't taking on Moscow and Beijing at the same time. By contrast, when America's leaders outline doctrines that require us to confront long lists of movements and regimes simultaneously, it's a sign that we've lost the capacity to prioritize. And that, in turn, is a sign that we think we are so powerful that we don't need to prioritize. And that, in turn, is a sign that we're flying too high.

These warning signs are a starting point. But in and of themselves, they are too crude. Because people are not particles, history is not physics. History does contain cycles—attitudes, behaviors, and outcomes that rhyme—but every moment is also unique and every event has multiple causes. Telling the story of the last century of American foreign policy as cycles of success leading to hubris leading to tragedy leading to wisdom, and then leading to more success, is like looking at a page covered with tiny, fuzzy dots and graphing the places they rise and fall. Reality is far messier than your smooth lines suggest. Others can map the evidence a different way.

For policymakers, the challenge is even tougher: to predict, on the basis of the dots you've already seen, where the next ones will fall. To do that well requires a trained eye. Just as a skilled radiologist can see patterns on an X-ray that may elude a medical student, the people who make foreign policy must draw meaning from a hazy landscape. In this effort, two kinds of knowledge can help. The first is a broad knowledge of history: not just American history, but the history of other great powers that in their day felt history's wind at their backs. This knowledge can provide a wider menu of analogies than the ones offered by America's recent past. Maybe Vietnam had more in common with the Spanish-American War than with Korea? Maybe our invasion of Iraq was more like Napoleon's invasion of Egypt than like the Gulf War? That American hawks relentlessly compare our adversaries to Hitler, and American doves relentlessly compare our wars to Vietnam, suggests that for much of our political class, the menu of available historical analogies is depressingly small.

If a broad sense of history can help policymakers navigate the fog, so can an intimate feel for one specific place. It is ironic that in recent years many American foreign policy commentators and practitioners, hoping to be the next George Kennan, have tried to formulate a universal foreign policy doctrine for the post–cold war, or post-9/11, age. In reality, Kennan, like William Graham Sumner, hated and feared universal doctrines. And after he saw containment swelling into one, starting in the late 1940s, he spent the next four decades screaming himself hoarse in opposition. What Kennan believed in, and what his writing on the Soviet Union exemplified, was the value of deep, local knowledge. By the time he wrote his famous "Long Telegram" from Moscow in February 1946, Kennan was on his third Russian posting and had also served in Riga and Tallinn, on Russia's northwest border. He spoke formal Russian better than Stalin (whose native language was Georgian), was writing a biography of the Russian writer Anton Chekhov, and would later author a prizewinning account of American foreign policy during the Russian Revolution. It was his profound understanding not only of Russia's politics but of its history and culture that allowed Kennan to conceive of containment, a strategy he developed not for dealing with "communism"—an abstraction he thought had little foreign policy meaning in and of itself—but for dealing with one communist regime at one moment in time.

In post–World War II America, where many saw the Soviet Union through the prism of America's last totalitarian foe, Nazi Germany, Kennan showed why the present was unlike the past; why Stalin, unlike Hitler, could be stopped by measures short of war. Like Daedalus, he crafted a set of conceptual wings appropriate to the specific problem at hand. And because he had built the wings, he could see when they were starting to melt.

It is no coincidence that a generation later, when the Johnson administration waded into its Indochinese bog, there were no high-level policymakers whose knowledge of Vietnam remotely rivaled Kennan's knowledge of Russia. On the contrary, America's best Asia specialists—people like Kennan's friend John Paton Davies—had been hounded from government by Joseph McCarthy, in part for daring to suggest that Asia's communist parties might not be clones of their Soviet counterparts. Similarly, there were no high-level policymakers in the Bush administration whose knowledge of Iraq approached the Kennan standard, either. Partly as a result, after 9/11, Bush and his supporters implied that all Middle Eastern terrorist movements and terror-supporting regimes were basically the same. They lumped together Iraq, Iran, Syria, Al Qaeda, Hezbollah, Hamas, and the Taliban into a single ideological construct called "jihadist terrorism" or "Islamofascism" and then outlined a set of responses—preventive war and coercive regime change—that they proudly equated with containment. The irony was that the Bush administration's one-size-fits-all approach to fighting terrorism *was* like containment: not Kennan's narrowly tailored strategy for containing Stalin's U.S.S.R., but Johnson's hubristic effort to contain communism everywhere in the world.

Having implied that national differences were irrelevant, the Bush administration then proceeded to appoint men too ignorant to spot them anyway. In May 2003, in the crucial first months after the overthrow of Saddam, Bush sent a man to run Iraq, L. Paul Bremer, who had never before been posted to the Arab world. To grasp the intellectual chasm between American foreign policy toward the U.S.S.R. in 1946 and American foreign policy toward Iraq in 2003, one need only try to envision Bremer writing a biography of an Iraqi writer, or, for that matter, being able to name one.

In important ways, America's wars in Afghanistan and Iraq echo the Icarus tale. But they differ in one crucial respect: Icarus dies from his hubris; America does not. In this sense, we are more like Xerxes in *The Persians,* who survives the consequences of his arrogance and returns home in rags. Tormented by the chorus, which demands an accounting for all the great soldiers who now lie dead, Xerxes at first wallows in self-pity. He blames the gods; he declares himself cursed; he says he wishes he too had died. But then he puts on a new robe, and rather than suffering the chorus' abuse, he begins to direct its actions. "We have been shattered by such a fate," he declares, but then tells the wailers to go home and prepare for the future.

As the play ends, there is at least the hint that although Xerxes has been transformed by tragedy, he has not in fact been shattered by it. He no longer feels sorry for himself. He can still lead his people. He turns their attention to the possibilities of a new day. Robert Kennedy, who like Arthur Schlesinger was transformed by Vietnam, and who late in life developed a fascination with the Greeks, once said that "tragedy is a tool for the living to gain wisdom, not a guide by which to live." If the last few years have acquainted a new American generation with tragedy and made us less inclined to believe that the course of history—or the power of the gods—is on America's side, we should see that as a cause not for despair, but for hope. For it is only when we abandon our false innocence and unearned pride that we can look around at the bits of feathers and wax that still surround us, and begin, carefully, to build wings.

PART I

———

THE HUBRIS
OF REASON

CHAPTER ONE

A SCIENTIFIC PEACE

Woodrow Wilson had very few friends, and that bothered him. People considered him "cold and removed," he groused. He wished journalists would write about his lighter side—his love of baseball, his gift for mimicry, his flair for limericks—but instead they depicted him as a bloodless "thinking machine." He longed for a nickname. Perhaps if he had kept his birth name, Thomas, he mused, people would call him "Tommy" and thus find him more approachable. Theodore Roosevelt was known as "Teddy," and no one ever called *him* cold and removed.

The problem was that while Wilson loved the idea of friends, he wasn't very friendly up close. One reporter compared his handshake to "a ten-cent mackerel in brown paper." A Baltimore ward leader said the president "gives me the creeps. The time I met him, he said something to me, and I didn't know whether God or him was talking." Wilson knew his reputation as aloof was partly his own fault. "I have a sense of power in dealing with men collectively which I do not feel always in dealing with them singly," he acknowledged, which may have had something to do with the fact that when he dealt with them singly, they often talked back.

There was one big exception, a genuine, honest-to-goodness presidential friend: Colonel Edward House. The two men took drives together, gossiped in their nightgowns, read poetry aloud to one another, shared secrets of their love lives. House, remarked Wilson, "is my second personality. He is my independent self. His thoughts and mine are one."

For Wilson, House had three endearing characteristics. First, he shared the president's background, having been raised in the genteel South. Second, he was a world-class sycophant. Third, he did the things that Wilson needed done but didn't like doing himself. The "Colonel" was not actually a military man at all; he was a fixer, highly skilled in the dark arts of patronage, conspiracy, and intimidation. "He could walk on dry leaves," remarked Oklahoma Senator Thomas Gore, "and make no more noise than a tiger." He always made you feel "intimate," noted another observer, "even when he was cutting your throat."

Woodrow Wilson, a politician with little patience for the more corporeal elements of his craft, needed someone like that, and he rewarded House with both affection and power. For a long spell in the middle of Wilson's presidency, House wielded more influence over American foreign policy than the secretary of state and the secretary of war combined. He was a kind of ambassador without portfolio: Wilson's emissary to the actually existing human race.

Walter Lippmann also seduced the powerful, but he did it through the front door. Disappointed by his own father, he acquired others, becoming the brilliant son that great men felt they deserved. At nineteen, as a Harvard junior, he took weekly tea with one world-renowned philosopher, William James, and gossiped over dinner with another, George Santayana. After graduating, he apprenticed for the famed muckraking journalist Lincoln Steffens. His first book, written at age twenty-three, drew praise from Sigmund Freud. After his second, published the following year, Theodore Roosevelt declared him "the most brilliant young man of his age in the United States." By 1916, at age twenty-seven, he was dining at the White House, and his editorials for a new progressive journal, the *New Republic*, were finding their way into Wilson's speeches. In the spring of 1917, while his contemporaries were being shipped across the Atlantic to the charnel house of World War I, he procured a draft exemption from Secretary of War Newton Baker, who was accumulating whiz kid assistants. Then, a few months later, as a damp Washington summer gave way to a frigid fall, Baker called Lippmann into his office. Colonel House wanted to meet him outside.

The fixer and the wunderkind began to walk, past Pennsylvania Avenue and the White House, and down Seventeenth Street toward the Potomac. In the distance spread the marshland that would become the National Mall, and beyond that the partially completed Lincoln Memorial, its white columns already standing, but empty of the seated figure who would reside within. As they walked, House described a project of the most fearsome secrecy. America had been at war for only six months, but already Wilson was designing a peace to redeem the slaughter. Three million Russians and five million Germans were either wounded or dead. In France, four thousand villages had simply ceased to exist. The European state system, long anchored by four vast overland empires, based in Berlin, Moscow,

Vienna, and Istanbul, was imploding. Like most Americans, Wilson had hoped against hope that the United States could escape the widening gyre. He had resisted fierce pressure to join the fray, even after hundreds of U.S. civilians were drowned by German subs. On the night he finally asked Congress to declare war, he returned to the White House and wept. But now that America had joined the battle, Wilson was determined to build a new world on the ruins of the old, to ensure that barbarism never overthrew civilization again.

When the slaughter stopped, the world would need a new map, which gave every nation the territory it deserved and not an inch more. And it would need laws, so that international affairs no longer followed the law of the jungle. For Wilson, it fell to America—which he considered the sole disinterested combatant, the only nation that wanted nothing out of the war save that it not happen again—to draw up the plans. And House was tapping the country's best minds for the task. He told Lippmann to join a group of experts working from a secret office in the New York Public Library. To disguise the project's import, they would call it simply "the Inquiry." From the ashes of war they would construct what House called a "scientific peace."

A scientific peace. The concept came straight from the belly of the progressive movement. Progressivism, historians insist, was not one thing. It was a swarm of impulses and interests, often colliding with each other. But if progressivism was an ideological cacophony, one note cut through the din: faith in human reason. This faith predated World War I. Indeed, it grew from progressivism's success in remaking the United States. Surveying their country at the dawn of the twentieth century, many progressives had seen a society hurtling toward a second civil war. Industrialization, urbanization, and immigration, they believed, were dividing the nation against itself, making it a battlefield of hostile tribes—business versus labor, urban versus rural, immigrant versus native-born—each trying to smash each other into submission, each trampling democracy along the way. It was up to government to impose order, not through brute force—then it would be just another selfish tribe—but through the force of reason. Government would provide answers to society's conflicts: answers so disinterested, so rational, so self-evidently fair that the tribes would lay down their arms. It was no accident that many progressives saw a scientific peace as the answer to Europe's war. For close to two decades, they had been fashioning a scientific peace inside the United States.

Rational, disinterested policies required rational, disinterested policy-makers: experts. And experts occupied a hallowed place in the progressive mind. Lippmann's boss at the *New Republic,* Herbert Croly, a homely, taciturn man—alternately likened to a yogi and a crab—suggested entrusting them with a fourth branch of government. The eccentric Norwegian-American economist Thorstein Veblen proposed handing them control over the economy. And no one gave the concept grander expression than House himself. Five years earlier he had anonymously published a strange little novel, in which a young "mastermind" named Philip Dru, watching the United States sink into civil war, seizes power and declares himself "Administrator of the Republic." He replaces Congress with a commission of five expert lawyers who decree all the high-minded reforms that selfish interests have long stymied. With order and reason thus enthroned in the United States, Dru sails off into the Pacific with his bride, determined to bring a scientific peace to the world beyond America's shores.

Philip Dru: Administrator was a lousy book, but a revealing one. Dru is not merely an expert. He is also an inspirational leader, able to rally the public to his cause. And as a great leader, he is largely an educator. While they revered experts, the progressives knew it took leaders to convince— to educate—the public to accept their conclusions. "The nation," wrote Croly, "like the individual, must go to school." And if the nation was a schoolhouse, the president was its schoolmaster. It was no coincidence that John Dewey, progressivism's most influential democratic theorist, was also a philosopher of education. To make democracy work, leaders had to educate Americans to embrace—indeed, to demand—the rational answers formulated on high.

This, then, was the essential progressive equation: Objective experts plus inspiring leaders plus educated citizens equal a society governed by reason, not force. And underlying it all was the most basic faith of all: in humanity itself. Against the late-nineteenth-century Social Darwinists who declared people fundamentally selfish, progressive thinkers like Dewey, Veblen, and the historian Charles Beard insisted that they were naturally generous and cooperative; it was the evil of anarchic capitalism that made them act like beasts. Against the theologians who saw humanity forever maimed by original sin, Walter Rauschenbusch and the champions of Social Gospel Christianity insisted that evil resided in the world, not in man.

Progressivism's great discovery, declared the historian Christopher Lasch, was the "lost innocence of the race." Progress was possible because deep inside, people were better—much better—than the world in which they lived. It was a stirring faith, nurtured by the progressive movement's success in bringing reason and order to the United States. And at a fevered moment at the height of World War I, after the planet itself had erupted in civil war, Woodrow Wilson decided to export it—via the sword—to the entire world. For a glittering constellation of American intellectuals, it was the crusade of a lifetime, a high-water mark of American optimism that would not be reached again until our own time. And it was shot through with hubris, hubris that America could make politics between nations resemble politics between Americans, hubris that the progressives could build a world governed by reason when, as it turned out, they weren't always that reasonable themselves. It was a hubris from which America would not fully recover until Wilson and House were dead, Lippmann was a weary, humbled man, and the world had endured more catastrophe than their innocent minds could even imagine in the brisk and intoxicating fall of 1917.

By the time House dispatched Lippmann to help draft a progressive charter for the postwar world, progressivism had already transformed public life in the United States. And it was because of their success rationalizing government at home that reformers dared imagine they could rationalize the entire globe. Until 1900, the tribes—especially the mega-corporations, or "trusts," which had amassed vast wealth since the Civil War—had dominated the impotent, corrupt federal government. But as the new century dawned, impartial experts and heroic leaders began to seize command.

They didn't come much more heroic than Theodore Roosevelt. A war hero, a boxer, a ferocious reader with a photographic memory, and a big-game hunter who did his own taxidermy, America's twenty-sixth president was also the author of a four-volume history of the American West; biographies of Gouverneur Morris, Thomas Hart Benton, and Oliver Cromwell; the standard work on the Naval War of 1812; a book on child rearing; and several on the birds of New York state. As deputy sheriff of Medora, North Dakota, in the 1880s, he had captured three men who tried to steal his boat, and guarded them single-handedly for forty hours, keeping himself awake by reading Tolstoy.

Roosevelt inaugurated the progressive age. In May 1902, 150,000 soot-covered men emerged from deep inside the earth in eastern Pennsylvania. They were anthracite coal miners, men at constant risk of suffocation, asphyxiation, or explosion, and they were on strike in pursuit of an eight-hour workday, a 20 percent wage hike, and recognition of their union. Past strikes had usually been resolved with buckshot, as the trusts and their government lackeys bloodied workers into submission. But Roosevelt didn't want a class war; he wanted a scientific peace, a settlement based on reason, not force. So he created what would become the quintessential progressive entity: an expert commission. Two engineers, a business specialist, a union leader, a judge, a priest, and the federal commissioner of labor heard testimony from 558 witnesses. They compiled statistics on the frequency of mine accidents, the quality of company-sponsored medical care, and the number of children working underground. In the end they gave the miners almost exactly half of what they asked for, along with a permanent commission to settle future disputes—all without anyone getting shot.

Using the coal settlement as a model, Roosevelt then proposed a Bureau of Corporations to gather data on business practices, and a Bureau of Labor to do the same for working conditions. When conservative senators balked, he took his case to the people, telling cheering crowds that selfish corporations must submit to rational control. In 1906, he signed the Hepburn Act, which empowered another group of experts—the Interstate Commerce Commission—to hold hearings on railroad pricing, and in case of abuse, set rates itself. Meanwhile, Upton Sinclair's bestseller about the revolting conditions in Chicago packing houses, *The Jungle*, was sparking a furor over the safety of American meat. Roosevelt invited Sinclair to the White House, then sent government investigators to confirm his findings. When they did, Roosevelt muscled through the Meat Inspection Act, which authorized Agriculture Department experts to ensure that America's chickens, hogs, and cows were not poisoning America's people, and to shut down slaughterhouses if they did.

From coal to railroads to beef, this was the ethic of reason sprung to life. Experts and muckraking journalists gathered data about the irrationality and injustice tearing America apart. A charismatic leader used it to educate the people, building a mighty tide of public opinion that overwhelmed the selfish tribes. And thus empowered by an active and reasoned populace, the leader built a permanent machinery of investigation, creating a virtuous cycle of objective information, public education, and scientific peace.

But these achievements were a mere warm-up for what progressives would accomplish under their greatest White House champion, the man with few friends but epic dreams, Thomas Woodrow Wilson. Even more than other progressives, Wilson feared a second civil war, mostly because he remembered the first one. His earliest memory was of standing outside his family's home in Augusta, Georgia, at age four and hearing people yell that Abraham Lincoln had been elected and there would be war. When war came, he watched Confederate soldiers moaning on gurneys inside his father's church, which had been turned into a makeshift hospital, and he peered at the Union soldiers milling in the courtyard outside, which had been turned into a makeshift jail. The war split the Wilson family asunder; its northern and southern branches never spoke again. "A boy never gets over his boyhood," Wilson later declared, and his left him with a gnawing anxiety that order might break down once more.

Wilson never truly identified with either Blue or Gray. In his mind, they were both selfish tribes. Even as a boy, he tried to stand outside the conflict, to find an impartial, rational vantage point. When he played soldier, his imaginary armies flew neither the Stars and Stripes nor the Stars and Bars, but rather the Union Jack. As he grew older, Wilson came to idolize Lincoln, not because Lincoln had led the North to victory, but because, in Wilson's mind, he had stood above the parochial interests of both sides and thus restored unity to a divided nation. Lincoln "was detached from every point of view and therefore superior," Wilson said of his political hero. "You must have a man of this detachable sort."

In his own mind, Wilson was such a man. He considered himself detached from narrow regional interests because although born in the South, he had spent his adulthood in the North and so understood both worlds. And he considered himself detached from America's looming second civil war—between rich and poor—because he was neither a worker nor a capitalist; he was a scholar. He believed government could be "reduced to science," a science in which he had particular expertise, since his own writings had helped establish public administration as an academic discipline in the United States. In 1910, when he ran for governor of New Jersey, his first foray into electoral politics, Wilson boasted that as the president of Princeton University he had run an institution dedicated to the production of unbiased expertise.

The job of impartial experts, in Wilson's view, was to write impartial rules. And in their grandest form those rules took the form of constitutions. Constitutions were precious to Wilson. The son and grandson of Presbyterian ministers, he was reared in a religious tradition that placed special emphasis on the covenant between God and his people. For

Wilson it was this basic compact, writes one biographer, that "imposed a comprehensible pattern—orderly, predictable, and permanent—upon the transient character of human affairs." Wilson saw the Bible as the greatest constitution of all, "the 'Magna Charta' of the human soul."

It was fitting, in Wilson's view, that the Civil War had ended in a re-written American constitution, a new, fairer set of rules under which the entire country could peaceably live. And in his own life, Wilson drafted constitutions wherever he went. As a kid, he wrote one for his local base-ball league. When he turned seventeen, he founded an imaginary yacht club and wrote a charter for it as well, complete with bylaws, regulations, and penalties for those who disobeyed. As an undergraduate at Davidson College, he transcribed the debating society's constitution by hand. At Princeton, where he transferred, he tinkered with the constitution of not one debating society, but two. From the University of Virginia, where he studied law, to Johns Hopkins, where he got his Ph.D., to Wesleyan, where he taught, it was more of the same. Like a legalistic Johnny Apple-seed, Wilson traveled from place to place replacing disorder with order, unfairness with fairness, anarchy with law. He even recommended to his wife that they draft a constitution for their marriage. Let's write down the basic principles, he suggested; "then we can make bylaws at our leisure as they become necessary." It was an early warning sign, a hint that perhaps the earnest young rationalizer did not understand that there were spheres where abstract principles didn't get you very far, where reason could never be king.

If Wilson's devotion to order and reason was classically progressive, so was his faith that leaders could educate people to want them. It was no coincidence that so many of his constitutions were written for debate societies. "Statesmen," Wilson declared, "must possess an orator's soul," and he certainly did: He was among the greatest orators in the history of American politics. If Wilson believed he was impartial, he believed just as strongly that his words could make ordinary people impartial, too. He possessed, his press secretary explained, "an almost mystical faith that the people would follow him if he could speak to enough of them." He knew that people sometimes believed irrational, selfish things. He just didn't think they would continue to believe them after listening to him.

Nothing in Wilson's early years as president shook that faith. On the contrary, his already considerable self-confidence swelled as he compiled one of the most dazzling first terms in American history. A month after taking office, he put his oratorical gifts to use, becoming the first president since John Adams to address Congress in person. His topic was the tariff on imports, which had more than doubled since the Civil War. In Wilson's view, the tariff was irrational: Many of the sheltered industries were perfectly capable of competing with their foreign competitors. They enjoyed government protection not because it served the public interest, but because they had members of Congress in their pockets. Outraged by Wilson's proposal, the selfish tribes descended on Washington to preserve their irrational advantage. But Wilson met them head-on. In a dramatic appeal to the American people, he denounced the "insidious" lobbyists seeking "to overcome the interest of the public for their private profit." Progressive senator Robert La Follette then shrewdly launched an investigation into whether the senators who opposed reform were benefiting personally from particular tariffs. Humiliated by what La Follette's investigation uncovered, several flipped their vote and the tariff bill passed. "I did not much think we should live to see these things," exclaimed Wilson's ecstatic secretary of agriculture. Three years later tariff rates were entrusted to an expert commission, ostensibly shutting out the selfish tribes forever.

From there, Wilson turned to the banks, which, according to a muckraking investigation by progressive House Democrats, used political rather than scientific criteria to make loans. The result was a Federal Reserve Board staffed with presidentially appointed expert economists to oversee America's regional banks. In the space of eight months, Wilson had passed the first tariff reduction in almost twenty years and the first banking reorganization in almost fifty.

Eleven months later, he passed the third jewel of his progressive agenda: a Federal Trade Commission empowered to investigate business fraud and abuse, ban unfair practices, and force the offenders to compensate their victims. And when Congress reconvened in December 1915, it passed seven smaller pieces of progressive legislation, including the creation of a Shipping Board to ensure safe working conditions for sailors, government loans to struggling farmers (the recommendation of a Wilson-appointed expert commission), a ban on child labor, workers' compensation for government employees, and an eight-hour day for railroad workers, which averted a potentially bloody nationwide strike.

Years later, after the war had made *progressive* an epithet—when Wilson was a pariah and reform lay flat on its back—critics would look back at these initiatives and see not impartial reason, but selfish interest in disguise. Entrenched industries learned how to turn the bodies designed to oversee them to their own advantage, often by helping to write regulations with which their smaller competitors could not afford to comply. No matter how tightly the progressives tried to lock the door, the inequities of power usually found a way in.

And the corruption did not only come from outside; the progressives themselves were not as disinterested as they liked to believe. Many of the experts appointed to the Tariff Commission, the Federal Trade Commission, and the Federal Reserve turned out to have their own biases and interests. Even Wilson himself, though convinced that he, like Lincoln, was detached from irrational prejudices, actually held quite a few of them, including a virulent antipathy toward blacks, an instinct that women shouldn't vote, periodic suspicions about the loyalty of America's newer immigrants, and an unshakable faith in the number 13. In expecting the public to be motivated by logic alone, Wilson was holding it to a standard that he could not meet himself.

This hubris of reason would become clearer when Wilson exported his progressive assumptions overseas. But in his triumphant first term, it was easy to be bullish about human progress. America had not experienced a traumatic war in fifty years, or a serious economic downturn in twenty-five. The country had been booming for so long that the influential University of Pennsylvania economist Simon Patten declared that America had transitioned from a "Pain Economy" to a "Pleasure Economy," in which scarcity would soon be a thing of the past. All this economic dynamism had produced class conflict, to be sure, but after more than a decade of progressive reform the government seemed to have that well in hand. Experts were steadily expanding their writ over the American economy, replacing selfishness with science and anarchy with order, and the American people—judging by their enthusiasm for first Roosevelt and now Wilson—were cheering them on. In the early years of a thrilling new century, America was fulfilling its destiny: It was becoming a republic of reason. "It was a happy time, those last few years before the First World War," Lippmann would later reflect. "The air was soft, and it was easy for a young man to believe in the inevitability of progress, in the perfectibility of man."

In 1910, the progressive-minded steel magnate Andrew Carnegie wrote a letter to the trustees of his newly formed Carnegie Endowment for International Peace. Their first project, he instructed, was to abolish war. Once that was accomplished, they should reconvene to "consider what is the next most degrading evil or evils whose banishment . . . would most advance the progress, elevation and happiness of man." He envisioned the process continuing on like this, with the trustees periodically checking off problems and thus assisting "man in his upward march to higher and higher stages of development unceasingly; for now we know that man was created, not with an instinct for his own degradation, but imbued with the desire and the power for improvement to which, perchance, there may be no limit short of perfection even here in this life upon earth." It was the kind of thing that, looked back upon from the other side of the abyss, could make you either laugh or cry.

For some Americans, the outbreak of World War I challenged this soaring optimism. Progressivism's success "was beginning to shake me in my very firm belief in original sin," wrote a friend of Lippmann's. "This war has restored it triumphantly. . . . How is Walter going to quench this fundamental and illogical passion in us all?" It was *the* question: If humans were rational creatures and progress was history's natural course, why were the most advanced nations on earth sending their young men to cower in trenches and gag on poison gas? But most progressives parried it in classic American fashion: by drawing a bright line between the enlightened new world and the degenerate old. Europe was the past; America was the future. Europe was the disease; America was the cure. "Evil and suffering did not of itself invalidate progressivism," notes the historian John Thompson; "on the contrary, its existence had always been the spur—so long as belief in the possibility of amelioration was retained." For Wilson, progressivism's success in fostering a scientific peace at home proved that amelioration was not just possible, but likely. "Europe is still governed by the same reactionary forces which controlled this country," he explained, but that "old order is dead . . . [and] the new order, which shall have its foundation on human liberty and human rights, shall prevail."

By "old order," Wilson meant, above all, "that unstable thing which we used to call the 'balance of power.'" For centuries, it had been the core principle of European statecraft. Nations looked out for their self-interest, and only their self-interest. They allied, schemed, armed, even warred, all to ensure that no adversary—or combination of adversaries—grew powerful enough to threaten their very existence. The balance of power was disorderly, unfair, and unscientific, everything Wilson disdained. At best, each country's efforts at self-preservation and self-aggrandizement produced an equilibrium in which no nation, or group of nations, grew so powerful as to launch an attack, and none grew so weak as to invite one.

The most anyone could hope for was a peace of drawn swords. The logic behind the balance of power mirrored the logic behind the unregulated free market. Given that nations, like people, were inherently selfish, a system based upon that selfishness was wiser than a system based upon the illusion that anyone gave a damn about anyone else.

For many European statesmen, who had learned from harsh experience to distrust their carnivorous neighbors, the balance of power was like gravity. You might not love it; but you defied it at your peril. "We too came into the world with the noble instincts and the lofty aspirations which you express so often and so eloquently," explained grizzled French premier Georges Clemenceau to Wilson. "We have become what we are because we have been shaped by the rough hand of the world in which we have to live and we have survived only because we are a tough bunch."

But to progressives such as Wilson, who had witnessed less tragedy than their European counterparts, and more triumph, the balance of power looked both immoral and archaic, the global equivalent of America's selfish tribes. Once upon a time, argued Lippmann, American politics had also been like Europe's, with the North and South "each trying to upset the balance of power in its own favor." But Lincoln had remedied that. The progressives also compared Europe's balance of power to the savage, anarchic capitalism of late-nineteenth-century America, which Social Darwinists had defended on the grounds that selfishness was the way of the world. Now, as the result of Roosevelt and Wilson's reforms, that too was being tamed.

Finally, many progressives pointed to the Western Hemisphere as a third model that proved that the balance of power could be overcome, that politics between nations, like politics between individuals, could be governed by impartial reason, not brute force. America's policy toward its southern neighbors, Wilson declared, "is known not to be [shaped by] a selfish purpose. It is known to have in it no thought of taking advantage of any government in this hemisphere or playing its political fortunes for our own benefit."

When Wilson said the United States had no selfish motives in Latin America, he believed it. Just as he considered himself a neutral arbiter of race, class, and regional interests in the United States—despite being a racist—he cast himself in the same role when it came to America's relations with its neighbors to the south. He proposed that the countries of the Western Hemisphere agree to a Pan-American collective security treaty under which they would protect each other from external aggression and internal disorder. On its face, it looked as objective as the Tariff Commission. But in reality, it was the United States that had the power to intervene to prevent disorder in its weaker southern neighbors, not the other way around. And while Wilson genuinely believed that such interventions were unselfish, it was no coincidence that where the United States did intervene—in Haiti, in the Dominican Republic, and in Mexico (twice)—it had the non-altruistic effect of securing U.S. investments and preventing meddling by European powers.

For Wilson, it was an article of faith that he wanted for the Latins only what they wanted for themselves. Unlike more naked U.S. imperialists—who would have been happy installing dictatorships in the countries to their south—Wilson was enough of an idealist to believe that the people of Latin America could achieve what the United States had achieved: democratic capitalist governments where property was respected and change occurred only via the law. But his idealism was marred by parochialism: He didn't understand Latin America well enough to realize that given the historical experience and economic realities of many of its countries, even necessary change would be far more chaotic than in the United States. Wilson wanted America's neighbors to be democratic, but only if they elected people like him—progressive capitalists—not radicals who might foment disorder or interfere with American business. Ever the believer in political education, he famously declared that "I am going to teach the South American republics to elect good men!"

When they did not, he convinced himself that the leaders he disliked didn't represent their people as well as he did. For Wilson, it was nearly axiomatic that selfish, irrational nationalism was not real nationalism at all. And so he sent U.S. troops to depose an anti-American dictator in Mexico, confident that the Mexicans would welcome the arriving U.S. soldiers as liberators. Instead they rose up in fury against the gringo invaders, and America quickly withdrew. Wilson was bothered by the experience, but not too bothered. He still saw his Latin America policy as a success. After all, U.S. economic and political influence in the hemisphere expanded dramatically during his presidency, largely because World War I cut off South American banks and businesses from Europe. Wilson's Pan-American treaty was never ratified, but it remained his model for the world.

During World War I, Wilson, who had never visited continental Europe before the war and spoke none of its languages, repeatedly recommended both U.S. domestic politics and U.S. relations with its Latin neighbors as templates for a world governed by reason, not force. His basic error was in not recognizing what was blindingly obvious to most African-Americans and Mexicans: that politics within the United States, and within the Western Hemisphere, was absolutely dependent on force. In the United States, it was the government's monopoly on military power that ultimately made possible the progressive agenda. No matter how rational Wilson's commissions, and how inspiring his rhetoric, they would not have mattered had Washington not been able to enforce its edicts via the sword. Similarly, in the Western Hemisphere, it was Washington's near monopoly on military power that undergirded Wilson's Pan-American visions. In the Americas, and in the United States, it was because Wilson had all the guns on his side that he could imagine that his authority did not really depend on them.

As Wilson saw it, the United States, given its lack of selfish motives, stood apart from Europe's balance-of-power clash in the same way government experts had stood apart from the clash between industry and labor in Pennsylvania's anthracite coal mines. America was not a player in Europe's game, but it could be an umpire, since its own domestic accomplishments suggested a better way. The United States, Wilson declared, must play "a part of impartial mediation." The war's "causes cannot touch us," he added, but its "very existence affords us opportunities for friendship and disinterested service."

But as was so often the case, Wilson was not as disinterested as he liked to believe. The supposedly peaceful, rational politics of the new world and the brutal, benighted politics of the old were deeply intertwined. The Monroe Doctrine, which for close to a century had kept the Americas free from European power politics, had been enforced for most of that time not by the United States, whose navy was comparatively puny, but by Britain, the greatest naval power on earth. For its own reasons, London had decided to allow the United States a safe haven in the Western Hemisphere, but its ability to do so depended on a balance of power in Europe. For centuries, the central goal of British foreign policy had been to prevent any one nation from dominating the European continent, since that nation could then threaten the royal fleet that protected the British Isles. Without a European balance of power, Britain could not control the high seas. And without British control of the high seas, the United States could not be sole master of its hemispheric domain. Wilson saw the Americas as an Eden free of balance-of-power politics, but he could only indulge that conceit by ignoring Beelzebub guarding the gates outside.

Britain entered World War I because it feared that if it did not, Germany might vanquish France and Russia and make itself overlord of all Europe. Since 1871, when Bismarck welded the German states into one, Germany had been casting an ever-longer shadow over the continent. In 1890, its population had been 11 million higher than France's; by 1913, the gap was 30 million. In 1890, it had produced about 2 million more tons of steel annually; by 1913, the gap was 13 million. In 1880, the British fleet had boasted more than seven times Germany's tonnage; by 1914, the ratio was less than two to one.

There were influential Americans who looked upon Germany's growing might the same way London did: with dread. Many of them were conservatives like Massachusetts Senator Henry Cabot Lodge, a dour, haughty, and cerebral Boston Brahmin, the first person ever to receive a Harvard history Ph.D. Lodge believed in the balance of power for the same reason he opposed most progressive reforms at home: He thought that trying to eradicate selfishness was a fool's errand, and a dangerous one. "We must deal with human nature as it is," he declared, "and not as it ought to be." But in his view of Europe's war, Lodge was joined by Theodore Roosevelt, who broadly shared Wilson's domestic views but, crucially, did not see them as a template for American policy overseas. Unlike Wilson, who had little experience abroad, Roosevelt had long been fascinated by America's role in the world and had for years maintained a lively correspondence with friends and associates in Europe. As early as 1910 he had traveled through Europe after a post-presidential African safari, and returned home worried that the Huns were growing too strong.

Temperamentally, Roosevelt also liked power politics. He liked—he was blunt about it—war. For Wilson, a domesticated world order, where decisions were made via education and law, not force, was heaven. For Roosevelt, who thought war made men virile and life exciting, a domesticated world order was hell. Roosevelt's bloodthirsty side was often ugly; in fact, it had produced a kind of hubris of its own in the Spanish-American War a decade and a half earlier. But his grasp of the fundamental difference between domestic and international politics made him more alive than Wilson to the role that force played—and would always play—in world affairs. And in Europe, at least, that understanding proved useful, since it helped him see the potential threat that Germany posed. "Do you not believe that if Germany won this war, smashed the English Fleet and destroyed the British Empire, within a year or two she would insist upon taking the dominant position in South and Central America?" he asked in 1915. Within months of the war's outbreak, Roosevelt and Lodge were demanding that America arm itself, if not to enter the war on Britain's side (something most Americans would not yet consider), then at least to ensure that if the Royal Navy did fall, America could defend its hemisphere if Germany went after that next.

Spurning Roosevelt and Lodge's demands, Wilson initially resisted building up American arms. He saw the United States as an impartial mediator—helping the Europeans evolve beyond power politics—not another selfish tribe. Impartiality, however, was harder than it looked. When Wilson declared America neutral, he demanded that the combatants respect its right to trade with both sides under international law. But enforcing that law against the world's two greatest military powers was not like enforcing it against the trusts, especially when America's military was so feeble. The British and Germans thumbed their noses at Wilson's request. London imposed a blockade on German ports, and Berlin countered with a blockade of its own, aimed at cutting the British Isles off from crucial transatlantic supplies. There was, however, a crucial difference. Lacking His Majesty's mighty fleet, Berlin employed a newer technology: submarines. Britain enforced its blockade by boarding and searching U.S. ships. Submarines, by contrast, relied on stealth. If they surfaced to inspect passing ships they would be easy prey for an armed merchant vessel. Subs could only enforce a blockade one way: by sinking whatever tried to pass.

In May 1915, a German U-boat blew a hole in the British ocean liner R.M.S. *Lusitania*, drowning 1,198 people, including 128 American civilians, in the icy waters off the Irish coast. To Americans, who had seen the war as a remote affair, the shock was seismic. A decade later, many still remembered where they were when they heard the news. For Roosevelt and Lodge, it just underscored the German threat. They knew the American people weren't ready for war, but in their hearts, Lodge and TR were already there. For Wilson, however, the problem wasn't German power; it was international law. "The rights of neutrals in time of war are based upon principle, not upon expediency," he wrote in a diplomatic note to Berlin, "and the principles are immutable."

It was a fateful choice. By insisting that Americans had a sacred right to travel the high seas, Wilson was putting the United States on a path to war with Berlin. But he was justifying it on legal and moral grounds, not geopolitical ones. As Wilson described it, Germany was dangerous not because it threatened to dominate the European continent and imperil the Western Hemisphere, but because it was violating America's neutrality rights. This rationale preserved the façade of objectivity that he cared about so much, the sense that America was acting from impartial principle, not selfish interest. But there was a cost: Wilson never used Germany's submarine warfare, as Lodge and Roosevelt urged him to, to tell Americans that they had a stake in a particular distribution of power on the European continent, that Germany was a danger not because it was violating international law but because it could threaten America. The United States was inching toward war, but with the illusion that it was still an umpire in the game of world politics, not a player on the field. It was an illusion that would prove extremely costly in the years to come.

The sinking of the *Lusitania* was like the starting of a stopwatch. Given Wilson's definition of neutrality rights, and Germany's need to blockade British ports, it was just a matter of time until U-boats drowned enough Americans to draw the United States into war. Well aware of this, Wilson flung himself into an effort to end the conflict before it engulfed his country. Colonel House shuttled between London, Paris, and Berlin trying to convince the belligerents to agree to a peace deal, any peace deal. Don't focus on "local settlements [such as] territorial questions, indemnities, and the like," Wilson instructed House. But for Europeans, those "local settlements" were what the war was all about. Germany wanted enough money and land to ensure that in tandem with its lackey, Austria-Hungary, it could dominate Europe; Britain, France, and Russia (and later Italy) wanted to thwart Germany's plans and aggrandize themselves at Berlin's and Vienna's expense.

In May 1916, Wilson tried to cut the Gordian knot. He announced that if the combatants laid down their arms, America would join a League to Enforce Peace, a global version of the Pan-American pact he had been pushing in the Western Hemisphere. Together, the league's members would guarantee freedom of the seas and repel aggression by any country against another. This system of collective security, in which countries respected and enforced rules of civilized behavior, would replace the balance-of-power system, in which countries—or alliances of countries— looked out only for themselves. It would ensure that international disputes were resolved based upon reason, not force. The implication was that

Europe's belligerents need not worry where they drew the armistice lines; the balance of power between adversaries would no longer matter.

Among American progressives, many of whom had been championing such a league for years, Wilson's speech was a triumph. Lippmann declared it "one of the greatest utterances since the Monroe Doctrine." But Europe's leaders were less enchanted. It was as if America were demanding that they call a halt to the game and insisting that the score no longer mattered. For statesmen who genuinely believed that they were fighting for survival and who, in pursuit of that belief, had sent millions of young men to be maimed and killed, there was something infuriating about Wilson's insistence that their death struggle for security was a triviality and a superstition. Wilson's vision of a world without power politics, declared French author Anatole France, was like "a town without a brothel." It was too innocent, too anti-septic, too inhuman. Wilson was casting off America's historic isolation and committing it to an active role in European affairs, but only conditionally: only if the old world became something radically different from what it was. The United States, Wilson declared, "can in no circumstances consent to live in a world governed by intrigue and force. We believe that our own desire for a new international order under which reason and justice and the common interests of mankind shall prevail is the desire of enlightened men everywhere." But what if it wasn't? Lurking behind Wilson's words was an implicit threat, which would come back to haunt him: If the world did not live up to its standards, America would pick up its marbles and go home.

So Wilson's mediation efforts failed, and slowly but surely, the war reeled America in. In March 1916, when a U-boat sank the French passenger ferry *Sussex*, injuring several Americans, Wilson issued an ultimatum: Unless Germany ceased attacking civilian ships, the United States would break off diplomatic relations. After a fierce internal debate, the kaiser's government backed down, partly because it was running low on subs anyway. But Wilson had placed himself in a diplomatic straitjacket: The moment Berlin resumed its submarine blockade, his own words would push him to the brink of war.

On January 31, 1917, Germany's ambassador informed Washington that his government was doing just that. The Germans now believed they had enough subs to enforce the embargo, and, with Russia crumbling on their eastern front, they saw a chance to rapidly win the war. "England will lie on the ground in six months," predicted German naval authorities, "before a single American has set foot on the continent."

Three days later, Wilson broke off diplomatic relations with Berlin. He still hoped for a miracle, but on March 18, Germany sank three U.S. merchant ships, killing fifteen Americans. Wilson's entire cabinet was now demanding war. Adding to the fever, the British had intercepted a telegram, sent from German foreign secretary Arthur Zimmermann to the government of Mexico, proposing a military alliance. If the United States declared war on Germany, and Mexico came to Germany's aid, Berlin would help it regain Texas, New Mexico, and Arizona. The telegram confirmed Roosevelt and Lodge's warning that if Germany won, its dark shadow would soon cross the Atlantic. When Wilson released Zimmermann's message to the Associated Press, it sparked a wave of public support for war.

If the Zimmermann Telegram demonstrated the dangers of a German victory, a very different event suggested the possibilities of a German defeat. On March 8, food riots broke out in Russia's imperial capital, Petrograd, and quickly grew into a general strike. When Czar Nicholas II tried to return to the city to reestablish control, soldiers refused to let his train pass. Seven days later, he abdicated. A dynasty that had ruled for three centuries had been toppled in a single week. And for the first time in its history, Russia was a republic, led by the moderate, pro-Western Alexander Kerensky.

For Walter Lippmann and John Dewey, Russia's liberal revolution was like a sign from God. Although aware that the carnage on the seas might force America into war, they yearned to enter it on behalf of something grander than the safety of America's merchant marine. If the United States had to descend into the valley of death, they wanted to believe it could lead Europe toward the light. Now, suddenly, the people of Russia, the most brutal, backward country on the continent, were breaking their chains, showing that the old world could indeed follow the path of the new. America could never join the war with its "full heart and soul," Dewey had written, until "the almost impossible happens . . . until the Allies are fighting on our terms for our democracy and civilization." Now the almost impossible had occurred: The most autocratic Allied power was embracing democracy. It is now "as certain as anything human can be," wrote Lippmann, "that the war which started as a clash of empires in the Balkans will dissolve into democratic revolution the world over."

Among pro-war progressives, the czar's overthrow was like the popping of a cork. In a great rush, the optimism that had been welling up during their years of domestic triumph bubbled over. Ambivalence, hesitation, even dread were replaced by giddy expectation. "We have been, as it were, a laboratory set aside from the rest of the world in which to make, for its benefit, a great social experiment," Dewey declared. "The war, the removal of the curtain of isolation, means that this period of experimentation is over. We are now called to declare to all the world the nature and fruits of this experiment." No longer was the war a mistake, or even a grim necessity; it was a crusade.

On the evening of April 2, 1917, as a light rain fell, mounted cavalry escorted Wilson's car from the White House to the Capitol. Soldiers saluted as he passed; thousands of well-wishers lined the route. Once he arrived, Wilson was taken to a congressional antechamber. Alone there for a moment, he began to shake uncontrollably, and had to stare hard into a mirror to regain his poise. At 8:32 P.M., the doors swung open and he walked into the House chamber, where he was greeted by a full two minutes of applause. With the exception of a tiny handful of war opponents, everyone in the audience wore or held an American flag.

"We are," Wilson declared, "at the beginning of an age in which it will be insisted that the same standards of conduct and of responsibility for wrong done shall be observed among nations and their governments that are observed among the individual citizens of civilized states." This was the crusade in a nutshell: America would make politics between nations resemble politics between Americans. Wilson went on to salute the Russian people, who "have been always in fact democratic at heart," for proving that America's democratic principles were becoming Europe's as well. And with the chief justice of the United States leading the crowd in raucous cheers, he ended by declaring that America "shall fight for the things which we have always carried nearest [to] our hearts—for democracy . . . for the rights and liberties of small nations, for a universal dominion of right by such a concert of free peoples as shall bring peace and safety to all nations and make the world itself at last free." As the crowd rose to its feet, members of Congress ripped the flags off their sleeves and lapels and waved them wildly. Five days later, America was at war.

To emphasize the conditional nature of America's commitment, Wilson declared that America would fight as an associate of Britain, Russia, France, and Italy, not an ally. Just as he believed Lincoln had done during the Civil War, he would fight and yet remain detached. Knowing (or at least suspecting) that London, Paris, Moscow, and Rome had made secret arrangements to feed on Germany's and Austria-Hungary's carcasses if they won the war, Wilson decided that America must establish a rational, disinterested process for designing the postwar world. To do that, in the fall of 1917, "the Inquiry" was born.

Lippmann was only one of the progressive luminaries whom House enlisted in the cause. Thorstein Veblen served as an informal adviser. So did the *New Republic*'s Herbert Croly, Harvard president A. Lawrence Lowell, and the famed historian of the vanishing frontier, Frederick Jackson Turner. Future Supreme Court justice Felix Frankfurter requested a role, as did Judge Learned Hand. Dewey considered heading the Inquiry's Russia division. The effort began in secrecy, at the New York Public Library on Forty-second Street, where only the head librarian and one aide knew the true purpose of the men working furiously in back. From there it moved to the American Geographical Society on Broadway and 155th, largely because of the society's vast storehouse of maps. Finally, a reporter for Philadelphia's *Public Ledger* broke the story, and the Inquiry was deluged with job applicants. Satellite offices were established at Harvard, Princeton, and Yale. Guards began patrolling the Inquiry's headquarters night and day.

The Inquiry was modeled on those great temples of progressive reform: the Tariff Commission, the Federal Trade Commission, the Interstate Commerce Commission, and the Federal Reserve. Just as they had established scientific principles for American industry and finance, the Inquiry would now establish them for entire nations—determining which countries should exist and where. In their effort to draw new European borders, Lippmann and his colleagues produced four types of studies. Historians determined which ethnic, religious, and linguistic groups had the best ancestral claim to a given patch of land. Geographers and sociologists searched for the right way of classifying the current residents. Economists allocated natural resources and industrial infrastructure so nations would be economically viable. And a final group—the smallest—handled politics.

Much of the work in this fourth category concerned domestic arrangements like federalism. Relations between states, by contrast, received little attention. How to put the genie of German dominance back in its bottle and rebuild Europe's shattered balance of power were questions that had no scientific answers, and were thus largely ignored. The whole idea behind the Inquiry, after all, was to replace borders imposed by force with borders conceived through reason. As Wilson's secretary of state, Robert Lansing, explained, "the fixing of frontier lines with a view to their military strength and in contemplation of war was directly contrary to . . . the policy of the United States."

The Inquiry produced more than two thousand reports, many filled with statistical tables and charts, often with no interpretation whatso-ever—the facts were meant to speak for themselves. But like Wilson himself, Lippmann and his colleagues were not as objective as they liked to believe. "In some respects the Koords [*sic*] remind one of the North American Indians," read one report on the Middle East. "Their temper is passionate, resentful, revengeful, intriguing, and treacherous." And it wasn't only racism that infected the Inquiry's efforts; power politics did as well. Just as the trusts had wormed their way into the tariff and trade com-missions, the British now influenced the Inquiry—plying the Americans with data that buttressed London's favored outcomes. So deeply enmeshed were they in the Inquiry's activities that British diplomats in Washington knew more about its work than did most members of Congress.

That, in and of itself, was not such a terrible thing. Had Wilson told Americans that Germany's power and ambition made it a potential threat to the United States, and that America was entering the war because Brit-ain, France, and Russia could no longer contain that threat alone, no one would have expected the Inquiry to be impartial. Had he forthrightly acknowledged that America was committing itself to a world it could not perfect, had he defined America's war aims as something less grandiose than the eradication of power politics, perhaps the postwar peace would not have proved such a crushing disappointment. Had he not allowed the hubris of reason to swell so dramatically, perhaps it would not have so violently popped.

In October, Russia's liberal experiment collapsed as Kerensky's govern-ment fell to the Bolsheviks of Lenin. Determined to discredit the capitalist powers and withdraw Russia from the war, Lenin published the Allies' secret treaties dividing up the spoils of war. Suddenly, for all his grand rhetoric, Wilson looked like an accomplice to a land grab. In mid-December, House summoned Lippmann to his Fifty-seventh Street apartment and told him that the Inquiry must show the world that America was still fighting for high principle, and fast. For the rest of the month, Lippmann and his fellow experts worked around the clock, often not even going home to sleep. We need "genius," Lippmann told his colleagues, "sheer, startling genius, and nothing else will do." Finally, on January 2, 1918, Lippmann gave House the Inquiry's final report. Three days later the president and his best friend met to mold it into a series of public recommendations. "We actually got down to work at half past ten," House noted in his diary, "and finished remaking the map of the world, as we would have it, at half past twelve."

The result, unveiled on January 8 before a joint session of Congress, was perhaps the most celebrated diplomatic initiative in American history: the Fourteen Points. It included free trade, freedom of the seas, disarmament, rights for colonial peoples, an end to secret treaties, and a League to Enforce Peace (increasingly called the League of Nations). When it came to borders, Wilson obscured some of the Inquiry's findings, employing vague language to avoid firm commitments, but the plan's eight territorial planks still veered toward national self-determination and the principle that borders should be determined by reason, not force. The boy who had once drafted constitutions for his baseball league and his imaginary yacht club had finally done it: He had drafted a constitution for the entire world.

Wilson knew that his wartime allies, especially the French, wanted a harsher peace. But he trusted in his capacity to teach. The experts had told him what was right, and now he would educate the people of Europe, rallying them against their own selfish regimes. That included the people of Germany. American planes dropped copies of the Fourteen Points behind enemy lines, and Lippmann joined a newly created Propaganda Commission, from which he planned "a frank campaign of education addressed to the German and Austrian troops, explaining as simply and persuasively as possible the unselfish character of the war, the generosity of our aims, and the great hope of mankind which we are trying to realize."

Finally, that August, after years of blood-soaked stalemate, the German military crumbled. With American troops now flooding into Europe, the Allies broke through enemy lines, destroying sixteen divisions in a matter of days. On the German home front, sailors mutinied; workers struck; starving women marched through Berlin holding their empty pots and pans. One of Germany's top generals, Erich Ludendorff, sank into depression; the other, Paul von Hindenburg, put on dark glasses and a fake mustache and sneaked across the border into Sweden. In November, the kaiser abdicated. A liberal government took power in Berlin and sued for peace on the basis of the Fourteen Points. The German people, it appeared, had heard Wilson's appeal. Now they too were demanding a scientific peace.

The stage was now set. No sitting U.S. president had ever left the

Western Hemisphere. (According to some interpretations, doing so was actually illegal.) But overruling his top advisers, Wilson decided to oversee the peace negotiations himself, so no subordinate would steal the glory. On December 4, 1918, as gunships saluted, handkerchiefs waved, and pigeons were released into the sky, the passenger liner *George Washington* left the Hoboken, New Jersey, harbor, carrying the president and virtually everyone else of importance in the executive branch, a delegation 1,300 strong. It was flanked on its left and right by ten destroyers; a battleship sped ahead to smooth the waves. Nine days later, at just past 1:30 P.M. on December 13—a time and date chosen because the president considered it lucky (his name contained thirteen letters)—the armada, now sixty vessels strong, reached Brest, France, on the far side of the Atlantic. On the morning it arrived, the sun, which had been absent for days, suddenly cut through the gray sky.

When Wilson disembarked, Europe's battered masses gave him a greeting that one journalist called "inhuman—or superhuman." At 3 A.M. that night, on the train carrying the American delegation to Paris, Wilson's doctor looked outside and saw men, women, and children lining the tracks as far as the eye could see. When the Americans reached the French capital, two million admirers jammed the streets, the largest crowd in French history. In Rome, the mayor likened Wilson's visit to the Second Coming. In Milan, banners compared him to Moses. Italian soldiers knelt before his picture; families placed his photograph on their windowsills, surrounded by sacred candles. "For a brief interval," wrote H. G. Wells, "Wilson stood alone for mankind. . . . He was transfigured in the eyes of men. He ceased to be a common statesman; he became a Messiah."

The lonely preacher's son had wagered an epic gamble: that through force he could build a world governed by reason. And now, with Germany vanquished and Europe's masses screaming his name, he seemed on the brink of success. It was the progressive dream, unfolding on a global scale. "It might turn out well," wrote House in his diary, "and yet again it might be a tragedy."

CHAPTER TWO

THE FRIGHTENING DWARF

Randolph Bourne made people uncomfortable. Partly, it was his body. At birth, a brutal doctor, making promiscuous use of forceps, elongated his head and disfigured his face. At age four he developed a double curvature of the spine, which left him a hunchback. As an adult, he wore a long black cape, like something out of *Phantom of the Opera*. John Dos Passos called him a "tiny twisted bit of flesh." Theodore Dreiser called him "that frightening dwarf."

Bourne's mind was frightening, too. He began reading grocery labels as a toddler and the Bible by kindergarten. By the time he reached college, he had become a withering polemicist, perhaps Lippmann's only generational equal. But if Lippmann used his pen to impress and ingratiate, Bourne used his to cut—a harsher style engendered by a harsher life. A self-described "homesick wanderer," he felt little affection for his aristocratic but downwardly mobile mother, barely knew his alcoholic, absentee father, couldn't stand his teachers and his hometown, and spent his twenties flitting from one mindless, dead-end job to another. Then, in 1909, he won a scholarship to Columbia University, where he met Charles Beard and John Dewey, the men who changed his life.

Beard, the charismatic son of an Indiana gentleman farmer, was progressivism's greatest historian. Dewey, a shy mumbler from Burlington, Vermont, was its greatest philosopher. On the eve of World War I, they were high priests of the church of reason, and their effect on Bourne was almost spiritual. From Dewey, whom he likened to a prophet, Bourne learned that schools could unlock the human instinct for cooperation and thus breed citizens who transcended their selfish desires. From Beard he learned to see American history as an upward march from plutocracy to democracy and anarchy to organization. By the time war struck, Bourne was earnestly spreading the gospel of progress. It was young people like himself, he argued, in whom selfishness had not yet fully taken root, who could serve as "the incarnation of reason." In 1914, Beard got his young disciple a job at the fledgling *New Republic*. Not yet thirty, the brilliant hunchback was becoming one of progressivism's sharpest voices. And in those heady, innocent days before the war, his future—like his movement's—appeared bright.

At first the carnage across the Atlantic changed little. Like his *New Republic* colleagues, and like Woodrow Wilson, Bourne wanted no part of Europe's war. But as 1916 turned to 1917, he began to feel the ground shift. Lippmann and Croly were now visiting Colonel House's New York City apartment every week, and their editorials were speculating with mounting fervor about how America's entrance into war might transform the country and the world. For his part, Bourne was moving hard the other way—his antiwar militancy growing ever stronger as America crept toward war. Not coincidentally, fewer and fewer of his articles made it into print, and by April 1917, when America finally joined the fray, the *New Republic* was no longer publishing his political writing at all.

But Bourne's bitterest break came not with his employer, but with his mentors, Dewey and Beard, who became two of the war's loudest academic cheerleaders. Wilson's vision of a rationalized postwar world, wrote Beard, reflected the "slowly maturing opinion of the masses of the people everywhere in the earth." It heralded "a new epoch in the rise of government by the people and in the growth of a concert among the nations." Dewey rhapsodized about the war's potential impact at home. Wartime mobilization, he predicted, would shift power from private to public hands, from selfish tribes to unbiased experts—exactly the shift for which progressives had been striving since the century turned. It would usher in "a more integrated, less anarchic [political] system" marked by the "systematic utilization of the scientific expert." For decades, Dewey had seen education as the key to overcoming American selfishness and disorder. Now he had a new laboratory: war.

For Bourne, the analogy was not just wrong, it was monstrous. Schools were "rational entities," safe laboratories for tinkering with human behavior. Nations at war were entirely different. Far from being a mechanism for reasoned planning, he argued, "war is just that absolute situation . . . which speedily outstrips the power of intelligent and creative control." War "determines its own end, victory, and government crushes out automatically all forces that deflect, or threaten to deflect, energy from the path of organization to that end." For Bourne, Dewey was like a man who, having called a tiger a horse, thought he could restrict its diet to grass.

War progressivism, Bourne was coming to believe, rested upon a lie: Force, once unleashed, could not remain the servant of reason. If the Inquiry tested that proposition overseas, it found its domestic counterpart in something called the Committee on Public Information, which was tasked with building support for the war at home. Led by George Creel, a former urban reformer and muckraking journalist, it embodied progressivism's love affair with the fact. Among the men and women who stocked the committee—many of whom had spent the prewar years exposing the abuses of the trusts—it was a point of pride not to coerce Americans into supporting the war, nor even to play upon their emotions. "A free people cannot be told what to think," declared Creel. "They must be given the facts and permitted to do their own thinking."

Bourne's mentors nodded approvingly. Beard went to work for Creel's committee, editing some of its pamphlets on the causes and purposes of the war. Dewey applauded the committee's calm, deliberative spirit, which he considered appropriate for a country in which "very large numbers of our citizens . . . have systematically taught themselves to discount all of the more violent appeals to passion."

Progressives had long placed their faith in the rationality of the common woman and man. And Dewey had predicted that ordinary Americans would act rationally even—and perhaps especially—during war. But by 1917, that claim sat uncomfortably alongside the wave of anti-German frenzy spreading from Washington, D.C., across the American hinterland. After America entered the war, Congress passed espionage and sedition acts that outlawed "disloyal, profane, scurrilous, or abusive language" against the Constitution, the government, the military, and the flag. Cincinnati outlawed the sale of pretzels; Iowa's governor made publicly speaking German a crime. When a Wyoming man was overheard saying, *"Hoch der Kaiser"* ("Up with the kaiser"), a group of townspeople hanged him, cut him down while still alive, and made him kneel and kiss the American flag. In April 1918, a St. Louis mob abducted a young German-American, stripped him, dragged him through the streets, and then lynched him, while a crowd of five hundred cheered. At trial, the defense attorney called the murder patriotic, and it took a jury twenty-five minutes to acquit.

"The war," wrote Bourne's friend, Waldo Frank, "which drove all the world, including Dewey mad, drove Bourne sane." In a series of scalding essays, Bourne put his mentors on intellectual trial. It wasn't just America's bloodthirsty yokels who proved that rationality was impossible in a time of war, Bourne argued; it was the war progressives themselves. Though Dewey and Beard saw themselves as apostles of reason, Bourne insisted that their motives for supporting the war were as primal as everyone else's. They were bored. Their domestic campaigns, having brought so much success, had lost the capacity to thrill. Having come of age at a time of swelling optimism, and possessing no primary experience of political tragedy, they were conditioned to see their people, their leaders, and themselves as capable of boundless achievement. And as a result, when Russia's liberal revolution offered a glimpse of a remade world, they could not resist the psychic pleasure of a crusade. "Hesitations, ironies, consciences, considerations—all were drowned in the elemental blare of doing something aggressive, colossal," wrote Bourne. "There seems to have been a peculiar congeniality between the war and these men. It is as if the war and they had been waiting for each other."

Betrayed by his prophets, Bourne cast about for a new creed. At times he veered toward isolationism. The horrors on the home front, he argued, showed that Dewey was wrong: American progressivism was not a successful experiment ready to be unveiled to the world. Americans should instead embark upon a "stern and intensive cultivation of our garden . . . a turning within in order that we may have something to give without." At other moments, Bourne lurched toward pacifism. Wilson, he argued, had been right to seek to replace the balance of power with a world of reason and law. Where he had gone horribly wrong was in believing that this new world could be enthroned—and maintained—via the sword. There is no such thing as a "democratic and antiseptic war," Bourne insisted. "The pacifists opposed the war because they knew this was an illusion."

Had Bourne lived longer, he might have reconciled these competing strains. But he did not live long enough to try. Cast out from the *New Republic* in 1917, he began writing for a small literary journal called *Seven Arts*. But it soon folded, largely because of the furor over his antiwar polemics. He briefly served as a contributing editor at another minor publication, *Dial*. But the journal's benefactor was also wooing John Dewey, the man Bourne was savaging in print. Dewey agreed to join the staff on one condition: that his nemesis be fired.

Under investigation by the Justice Department, increasingly unable to publish, and virtually homeless—living in the apartments of friends— Bourne died of influenza just weeks after the war's end. A legend soon grew that he had starved because no journal would print his work. That was a fable, but its appeal testified to Bourne's hold over the postwar imagination. In his final, fevered days, his assault on the hubris of reason offered a template for many of America's leading intellectuals and politicians in the two decades to come. Isolationists in the 1920s and '30s demanded that America never again bind its fate to the alien civilization across the Atlantic. Pacifists kept alive the dream of a rationalized world but insisted that achieving it required America not to fight wars, but to abolish them. And leading these two intellectual currents, oddly enough, were Charles Beard and John Dewey. In death, the protégé became mentor, and the fathers carried on the work of the prodigal son. Had Bourne's ravaged body endured into the postwar age, the homesick wanderer might have found himself, finally, at home.

If the hubris of reason was cracking domestically, it would soon crack overseas as well. On January 18, 1919, five weeks after Wilson's triumphant arrival in Europe, the Paris Peace Conference began. The date was no accident. It had been on that same day in 1871 that Wilhelm I was coronated kaiser of a newly unified Germany. French premier Georges Clemenceau, who orchestrated the timing, remembered 1871. Back then, as a young man, he had helped barricade his neighborhood against the German siege of Paris. Now, in his seventies—suffering from diabetes, insomnia, and eczema on his hands so severe that he was forced to wear gloves—he had seen the Germans invade again. Germany, he told a journalist, was his "life hatred." His one overriding goal before he died—for which he was willing to beg, lie, cheat, steal, and kill—was to ensure that France was never raped again. According to legend, he had asked to be buried upright, facing Germany, so he could keep vigil even from the grave.

Clemenceau wanted Wilson to see Germany as he saw it, to understand life in the lion's den. In January 1919, France was a ravaged nation. The war had taken one quarter of its men between the ages of eighteen and thirty. Double that number had been wounded; they were everywhere, pitiful, limbless creatures begging in the streets. On paper, France had won the war. But Clemenceau knew that it was less a victory than a stay of execution. France had lost a higher percentage of men than Germany, and since the fighting had occurred largely on its soil, it had suffered more damage to its infrastructure. By sparking the Bolshevik Revolution in Russia, the war had also removed France's most powerful ally—and Germany's most powerful foe—in Eastern Europe.

Structurally, therefore, France finished the war even more vulnerable to its bigger, more productive neighbor than it had been at the start. Even the souvenir penknives that French shopkeepers sold to commemorate victory were made in German factories.

Clemenceau urged Wilson to see the destruction firsthand, the six thousand miles of French territory that the war had laid waste. "There are hundreds of villages into which no one has yet been able to return," explained one French minister. "Please understand: it is a desert, it is desolation, it is death." But Wilson resisted French invitations to tour the battlefields; he did not want to be swayed by emotion. He hadn't even wanted to hold the peace conference in Paris, preferring Geneva, where there would be fewer "hysterical" French. As always, Wilson saw himself and his nation as impartial, detached. Americans, he explained, were "the only disinterested people at the Peace Conference." Asked by foreign correspondents what he would say to his fellow leaders once the conference began, he replied, "[W]e come here asking nothing of ourselves and we are here to see you get nothing."

Clemenceau found Wilson's detachment maddening. He grumbled about the American president's "cold reason" and "mathematical justice." Wilson, he moaned, "believed you could do everything with formulas and his fourteen points. God himself was content with [only] ten commandments." For Clemenceau, there could be no objectivity between the lion and the lamb. In his mind, the purpose of the peace conference was not to repeal the balance of power—a system he pointedly endorsed just before the negotiations began—but to rectify it. France must be made stronger; Germany weaker. There must be a redistribution of force. "I have come to the conclusion that force is right," he told a dinner guest. "Why is this chicken here? Because it was not strong enough to resist those who wanted to kill it. And a very good thing too!"

France needed security; in that pursuit Clemenceau was unyielding. But he was flexible about how to attain it. The most obvious way was to ravage Germany as it had ravaged France: to strip it of as much land and money as possible. In addition to Alsace-Lorraine, whose return to France everyone took for granted, Clemenceau suggested annexing the Rhineland, a strategically vital, mineral-rich slice of land that hugged Germany's borders with Holland, Belgium, and France, or establishing a puppet state there. That way, if Germany struck again, at least France would have a territorial buffer to absorb the blow. He also proposed reparations so severe that they would not only help France rebuild but shift the economic balance of power with Berlin.

But France needed more than money and land. It needed allies, since without them it couldn't enforce a peace agreement anyway. Clemenceau was prepared to go softer on Germany in return for ironclad guarantees that America and Britain would come to France's aid if the Huns stirred again. But here his desires and Wilson's progressive vision clashed. Clemenceau wanted that most old-fashioned of things: an alliance against a potential foe. Wilson wanted a world without alliances, where nations no longer banded together in selfish blocs but instead championed the common interests of all humankind.

For five months, Wilson and Clemenceau bickered, with British prime minister David Lloyd George usually somewhere in between. Clemenceau embraced Wilson's call for a League of Nations but tried to turn it into an anti-German alliance. The League, he proposed, should exclude Germany and establish a standing army dominated by America, Britain, and France. When the Americans and British refused to create a League army, Clemenceau suggested a common military planning staff instead, so the League could organize quickly in response to German aggression. But Wilson rejected that, too. For Clemenceau, who assumed that Germany would resist whatever peace settlement the Allies imposed, the League was only useful as a tool of Allied enforcement. For Wilson, by contrast, the whole purpose of the peace settlement and the League was to establish principles so rational and fair that they would be accepted by all countries, Germany included. Wilson was no pacifist; he conceded that in theory the League might have to enforce its rulings at gunpoint. But he insisted that military action would be rare. As in his crusades for progressive reform at home, and for collective security in the Western Hemisphere, he claimed that it was education, not force, that would convince people to accept the scientific principles formulated on high. In the new world being born, he insisted, stopping aggression would depend "primarily and chiefly upon one great force, and that is the moral force of the public opinion of the world."

In the end, the French and Americans struck a compromise: The Rhineland would stay in German hands but the Allies could occupy it for fifteen years. Washington and London also offered Paris security guarantees: They would come to its aid if the Germans attacked again. On paper Clemenceau had what he wanted. But the alliance was built on sand. Wilson downplayed it, saying it did not commit the United States to do anything it hadn't already pledged to do under the League. That made Clemenceau shiver. The League, in his mind, was utopian precisely because it bound its members to defend the borders of *all* nations—whether those borders mattered to their security or not. In the real world, Clemenceau believed, that made America's pledges under the League virtually meaningless—a point underscored by Wilson's insistence that upholding those pledges would rarely require force. Clemenceau wanted a French alliance with the United States to be everything the League was not: a binding, narrowly tailored commitment made for reasons of national security, not universal principle. He wanted Wilson to acknowledge that aggression against France mattered to America in a way that aggression against Latvia did not, to admit that America had entered the game of global politics not to protect the rights of all nations, but to protect the rights of those crucial few—like France—upon which America's own security relied.

But that is not how Wilson had sold the war back home. He had told Americans that the United States was an associate of Britain and France, not an ally—that it was fighting not on their behalf, but on behalf of all humankind. The clear implication was that America's goal at the peace conference was not to buttress any one group of nations but to ensure that all nations, the victors as well as the vanquished, were treated rationally and fairly.

In Wilson's mind, his rapturous welcome across the continent proved that ordinary Europeans wanted this, too. They wanted to transcend their selfish, tribal desires; they just needed a great educator like himself to show them the way. "National purposes have fallen more and more into the background," he declared, "and the common purpose of enlightened mankind has taken their place." Mastodons like Clemenceau didn't understand that; they still hungered for alliances, reparations, and land. But Wilson told aides that Europe's leaders did not represent their people. "If necessary," he insisted, "I can reach the peoples of Europe over the heads of their rulers."

But the peoples of Europe let Wilson down. Time and again during the Paris negotiations, when Europe's leaders resisted his designs, Wilson threatened to take his case to their people, in a campaign of public education. And time and again, ordinary Europeans refused to be educated. When Wilson said he would appeal to the British to overturn their government's opposition to freedom of the seas, one of the Fourteen Points, Lloyd George urged him on, no doubt amused at the prospect of an American telling Britons to place their trust in his abstract principles rather than the Royal Navy. When Italian prime minister Vittorio Orlando walked out of the peace conference after his nation was denied the Adriatic city of Fiume, Wilson promised to explain the decision to the Italian people himself. Unmollified, the Italians, who had erected plaques to Wilson only weeks before, covered them over in disgust. If ordinary Americans weren't as rational in wartime as Wilson and his fellow war progressives had assumed, ordinary Europeans, it turned out, weren't, either.

Wilson was hoisted on his own petard. Against Roosevelt and Lodge's advice, he had denied that America was entering the war because its own security required preventing Germany from overthrowing the European balance of power. Instead he had told Americans that they were entering the war to abolish the European balance of power, to build a world in which reason governed force. The implication was that America's participation was conditional: If the old world refused to change, America could always retreat to its own hemisphere, since it could "in no circumstances consent to live in a world governed by intrigue and force." Now, as the terms of the Paris Peace Conference trickled out, it became clear that the postwar world would indeed be governed by intrigue and force. The Treaty of Versailles was not as punitive as legend suggests. It was considerably less onerous, for instance, than the treaties that Germany had imposed upon Romania and Russia when they exited the war in May 1918. But it was far from a scientific peace. German territory was parceled out to France, Belgium, Denmark, Czechoslovakia, and Poland, sometimes in flagrant violation of the principles of national self-determination laid out by the Inquiry and the Fourteen Points. Italy seized several choice parts of the former Austro-Hungarian Empire; Japan grabbed German islands in the Pacific and a slice of China; Britain and France snatched Berlin's colonies in Africa and the Middle East. The settlement reflected the distribution of postwar power, not objective standards of right and wrong. The peace treaty failed that classic progressive test: It was not seen as impartial by all sides.

Wilson did his best to sugarcoat the results. He had convinced the Europeans to join a League of Nations, he explained, and over time the League would smooth out the treaty's rougher edges. But many of Wilson's progressive allies refused to be placated. Outraged that his work for the Inquiry had been in vain, Lippmann warned the Europeans that "if you make it a peace that can be maintained only by the bayonet we shall leave you to the consequences and find our own security in this hemisphere."

Wilson had made America's entry into the war sound altruistic: The United States had joined the fray to lift up the benighted Europeans, not because its security was intertwined with theirs. Now the Europeans had refused to be lifted up. And so for many disillusioned progressives—and many Americans more generally—there was only one thing to do: return home.

On July 10, 1919, Wilson submitted the Versailles Treaty—which included American membership in the League—to the Senate. There he faced three rough blocs. First were the forty-seven Senate Democrats, who as loyal partisans nearly all backed U.S. entrance into the League. But by themselves they lacked the two-thirds votes necessary for ratification. Next were a group of Republicans called "Irreconcilables," led by Idaho's Senator William Borah, who for various reasons were certain to vote no. They numbered only about fourteen, but their voice was amplified by disillusioned war progressives like Lippmann, who now opposed the League with the fury of lovers scorned.

The opposition of the Irreconcilables meant that to win Senate ratification the treaty required votes from the third bloc: a group of roughly thirty-five Republicans called "Reservationists." The Reservationists were prepared to swallow America's entrance into the League, but only if certain "reservations" were added to the treaty, the most important of which was that only a vote of Congress—not a vote of the League—could take America to war. Their patron saint was Theodore Roosevelt, who six months earlier had dropped dead of a heart attack. Their leader was Roosevelt's soulmate, Henry Cabot Lodge, who saw his battle with Wilson as a tribute to his dead friend.

At first glance, Lodge stood in the ideological middle, with the militantly pro-League Wilson and the militantly anti-League Borah occupying the two poles. But in a deeper sense he stood apart from both. Wilson and Borah were, in the political scientist Hans Morgenthau's words, "brothers under the skin." Wilson claimed that the League would rationalize international affairs and therefore must be embraced. Borah insisted it would not and thus must be spurned. Wilson said America should engage with the world in order to perfect it; Borah said the world could not be perfected and so should not be engaged. Only Lodge articulated a realistic internationalism: He acknowledged that the world could not be perfected—that reason would never fully govern force—but urged that America bind itself to it anyway.

Lodge did not believe in detachment. He was a fierce partisan of his city, Boston; his political party, the Republicans; and his intellectual tradition, which he traced to Alexander Hamilton and the Federalists. Like Wilson, he revered Lincoln—but Lincoln the partisan and Lincoln the conqueror, not Lincoln the impartial reconciler. When World War I broke out, Lodge became a partisan of Britain and France. The United States, he insisted, "should act, not as an umpire between our allies and our enemies, but as one of the allies." Now, with the war over, his goal mirrored Clemenceau's: to put Germany in a box from which "it will be physically impossible for her to break out again."

For Lodge, the League—as Wilson described it—was a dangerous delusion. America, in his view, would never go to war for Latvia, and he did not like making promises that his nation could and should not keep. He was willing to support U.S. membership, but only once Congress made clear that the League was a mere debating society, which would change nothing fundamental about international affairs. Lodge was not sentimental; he did not believe that international politics would ever be much more than a jungle. He urged Wilson to eschew "efforts to reach the millennium of universal and eternal peace," to stop trying to make "mankind suddenly virtuous by a statute or a written constitution." Better to seek a balance of power against Germany, anchored by an alliance with France. Better to fly low and steadily than be seduced by the clouds above.

In theory, Wilson and Lodge should have been able to come to terms. Wilson desperately wanted the League; Lodge desperately wanted the treaty with France. Each man had the power to grant the other's wish. But to convince Lodge to support U.S. membership in the League, Wilson had to admit that the emperor had no clothes: that America would never go to war at the League's behest. Even more humiliatingly, he had to separate the French security treaty from the League itself, to show that the former, unlike the latter, constituted a binding commitment. He had to acknowledge, in other words, that his dream of a rationalized world would not come true anytime soon.

This Wilson would not do. Instructing his Democratic Senate allies not to compromise, he instead set out to speak to the people, to rally them against selfishness and call them to reason in one last, great educational campaign. His longtime doctor, Cary Grayson, begged him not to go. Wilson had a history of strokes, and that April in Paris, while running a high fever, one side of his face had gone numb and his eye had twitched uncontrollably. Since the president's return to the United States, Grayson noticed, his memory had deteriorated; he seemed confused by simple things. But Wilson brushed off the warning. "I know I am at the end of my tether," he declared, "but . . . in the presence of the great tragedy which now faces the world, no decent man can count his personal fortunes in the reckoning."

On September 3, 1919, he left Washington in a custom-designed train. From Ohio to Nebraska to California and back through the interior West, he delivered thirty-seven speeches in twenty-two days over nearly ten thousand miles. The crowds often swelled into the thousands, and, in every speech save one, Wilson addressed them without a microphone. Finally he began to buckle under the strain. On September 25, during a speech in Pueblo, Colorado, he lost his bearings, and then, upon recalling America's fallen soldiers, burst into tears. That night he coughed so violently that he had to sleep upright to breathe and suffered a headache so painful he could barely see. His train raced back to Washington, but, on the car ride from Union Station to the White House, as his motorcade drove past deserted sidewalks, he began doffing his hat to imaginary crowds, fueling rumors that he had lost his mind. Within a week the left side of his body was paralyzed. He spent most of the rest of his presidency in seclusion, tended to by his wife, Edith. Neither his vice president nor any member of his cabinet would see him for the next year and a half, until he emerged to watch his successor sworn in.

The people had failed Wilson yet again. Even Democrats detected little shift in public opinion as the result of his trip; if anything, anti-League sentiment was building as more and more Americans came to see Wilson's European crusade as one big waste. Colonel House—long Wilson's anchor to the messy, real world—sent him two letters urging him to accept Lodge's changes. "Your willingness to accept reservations rather than have the treaty killed," he implored, employing the flattery that had long greased their relationship, "will be regarded as the act of a great man." But even before his stroke, Wilson had turned on his old friend, convinced that House had made concessions in Paris aimed at sabotaging his dream. Edith Wilson, who had long resented House for competing with her for the president's meager affections, refused the colonel's pleas for a meeting. So Wilson spent the disastrous fall of 1919 walled off from reality, barricaded in what one observer called "the lonely citadel of his soul."

With Wilson a virtual cadaver, and the politically primitive Edith in firm control, the vultures circled. In November, the Irreconcilables and Reservationists joined forces to defeat the unamended treaty. Then, when

Lodge's reservations were added, the treaty was voted down again, this time by a coalition of Irreconcilables and Senate Democrats, whom Wilson had instructed to oppose anything but a clean draft. The following March, when the treaty came up for one final vote, twenty-one Senate Democrats defied the president and backed the amended draft. But they still fell seven votes short of the two-thirds needed for ratification. In retaliation, Wilson refused to urge Senate Democrats to support the treaty with France, and so it died as well.

The tragedy was not, as Wilson's supporters claimed, that Lodge and the selfish Republicans had killed his dream of a rationalized world. That dream was always a fantasy, born of a fundamentally flawed analogy between politics in the United States—where leaders possessed a near monopoly on force and thus could imagine that force had given way to reason—and politics in the world at large, where no such monopoly existed and force stood naked as the arbiter of events. The real tragedy was that Wilson could not himself abandon his dream, even after Paris, and thus help bring Americans to the painful, crucial realization that they must commit themselves to a world they could not perfect. As a result, in the years after World War I, the hubris of reason gave way not to a sober and dogged realism, but to isolationism and pacifism—new flights of fantasy that evaded the central truth that Wilson would not help his country learn: that America must bind itself to the world, even though the world would always let America down.

For John Dewey and Charles Beard, it had been a very bad war. Before America joined the fighting, Dewey had predicted that wartime Americans would be models of rationality. We will be "lenient and amiable in our judgments," he prophesized, and "discount all the more violent appeals to passion." But even on Dewey's own campus, violent passions stirred. Two months after America entered the fray, Columbia University president Nicholas Murray Butler told a commencement audience that "what had been tolerated before becomes intolerable now. What had been wrongheadedness was now sedition. What had been folly was now treason." Four months later, Butler fired psychology professor James McKeen Cattell, the man who had recruited Dewey to Columbia, for writing a letter on university stationery claiming that draftees had the right to conscientiously object. Dewey protested, but most of his colleagues considered the firing entirely justified. Soon Dewey was being investigated by the Justice Department for "pro-German" leanings himself.

Beard's experience was no better. In the spring of 1916, after publicly defending a man's right to shout "To hell with the flag" in a New York City public school, he was summoned by the Columbia trustees. For thirty minutes they questioned him about his allegedly subversive political and historical views before sending him away with a warning not to continue to "inculcate disrespect for American institutions." Cattell's firing the following year was the final straw. On the afternoon of October 9, 1917, Beard finished his afternoon lecture, then told his students, "This is my last lecture in Columbia University. I have handed in my resignation this afternoon to take effect at 9 A.M." Butler and the trustees, Beard later explained, were using "the state of war to drive out, or humiliate, or terrorize every man who held progressive, liberal, or unconventional views in political matters." After a moment of stunned silence, Beard's students rose and applauded for twenty-five minutes straight, while their teacher wept.

Soon Beard himself—like Dewey—was under investigation, in his case by the Senate Judiciary Committee for associating with groups deemed unfriendly to the war effort. The army banned some of his books from its training camps.

Politically, neither man was ever the same. The war, Dewey admitted, had fostered a "cult of irrationality." The "small voice of reason" had been utterly silenced "amid howling gales of passion." Equally disillusioned by the war's fruits overseas, he joined Lippmann and the Irreconcilables in urging the Senate to reject the Versailles Treaty. "Were not those right who held that it was self-contradictory to try to further the permanent ideals of peace by recourse to war?" he wrote in early 1919. It was a veiled apology to Randolph Bourne, the disciple he had betrayed, who had died just weeks before.

In what one friend called "a state of tension that in most people would have been an illness," Dewey left for Asia, where he spent the next two years. He returned to the United States a devout pacifist, insisting, as Bourne had in his final days, that instead of using force to build a world of reason, America should use reason to build a world free of force. He would never support another war.

For his part, Beard carried on the other half of Bourne's revolt against the hubris of reason: isolationism. In 1922, he called for a foreign policy of "continentalism," under which the United States would divest itself of all overseas territories and commitments, refuse all military and diplomatic efforts to protect U.S. investments abroad, and build a military solely devoted to protecting the homeland from foreign attack. Henceforth, Beard pledged, he would "center my efforts on the promise of America rather than upon the fifty century-old quarrels of Europe." The United States, he declared, should "set its own house in order under the stress of its own necessities and experiences."

On paper, Dewey's pacifism and Beard's isolationism were worlds apart. Dewey's vision remained every bit as hubristic as Wilson's; he was just no longer willing to kill to achieve it. Beard, by contrast, now rejected not merely Wilson's means, but his ends: He no longer wanted to rationalize the world. But despite this theoretical gulf, Dewey and Beard shared one core belief, which would define American foreign policy in the interwar years: that America should never again make any overseas commitment that would require enforcement via the sword.

For the French, who desperately needed America's drawn sword to keep the Germans at bay, it was a slow-motion tragedy. In August 1921, Wilson's successor, Warren Harding, signed a separate peace with Germany that freed the United States of all obligations to enforce the Treaty of Versailles. Since Britain had conditioned its own security alliance with France on America's, Paris was now on its own.

A year later, Germany signed the Treaty of Rapallo, normalizing relations with the Soviet Union. For France, the rapprochement was ominous. Paris, which had long pursued close ties with the nations to Germany's east so that in the event of war Berlin would have to fight on two fronts, was now banking on alliances with the newly created nations of Eastern Europe, some of which had been carved from German or Russian soil. Rapallo was the first step in Germany and Russia's efforts to take that land back. Even worse, the treaty contained a secret clause allowing Germany to train its forces on Soviet soil—a direct violation of Versailles, which required the Germans to substantially disarm. Less than three years after the peace treaty, Berlin was already breaking free.

By 1923, the Germans were no longer paying their war reparations, either. Ever since 1921, when an Allied commission had set the reparations bill at 132 billion marks, Berlin had been screaming about the insane, intolerable burden under which it labored. But in reality the burden wasn't as insane and intolerable as it appeared. Given the way the reparations were structured, the bulk of the 132 billion would never have to be paid; the real figure, by some estimates, was only 25 billion marks (about $550 million). Before the Allies presented the bill, the Germans had actually proposed paying more than that themselves.

The real problem was not the amount of reparations Germany had to pay; it was the fact that most Germans did not believe they should pay any reparations at all, just as they did not believe they should forfeit territory, armaments, or their right to immediately join the League. The Versailles peace, which the Germans called a virtual crime against humanity, was in many ways milder than the one the Allies would impose in 1945, and which Germans swallowed with barely a whimper. The difference was that after World War I—unlike after World War II—the Germans did not feel defeated. They had never seen foreign troops marching through their razed cities. Since the Allies, in a terrible blunder, had made Germany's postwar civilian leaders sign the hated Versailles Treaty, rather than forcing the generals who had launched the war to take ownership of the peace, German militarists and hypernationalists found it all too easy to claim that the war had not been lost on the battlefield. It had been lost because liberals, socialists, and Jews stabbed Germany in the back.

Clemenceau was right. The Germans had not abandoned their dreams of European hegemony. And his nightmare was coming true: France was being left to face them alone. It was a vicious cycle. The more abandoned France felt, the more it lashed out against Berlin in a desperate bid to strangle Germany's growing postwar power in the crib. And the more France lashed out, the more it alienated public opinion in America, where isolationists and pacifists alike insisted that the selfish, irrational French were as bad as their German foes, if not worse.

In January 1923, after Germany, which was paying some of its reparations in coal, defaulted on its thirty-fourth shipment in thirty-six months, France and Belgium sent troops into the Ruhr Valley to seize German mines and steel plants as collateral. German inflation had already been running dangerously high as the new Weimar-based government printed currency to fill the gap between what it received in taxes and customs duties (which had fallen sharply) and what it spent (which remained high). The Ruhr occupation tipped that problem into a catastrophe. Furious that French troops were squatting on German soil, Weimar encouraged passive resistance: paying miners and factory workers not to work. To do that, Weimar printed even more money. By fall, it required thirty paper mills and 150 presses all operating twenty-four hours a day to generate all the marks that Germany's government was sending into circulation. To purchase the goods that one mark would have bought in 1914 now required one trillion. Its savings rendered virtually worthless by hyperinflation, the German middle class was now easy prey for a Nazi Party whose popularity was beginning to build.

Americans mostly responded by blaming France. The Ruhr invasion, declared William Borah, was "utterly brutal and insane." Oswald Garrison Villard, editor of the *Nation*, called it every bit as despicable as Germany's invasion of Belgium during World War I. Thousands assembled at New York's Madison Square Garden, where a litany of senators denounced French militarism. To register his disapproval, President Harding withdrew all remaining U.S. troops from the Rhineland, abandoning the final element of American enforcement of Versailles. Four years earlier, when they signed their names at Versailles, French leaders believed they had an American commitment to defend their borders and to help keep Germany from arming in the Rhineland. By 1923, they had neither.

The Ruhr occupation did not just offend American sensibilities; it hurt America's bottom line. World War I had turned the United States into the world's largest creditor, and U.S. investors were keen to sink their money into postwar Germany, whose reviving industries they believed would offer fat rates of return. But as long as the French occupied the Ruhr, Germany's economy could not revive. So in 1924, President Calvin Coolidge sent banker Charles Dawes to convince Berlin to start paying a reduced reparations bill, and to convince Paris to withdraw its troops. Dawes succeeded. The Germans began paying, France pulled back, and Americans made money. But for France, it was yet another defeat. By agreeing to the Dawes Plan, the French effectively conceded that it would be Washington—not Paris—that determined whether Germany was paying sufficient reparations. And since Dawes and Coolidge didn't much care whether Berlin paid its reparations to France, so long as it repaid its loans to America, Weimar could now default with virtual impunity. Just as significant was the way Dawes approached the two sides: as an impartial mediator. As Coolidge put it, "We are independent, detached and can and do take a disinterested position in relation to international affairs." Wilson had at least made America an "associate" of the French cause. Coolidge went further: America was now strictly neutral between the nation it had fought alongside and the nation it had fought against, between France, which had no capacity to dominate Europe, and Germany, which still did.

Soon the impartiality was enshrined in international law. In 1925, at the Swiss resort of Locarno, Germany, France, and Belgium pledged to respect each other's borders, with Britain and Italy serving as guarantors. It all seemed wonderfully hopeful: The wartime belligerents were finally starting to reconcile. Europeans began speaking of a "spirit of Locarno" breaking out across the continent; the British, German, and French negotiators all won the Nobel Prize. But there were darker forces at play. At Locarno, the final, slim prospect of a British alliance with France died. Instead of pledging to protect France against German attack, Britain was now pledging to protect Germany against French attack as well, which presumably meant that if the French again sent troops into the Ruhr to enforce Versailles, then London and Paris would be at war. At Locarno, the Germans also conspicuously refused to pledge to respect their borders with Austria, Czechoslovakia, and Poland. It was yet another sign that Berlin intended to break out of its Versailles straitjacket in the east, a move that would in turn imperil the two-front strategy on which French security relied.

Slowly and artfully, using progressivism's language of impartiality as their pretext, the Germans were wriggling free. In 1926, they were admitted to the League and even given a seat on its council. Now any French bid to use the League to enforce Versailles was subject to Berlin's veto, and any League-sponsored effort at disarmament would have to apply equally to both Germany and France. Since France, given its visceral fear of its eastern neighbor, was unwilling to disarm, Germany no longer had to, either. In January 1927, the Inter-Allied Military Control Commission, which had been (halfheartedly) monitoring Berlin's compliance with its disarmament obligations, left German soil. In theory, under the Versailles Treaty, Germany still had only one hundred thousand troops and no tanks, heavy weaponry, or air force. Meanwhile, France began building a chain of military defenses across the length of its border with Germany: the Maginot Line.

For the pacifists and isolationists who shaped American foreign policy in the 1920s, it was France—more than Germany—that exemplified the selfish, irrational tendencies they disdained. The Germans, with their calls for international equality, mutual disarmament, and self-determination, spoke America's language. The French, by contrast, kept going on about fortifications, punishments, alliances, and the balance of power—all grating to the American ear. France, wrote Dewey, "affords a striking example of the fact that" in international affairs, "the old policies and the old type of politicians are still absolutely in control." When it came to security, the French were like a man desperate for food, and every time they lunged for it, Americans reprimanded them for their poor table manners.

While France grimly prepared for another war, Americans in the 1920s fell madly in love with the idea of peace. From Ernest Hemingway to John Dos Passos to William Faulkner to H. L. Mencken, American novelists and critics savaged the idea that there was nobility in war. Harding and Coolidge's influential secretary of state, Charles Evans Hughes, began calling the State Department "the Department of Peace." In 1922, publisher Edward Bok offered one hundred thousand dollars to anyone who could outline a scheme for world peace in less than five thousand words, and a staggering 250,000 Americans inquired about taking part. The *Nation* demanded "a burning, an unconquerable, an undeviating hatred of war, any war for whatever reason." On Armistice Day 1932, Nicholas Murray Butler, the Columbia University president whose wartime jingoism had so traumatized Dewey and Beard, called on the nations of the world to destroy their weapons, but leave a few to put in museums so future generations could see how barbaric human beings had once been.

Influential Americans actually said such things in the 1920s, in full sincerity. What's more, they tried to turn their words into action. In 1920, Borah introduced legislation proposing a conference with Britain and Japan aimed at cutting each nation's naval expenditures in half. It passed the Senate, 74–0. The following fall, Secretary of State Hughes convened the Washington Naval Conference, which produced an agreement between America, Britain, Japan, France, and Italy on a ratio limiting their battleship fleets. Hughes, one British observer quipped, had destroyed more royal vessels than all the admirals in history. Scattered voices in the U.S. departments of Navy and War muttered that the ratios did not allow America to keep pace with Japan in the Pacific, but they were missing the larger point: Congress wouldn't have appropriated enough money to keep pace even without the treaty. Between 1922 and 1926, American naval spending dropped by one-third.

By 1927, the Coolidge administration was itching for another disarmament conference. But the French, who had been dragged unhappily into the first one, refused to play along. Instead, French foreign minister Aristide Briand—a man renowned for his wit and great mop of disheveled gray hair—proposed that the United States and France pledge never to go to war. It was a sly move. Since Versailles, Paris had been busily encircling Germany with alliances: with Belgium in 1920, Poland in 1921, Czechoslovakia in 1924, and Romania and Yugoslavia in 1926. However, in the rational, unselfish new world supposedly born from World War I, *alliance* was a dirty word. So the French didn't call their treaties "alliances"; they called them pledges never to go to war (sometimes with secret addenda promising military aid in case either signatory went to war with anyone else). Now Paris was proposing such a pledge with the United States.

Hughes's replacement as secretary of state—an elderly, profane, former corporate lawyer named Frank Kellogg—was not fooled. He knew that if America and France pledged never to go to war, America would find it much harder *not* to go to war with whomever France was fighting against. With a U.S. nonaggression pact in its back pocket, France could infringe upon U.S. neutrality rights—as the British had done during World War I—secure in the knowledge that America could not retaliate. This would make the United States a de facto belligerent on France's side.

The American peace movement, however, was enthralled by Briand's idea, and so Kellogg needed a counterproposal. In that effort, he turned to the work of John Dewey. Upon his return from self-imposed exile in Asia, Dewey had thrown himself into a movement called the Outlawry of War. War, Dewey noted, was not currently illegal under international law. There were laws *of* war—rules about treating prisoners, and the like—but no laws *against* war. This, Dewey declared, made international politics barbaric and archaic. Once upon a time, he claimed, societies had not outlawed murder, either; they had just regulated it. Dueling was once a legal method of settling disputes within nations. But now humanity had progressed. Just as dueling had been outlawed between individuals, war should now be outlawed between nations.

It was a deeply progressive argument: History was marching upward from barbarism to civilization, and, in that effort, the triumph of reason and law *within* nations should serve as a template for the triumph of reason and law *between* nations. It was similar in spirit to what Dewey had believed in 1917, except that now Dewey insisted that this upward march must proceed not via war, but via its abolition. As always, the key was public opinion. Dewey, like Wilson during the Paris Peace Conference, insisted that ordinary people were more morally advanced than their leaders. Across the world, he insisted, a "community of moral feeling" against war already existed. By helping to outlaw war, America would nurture that moral feeling until it became a stigma so profound—like the stigma against slavery or incest—that war became almost unthinkable. Dewey's pacifism was the ethic of reason turned inside out, a kind of hubris of the impotent. America should insist on an utterly remade world, governed by reason, not force. But the means should be as pure as the end: To avoid sullying the dream of a rationalized world, America should shed no blood in its pursuit.

In 1921, Dewey had written to Borah, the flamboyant leader of the Senate Irreconcilables, urging him to draft legislation committing the United States to the abolition of war. Two years later, Borah did just that, and in 1927, when Kellogg needed an answer to Briand, Borah's legislation was still sitting there like an unlit match. When Kellogg struck the fuse and proposed a treaty banning war not just between the United States and France but between all nations, the peace movement erupted in even greater exhilaration. Carrie Chapman Catt's Committee on the Cause and Cure of War convened ten thousand town meetings to pass pro-Outlawry resolutions; side by side the resolutions stretched for two miles. Letters endorsing Kellogg's proposal flooded the State Department at the rate of six hundred per day.

The French were furious. Once again they had tried to drag Americans down into the harsh, real world only to see them float back up into the haze of dazzling abstraction. There was nothing left to do but smile grimly and play along. At the signing ceremony, where sixty-four nations pledged to forever abstain from war, Briand—a glorious bullshitter when necessity required—solemnly dedicated the treaty to the World War I dead. President Coolidge, with greater sincerity, declared that the Kellogg-Briand Treaty held "greater hope for peaceful relations than was ever before given to the world." The Senate passed the treaty, 85–1. Dewey called it one of the most important events in the history of international affairs. Kellogg won the Nobel Prize.

Some of the senators who voted for the treaty were isolationists. They didn't believe for a minute that the blood-soaked Europeans would refrain from war. But the treaty was a harmless sop to their pacifist constituents, a meaningless commitment that substituted for a real one. Pacifists, by contrast, supported the treaty in deadly earnest; they considered Kellogg-Briand the dawn of a virtual messianic age. Between them, the two factions embodied the same dual impulses that had animated Randolph Bourne in his final days: to avoid war by fleeing it and to avoid war by banning it. Wilson, for all his self-delusion, had actually risked something for his vision of a rationalized world. Beard and the isolationists abandoned that vision; Dewey and the pacifists thought they could achieve it by proclamations alone. What they all shared—Wilson and those who rebelled against him in the giddy, unreal 1920s—was a refusal to meet the world on its own terms, to accept that politics between nations would never resemble politics between Americans, yet must be embraced nonetheless. The real answer to the hubris of reason was still waiting to be born.

CHAPTER THREE

TWICE-BORN

On a fall day in 1932, John Dewey ascended a Manhattan podium to endorse a candidate for Congress: a theologian thirty years his junior named Reinhold Niebuhr, who was running on the Socialist ticket in New York state's 19th District. It was a decision he would regret. Unbeknownst to Dewey, Niebuhr had just completed a book that attacked him by name. Dewey, Niebuhr charged, placed too much faith in reason, forgetting that "reason is always, to some degree, the servant of interest." And his pacifism was naïve, since "conflict is inevitable, and in this conflict power must be challenged by power." Niebuhr's book, *Moral Man and Immoral Society*, was a broadside against the faith to which Dewey had clung in both war and peace: that through education humanity could be fundamentally improved. For the aging philosopher, it would prove little consolation that the young cleric won only 4 percent of the vote.

Like Randolph Bourne, Niebuhr had once been a fervent admirer of Dewey, and like Bourne, he had scores to settle from World War I. Niebuhr had been twenty-four years old when America plunged into the Great War, and his war progressivism had an almost desperate tinge. Niebuhr, after all, was German, the son of an immigrant from the province of Lippe-Detmold. When the war broke out, he was a student at Yale Divinity School, struggling to perfect his English. And when the United States entered the fray, he was a minister at a German-language church in northwest Detroit, having only recently convinced the board to allow English-language services every week. As war hysteria gripped the nation—and publicly speaking German became in some places a crime and in others an invitation to mob attack—Niebuhr refused to defend his besieged flock; to the contrary, he added to their torment. Eager for literary fame, he broke through with a cover story in the *Atlantic* implying that when it came to German-Americans, the public's "suspicions of disloyalty" were wholly justified. He demanded removal of the word *German* from his denomination's letterhead, and when a church newspaper gently criticized wartime censorship, Niebuhr chastised its editor, insisting that the paper express "out-and-out loyalty."

The ambitious young preacher justified all this by declaring World War I a noble crusade: "the inspiring spectacle of a nation making every sacrifice of blood and treasure for aims which do not include territorial ambitions or plans for imperial aggrandizement." He was gambling that Wilson and Dewey were right: that the war would midwife a nation and a world governed more by reason than force. When that hope died at Versailles, Niebuhr announced—like Randolph Bourne and Walter Lippmann—that he had been betrayed. He called Wilson a dupe. He denounced the wartime pastors so "anxious to prove that they liked the smell of blood." He fretted that "no one will ever know the tragedy in the millions of lives of German immigrants" treated "as potential criminals" by their adopted nation. Niebuhr understood the tragedy all too well. He had seen it up close, not as a victim, but as a victimizer. Like Lippmann, he responded to the collapse of his war progressive dreams with rage when the more honest reaction would have been shame.

For a time after the war, Niebuhr called himself a pacifist. But his heart was never in it. Even when he supported outlawing war, he kept rebuking his fellow outlawers for their excessive idealism, for their failure to see that war was a product not of human ignorance, which could be eradicated, but of human sinfulness, which could not. World War I, he declared, "made me a child of the age of disillusionment." And that disillusionment applied not merely to the belief that the world could be rationalized through war, but to the belief that it could be rationalized through peace as well.

What exactly all this disillusionment meant for Niebuhr's foreign policy views was not yet clear, since in the 1920s foreign policy was not his major concern. But when it came to domestic affairs, he kept pointing out the limits of reason and the centrality of power, taking aim at core principles of the progressive creed. One of his favorite examples was Henry Ford, the auto titan who dominated Niebuhr's adopted city of Detroit. Ford made a great show of his Christian piety, thus suckering some progressive-minded ministers into believing that by citing the ethical teachings of the gospel they could convince him to treat his workers more fairly. Niebuhr called that a delusion. Ford didn't exploit his workers because he was ignorant of the Bible; he exploited them because it made him rich. The problem was not inadequate education; it was naked self-interest. And the answer, in Niebuhr's view, was to confront Ford with the organized self-interest of his workers, who should fashion themselves into a labor union and refuse to work at starvation wages. Unlike the progressives, Niebuhr didn't think the clash between Ford and his workers had an objective solution. He wasn't seeking a scientific peace. He was seeking a new industrial balance of power, in which the rich—acting less from reason or love than from fear—satisfied the selfish desires of the empowered poor. In 1929, Niebuhr made it official, joining the Socialist Party.

Then the Depression hit. By 1933, America's gross national product was less than one-third its level in 1929. Food riots grew common; families foraged in garbage dumps for something to eat. In this bleak and savage new America, Niebuhr lurched further left. Once again he took aim at progressives, whom he accused of trying to persuade the rich to treat the poor more justly. The old progressive faith in education, and in the ultimate harmony of society, he argued, was naïve. The class war had begun, and it was a contest not of reason, but of might. Dewey, Niebuhr claimed, would not stare this harsh reality in the face. He could not see that when oppression forces you to the barricades, the time for pedagogy is done.

In his attack on Dewey, Niebuhr reflected his times. The Depression politicized intellectuals and radicalized them, leaving many contemptuous of the claim that the struggle between business and labor could culminate in a scientific peace. In 1931, Lippmann's first boss, the famed muckraker Lincoln Steffens, published an autobiography mocking his earlier belief that by exposing society's crimes, journalists could educate those in power to remedy them. In his influential 1932 polemic, *Farewell to Reform*, John Chamberlain took stock of three decades of progressive good intentions and declared them a failure; revolution was the only option left. That same year, a host of prominent writers, including the novelist John Dos Passos, the philosopher Sidney Hook, and the literary critics Malcolm Cowley and Edmund Wilson, announced that they were voting communist for president.

Looking at the election results, one might have wondered whether any of this mattered. American politics did move left in the 1930s, but the result was not revolution, but the New Deal—not a farewell to reform, but its spirited revival. At the very moment leftist intellectuals were burying progressivism, Franklin Roosevelt seemed to be dredging it up.

But the intellectual rebellion that Niebuhr led—against Dewey's faith in the inevitability of progress, the efficacy of education, and the inherent goodness of man—would by the early 1940s profoundly affect American foreign policy, as the United States climbed out of the Depression and found itself standing at the precipice of a second world war. As in 1917, a vision of America would become the template for a vision of the entire globe. And a generation less intoxicated by success than its predecessor, and more toughened by tragedy, would do what the pacifists and isolationists of the 1920s and 1930s could not: bury the hubris of reason and reconcile America to a fallen world.

All this, however, was still a long way off when Dewey endorsed Niebuhr on that fall day in 1932. America's president was Herbert Hoover, a man who neatly blended the pacifism and isolationism of the interwar years. Hoover believed in reason, not force, and understandably so: He had seen the former transform his life and the latter devastate the world. His biography would have made Horatio Alger blush. Orphaned at age nine, he was sent to live with his uncle in Oregon, who forced young Herbert to labor such long hours in the family real estate office that he was unable to attend high school. Undaunted, the young striver took classes at night, where he mastered bookkeeping, typing, and math and so impressed his teachers that he gained admission to Stanford University's inaugural class. There he trained as a mining engineer, and by age forty his genius at exploiting inaccessible deposits had made him a multimillionaire. (To this day, Hoover's entry in the Australian Prospectors & Miners Hall of Fame does not even mention that he became president.)

In early-twentieth-century America, engineering was a quintessentially progressive profession, and Hoover was, in one colleague's words, the "engineering profession personified." Engineers saw themselves as scientifically trained experts standing above the selfish clash between business and labor. They occupied, in Hoover's words, a "position of disinterested service," "want[ing] nothing . . . from Congress [except] efficiency." And this, in Hoover's mind, made them just the kind of people America needed to run its government.

World War I provided Hoover his chance. In London on business, he was asked by the U.S. Embassy to help find food, lodging, cash, and safe passage back to the United States for the thousands of Americans marooned in Europe. This he did with dazzling efficiency: Of the $1.5 million in loans he arranged for roughly 120,000 stranded Americans, all but $300 was paid back. When the job was done, the U.S. ambassador called Hoover "a simple, modest, energetic man who began his career in California and will end it in heaven."

After that, Hoover was charged with aiding the people of Belgium, who were suffering both from German occupation and a British blockade. This time he oversaw the shipment of more than $5 billion in food and supplies while spending less than half a percent on overhead. By war's end, close to four million Europeans had written letters or signed petitions thanking him for his work. The people of Finland invented a new verb: *hoover*, meaning "to help." Now something of an international celebrity, he was tapped by Wilson to head the American Food Administration, where he ran a vast public relations campaign aimed at convincing Americans to ration food and other staples so there would be enough for the troops overseas. Soon, the word *Hoover* entered the American lexicon as well, as a verb meaning "to conserve." "I can Hooverize on dinners and on lights and fuel too," read a popular Valentine's Day card in 1918, "but I'll never learn to Hooverize when it comes to loving you!"

Hoover's wartime experience left him with foreign policy views that blended the pacifism of John Dewey and the isolationism of Charles Beard. Already predisposed to nonviolence by his Quaker upbringing, Hoover now contrasted his accomplishments as an international engineer and relief worker—using science, commerce, and moral appeals to improve humanity's lot—with the savagery and waste of Europe's war. But if Hoover, like Dewey, believed that America should try to build a world free of force, his wartime experience also inclined him toward Beard's pessimism that Europeans would ever shed their murderous ways. The result was a foreign policy ideology best described as "pacifism if possible, isolationism if necessary." The U.S. government, in Hoover's view, should appeal to other nations to cut armaments and renounce aggression, just as he had appealed to Europeans to aid Belgium and to Americans to stop eating meat. And the United States should spread capitalism across the globe, just as he had in his mining days. But America should always retain the freedom to retreat across the oceans if its efforts were rebuffed. The United States, Hoover believed, was militarily and economically self-sufficient. For him, as for Harding, Coolidge, and even Wilson, America's international engagement was conditional: If the world lived up to our standards, great. If not, we could survive perfectly well on our own.

This was exactly the attitude that had been making French leaders twitch ever since Versailles. In 1929, soon after entering the White House,

Hoover, along with British prime minister Ramsay MacDonald, hatched plans for another naval disarmament conference, to extend to smaller ships the restrictions put in place in 1921. Thrilled, peace activists sent the State Department so many letters that clerks complained they could not process all the mail.

In London the following spring, the United States, Britain, and Japan agreed to limit their fleets of destroyers, cruisers, and submarines. But the French—confirming their reputation as the troglodytes of Europe—would not join in. Pressed to accept naval parity with Benito Mussolini's Italy, Paris responded with that most tedious and hoary of demands: a security alliance with America and Britain. When Hoover refused, the French walked out. France, Hoover commented, was "rich, militaristic, and cocky; and nobody can get on with her until she has to be thrashed again."

The Germans could not have agreed more. Berlin's bid to free itself from the fetters of Versailles, which it had once pursued with stealth and good manners, was becoming more blatant. A big reason was the Depression, which further devastated a German middle class still reeling from the hyperinflation of 1923. In September 1930, with unemployment nearing 15 percent, the Nazis—who had won roughly 800,000 votes two years earlier—garnered more than 6 million, making them the Reichstag's second-largest party. Six months later, Berlin announced a customs union with Austria, the first step toward unifying the two German-speaking nations, which Versailles explicitly banned. The French were apoplectic and ultimately squashed the move, but Hoover, who saw the customs union as economically rational, not geopolitically threatening, gave his blessing. "Evidently you don't approve of the Versailles Treaty," remarked Secretary of State Henry Stimson. "Of course I don't," Hoover replied. "I never did."

"In the realm of reason," observed Walter Lippmann, Hoover "is an unusually bold man." But "in the realm of unreason he is . . . easily bewildered." Unfortunately for the Great Engineer, unreason was on the march in the Depression years, and not just in Germany. In September 1931, a cadre of younger Japanese army officers swearing fealty to *bushido*, the ancient Japanese warrior code, engineered a massive invasion of the northern Chinese province of Manchuria. Hoover was stunned: He had thought American morality and commerce were taming the Japanese. Just a year earlier, Tokyo had signed the London Naval Treaty, and Japan's economy was deeply enmeshed with America's own. But the Japanese prime minister who signed that treaty was now dead, shot in the abdomen by a young hypernationalist. And during the Depression, America's own protectionist policies changed Tokyo's economic calculus. In an increasingly autarkic world, Japan could no longer depend on financial and commercial links across the oceans; it needed an economic bloc of its own. That required securing Manchuria, its primary source of iron and coal, and the recipient of 90 percent of its foreign investment. By year's end, Japanese troops had pushed several hundred miles into China. In Tokyo, yet another civilian prime minister lay dead, and the army and navy were firmly in control.

In response, Hoover turned to the methods that, in his own life, had served him so well. He issued a moral appeal, informing Tokyo that since it had violated the Kellogg-Briand Treaty banning war, America would not recognize the fruits of its aggression. But Japan's war-frenzied regime paid little heed. Hoover hoped the strain of conquest might spark Japan's economic collapse, but instead its economy improved. He backed a League of Nations investigation into the incursion, but when the world body issued its condemnation, Japan walked out. Finally, Secretary of State Stimson—a blue-blooded Theodore Roosevelt protégé who feared war less than his boss, and understood power politics more—suggested threatening sanctions. But Hoover refused; his experience in Europe had bred in him a deep aversion to blockades. "The United States," he declared, "has never set out to preserve peace among other nations by force."

In the 1920s, America's refusal to help uphold the European or Asian balance of power had been more understandable. Given the lack of obvious foreign threats, and the American public's hostility to overseas obligations, it would have taken an unusual leader to see that Germany's potential strength and undaunted ambition necessitated an American commitment to France. Theodore Roosevelt and Henry Cabot Lodge might have been such leaders, but by the mid-1920s both were dead.

Harder to defend was Hoover's unwillingness to build up American strength in the early '30s, with the Japanese rampaging through China, and Germany growing increasingly aggressive in its territorial demands. Even after the Manchuria invasion, Hoover still did not commission a single new warship. To the contrary, he began planning the grandest disarmament initiative yet. For years peace groups had been agitating for a meeting to address not merely limitations on a few classes of ships among a few powers, but the disarmament of the entire globe. And in 1932 they got their wish, when representatives from fifty-nine nations arrived in Geneva for the World Disarmament Conference. U.S. officials quickly electrified the delegates by proposing that all "offensive" weapons be banned, and all other armaments be cut by at least a quarter. A peace convoy of more than two hundred cars, carrying a petition with 150,000 signatures in support of Hoover's efforts, drove from Los Angeles to Washington, where it was taken to the White House by police escort.

Predictably, the French were hostile. Naval disarmament was bad enough, but slashing the size of their army—when intelligence reports showed that Germany was rapidly rearming—was downright terrifying. Yet again Paris demanded security guarantees from Washington and London, and yet again Paris was rebuffed. "When would the American government learn," exclaimed French foreign minister André Tardieu, "that it could not brutalize France into concessions which it was not prepared to make?" In July, the disarmament conference temporarily adjourned. Four months later, Germany's parliamentary elections produced an upset: the Nazis had won a plurality of the vote.

If Hoover clung to the foreign policy principles of a dying age, his successor, Franklin Roosevelt, was widely thought to lack any firm principles at all. Even as a child, his cousins, Theodore's children, considered him a toady, eager to tell adults whatever they wanted to hear. At Harvard, classmates called him a "false smiler" and "two-faced." When FDR ran for president, H. L. Mencken quipped that if he "became convinced tomorrow that cannibalism would get him the votes he so sorely needs, he would begin fattening up a missionary in the White House backyard." The charges of spinelessness and opportunism would never entirely stop. "Every great leader had his typical gesture," declared the writer and Republican congresswoman Clare Boothe Luce years later. "Hitler the upraised arm, Churchill the V sign. Roosevelt?" She licked her index finger and raised it to the wind.

But FDR did have foreign policy principles, or at least foreign policy instincts, instincts that would eventually help him overcome the hubris of reason. Those instincts were just difficult to see early on because one of them was that a politician should conceal his views when they clashed with the public's, and for most of the 1930s, FDR's did. Had he been more forthright—less two-faced and opportunistic—it is unlikely he would have been elected, or reelected, president of the United States.

FDR's foreign policy instincts were shaped by two mutually reinforcing passions: for Teddy Roosevelt and for the sea. As a boy, the young

Franklin hero-worshipped his fifth cousin, whom he called the greatest man he had ever known. At fourteen, he bought a pair of TR's trademark pince-nez eyeglasses. At twenty-three, he married TR's favorite niece, Eleanor. He slavishly patterned his political career on his famous relative's: winning election to the New York state assembly, getting himself appointed assistant secretary of the navy (and trying to leave that job to raise a regiment to fight in Mexico, just as TR had done in Cuba during the Spanish-American War), and then becoming governor of New York.

FDR mimicked his cousin ideologically as well. Although he was working in Wilson's Navy Department when World War I broke out, FDR's views more closely resembled those of TR and Lodge. Like them, he made little pretense of objectivity; he wanted Britain and France to win. And like them, and in defiance of his superiors in the administration, he urged as early as 1915 that the United States build up its military to deal with a potential German threat. FDR's wartime statements bristled with TR's belief that war between nations was inevitable, necessary, even fun, and that it would do America good to get into the fight. He mocked his boss at the Navy Department, a pacifist named Josephus Daniels, whose "faith in human nature and civilization and similar idealistic nonsense was receiving such a rude shock" from the war. And he derided the "soft mush about everlasting peace which so many statesmen are handing out to a gullible public." When the war ended, Roosevelt supported American membership in the League of Nations but added that "I don't care how many restrictions or qualifications are put on [it]." The League, he explained, was "merely a beautiful dream, a Utopia." In the harsh and exhilarating real world, the struggle for power would rage on.

If FDR's muscular foreign policy views echoed TR's, they also stemmed from his almost mystical connection to the sea. Growing up on an estate alongside the Hudson River, he built toy boats before he could read, sailed his first real one at age nine, and owned a vessel by age sixteen. At his prep school, Groton, he filled his room with prints of sailing ships and tried to run away and join the navy during the Spanish-American War. At Harvard he began collecting manuscripts about naval history and developed a lifelong fascination with geography. He also read the works of the famed Admiral Alfred T. Mahan, who argued that a strong navy, capable of protecting commercial shipping, securing colonies, and protecting home waters, was essential to a nation's power.

FDR's love of the sea reinforced his cosmopolitanism: As a child, he spent so much time learning German and French from foreign-born governesses that when he entered Groton he spoke with a slight European accent. He traveled to Europe for the first time at age three, and then every summer from ages seven to fifteen. Unlike Wilson, whose image of the world derived largely from theories developed in the United States, FDR was used to seeing events through foreign eyes. And unlike Hoover, whose international mining and relief work inclined him to see foreign nations as consumers or charity cases, FDR, because of his naval obsession, could more easily see them as competitors. Expert sailor that he was, FDR knew that the Atlantic and Pacific were no longer America's moats—that they grew smaller with every naval advance. Even before World War I, he urged Daniels and Wilson to build a fleet capable of patrolling one thousand miles off American shores. If "an enemy of the United States obtains control of the seas," he warned, the Western Hemisphere would offer no refuge.

Finally, it was the sea that introduced Roosevelt, just before his fortieth birthday, to tragedy and forced him to summon a previously untapped, inner strength. As a young man, he had indeed been—as his detractors whispered—pliable, smug, and light. For all his mimicry of his famous relative, the contrasts spoke louder than the similarities. While an undergraduate at Harvard, TR had begun writing his authoritative, five-hundred-page history of the Naval War of 1812. FDR, by contrast, wrote editorials for the *Harvard Crimson* urging louder cheering at football games. TR had actually waged war; FDR just pretended to. In July 1918, after much pestering, Daniels let him travel to wartime Europe to inspect the U.S. fleet. Although he later giddily recounted his voyage through (allegedly) sub-infested waters and described the artillery fire he heard on the battlefields of northern France, FDR never encountered any real hardship or danger. To the contrary, traveling with his old Harvard friend and golfing partner Livingston Davis, he stayed in the fanciest hotels, dined with foreign dignitaries, and drank himself sick. (After one particularly wild night of carousing in Scotland, he awoke to find himself sharing a bed with a fox.) Although already sporting a fever, he stayed out until 4:30 A.M. on his final night in Europe, and then promptly contracted double pneumonia on the voyage home. When the boat docked in New York City, he was carried by stretcher to his mother's house, where Eleanor, while unpacking his clothes, discovered his love letters to her former assistant, Lucy Mercer. Eleanor offered him a divorce, but when his mother threatened him with disinheritance, FDR declined. For Roosevelt's critics, the trip just illustrated his true nature: boastful, spoiled, irresponsible, dilettantish, and weak—"the featherduster," as college classmates called him behind his back.

But something changed in his thirty-ninth year. After a long and exhausting deep-sea fishing trip near the Roosevelt vacation compound off the New Brunswick coast, FDR fell into icy waters and soon began to shake uncontrollably. Within days, he was paralyzed from the waist down by polio. As his muscles atrophied, his shriveled legs began to bend backward behind his knees. To keep them straight, a doctor pressed them into heavy plaster casts, where they were forced upright by iron rods, in a procedure that one biographer compared to "being stretched on a rack in a medieval torture chamber." Years later, FDR told a friend that in those days of agony, for the first time in his life he lost his faith in God, and then, somehow, found it again.

Outwardly, FDR remained much the same. He rarely complained, and he retained his usual breezy demeanor. But those close to him noticed a profound change. Polio, noted Eleanor, was the "turning point" in her husband's life. It "gave him strength and courage he had not had before." Roosevelt's close friend and future labor secretary Frances Perkins described "a spiritual transformation" and a new "humility of spirit." FDR's uncle called him a "twice-born man."

This can be taken too far. In important ways, the foreign policy realism that FDR smuggled into the White House antedated his disease. It had its roots in his adoration of TR, his reading of Mahan, his travels in Europe, and his image of the sea. But in some inchoate way, polio reinforced his view of the world as an irrational, unknowable, unyielding place. By 1932, FDR understood, in a way Wilson and Hoover never did, that politics, like life, did not permit tidy constructs or linear progress. His personal story—brash optimism and easy success followed by sudden calamity and a struggle against the limits it imposed—was a kind of metaphor for his entire generation. Like Niebuhr, he was also, in his way, a "child of the age of disillusionment." As Roosevelt campaigned for president on crutches, rocking his stiff, steel-encased legs and hips forward in order to move a few excruciating feet, his old drinking buddy, Livingston Davis, shot himself. It was yet another sign of an end to childlike things.

At first, Charles Beard was enthusiastic about Roosevelt's presidency. Tacking as he so often did with the political wind, FDR had spent the 1920s downplaying his internationalist foreign policy views. At the 1932

Democratic National Convention, when the isolationist publisher and power broker William Randolph Hearst demanded that he renounce his support for the League, Roosevelt quickly complied, thus helping secure his party's nomination. In his first year in office, he torpedoed efforts to coordinate a global response to the Depression and his delegate to the London Disarmament Conference announced that America would make "no commitment whatever to use its armed forces for the settlement of any dispute anywhere." FDR did triple the budget for new ship construction, the biggest increase since 1916, but even this he justified on isolationist grounds: as a way to fight unemployment and defend America's home waters. Impressed, Beard applauded Roosevelt's renunciation of "any duty owed by the United States to benighted peoples." By the fall of 1933, Beard was dining at the White House and reporters were calling him "one of the intellectual parents of the New Deal."

Hitler was now in power. Four days after being appointed chancellor, he ordered a massive rearmament effort. A few months later, he pulled Germany out of the London Disarmament Conference and the League. In early 1934, he made his rearmament efforts public, in bald defiance of Versailles. Seeing no response from Britain or France, he introduced conscription the following year. Then, on the morning of Sunday, March 7, 1936, German troops marched into the Rhineland. France asked if it could expect Britain's support if it met Hitler's move with force, support that Britain had twice pledged to provide, at Versailles and Locarno. But His Majesty's government refused, with its secretary of state for war telling Hitler's ambassador that the British public "did not care 'two hoots' about the Germans reoccupying their own territory." For his part, FDR's secretary of state, Cordell Hull, declared that since the United States had signed neither Versailles nor Locarno, Hitler's actions were none of its concern. Only later would historians learn that Hitler, whose military buildup was still in its early stages, had instructed his generals to retreat at the first sign of foreign resistance.

Privately, FDR called Hitler a madman and mused about offering the Fuhrer a disarmament deal, and imposing a naval blockade if he refused. But publicly he said nothing of the sort. An astute student—and often slavish follower—of popular opinion, Roosevelt knew exactly how his fellow citizens felt about German aggression: Far from undermining American isolationism, Hitler's moves were bolstering it. During the 1920s, when the prospect of another European war seemed remote, pacifism had been as strong a public force as isolationism, although the line between the two sometimes blurred. But by the mid-1930s, with militarism rising in Germany, Italy, and Japan, Americans found it harder to imagine that through commerce, disarmament deals, and nonaggression pacts they could build a less violent, more rational world. And so pacifism, with its optimistic, outward-looking spirit, receded before isolationism: the sense that with storms brewing on the horizon, America should shut its windows and lock its doors. "Let us turn our eyes inward," declared Pennsylvania governor George Earle in 1935. "If the world is to become a wilderness of waste, hatred, and bitterness, let us all the more earnestly protect and preserve our own oasis of liberty."

American efforts to do that were shaped, to a striking degree, by the memory of World War I. From 1934 to 1936, North Dakota's Gerald Nye oversaw a vast Senate investigation, involving ninety-three hearings and more than two hundred witnesses, into whether bankers and arms dealers had lured America into the Great War. Bestselling books such as H. C. Englebrecht and F. C. Hanighen's *Merchants of Death*, Walter Millis's *Road to War*, and Beard's *The Open Door at Home* and *The Idea of the National Interest* hurled essentially the same charge. On April 6, 1935, the eighteenth anniversary of America's entry into the war, fifty thousand veterans marched for peace in Washington, laying wreaths at the graves of three congressmen who had voted no. A January 1937 poll revealed that 71 percent of Americans considered U.S. intervention in World War I a mistake.

The result of all this historical remorse was the Neutrality Acts, a series of laws that between 1935 and 1937 prohibited the sale of weapons or the lending of money to countries at war, and warned Americans that if they traveled on belligerents' ships, they did so at their own risk. This time reckless tourists would not be permitted to get themselves drowned, and weapons manufacturers and Wall Street speculators would not be allowed to sign contracts that gave them a financial incentive in the spilling of American blood.

The Neutrality Acts represented the final, tragic, unintended consequence of Wilson's justification for U.S. entry into World War I. Wilson had justified that entry by invoking America's sacred rights under international law. Now the historical revisionists were insisting that those rights hadn't really been sacred at all; they had masked the pecuniary interests of arms dealers and Wall Street fat cats—fat cats who were none too popular with American capitalism on its knees. The war progressives had offered an idealistic rationale for war, not a geopolitical one, and so once Americans decided that this idealism was a fraud, no compelling argument for international engagement remained. In this sense, the isolationism of Charles Beard was war progressivism's bastard child.

Franklin Roosevelt had World War I on the brain as well. "From 1913 to 1921, I personally was fairly close to world events," he told the nation in a radio address, "and in that period, while I learned much of what to do, I also learned much of what *not* to do." FDR didn't say so out loud, but unlike 71 percent of his fellow citizens, he still believed the war had been worth fighting. In his mind, Wilson's mistake was in how he justified it. Wilson had portrayed it as a war for high principle when it was actually a war to keep America safe.

Through the end of 1936, FDR kept his head down, signing the Neutrality Acts and promising to "isolate America from war," thus helping ensure a reelection landslide. But in October 1937, when Tokyo resumed its attack on China, he began to do what Wilson had not: argue that the shifting balance of power on far-off continents endangered the United States. Condemning "terror and international lawlessness," FDR warned that "if those things come to pass in other parts of the world let no one imagine that America will escape, that it may expect mercy, that this Western Hemisphere will not be attacked." Roosevelt wasn't arguing that Japanese expansionism merely threatened principles like self-determination that Americans held dear. He was arguing that Japanese expansionism threatened the United States itself, that stopping it was a matter of necessity, not choice.

FDR didn't say how the United States should stop Japan, except to speak vaguely about an international "quarantine." Still, isolationists got the point. He was taking aim at their central argument: that America could live apart. Members of Congress threatened impeachment. "Stop Foreign Meddling; America Wants Peace," screamed the *Wall Street Journal*. The quarantine speech, FDR later claimed, "fell upon deaf ears, even hostile and resentful ears." The following January, after Japanese planes sank an American gunboat on the Yangtze River, the House of Representatives came within twenty-one votes of passing the Ludlow Amendment, which would have required a national referendum on taking the country to war except in cases of foreign attack. For the moment, FDR's bid to convince his fellow citizens that they had a stake in the global balance of power had failed. "It's a terrible thing," he told an aide, "to look over your shoulder when you are trying to lead and to find no one there."

In March 1938, Hitler swallowed Austria, with Washington offering only the mildest of protests. Then in September, in his most brazen
move yet, he demanded the Sudetenland, the German-speaking chunk
of Czechoslovakia. The Czechs asked the French, with whom they had
a security treaty, if they could expect support in the event of war. The
French in turn asked the British, and were told—as Neville Chamberlain later famously put it—that the Sudetenland crisis was "a quarrel in a
faraway country between people of whom we know nothing." Not only
would Britain not fight to defend Czechoslovakia's border with Germany,
it would not even commit to defending *France's* border with Germany
(despite having pledged to do so at Locarno), if in trying to defend its
imperiled Eastern European ally, Paris incurred Hitler's wrath. On paper,
the U.S.S.R. was also committed to France and Czechoslovakia's defense.
But with Britain fleeing its commitments, Moscow would not act alone.

The French were actually relieved. By the late 1930s, Hoover had gotten his wish: French leaders were no longer militaristic and cocky. To the
contrary, they were almost morbidly fatalistic, like a man who knows he
will soon face the gallows and has spent so many hours shouting himself
hoarse from his prison cell, with no response, that he can no longer muster
the breath. The French knew their best chance at stopping Hitler had been
to act early, before he converted Germany's massive industrial capacity
into military might. By 1938, it was too late: France was producing forty-
five warplanes per month; Germany was producing 450. When the U.S.
ambassador in Paris asked French premier Edouard Daladier if France
would go to war for Czechoslovakia, Daladier replied, "With what?"

On September 15, Chamberlain flew to Hitler's mountain castle in
Berchtesgaden—a remote location that the Fuhrer had chosen largely to
humiliate the British leader—and told him that London and Paris would
back a plebiscite in those parts of Czechoslovakia where ethnic Germans
constituted a majority. But Hitler soon upped the ante. The Czechs, he
demanded, must evacuate the Sudetenland by October 1 and hand the
territory—and all its military installations—over to the German army,
which would then supervise the vote. Hitler also insisted that Prague surrender its Polish- and Hungarian-speaking provinces to those neighbors,
just to ensure that the rump Czech state would be utterly defenseless. At
first Chamberlain refused. But his generals, long starved of resources by
governments committed to disarmament, said they lacked the planes to
defend London. So on September 28, in a last-ditch bid for peace, Chamberlain flew to meet Hitler again, this time in Munich. For thirteen hours,

the leaders of Britain, Germany, Italy, and France negotiated, while Czech representatives waited at a nearby hotel. In the end, Hitler got the Sudetenland, and Chamberlain got a pledge that Germany and Britain would remain forever at peace. The British leader returned home triumphant, carrying his trademark umbrella, and telling a cheering crowd that he had achieved "peace with honor . . . peace for our time." Privately, FDR compared Britain's treatment of the Czechs to Judas Iscariot's betrayal of Jesus. But he didn't say so publicly. To the contrary, he sent Chamberlain a two-word cable: "Good man."

As 1938 turned to 1939, FDR kept telling Americans that their fate and Europe's were intertwined. But as his actions in the Munich crisis showed, he held out little hope of translating those words into actions anytime soon. There was something ironic in the fact that this master communicator, whose radio addresses had rallied the nation during the depths of the Depression, did not believe his oratory could rouse isolationist America from its slumber. But this was part of the lesson he had drawn from World War I. Wilson's failure during the treaty fight had left Roosevelt dubious about the degree to which a leader, even an articulate one with the facts on his side, could sway the public on questions of war and peace. In this way, FDR's cautious response to the darkening international scene wasn't mere timidity; it reflected a retreat from the progressive belief that citizens were essentially pliable. In FDR's mind, political education could only do so much. Events would have to speak for themselves.

On September 1, 1939, international dignitaries gathered in Geneva, home to the League of Nations, to unveil a long-planned statue of Woodrow Wilson. On the same day, hundreds of miles to the northeast, German soldiers poured across the Polish border. At last, Britain and France decided to fight. At 2:50 A.M. Washington time, Roosevelt was awakened by a call from his ambassador in Paris, William Bullitt, informing him that World War II had begun. The call, FDR told his cabinet later that week, provoked "a strange feeling of familiarity, a feeling that I had been through it all before." The two decades since World War I suddenly seemed like a long intermission. Now he was "picking up again an interrupted routine."

Addressing the nation on September 3, FDR assured Americans that he had no intention of taking them to war. But neither did he ask them to be detached observers in the struggle between Nazism and its foes. "I cannot ask that every American remain neutral in thought," he declared, in an oblique reference to Wilson's call for impartiality in the early days of World War I. "There is a vast difference," he added in his State of the Union address the following January, "between keeping out of war and pretending that war is none of our business. . . . For it becomes clearer and clearer that the future world will be a shabby and dangerous place to live in—yes, even for Americans to live in—if it is ruled by force in the hands of a few." With Europe at war, Congress finally amended the Neutrality Acts to allow Britain and France to buy U.S. arms, so long as they paid cash and carried them away in their own ships.

As 1939 turned to 1940, Hitler's Germany and Stalin's U.S.S.R.—which had signed a nonaggression pact the previous summer—carved up Eastern Europe. Then, in April, Germany conquered Denmark and Norway, and then Holland and Belgium in May. Keeping up his rhetorical campaign, FDR warned that if Hitler gained control of Europe, Americans would be a "a people lodged in prison, handcuffed, hungry, and fed through the bars from day to day by the contemptuous, unpitying masters of other continents." But these were just words. "If you cannot give France in the coming days a positive assurance that the United States will come into the struggle within a short space of time," wrote French premier Paul Reynaud on June 14, "you will then see France go under like a drowning person after having thrown a last look towards the land of liberty from where she was expecting salvation." But with more than 80 percent of Americans still against entering the war, FDR let France drown. Only the sight of it, he reasoned cruelly but correctly, would shift the public mood. Eight days after Reynaud's plea, the Nazis accepted France's surrender on the same spot, in the forests of Compiègne north of Paris, where Germany had surrendered twenty-two years before. Mercifully, Georges Clemenceau was not alive to see his nightmare come true.

France's fall jolted American opinion like no presidential speech ever could, and FDR quickly capitalized. In August, with Winston Churchill now Britain's prime minister, and bracing his nation for a German assault, FDR sent the United Kingdom fifty destroyers in exchange for military bases in the Western Hemisphere. The following month, he convinced Congress to institute the draft, although mail to the White House ran ten to one against. And in December, with the British no longer able to pay for American arms, he pushed through "lend-lease," which essentially provided them for free.

As he edged closer to war, FDR began laying the foundations that would undergird American foreign policy for the next quarter century. First, he convinced former Republican secretary of state Henry Stimson to become his secretary of war, and Frank Knox, who had been the GOP's vice presidential nominee in 1936, to serve as secretary of the navy. Stimson's former boss, Herbert Hoover, was now a rabid isolationist. But Stimson, who had quarreled with Hoover almost a decade earlier on Manchuria, now broke with him decisively and became the anchor for a revived Republican internationalism, which helped FDR push lend-lease and the draft through Congress. In choosing Stimson, FDR—the country squire whose New Deal domestic policies had made him a hated figure in conservative East Coast financial circles—was also reconciling with his class. American foreign policy would henceforth be led by a bipartisan, hard-nosed elite entirely comfortable with power politics. In America's hour of peril, the children of Theodore Roosevelt—who espoused neither Woodrow Wilson's ethic of reason nor the pacifism of John Dewey nor the isolationism of Charles Beard—were closing ranks. They would guide American foreign policy for a generation, until discredited by hubris of their own, in Vietnam.

At the same moment he embraced bipartisan internationalism, FDR also began dramatically expanding presidential power. The deal to provide Britain fifty destroyers violated the Neutrality Acts and would have met fierce resistance on Capitol Hill. But the White House, advised by a corporate lawyer named Dean Acheson, decided it had the legal authority to transfer them without congressional approval. Then, in May 1941, FDR declared an "unlimited national emergency," which gave him the power to take any action he deemed necessary to protect the Western Hemisphere. In July, he used it to send four thousand Marines, without congressional approval, to Greenland and Iceland, to prevent them from falling into Nazi hands. It was an historic shift. Ever since Versailles, Congress had held foreign policy tightly in its grip, giving the president little scope for overseas adventures. Now executive power—nurtured by America's bipartisan internationalist elite—was starting to swell. It would swell unabated for a generation, until it too collapsed in the jungles of Vietnam.

In May 1941, FDR had a dream that Hitler's Luftwaffe was bombing New York. Privately, some historians believe, he had already decided to enter the war. By fall, U.S. warships were escorting British merchant vessels across the Atlantic, ranging hundreds of miles in pursuit of German subs, doing everything possible to provoke an incident. They soon got one. In September, a German U-boat fired torpedoes at the U.S.S. *Greer* near Iceland. In October the U.S.S. *Kearny* was hit, and later the U.S.S.

Reuben James. Public opinion swung toward war, particularly since FDR claimed that the ships had been attacked without provocation, which was a lie. But still, Roosevelt waited. Calling for war now, he told Churchill, would trigger a bitter three-month congressional debate. And if events on the battlefield or at the subsequent peace conference turned sour, that initial disunity would return to haunt the nation.

The basis upon which America entered the war, FDR believed, would make all the difference once the shooting stopped. When subs attacked the *Greer,* he did not say—as Wilson had—that Germany had violated America's neutrality rights. Instead he warned that Hitler was seeking "absolute control and domination of the seas." Roosevelt was groping for a way to convince the public that this time Germany threatened not just international law, but American security. Only then—once Americans saw international engagement as survival, not social work—would they sustain it even when the world let them down. "Our policy is not based primarily on a desire to preserve democracy for the rest of the world," FDR added. "It is based primarily on a desire to protect the United States and the Western Hemisphere from the effects of a Nazi victory." This, he believed, was what Wilson should have said in 1917.

It was easier in 1941, of course, because Hitler was a more terrifying figure than the kaiser, with more terrifying weapons and more grandiose plans. But FDR still needed an event to dramatize his words. It came, as it happened, not in the Atlantic but in the Pacific. When France fell to Germany in 1940, Japan conquered its colonies in northern Indochina, leading the United States to halt sales of scrap iron and steel in protest. The following summer, Japan took the rest of France's Asian empire, prompting FDR to take the much graver step of cutting off Tokyo's supply of oil. That set the clock ticking. With no fuel supply of its own, Japan faced a choice: ignominious withdrawal or a push farther south into the oil-rich Dutch East Indies. By the first week of December, Roosevelt knew that Tokyo was planning the latter course, which he feared would involve an attack on U.S. forces in the Philippines. His close aide Harry Hopkins suggested that America strike first and thus gain the element of surprise, but FDR refused, partly because he believed that only a Japanese attack would unify the country behind war.

At 7:55 A.M. Hawaii time on December 7, carrier-based Japanese aircraft began appearing in the Oahu sky. For two hours, 360 planes turned Pearl Harbor into an inferno, killing more than two thousand Americans and destroying most of America's Pacific fleet. The decision for war, FDR told Hopkins, was now out of his hands. Events, not mere words, had shown that America could not live apart. "Franklin," commented Eleanor, "was in a way more serene than he had appeared in a long time."

On May 22, 1940, Reinhold Niebuhr received a letter from the executive secretary of the American Socialist Party. The party had just held its annual convention, which reaffirmed its devotion to class struggle and its indifference to the struggle between nations currently bloodying much of the world. World War II, according to party doctrine, was a contest between capitalist powers in which Socialists should have no allegiance. This placed the party in direct conflict with groups like the Committee to Defend America by Aiding the Allies, which insisted that the United States had a strategic and moral interest in fascism's defeat. Niebuhr, the secretary noted, was a prominent member of both organizations. "I am interested," he wrote, "in getting your reaction on this matter."

Two days later, Niebuhr resigned his party membership. "The Socialists have a dogma that this war is a clash of rival imperialisms," he wrote in the *Nation*. "Of course they are right. So is a clash between myself and a gangster." The party was so wedded to "purely ideal perspectives"—it viewed events from such Olympian heights—that it could see no important difference between the capitalism of Winston Churchill and the capitalism of Adolf Hitler. That, Niebuhr argued, was "utopianism." For him, few epithets were worse.

Niebuhr's break with the left had been several years in coming. By the mid-1930s, with the New Deal in full swing, his fury over economic oppression had begun to cool. He still believed that ultimately the proletariat would rule. But FDR's "amiable opportunism," he admitted, was "arresting the decay of capitalism as effectively as that can be done." His sympathy for Marxism also faded as he learned more about its Soviet incarnation. In 1939, when Nazi Germany and Soviet Russia became allies, he began to stress their common totalitarian features. If the Socialists saw little difference between Churchill and Hitler, Niebuhr saw less and less difference between Hitler and Stalin.

But there was something odd about Niebuhr's ideological transformation. Despite moving from left to center, he kept sounding many of the same themes. In 1932, he had attacked progressives like Dewey as naïve for thinking that through reason they could achieve justice without conflict. Now he accused the communists of naïveté as well. They too, he argued, championed a supposedly scientific vision of inexorable progress.

They too believed that if you overhauled society's institutions you could eradicate evil. Because they would not acknowledge that human nature was inherently sinful, they could not see that it wreaked havoc upon all efforts to rationalize society, including their own. Whereas Niebuhr had once applauded communism for its hard-boiled embrace of class conflict, he now lumped it with progressivism as yet another utopian creed.

Once again Niebuhr reflected his times. By the mid-1930s, many American intellectuals were shifting anxieties, worrying less about the injustice of Depression-era capitalism and more about the menace of fascist aggression. At first communism seemed a sturdy bulwark against this rising threat, and to underscore the point, in 1935 the American Communist Party ceased its revolutionary agitation and joined liberals in an antifascist Popular Front. But between 1936 and 1938, Stalin's show trials, which culminated in the murder of dozens of his political rivals, prompted some American intellectuals to reconsider their communist sympathies. And when Stalin and Hitler joined forces the following year, the trickle of anticommunist disillusionment became a flood. One after another, intellectuals who had embraced some species of communism during the Depression left the Marxist faith. The disillusioned communists—a remarkable group that included the philosophers Sidney Hook and James Burnham; the journalists Max Lerner, Louis Fischer, Irving Kristol, and Max Eastman; the critics V. F. Calverton and Irving Howe; and Philip Rahv and William Phillips, editors of the new, fiercely anti-Stalinist journal *Partisan Review*—did not all end up in the same place politically. But they shared a sensibility. Like Niebuhr, they had lost faith in grand schemes for remaking the world because they had lost faith in grand schemes for remaking human beings. The 1930s taught that evil did not stem merely from failures of education and expertise, or from one economic system. Its roots lay deeper, inside humanity itself. A whole generation of intellectuals carried this tragic insight with them as they journeyed back to the democratic fold. They approached American democracy not as virgin enthusiasts, like the progressives, but as weary travelers. Noting that they had gained wisdom through pain, the sociologist Daniel Bell called them "twice-born," which, as it happens, was the same phrase FDR's uncle had applied to him.

It was these "twice-born" intellectuals who in the late 1930s provided intellectual ballast for Roosevelt's moves toward war. Unlike the war progressives, they held out little hope that through military action America could eradicate power politics. They conceded that Americans would die and kill for racist, colonial powers like Britain and France and for an immoral economic system at home. But they were comfortable fighting for lesser evils. In their minds, that was what politics was all about. "The civilization we are called on to defend," wrote Niebuhr, "is full of capitalistic and imperialistic injustice . . . [but] it is still a civilization."

In their argument for war, the ex-communists were joined by another child of the age of disillusionment: Walter Lippmann. Lippmann had walked his own tortured intellectual path since Versailles. Obsessed with the public irrationality that World War I had exposed, in the 1920s he had questioned whether ordinary people were competent to govern themselves at all, whether democracy could ever really work. But like the ex-Marxists, he became a partisan of liberal democracy in the late 1930s, not because he had regained his youthful optimism about what it could accomplish but because the alternatives were so horrifying.

Early in FDR's presidency, fearful of repeating the mistakes of 1917, Lippmann had urged America to steer clear of Europe's troubles. But as the Nazi menace grew, he came to terms with the war that as a young man he had championed and then spurned. He did so by replacing the idealistic arguments of his youth with realist ones. America had not entered World War I to rationalize the world, he now argued. It had fought to ensure "the safety of the Atlantic highway": to prevent Germany from smashing the British fleet that guaranteed American security and prosperity. This, he argued, "is something for which America should [again] fight." He had, it turned out, discovered the same rationale for war as FDR, the man he once called an "amiable boy scout" who "just doesn't happen to have a very good mind."

By the early 1940s, John Dewey and Charles Beard were old men. They could see war coming and had long ago resolved to resist it with everything they had so as never to repeat the sins of World War I. In 1940, when the City University of New York revoked a job offer to the British philosopher Bertrand Russell because he was a pacifist, Dewey felt he was reliving a nightmare. Beard, who had long since stopped receiving invitations to the White House, grew convinced that FDR's power grab was leading to dictatorship. He even alleged that Roosevelt had known the Japanese might bomb Pearl Harbor but did nothing because he needed a pretext for war. Both Dewey and Beard felt in their bones that this war would be worse than the last, that it might destroy American democracy once and for all. "Day and night I wonder and tremble for the future of my country and mankind," Beard wrote to a friend. "An epoch has come to an end," Dewey declared, "but what is beginning is too much for me."

In their opposition to American entry into World War II, Dewey and Beard were fulfilling their silent oath to their martyred student, Randolph Bourne. But ironically, they now came under assault from a new group of younger intellectuals, for whom the Bourne legend held no appeal. To these sad-eyed realists, who in the 1930s had grown old beyond their years, Dewey and Beard were moral perfectionists, men who would allow the greatest of evils to triumph for fear of implicating themselves in the lesser evil of war. Dewey, Niebuhr charged, would not "defend democracy because it is not pure enough." The historian Lewis Mumford, once a Beard disciple, now called him "a passive—no, active—abetter of tyranny, sadism, and human defilement."

In 1941, the American Communist Party, now in bed with Nazism via the Hitler-Stalin pact, created a Randolph Bourne Award to honor an intellectual who opposed America's march to war. They chose the novelist Theodore Dreiser, a raving anti-Semite who considered FDR's foreign policy a Jewish plot. Pacifism and isolationism, which sought to keep America free from the barbarism of the old world, had become complicit in its greatest barbarism ever. The age of Randolph Bourne—which had begun, oddly enough, when influenza took his life—was now dead. In its wake, the real answer to the hubris of reason was being born.

CHAPTER FOUR

I DIDN'T SAY IT WAS GOOD

On Friday, September 15, 1944, Franklin Roosevelt went to the movies. The film was *Wilson*, a lavish, syrupy, star-studded account of the twenty-eighth's president's life that cost $5 million to produce, making it the most expensive film yet made. A decade earlier, when World War I revisionism ran rampant and Wilson was a figure of public scorn, such a project would have been unimaginable. But World War II rescued Wilson's reputation. In 1941, Otto Preminger directed *In Time to Come*, a Broadway play about the Paris Peace Conference, which painted Clemenceau and Lodge as villains and Wilson as a man too good for his time. Three years later, the magazine *Look* published a glossy pictorial biography of Wilson subtitled "The Unforgettable Figure Who Has Returned to Haunt Us." At their 1944 convention, Democrats voted a resolution of tribute to the former president. And by the time *Wilson* hit the theaters, the nation was in the midst of a full-blown Wilson revival. At the Roxy in New York, twenty thousand people stood in line for a ticket; one million saw the film in its first five weeks alone. The movie, which noted Wilson's love of football, golf, and vaudeville, even made him seem friendly. But FDR was less charmed by the experience than stressed. While he was watching it, his blood pressure spiked to 240 over 130.

For Roosevelt, Wilson's failure was a near obsession. When writing speeches in the Cabinet Room of the White House, he often glanced up at Wilson's portrait, which he had installed above the mantelpiece. Versailles, wrote FDR's speechwriter Robert Sherwood, was "always somewhere within the rim of his consciousness." But unlike the creators of *Wilson*, FDR did not exactly consider his old boss a role model; more like a cautionary tale. Wilson, he believed, had dug his own political grave by imagining that other countries and his own citizens were more malleable than they really were. In response, in the years before Pearl Harbor,

FDR had labored mightily to convince Americans by word and deed that saving Europe and Asia from fascism was self-defense, not philanthropy. And now, with America in the fight, he had two big and rather un-Wilsonian ideas about the postwar world. First, it should be built on power, not reason, because only a coalition of the powerful could keep the peace. Second, this must be concealed—as much as possible—from the American people, who, if confronted with the ugly truths of international affairs, might scurry back to their shores, in which case the world would again go to hell. It was in his messy, often unsentimental, sometimes duplicitous efforts to square these dual imperatives that FDR, in his final years on earth, began formulating an answer to the hubris of reason. That answer—which his successor, Harry Truman, amended and completed— was far from perfect. In some ways, it was downright tragic. But it laid the foundation for a fundamentally better world.

If FDR had an historical inspiration for his postwar vision, it was not America's twenty-eighth president; it was the twenty-sixth. Theodore Roosevelt had also envisioned an international order built on power. As early as 1910, he had offered his own conception of a League of Nations. What distinguished it from Wilson's was that TR wanted "to create the beginnings of international order out of the world of nations *as those nations actually exist.*" In other words, power in the League should reflect power in the world: TR had little patience for an international body based upon the principle that strong and weak nations were equal in the eyes of international law, since for him law was only as good as the gun that enforced it. When the United States entered World War I, he saw the Allies as the core of this new body. He wanted a league of victors, built upon security treaties between America, Britain, and France. The victorious powers would be allowed their spheres of influence, no matter how unscientific and unfair, as long as they stuck together and thus kept the defeated Germans from again disturbing the peace.

FDR's vision was strikingly similar. In August 1941, six months before America entered the war, he met Churchill off the coast of Newfoundland. They agreed to formulate an "Atlantic Charter" that outlined their hopes for the postwar world. But when the British prime minister proposed that this vision include an "effective international organization," FDR balked, saying the League had been tried and had failed. Instead, he suggested, Britain and America should police the world themselves. "The time had come," he declared, "to be realistic."

The following May, Roosevelt broadened the idea, telling Soviet foreign minister Vyacheslav Molotov (whose country was now at war with Hitler) that the postwar world should include "Four Policemen": the United States, Britain, the U.S.S.R., and China, the four nations that would emerge as most powerful once Germany and Japan were crushed. Each great power would walk a particular beat: Britain would control Western Europe; the Soviet Union would control Eastern Europe; the United States would guard the Western Hemisphere and the Pacific; and China (with America's help) would control the Asian mainland. Together, these Four Policemen would keep down Germany and Japan, so that they couldn't stir up trouble as they had after World War I. Such an arrangement, FDR told the "cannonball"-headed, "slab"-faced Soviet foreign minister, was far preferable to another League, which would include "too many nations to satisfy." As the *Saturday Evening Post*'s Forrest Davis reported the following April, "Mr. Roosevelt primarily is concerned not with aspirations toward a better world such as he articulated in the Atlantic Charter, but with the cold, realistic techniques, or instruments, needed to make those aspirations work. This means that he is concentrating on power. . . . The problem of security rests with the powers who have the military force to uphold it."

This time there would be no Inquiry poring over maps in the New York Public Library, scientifically determining which nations should exist, and where. Unlike Wilson, FDR clearly envisioned spheres of influence, in which great-power "policemen" made the rules and small nations did as they were told. And unlike Wilson, who had imagined a peace settlement so impartial that it was embraced by victor and vanquished alike and thus required little enforcement, FDR wanted the victors to impose peace and sustain it by confronting Germany and Japan with overwhelming force. His greatest desire was that the wartime *allies* stick together to uphold the peace. (Unlike Wilson, FDR always called America an ally of Britain, France, and the U.S.S.R., never a mere "associate.") Ideally, in Roosevelt's view, the Germans and Japanese would see the justice of this new postwar order. But he wasn't relying on public opinion to keep the postwar peace; he was relying on a powerful alliance. All this would have brought a smile not only to the faces of Teddy Roosevelt and Henry Cabot Lodge, but to that other mustachioed devotee of power politics: Georges Clemenceau.

Secretary of State Cordell Hull and his influential deputy, Sumner Welles—both unreconstructed Wilsonians—were appalled. "During the spring of 1943," wrote Hull in his memoirs, "I found there was a basic cleavage between him [FDR] and me on the very nature of the postwar organization." Hull and Welles tried and failed to convince FDR that opposing another League of Nations was wrong on the merits. But fortunately for them, Harry Hopkins, the president's most influential adviser, convinced him he was wrong on the domestic politics. Roosevelt had assumed that the American people did not want another League; he had told Churchill as much at Newfoundland. Since America entered the war, however, public sentiment had shifted, as the reaction to the movie *Wilson* illustrated. In 1941, only 38 percent of Americans had backed U.S. entrance into a new international organization; by 1944, 72 percent did. This was, in a sense, a testament to FDR's own strategy: By waiting until Pearl Harbor to enter the war, he had convinced most Americans that isolationism was impossible. But now, having decided to participate in the world, Americans were again determined to bend it toward their ideals. For FDR, that was a recipe for disillusionment, since he suspected that the world would not bend very quickly or very much. Hopkins, however, convinced him to conceal those thoughts from public view. Henceforth FDR would continue crafting a postwar settlement based on power politics, but he would give it a Wilsonian face, especially when addressing audiences at home. And he would try, in this way, to do what Wilson and the war progressives could not: reconcile America's missionary impulse with the realities of an imperfect world.

In keeping with this political strategy, FDR in the spring of 1943 agreed to encase his Four Policemen idea within a Wilsonian shell. In a meeting with British foreign secretary Anthony Eden in March, he suggested a three-tiered international body including an executive committee, comprised of the Four Policemen, which made all important decisions, an advisory council including six or eight other nations, which handled smaller, nonmilitary matters; and a general assembly with universal membership that met once a year so that small countries could, in Eden's words, "blow off steam."

In November, Roosevelt presented his vision to Churchill and Stalin. To meet them, he flew to Tehran, a trip of more than seventeen thousand miles, undertaken in an age when aircraft cabins often lost pressurization, by a man suffering serious respiratory problems and in chronic pain. His fellow leaders embraced the Four Policemen concept, but they were amused by FDR's insistence that they make no public reference to it and his claim that the American people would never accept great-power spheres of influence. It was as if, having drawn them a picture of a wolf, he was demanding that they call it a sheep.

The contradiction went to the heart of Roosevelt's political dilemma. He desperately wanted good postwar relations with Stalin. Otherwise, he feared, the Germans would exploit divisions between the wartime victors, as they had after World War I. And he knew Moscow's price for such a partnership: a sphere of influence in Eastern Europe, which could serve as a buffer in case Germany ever invaded again.

On the battlefield, that sphere was already being demarcated. By the time FDR went to Tehran, the U.S.S.R.—having turned the tide against Hitler's forces at Stalingrad and Kursk, two of the bloodiest battles in the history of war—was already emerging as the most powerful player in Eastern Europe. That was no accident. From the beginning, FDR had adamantly opposed spilling American blood in the lands between the Baltic and the Adriatic seas. When his adviser William Bullitt, and then Churchill, suggested that Washington and London alter their European invasion plans, choosing a route that took longer to defeat Hitler but put American and British troops on the ground in Eastern Europe to contain Soviet influence, FDR refused. Fighting in Eastern Europe, he knew, would require mobilizing far more American troops, which could disrupt the economy back home. The military warned that it would prove a logistical nightmare, since the region lacked good supply routes from the sea. And finally, Roosevelt feared that Americans would revolt against the additional loss of life. He had sold them not a war for the rights of small nations, but what he called "The Survival War"—a war to defeat Hitler, Mussolini, and Tojo. He knew Americans didn't want Eastern Europe to fall under Stalin's thumb, but his gut told him that they were even more opposed to sacrificing their sons to prevent it. There was a gap, he believed, between the world that Americans wanted to create and the blood-price they were willing to pay to create it. And when that gap was exposed, he feared, a foul, isolationist wind might sweep across the American political landscape. His strategy, therefore, was to hide the discrepancy between ideals and reality as much as possible until America was firmly ensconced in a system of postwar security. Then, no matter how disillusioned Americans became, they would be unable to again retreat from the world.

For his strategy to work, FDR needed his fellow leaders to lie. That was a major reason he had traveled those forty excruciating hours on a primitive plane—not merely to get Stalin and Churchill to agree to his scheme for postwar cooperation, but to enlist their aid in convincing his own people. In Tehran, FDR told Stalin he "fully realized" that Russia planned to reoccupy the Baltic countries, which it had ruled for roughly a century until they gained independence at Versailles. But since "the question of referendum and self-determination" would constitute a "big issue" in America, it would be helpful to him "personally" if Stalin gave Lithuania, Latvia, and Estonia some say over their fate. On Poland, Stalin informed Roosevelt and Churchill that he intended to annex an eastern slice of the country while compensating the Poles with a slice of eastern Germany. FDR said that this was fine but that he "could not publicly take part in any such arrangement," since he needed the votes of "six to seven million Americans of Polish extraction" in his upcoming reelection bid.

Upon returning to the United States, Roosevelt delivered a Christmas Eve radio address about his trip. "The rights of every nation, large or small, must be respected and guarded as jealously as the rights of every individual within our own republic," he announced. "The doctrine that the strong shall dominate the weak is the doctrine of our enemies—and we reject it." Just back from a whorehouse, he was insisting that America remained chaste. One day, he knew, reality would become too graphic to ignore.

As 1943 turned to 1944, the particulars of FDR's vision grew clearer. By March, the State Department had hammered out an outline of the new international organization to be born after the war. Its very name—the "United Nations Organization" (later referred to as simply the "United Nations")—hinted at its differences with the League. "United Nations" was the appellation that America, Britain, the Soviet Union, and their smaller partners had given their alliance against Germany, Italy, and Japan. It suggested a continuation of the wartime partnership, an organization comprising (or at least dominated by) the alliance that had won the war.

In crucial ways, the UN was also structured differently from the League. The League had also contained a council composed of big countries, but it could merely offer recommendations to the General Assembly, which made the final decision. What's more, every member of that General Assembly could exercise a veto, which created a significant equality of power between members. At the UN, by contrast, the General Assembly would make recommendations to the Security Council, not the other way around. In fact, the Security Council would not even need a General Assembly recommendation; it could simply act on its own authority. Moreover, only the Security Council's permanent members—the Four

Policemen (with France later added as a fifth)—enjoyed veto power. The result, observed British historian Evan Luard, was "a revolutionary transformation of the existing international system in favour of the great."

With a draft of the prospective United Nations in hand, Secretary of State Hull began corralling congressional support, taking particular pains to woo prominent Senate Republicans, which Wilson had failed to effectively do while negotiating the League. Key Republicans such as Arthur Vandenberg, the GOP's ranking member on the Senate Foreign Relations Committee, were pleased by the Security Council veto and the absence of an international police force, both of which allayed their fears that the nascent UN would infringe upon American sovereignty. They were happier with the organization's structure, in fact, than some Wilsonians, who objected to the outsize power that the body afforded large nations. But politically, the UN's real vulnerability wasn't its structure but the territorial settlement it would implicitly ratify, since both Wilsonians and anticommunist conservatives strongly opposed granting the Soviets a sphere of influence in Eastern Europe. In May 1944, when Forrest Davis penned an article titled "What Really Happened at Teheran," which claimed that, in his dealings with Stalin, FDR was "willing to settle for the lowest common denominator . . . peace among the great powers," prominent senators began to suspect that the president was cutting secret deals on Eastern Europe. "We all believe in Hull," wrote Vandenberg in his diary, "but none of us is *sure* that Hull *knows* the whole story."

They were right; he didn't. That fall, Churchill traveled to Moscow without FDR, who authorized the U.S. ambassador to the U.S.S.R., Averell Harriman, to represent him. At 10 P.M. on October 9, at a meeting to which Harriman was not invited, Churchill took out half a piece of paper, wrote down percentages alongside the names of various countries, and handed it to Stalin, who made a check mark in blue pencil. It was a parody of sinister old-world power politics: The two leaders had just divided up southeastern Europe. Britain's sphere of influence would include 90 percent control in Greece, while Russia's would give it 90 percent in Romania and 75 percent in Bulgaria and Hungary. They would split Yugoslavia 50-50. Churchill told Stalin that this "naughty document" should be kept secret "because the Americans might be shocked." That was true. Many Americans, including Vandenberg and probably also Hull, would indeed have been shocked. But FDR was not one of them. At Tehran, Roosevelt later confided, he had "already told the Russians they could take over and control [Eastern Europe] completely as their sphere."

The president, commented Harriman several weeks after the Moscow summit, "consistently shows very little interest in Eastern European matters except as they affect sentiment in America."

FDR spent 1944 in a race against time: hoping to win reelection to a fourth term before voters learned of his compromises on Eastern Europe. In November, he successfully crossed the finish line, after a campaign in which his Republican opponent, Thomas Dewey, rarely challenged his postwar efforts. But the 1944 campaign was just one leg of a longer contest, in which Roosevelt struggled to entrench America in the UN before the tragedy in Eastern Europe made Americans recoil. And within that competition lay another, more intimate one: against the limitations of FDR's own body. Since Tehran, Roosevelt's health had deteriorated. A medical examination in early 1944 revealed hypertension, an enlarged heart, and signs of cardiac failure. Over the course of the year, he lost twenty pounds, and a picture published in July, which captured his withered, dazed visage, briefly threatened his reelection bid. In September, when a former aide saw Roosevelt for the first time in eight months, he was appalled by "the almost ravaged appearance of his face."

Nevertheless, a month later, FDR flew to Yalta to meet Churchill and Stalin for what would be the last time. The flight was again arduous, and the accommodations were worse. The leaders stayed in Livadia Palace, a fifty-room czarist castle perched on a cliff above the Black Sea. But the building, which like most things in Crimea had been ravaged by Hitler's armies, contained barely any working bathrooms and teemed with typhoid and lice. Churchill muttered that it must have taken the Russians ten years to find a site so ghastly, and endured the stay by fortifying himself regularly from a private supply of whiskey. FDR—his face pallid, his lips blue, his hands trembling—wrapped himself in a large cape to stay warm.

Still, Roosevelt pressed his agenda: to secure Moscow's pledge to enter the war against Japan, to decide how to govern postwar Germany, to nail down the structure of the new United Nations, and to again fudge the question of Eastern Europe so anger over Soviet influence there didn't turn Americans against the UN. On the first issue, he succeeded: Stalin agreed to enter the Asian theater three months after the European war was won. (In return the U.S.S.R. expected to replace Tokyo as the dominant power in Manchuria and northern China.) The UN's organizational structure wasn't much of a problem, either. For six months Stalin had been demanding that the sixteen Soviet republics each have its own General Assembly vote, presumably to counter the pro-Western Latin American and British Empire blocs. At Yalta, he agreed that between them the republic would have two or three votes, and FDR said he would demand the same number for the United States if the Senate raised a fuss. But the issue, Roosevelt insisted, "is not really of any great importance"; the UN was not going to make decisions by majority vote.

The tough nut was Poland. By the time of the Yalta Conference, Soviet troops were already occupying Warsaw, Krakow, Lodz, and Posen, while their American and British counterparts remained one thousand miles away. FDR had already agreed to shift the Polish border west, and Churchill had traded away Romania, Bulgaria, and Hungary on a single piece of paper. But the composition of Poland's postwar government was harder. The issue had special meaning for Churchill, since the destruction of Polish independence had been Britain's trigger for war. Roosevelt was less personally invested, but he knew that of all the Eastern European nations, Poland's fate was the most politically explosive back home. Weeks earlier, Moscow had installed a puppet regime based in Lublin. The United States and Britain, by contrast, still recognized the exiled prewar Polish leadership in London. At Yalta, Roosevelt and Churchill pushed Stalin to bring the London Poles into a broad-based government. In return they got words. In a Declaration on Liberated Europe, all three leaders called for "broadly based" governments and "free elections" in the lands liberated from Hitler. It was a lovely vision: a "Crimean Charter" that echoed the Atlantic one penned by FDR and Churchill in 1941. But with the Red Army squatting on Polish soil, Roosevelt knew he had no way of making the rhetoric real. All he could do was plead. "I want this election in Poland to be the first one beyond question. It should be like Caesar's wife. I did not know her but they said she was pure," he told Stalin. "They said that about her," replied the Soviet leader, his smallpox-scarred face widening into a grin, thus revealing his yellowish-brown teeth, "but in fact she had her sins."

On March 1, Roosevelt went before Congress to report on what he had achieved. One aide said he looked "ghastly, sort of dead and dug up." But his words stirred the chamber. At Yalta, he declared, he and his fellow leaders had put an end to "the spheres of influence, the balances of power . . . that have been tried for centuries—and have always failed."

Within weeks, events in Poland were mocking his words. Not only were the Soviets exercising a sphere of influence, but they were enforcing it with characteristic brutality: systematically silencing, deporting, or murdering Poles who wanted a government free from Moscow's grip. Furious, Churchill sent FDR a stream of missives proposing that they make a joint protest, thus exposing Stalin's violation of the principles to which he had signed his name. But Roosevelt did not want a public row. The creation of the United Nations, its ratification by the Senate, and the exorcism of the ghost of Versailles were within reach. He knew the realities in Eastern Europe would be nastier than the proclamations in the Atlantic and Crimean charters implied, but he believed that he had laid down a moral marker, a standard by which the world could judge. Over time, he hoped, things would improve, as long as America did not pull away from the world in disgust. "Perfectionism," he declared in his 1945 State of the Union address, perhaps with an eye toward the realities Americans would soon discover, "no less than isolationism or imperialism or power politics may obstruct the paths to international peace. Let us not forget that the retreat to isolationism a quarter of a century ago was started not by a direct attack against international cooperation but against the alleged imperfections of the peace. . . . We gave up the hope of gradually achieving a better peace because we had not the courage to fulfill our responsibilities in an admittedly imperfect world." For FDR, this was the core lesson of World War I, the true answer to the hubris of reason. America must dirty its hands. To preserve its freedom, it must practice power politics, even though power politics would threaten American freedom in other ways. And it must do so while recognizing that despite its efforts, the world would remain an ugly place, just not quite as ugly as if America tried to remain pure.

Those close to the president knew the end was near. "He is slipping away from us," said his scheduler, William Hassett, "and no earthly power can keep him here." On April 12, as he sat for a portrait, FDR suddenly complained of a "terrific headache," then slumped over unconscious in his chair. At 3:35 in the afternoon he was pronounced dead of a cerebral hemorrhage. By his side was Lucy Mercer, the mistress he had told Eleanor he would jettison more than twenty-six years earlier.

Twelve days later, a U.S. platoon made contact with Russian troops in eastern Germany, near the city of Torgau, on the Elbe. The Americans followed the Russians back to the farmhouse where they were encamped and drank toasts to the Allied cause, its triumph over the Nazis now virtually assured. A few hours later, half a world away, 282 delegates entered the San Francisco Opera House for the opening ceremony of the United

Nations. By late June they had approved a structure that closely resembled the one envisioned at Yalta. It would comprise a General Assembly, where all nations talked, and a Security Council, where the great powers— vetoes in hand—acted. The UN's Universal Declaration of Human Rights spoke about "the equal and inalienable rights of all members of the human family," but once again reality could not match the soaring words. Despite some vague language about trusteeship, Britain and France kept their colonial possessions. On Poland, delegates deferred seating Russia's puppet government. But the ultimate outcome was already clear: With Soviet troops in Eastern Europe, and a Soviet veto on the Security Council, Poland's awful fate was sealed.

A month later, the Senate took up American membership in the UN. Especially among Democrats, Wilson's ghost was summoned again and again. It is rare for a man "to be given a second chance in his lifetime to correct a great mistake," declared Connecticut's Brien McMahon. "It is even more seldom that that chance comes to a nation." Senator Charles Andrews informed his colleagues that he had recently visited Wilson's grave for inspiration. Florida's other senator, Claude Pepper, said that specters hovered over the proceedings, watching through the "veil which their immortal eyes can always penetrate. . . . One of them is Woodrow Wilson, and the other is Franklin D. Roosevelt."

In the end, the vote wasn't close. Pearl Harbor had killed American isolationism, and the realities in Eastern Europe—which FDR had helped mask—hadn't yet revived it. On Saturday, July 28, with only two die-hard isolationists voting no, the Senate overwhelmingly ratified the Charter of the United Nations. Upon hearing the news, one aging League sup- porter said he "almost shouted and cried with joy." Another wrote, "The faith of Woodrow Wilson has been vindicated. The record of the United States of 1920 has been expunged." But in the Senate chamber, the mood was oddly muted. Spectators packed the galleries, but the vote sparked no cheering or applause. In truth, Wilson's spirit had not been entirely re- vived; his faith had not been fully vindicated. The world had been saved, but not exactly transformed. As average Americans would gradually come to realize, Eastern Europe was slipping away. And there were other hor- rors. As U.S. troops occupied Germany, they discovered a vast machinery of death, whole cities devoted to race murder. In August, Americans— who a few years earlier had been shocked by the Spanish bombing at Guernica, which killed roughly two hundred people—watched their own nation drop an atomic bomb on Hiroshima, killing eighty thousand in a single night. Meanwhile, from Arkansas to Arizona, one hundred thousand American citizens whose only crime was their Japanese ancestry were slowly exiting the camps where they had been imprisoned for most of the war. In such a world, who could declare, even in victory, that humanity was marching toward a bright dawn? A few idealists attacked the UN as a betrayal of the Wilsonian dream, "a mere camouflage," in the words of one church group "for the . . . exercise of arbitrary power by the Big Three for the domination of other nations." But many Americans seemed to intuitively grasp the lesson that FDR believed the war progressives and their pacifist and isolationist successors had never learned: that the world was a tragic place, which America could neither escape nor fully redeem. What the Senate had ratified, wrote *Time* magazine, was "a charter written for a world of power, tempered by a little reason." Or as FDR exclaimed about the Yalta deal, in words that could have summed up his entire foreign policy, "I didn't say the result was good. I said it was the best I could do."

FDR wasn't much of an intellectual; he relaxed by building model ships, not reading highbrow journals. But the sentiment he expressed—that people must do their best in the world while not expecting too much of it—captured the intellectual spirit of the age. Even before World War II, the Great War, the Depression, and Stalin's crimes had darkened the mood of many American writers and thinkers. And now, the Holocaust, Hiroshima, and the Iron Curtain joined the register of human horror. Looking back, the intellectuals of the 1940s tended to lump together the political faiths that had come before—war progressivism in the 1910s, pacifism in the '20s, Marxism in the '30s—and call them the age of optimism and reason. Their own challenge, as they saw it, was to practice politics in a world that had reduced all utopian visions to rubble. "Frustration is increasingly the hallmark of this century—the frustration of triumphant science and rampant technology, the frustration of the most generous hopes and of the most splendid dreams," wrote a precocious Niebuhr disciple named Arthur Schlesinger, Jr., in his 1949 book, *The Vital Center*. "Nineteen hundred looked forward to the irresistible expansion of freedom, democracy and abundance; 1950 will look back to totalitarianism, to concentration camps, to mass starvation, to atomic war."

For many intellectuals, coming to terms with these frustrations meant acknowledging that their roots lay not in any economic or political system, but in human nature itself. "Our modern civilization . . . was ushered in on a wave of boundless social optimism," wrote Niebuhr in a series of

1944 lectures that became the book *Children of Light, Children of Dark-ness.* "Modern secularism is divided into many schools. But all the various schools agreed in rejecting the Christian doctrine of original sin." This was a little overstated, as Niebuhr's generalizations often were. But it was a feature of intellectual life in the 1940s to see the past in prelapsarian terms.

Niebuhr's vision of man as a fallen creature was rooted in his Christian theology. But his views penetrated far beyond the community of faith. Lionel Trilling, a secular Jew and America's most influential literary critic, was drawn to writers like Nathaniel Hawthorne and T. S. Eliot, whose work evoked a sense of tragedy about the human condition. In *The Vital Center,* Schlesinger attacked the progressives for their "soft and shallow concept of human nature" and their "unwarranted optimism about man." There was a self-consciously jaded quality to many wartime intellectuals. They were like Rick, the hard-boiled nightclub owner in the 1942 film *Casablanca,* who seems utterly cynical about the world yet risks his life to resist the Nazis nonetheless. Niebuhr and Schlesinger loved calling their adversaries—be they communists or old-style progressives—idealistic and innocent. For their part, they prided themselves on seeing the world in all its awfulness without succumbing to despair.

In addition to assaulting the progressives for their overly sunny views of human nature, intellectuals in the 1940s attacked the claim, made by pro-gressives and Marxists alike, that politics could be scientific. The wartime intellectuals were not hostile to science and technology per se. How could they be, since without radar and penicillin and the factories that churned out ships and planes faster than the Nazis, America might not have won the war? But they denied that the scientific method—the gathering and testing of objective facts—could underpin government, since people, un-like machines, did not always act in rational ways, and since politics, un-like engineering, was fundamentally a moral pursuit. Once again, John Dewey served as whipping boy. The University of Chicago philosopher Mortimer Adler, yet another student turned assailant, accused him of cre-ating an educational philosophy based on the principle that "science . . . would eventually supply all the guidance necessary for human conduct." That made a Deweyian education ultimately amoral, Adler alleged, since science "cannot make a single judgment about good and bad, right and wrong."

If Adler accused Dewey of thinking America should brew its politics in a beaker, no one applied the critique more systematically to foreign policy than Hans Morgenthau, who took deadly aim at Colonel House's old idea of a scientific peace. A Bavarian-born Jew, Morgenthau spent his youth at the knee of scholars steeped in the tradition of Bismarckian realpolitik. He admired that conservative, Germanic tradition, with its reverence for the national interest and the balance of power, because he believed it respected limits, limits of both human reason and national power. He mourned its replacement by Nazism, and when Nazism chased him to America's shores, he used the realpolitik tradition of his youth to challenge the foreign policy culture of his adopted land, in hopes of better preparing it for its confrontations with the old world.

For too long, Morgenthau argued in his 1946 book, *Scientific Man vs. Power Politics*, Americans had succumbed to the "misconception of international affairs as something essentially rational, where politics plays the role of a disease to be cured by means of reason." This, in Morgenthau's view, was a profound mistake; human behavior could not be rationalized, because human beings possessed an irrational and incurable lust for power. That was particularly true in international affairs. In the domestic arena, Morgenthau observed, liberal democracy camouflaged the harsh realities that underlay all politics, redirecting the struggle for power from the bullet to the ballot. As a result, people who took democracy for granted—and no people did so more than Americans—tended to believe that international politics could be tamed in the same way: by creating global parliaments that put international disputes to a vote or global courts that ruled aggression illegal. But the domestic camouflage of power, Morgenthau insisted, depends upon the fact that one sovereign possesses a monopoly of it. It is precisely because the state's power is so uncontested that it can sheath its sword, thus creating the illusion that its authority doesn't ultimately rest on force. In the international arena, by contrast, where there is no sovereign, the contest is naked. In a neighborhood with no policeman, no one can pretend that safety does not rely on the barrel of a gun. And thus, Morgenthau posited, it was in international affairs that the illusion that politics could be made rational and scientific—made into something other than the self-interested search for power—was most dangerous. Both the war progressives and their pacifist successors had relied heavily on analogies between domestic and international affairs. Those analogies, Morgenthau insisted, were dead wrong.

Morgenthau urged Americans to abandon the progressive illusion that through scientific knowledge and charismatic leadership they could educate other nations to desire something besides power. And he urged them to abandon that illusion about themselves as well, recognizing what other nations already knew: that beneath America's moralistic rhetoric lay its own self-interest. Instead of trying to create a rational system to govern the world, he argued—be it Wilson's collective security or Dewey's pacifism—Americans should see themselves as guardians not of universal morality, but of their own national interest. For Morgenthau, a foreign policy based on national interest was not amoral; to the contrary, it was the most ethical response to an anarchic world. After all, in a town with no policeman, what better keeps the peace: the household that in the name of nonviolence disarms itself, and thus tempts others to attack; the household that equates its own interests with everyone else's, and thus seeks to deny others the right to self-defense; or the household that carefully delineates its domain and maintains sufficient strength to defend it? In foreign policy, Morgenthau argued, where precepts drawn from domestic affairs hold little sway, a concern for the national interest was most likely to establish a balance of power, an equilibrium that confronted potential aggressors with countervailing force. And a balance of power could effectively mitigate international conflict because rather than requiring nations to transcend their self-interest, it relied upon it. It appealed to something more solid than reason: fear.

In the 1940s, Morgenthau was not alone. He was part of a remarkable crop of foreign policy intellectuals determined to slay the hubris of reason once and for all. One of those intellectuals, ironically enough, was the former boy wonder of war progressivism, Walter Lippmann. Having already recast World War I as a war of self-defense, Lippmann, now America's most famous columnist, argued that after World War II America should not again try to create an institution like the League of Nations based upon "a moral code which was beyond human capacity to observe and which the United States had been unable to honor." Instead, he argued, sounding a lot like FDR, that the United States should focus on maintaining cooperation among the big powers that had vanquished Germany and Japan. And if that required granting Moscow a sphere of influence in Eastern Europe so it would feel less threatened, so be it. After all, didn't America have its own sphere in Latin America? Here Lippmann picked up a theme that both Niebuhr and Morgenthau stressed: that America must acknowledge the self-interest in its own actions rather than demanding that other nations comply with moral norms it did not uphold itself.

By the end of World War II, Lippmann was no longer a wunderkind. Nearing sixty, he had enjoyed dazzling professional success. But he had also experienced not only national tragedy, but personal tragedy as well, having seen his marriage collapse in a blaze of public humiliation after an affair with the wife of one of his best friends. He was no longer the optimist and rationalist he had been when he met Colonel House for that autumn 1917 stroll. His views, he wrote in 1943, had evolved "slowly over thirty years, and as a result of many false starts, mistaken judgments and serious disappointments. They represent what I now think I have learned, not at all what I always knew." It was a measure of the age that when it came to foreign policy, even a man as haughty as Walter Lippmann was in a chastened mood.

If Niebuhr was realism's theologian, Morgenthau its political scientist, and Lippmann its journalist, its greatest postwar practitioner was an introverted hypochondriac from Milwaukee named George Frost Kennan.

Kennan was too young to be a disillusioned progressive and too conservative to have ever been a communist, but he didn't need grand public embitterments to decide that the world was a grim place. Private embitterments more than did the job. He was born in a house built backward, so it received little natural light. His mother died when he was an infant, leaving him to be raised by a repressed, distant father and a stepmother who favored her own boy over him. At twelve they dispatched him to military school, where he was tormented by bigger boys and his class yearbook read, "Pet peeve—the universe." Fascinated with the American elite ever since reading F. Scott Fitzgerald, he somehow made it to Princeton, where he was even more miserable than he had been in high school. A poor, unsophisticated son of the Midwest with weak social skills and something of a persecution complex, he entered the Ivy League expecting to be looked down upon, and was not disappointed. In his first week, he asked a fellow student for the time and had smoke blown in his face. On a campus dominated by eating clubs, he had neither the money nor the cachet for sustained membership, and ate most of his meals alone. After graduating, he joined the foreign service, a profession he loved but tried to resign from eight times in response to imagined slights. In his time off he often checked himself into sanitariums to cure an array of real and perceived maladies. He enjoyed the company of nurses, he later admitted, because they conjured the mother he had never had.

Kennan once wrote that he felt like "a guest of one's own time and not a member of its household." This alienation from the modern world was sometimes quaint: Kennan loathed cars and proposed that candidates for the foreign service be required to till the soil. And it was sometimes appalling: Kennan's love affair with the eighteenth century often left him unsympathetic to blacks, Jews, women, and immigrants who demanded liberation from ancient hierarchies. Kennan's affection for hierarchy, in fact, sometimes left him dubious about democracy itself and led him to muse about the benefits of a government run by unelected wise men like, well, himself.

As an observer of his own country, Kennan was often unfair and unreliable. But it wasn't his analysis of the United States that in the 1940s made him famous: It was his analysis of the U.S.S.R. Upon joining the foreign service, he was sent to study Russian at the Oriental Seminary in Berlin, a citadel of realism founded by Otto von Bismarck, and then to the Baltic capitals of Riga and Tallinn, where he was introduced to Russia by its czarist émigrés, who bred in him an abiding love for its culture and a fierce hostility to its communist rulers. As the years passed, and his foreign policy views jelled, Kennan's realism and his anticommunism came together in a critique of what he called "universalism," a pathology he believed that American progressives like Wilson and Soviet communists like Lenin shared. Both groups, Kennan argued, wanted to remake the world according to supposedly objective principles and promised that once they had, "the ugly realities [of world politics]—the power of aspirations, the national prejudices, the irrational hatreds and jealousies—would be forced to recede" and nations would live in harmony. The nature of those principles differed, of course. Wilson wanted a world of reason and law, a world that looked like America in the progressive age; Lenin wanted a global dictatorship of the proletariat, a world that looked like the U.S.S.R. But the fundamental problem with both visions, Kennan insisted, was that they imposed an abstract, universal vision on a messy, diverse world. For Kennan, like Morgenthau, foreign policy was not a science. Practitioners should not build logical models; they should tend to organic matter. They should be not mechanics, but gardeners, cultivating each plant or flower based on its own circumstances and needs. In foreign policy, as in marriage, Kennan argued, general principles of conduct are far less useful than an intimate understanding of the participants. Had he known that Wilson proposed drafting a constitution to govern his union, he would have laughed and laughed.

When it came to Russia policy, Kennan saw two imperatives: to prevent Soviet universalism from threatening America and to prevent America from threatening itself by lapsing back into a Wilsonian universalism of its own. Kennan didn't begrudge FDR's willingness to grant the Soviets a sphere of influence, since preventing it was beyond America's power. But he considered Roosevelt naïve for thinking that granting the U.S.S.R. such a sphere would make it an ally in policing the postwar world. Russian history, Kennan argued, inculcated a suspicion of outsiders, and Marxist-Leninism, with its ideological hostility to capitalism, confirmed the tendency. Given the Kremlin's revolutionary aspirations, he believed, FDR's efforts to enlist it as a partner in upholding the postwar status quo were doomed. Even worse, America's efforts at cooperation would likely be taken in Moscow as signs of weakness, thus increasing rather than reducing the chances of war.

Kennan did not consider Stalin another Hitler; he thought the Soviet leader was basically cautious. But if America, in its zeal to maintain the wartime alliance, left the door to expansion open, Stalin would walk through. Russia, wrote Kennan, is like a "toy automobile wound up and headed in a given direction, stopping only when it meets with some unanswerable force." America needed to carefully choose which chunks of the globe were essential to its security, and in each one establish a balance of power sufficient to deter Soviet influence. Such balances would not require governments that resembled America's: that would be Wilsonian universalism, Kennan's other bête noire. They could be democratic or dictatorial, capitalist, socialist, or even communist. All America should ask of its allies was the will to resist Soviet domination. With this strategy, Kennan modified FDR's response to the hubris of reason. The final answer would not be called the Four Policemen; it would be called containment.

By 1946, Washington was finally ready to hear what Kennan had to say. During the war, given FDR's obsession with maintaining the anti-German alliance, Kennan and the other anti-Soviet hard-liners in the State Department's Eastern European division had been mostly ignored. But in his final days, even Roosevelt, by some accounts, was losing patience with Moscow. While he sympathized with Stalin's desire for friendly governments on Russia's border, he had grown increasingly annoyed by the Soviet leader's inability to orchestrate them with a light and subtle hand so as not to embarrass him at home. Roosevelt may have genuinely believed that the nations of Eastern Europe could elect governments that were both democratic and deferential to Moscow—a wildly unrealistic hope in a country like Poland, which seethed with hatred toward its large and brutal neighbor. Or he may simply have wanted to delay any conflict with Stalin until

Germany and Japan were defeated and America was ensconced in the UN. Either way, when Harry Truman succeeded him, relations between Washington and Moscow began to chill. Truman—a failed businessman, a machine politician, the vice president largely because special interests had blocked more qualified contenders, and now the first president in a half century without a college degree—was staggeringly unprepared for the challenges he would face. FDR hadn't even bothered to tell him that America was building an atomic bomb. On April 12, 1945, when Eleanor Roosevelt broke the news that the president was dead, Truman was silent for a long time; then asked, "Is there anything I can do for you?" To which she replied, "Is there anything *we* can do for *you*? For you are the one in trouble now."

One consequence of Truman's ignorance was that he viewed U.S.-Soviet relations from the perspective of an outsider. Roosevelt had felt grateful to the U.S.S.R., which had lost ninety times as many lives in the war as had the United States. He knew that by delaying America's invasion of France until 1944 he had purchased victory cheap, and that he had been able to do so because for almost three years Soviet ground troops had faced the Nazi meat grinder largely alone. At Tehran and Yalta, FDR had also given Stalin the winks and nods that undercut Allied declarations about democracy and self-determination. Truman hadn't been in those back rooms, nor even told what transpired there, and so he saw the Allied victory the way most of his countrymen did: as the story of America saving Europe, and especially Russia, which Truman assumed had only survived Hitler's onslaught because of American aid. It was Stalin, in his view, who should be grateful, and should express that gratitude by upholding his public pledges to safeguard democracy in the nations he controlled. Over the course of 1945, as it became clear that Stalin was doing no such thing, Truman began to get tough.

Convinced that FDR's diplomacy had involved too much carrot and too little stick, Truman spent much of 1945 trying to muscle Moscow into cooperating on America's terms. He brought Soviet foreign minister Molotov into his office and gave him "the straight one-two to the jaw." He halted wartime aid and made it clear there would be none in the future unless the Kremlin cleaned up its act. He threatened to publicly denounce Soviet behavior in Eastern Europe and he brandished America's new atomic bomb, thinking it would scare Stalin into complying with U.S. demands. But all these efforts failed: The bomb was too crude an instrument to be wielded for geopolitical effect, and while the Soviets didn't relish the cutoff of U.S. aid, having pliant governments on their border mattered more. As for Molotov, it wasn't easy to intimidate a man who slept with a hunk of brown bread and a pistol under his pillow. So Truman and his advisers began a more fundamental reassessment. Perhaps the problem wasn't that FDR's efforts at cooperation had involved too many carrots. Perhaps cooperation wasn't possible at all.

That's where Kennan came in. In February 1946, in his capacity as deputy head of the U.S. mission in Moscow, he was asked by the State and Treasury departments to explain Stalin's increasingly anti-American rhetoric and his refusal to join the newly created International Monetary Fund and World Bank. Kennan read the request and groaned. He had explained the Kremlin's behavior to his superiors in Washington many times before, and received no response; it was like "talking to a stone." What's more, the "sunless, vitaminless" environment of Stalinist Moscow was affecting his health. Suffering from a fever, cold, and ulcers, he retired to his sickbed and dictated an eight-thousand-word telegram, the longest ever sent from the embassy in Moscow. He expected that this time too it would disappear into the bowels of the Washington bureaucracy with barely a trace.

In his missive, Kennan took aim at the key assumption underlying both Roosevelt's and Truman's policies: that the United States—either through carrots or sticks—could maintain its alliance with the Soviet Union now that the war was won. Moscow, argued Kennan, was not eschewing cooperation because American policy was too tough or too soft; American policy was irrelevant. Stalin was not Georges Clemenceau, a prickly character eager for a postwar partnership with the United States. Stalin *wanted* conflict with the capitalist world because he ruled by repression, and to justify that repression he needed an external threat. With Hitler vanquished, and America the most powerful capitalist country in the world, we were now it.

Kennan's answer was containment. If Roosevelt had transgressed the Wilsonian faith by accepting spheres of influence, Kennan now committed an even greater apostasy: He blessed the balance of power. Even FDR, Kennan argued, had been dewy-eyed. The Soviets would not help America police the world. Rather, they would pocket their Eastern European spoils and seek new ones, foraging in places such as Western Europe, Japan, and the Middle East, which America must not allow a hostile power to control. To prevent this, Kennan argued, the United States would have to seek regional balances of power, alliances that met the U.S.S.R. and its clients with countervailing force.

This time, Kennan's telegram did not reach Washington and disappear. To the contrary, it detonated with near-atomic force. Kennan's boss, Ambassador to Russia Averell Harriman, gave it to Secretary of the Navy James Forrestal, who sent it to Truman, the entire cabinet, top military officers, key members of Congress, and influential columnists and publishers. The State Department distributed it to every U.S. Embassy in the world. Kennan had provided a rationale for the policy Washington was already groping toward. The central reality of the coming age, it was now clear, would be not great-power cooperation, as FDR had hoped, but great-power conflict. As Kennan's fellow Soviet expert Charles Bohlen noted, there would be not one postwar world, but two.

As containment became the prism through which Washington saw the world, other aspects of the postwar settlement fell into place. Originally, FDR had wanted Germany crushed. The more emasculated it was, he assumed, the less it could challenge the Four Policemen who patrolled the postwar world. In 1944, he endorsed Secretary of the Treasury Henry Morgenthau's plan to exile Germany to the nineteenth century, dismantling its industry and carving it into small agricultural states. When FDR talked about how to treat Hitler's vanquished nation, his favored image was castration, and it wasn't always clear that he was speaking entirely metaphorically.

By 1945, Roosevelt had ditched Morgenthau's plan, in part because of opposition from Secretary of State Hull, who feared inflaming the Germans. But what sealed the decision was containment. Preventing Soviet expansion, Kennan argued, required building up the capacity of key regions to resist. And unless Germany's economy revived (at least in that part of Germany under Western control), neither would Western Europe's as a whole, leaving it ripe for Soviet subversion. Ironically, then, it was the recognition that great-power cooperation was impossible— that a balance of power was the best America could achieve—that helped convince the United States to treat the defeated Germans more generously. The same principle underlay the Marshall Plan, America's initiative to alleviate the economic disorder and despair stalking postwar Western Europe. That noble effort would never have passed Congress had Americans not feared that Moscow would turn Europe's suffering to its advantage. Such ironies fit the intellectual ethos of the age: By

lowering America's moral sights and substituting a more realistic vision
(containment) for a more optimistic one (U.S.-Soviet cooperation),
America was actually doing more good.

The contrast was clear. After World War I, the absence of foreign
threats had contributed to America's withdrawal. After World War II,
by contrast, the specter of Soviet power drew America further and further into European affairs. Soon the United States wasn't only paying
to rebuild Western Europe economically; it was promising to defend it
militarily. If in 1940, when France fell, Clemenceau's nightmare had come
true, in 1949 Washington and Paris realized his dream, a binding security
guarantee, the NATO pact.

What held the postwar settlement together, then, was restraint on American universalism and fear of the universalism of the Soviet Union. FDR
granted Moscow its sphere of influence; Truman contained it. Together
they acknowledged the limits of America's capacity to remake the world
yet convinced Americans to sacrifice so that Hitler and Stalin could not
remake it, either. Roosevelt and Truman set their ideological ambitions
lower than Wilson and paid a higher price to achieve them; they deployed
greater power in pursuit of narrower goals. By eschewing a scientific peace,
they forged one better able to endure in an imperfect world.

And yet, while America's leaders often acted like realists, they didn't
sound like realists. They outfitted their Lodgean peace with Wilsonian
clothes. From Newfoundland to Yalta to San Francisco, FDR and Truman often spoke in universalistic, missionary terms. Partly it was politics.
Kennan, Morgenthau, and Lippmann wanted to keep the citizenry out of
foreign policy, since they believed that public participation would hamper
the subtle and ruthless adjustments that an effective balance of power
required. FDR and Truman, however, could not afford such an elitist
view. They had to speak to Americans in their own language; they had to
get legislation through Congress. They had to worry about the balance of
power at home as well.

But there was more to it than that. The words were window dressing,
but they were not only window dressing. They reflected a discomfort
with brute, unreasoning power even as America accommodated itself
to it. If FDR and Truman lacked Wilson's faith in a world governed by
reason, neither were they fully reconciled to its opposite. Had they not
at least partially believed their own words, they could not have spoken
them so convincingly. In this sense, they were less like Kennan, Mor-

genthau, and Lippmann than like Niebuhr, who refused the realist label because while he acknowledged that self-interest powered human affairs, he did not want to sanctify it. He believed that while both nations and individuals were inherently sinful, they could still be redeemed through the mystery of God's grace. And it was this faith that ordinary individuals could somehow be redeemed that kept Niebuhr from the dark and sometimes ugly skepticism of democracy to which Kennan fell prey. "Realities are always defeating ideals," he wrote, "but ideals have a way of taking vengeance upon the facts which momentarily imprison them."

This is where Niebuhr and Roosevelt met. Perhaps it is no surprise that a man who for decades insisted, in the face of all reality, that he would cast off his own momentary imprisonment and walk again would insist on seemingly fanciful declarations about what was possible in the world. With the UN charter, as David Fromkin puts it, FDR created a sleeping beauty. Like the Declaration of Independence, its words rebuked their authors and offered the prospect, however distant, that justice and reason would one day rise from their slumber.

It was a treacherous balance. If the ideals grew too intoxicating they could blind Americans not only to the limits of what they could expect from the world, but of what they could expect from themselves. Missionaries unable to recognize their own sinfulness, Niebuhr noted again and again, were worse than merely naïve; they were often monstrous. Yet if the ideals disappeared altogether, as the realists seemed to desire, the result might be utter cynicism, a world grown numb to evil.

There was no abstract formula for weighing these competing imperatives. Abstract formulas, in fact, had been part of the problem. Much of the time, FDR and Truman were simply feeling their way. But overall, their generation—which was more aware than the war progressives and pacifists of the tragic nature of world affairs and more aware than the isolationists that America must struggle to improve it nonetheless—found a balance that served their nation astonishingly well. For an extraordinary period during and immediately after World War II, they not only built a stable balance of power but swathed it in the garb of reason. They saved Asia from Japanese imperialism, and Western Europe from not one species of totalitarianism but two. They took a nation that wanted no commitments in the world, especially if they involved carrying a gun, and made it the leader of the mightiest wartime and peacetime alliances in human history. They did all this while strengthening America's economy and preserving its democracy, in the face of Beard's and Dewey's predictions that another world war would destroy both.

And then, as the years passed, and Americans increasingly took the fruits of their labor for granted, a younger generation of thinkers and leaders began to suspect that they had set their sights too low.

PART II

THE
HUBRIS OF
TOUGHNESS

CHAPTER FIVE

THE MURDER OF SHEEP

For all his misanthropy, George Kennan had to admit that the world was treating him pretty well. Before his February 1946 "Long Telegram" from Moscow, he had been an obscure, miserable functionary; perhaps two hundred people in the entire government had known his name. Now he was the State Department's resident genius. In April he was recalled to Washington and made deputy commandant of the National War College, where he taught officers and diplomats about the U.S.S.R. A year later he became the State Department's first-ever director of Policy Planning, which put him in charge of foreign policy big-think and gave him direct access to Secretary of State George Marshall. Even his health began to improve.

The key figure in Kennan's ascent was Navy secretary James Forrestal. Like Kennan himself, Forrestal was an outsider in the patrician, East Coast, WASP-dominated foreign policy elite. Born to a working-class Catholic family, he had twice broken his nose boxing, which added an air of menace to his rugged good looks. Some people said he resembled the actor James Cagney, who was famous for playing gangsters. Like Kennan, Forrestal had made it—against the odds—to Princeton. But unlike Kennan, who was miserable there yet stayed true to himself, Forrestal used the Ivy League to don a new identity as an upper-class dandy. He shed his Catholicism, cut ties to his parents, made millions on Wall Street, and romped with the Jazz Age Long Island set like a character from a Fitzgerald novel. Yet he never seemed entirely at ease with his new self. His many business associates, his many drinking buddies, his many lovers, even his ill-fated wife, all admitted that beneath the high-achieving, high-living exterior lurked a man they didn't really know at all.

Politically, Forrestal was a realist of sorts. He certainly saw the world as a tough and irrational place. But for Kennan, Niebuhr, Morgenthau, and

Lippmann, the recognition that the world was tough and irrational represented a reaction against Wilson's messianic fervor, a kind of anti-ideology. Forrestal, by contrast, invested toughness with an ideological fervor of its own. He saw the Soviets as Nazis, ruthless fanatics determined to snuff life from the free world. And he believed that only brute, unyielding force could keep them at bay. In his use of Nazism as a template for understanding international affairs—his belief that evil was a permanent, pervasive force in the world and that any concession to it would unleash the same horror as Chamberlain's had at Munich—Forrestal was, in 1946, a man ahead of his time, a harbinger of a new hubris gradually taking root in a new age.

Forrestal was morbidly fascinated by the writings of Marx and Lenin, which he saw as the postwar equivalent of *Mein Kampf*: a Rosetta stone for understanding America's new totalitarian foe. In late 1945, he asked an economist on his staff named Edward Willett to do a study of "dialectical materialism," a term Willett had never heard before. Willett, a man whose intellectual reach often exceeded his grasp, read two books on Marxism and then produced a paper titled "Random Thoughts on Dialectical Materialism and Russian Objectives," which lived up to its name. Forrestal distributed the paper to some Russia experts for review, but the general response was that Willett didn't know what he was talking about, and that in any case, Soviet foreign policy was motivated more by national interest than by Marxist theory. Forrestal was crestfallen. He had commissioned the paper precisely because he believed the Kremlin *was* motivated by Marxist theory and thus was hell-bent on global revolution. "I realize it is easy to ridicule the need for such a study," he wrote to one associate, "but I think in the middle of that laughter we always should remember that we also laughed at Hitler."

Just then, like an answered prayer, Kennan's Long Telegram landed on his desk. It was well reasoned, well informed, articulate—everything Willett's paper was not. Kennan, like most other Russia experts, believed that Stalin's foreign policy was motivated more by Russia's historic insecurity and animosity toward the West than by the writings of Lenin and Marx. But in his effort to puncture the still-prevalent view that America could charm Moscow into postwar cooperation, Kennan had described the Kremlin in harsh, pungent terms. And he did acknowledge that communism exacerbated Russian suspicions of the outside world. This was good enough for Forrestal, who now had a new horse to ride.

After summoning Kennan back to Washington to teach at the National War College, which he had helped create, Forrestal asked him to revise Willett's article, hoping to invest its argument with greater intellectual heft. Kennan demurred, admitting that he didn't share Willett's belief that Stalin, like Hitler, wanted a world war. But Forrestal countered with a new request: Kennan should write a paper of his own. This put Kennan in a bind. He wanted to keep Forrestal, his most powerful patron, happy, yet he didn't really share his views. He tried to square the circle by drafting an essay that claimed Moscow's actions were motivated by both nationalist and ideological impulses, without suggesting which was more important. That was good enough for Forrestal, who distributed Kennan's essay with a cover note stating that "nothing about Russia can be understood without also understanding the implacable and unchanging direction of Lenin's religion-philosophy." For Kennan, who later admitted that he had written the essay to serve "what I felt to be Mr. Forrestal's needs at the time," it was a small, strategic breach of intellectual honesty. And it worked. Three months later, with a push from Forrestal, he got his dream job: head of the State Department's Policy Planning staff.

But Kennan's breach came back to haunt him. In January 1947, he gave a talk on Russia at the Council on Foreign Relations, which the editor of the Council's journal, *Foreign Affairs*, asked him to turn into an article. Kennan didn't have time to write something new, so he submitted the essay he had penned for Forrestal. The State Department gave him permission to publish it as long as it did not bear his name. Under the article's title, the byline read simply "By X."

On the surface, the X article brought Kennan another burst of good fortune. It was excerpted in *Reader's Digest* and *Life*, and quoted in *Newsweek*. After Forrestal tipped off *New York Times* columnist Arthur Krock about the author's identity, Kennan was named one of *U.S. News & World Report*'s people of the week. Suddenly he was deluged with speaking requests. At Radcliffe, where his daughter attended college, friends began calling her "Miss X."

But beneath all the adulation lay a problem, one that would shadow Kennan until he was an old man: the problem of what containment actually was. For Kennan, it was something modest, something significantly less than a global doctrine. First of all, it was a strategy against a country, the Soviet Union, not an ideology, communism. If communist movements were pawns of Moscow, Kennan conceded, then their ascent to power in strategically important countries should be vigorously opposed. (He even supported covert action to undermine Soviet control in Eastern

Europe.) But unlike Forrestal, he suspected that some communist regimes might refuse orders from the Kremlin and perhaps even grow openly hostile, in which case they themselves might become vehicles for containing Soviet power. For Kennan, it was critical that while the Soviet Union pursued "universalism"—a world remade in its ideological image—the United States should not. That, in his view, had been Wilson's mistake. The United States might be partial to democracy, but it should condone ideological diversity. It should tolerate governments of any type as long as they were not agents of a threatening power.

Secondly, for Kennan, containment was primarily a political strategy, not a military one. He had seen firsthand the way World War II shattered Soviet society, and he believed the Kremlin had little appetite or capacity for another world war. Unlike Forrestal and Willett, therefore, he put little stock in the Hitler analogy. The real danger, in his mind, was not that the Red Army would sweep across Western Europe. It was that the people of Western Europe, made desperate by the continent's postwar misery and chaos, would elect Kremlin-controlled communist parties and thus unlock the door to Soviet domination from the inside. That's why for Kennan, containment's crown jewel was the massive program of American economic assistance dubbed the Marshall Plan. NATO, by contrast, which committed the United States to Western Europe's military defense, struck him as unnecessary and probably counterproductive.

Finally, Kennan did not believe that containment applied everywhere. Even the spread of Soviet influence did not bother him if the spread was to countries that lacked the industrial capacity, natural resources, or strategic location to tip the geopolitical balance in Moscow's favor. If the Kremlin wanted to waste its money and blood in irrelevant backwaters, he believed, that was no reason for America to follow suit. Kennan was not always precise or consistent in defining which chunks of the globe he considered strategically important and which he did not, but he was adamant that containment have geographic limits. The central point, in his mind, was that America must have a prior conception of its interests, which it defended against Soviet threats. To oppose Soviet power everywhere would flip that on its head: It would mean deciding that anywhere the U.S.S.R. probed suddenly became in the interest of the United States to defend, in which case Moscow, not America itself, would determine which areas of the globe mattered to America.

By the summer of 1947, Kennan had articulated these principles many times. What he had not done—on purpose—was clearly articulate them in his paper for Forrestal, the paper that, to his surprise, made *containment* a household world.

In his summer home, surrounded by pine trees on an island off the coast of Maine, Walter Lippmann read the X article and decided that containment was a dangerous idea and Kennan was a dangerous man. He proceeded to write fourteen straight columns attacking the concept, which he packaged into a short book. He took its title from a French phrase describing the standoff between Berlin and Paris before World War II: *la guerre froide*, the cold war.

For Lippmann, Mr. X's containment doctrine had two big and interrelated problems. First, it assumed that the Soviets were motivated by ideology, not nationalism. As a result, it interpreted the Kremlin's insistence on an Eastern European sphere of influence not as the fulfillment of an historic Russian desire for a buffer against attack from the West (a desire that Hitler's invasion had graphically underscored), but as the first step in a bid to take over the world. Second, because Mr. X had defined Soviet ambitions as global, he was insisting that America's response be global, too. Containing Moscow everywhere, Lippmann warned, would require the United States to commit itself to a motley array of "satellites, puppets, clients, agents about whom we can know very little." It would require "a drawing account of blank checks both in money and military power."

The irony was exquisite. To please Forrestal, Kennan had described Soviet policy as more ideological than he thought it really was, and containment as more global than he believed it needed to be. To make a polemical point, Lippmann had put the most extreme gloss on Kennan's purposefully ambiguous pronouncements. The result was a grand, public feud between two men who basically agreed. Kennan called it a "misunderstanding almost tragic in its dimensions." He checked into Bethesda Naval Hospital complaining of ulcers and began composing a letter to Lippmann expressing his "bewilderment and frustration" that the columnist had "held against me so many views with which I profoundly agreed." Kennan considered publishing the letter, but Secretary of State Marshall, who believed his planning chief had gotten too much publicity already, refused. And Kennan, for reasons that remain obscure but probably stemmed from his generally neurotic personality, never sent it, either. Two years later, finding himself on a train with Lippmann from New York to Washington, he launched into an impassioned explanation of why the columnist had gotten his views all wrong. But by then it was too late.

* * *

It was too late because between 1946 and 1950, containment swelled from the limited doctrine Kennan espoused into the hubristic one Lippmann condemned. And the vehicle for that expansion, as usual, was success.

What expanded was not merely a foreign policy doctrine, containment, but a foreign policy mind-set, toughness, which had its roots in World War II. For virtually everyone in the Truman administration, Munich had been a totemic trauma, a searing, life-altering event. And to most policymakers its lessons were clear: The world contained evil people and evil regimes, which were impervious to reason. If their aggression was not met, early on, by force, the result might be death, destruction, and dishonor on an epic scale.

But despite this, America in the early postwar years did not go searching the world for Hitlers to destroy. In Western Europe and Japan, to be sure, Truman administration officials generally agreed that America must contain Soviet power; World War II had shown we had vital interests there. There were other chunks of the world, however, where the United States had never before projected much military power and whose political fate had never seemed tightly bound up with its own. The United States in the late 1940s did not have global military commitments, and most politicians, on both sides of the aisle, wanted to keep it that way. In 1948, with the Truman administration negotiating the treaty that created NATO, thus obligating America to defend Western Europe, Arthur Vandenberg and John Foster Dulles, two of the Republican Party's most influential spokesmen on foreign affairs, both demanded Secretary of State Marshall's assurance that America not go around making similar pledges in other corners of the globe.

One reason they feared such pledges was financial. With memories of the Depression still fresh, many politicians and policymakers considered the U.S. economy a fragile organism, vulnerable to sudden collapse. Containing communism across the entire globe would require vast sums for defense, which would likely lead to budget deficits and their ugly cousin, inflation, a prospect almost as frightening to government officials in the early Truman years as the Red Army.

As a result, to Forrestal's dismay, Truman put sharp restrictions on the amount he would spend containing the Soviets, restrictions that buttressed Kennan's insistence on containment's geographic limits. Since "our nation's economy under existing conditions can afford only a limited amount for defense," announced Omar Bradley, chairman of the Joint Chiefs of Staff, in October 1949, "we must look forward to diminishing appropriations for the armed services." That same month, Truman unveiled his budget, which slashed military spending by 15 percent. In the late 1940s, America's leaders did not know how much military spending the U.S. economy could bear, and they were not eager to find out.

But success turned Kennan's version of containment, with its keen appreciation of America's financial and geographic limits, into something more like Forrestal's. It began in the Near East, where Britain, which had long checked Russian ambitions, was now too bankrupt and exhausted to continue the job. In 1946, Iran's shah complained of bullying from Moscow, which had stationed troops on his nation's soil during the war and was refusing to remove them until it got a share of Persian oil. Truman demanded that the Red Army leave and dispatched $10 million in military aid, which Tehran used to fight a separatist uprising that the Kremlin had stoked in the country's north. Within a year, the Soviet troops were gone; the separatists had been crushed; and the United States, not the U.S.S.R., had won lucrative new rights to drill for Iranian crude. It was Munich in reverse: by standing tough—and not abandoning Iran, as Chamberlain had abandoned Czechoslovakia—America had nipped aggression in the bud. Iran had been a "test case," declared the State Department, of a "small state victim of large state aggression." And this time, America passed.

It was, it turned out, just a warm-up. A year later, in February 1947, a British embassy official drove to the State Department to inform U.S. officials that His Majesty's government, now liquidating its once-mighty empire in a kind of geopolitical fire sale, could no longer afford to defend Greece or Turkey, either. In Greece, a right-wing government was fighting for its life against communist rebels. In Turkey, the Soviets were demanding bases near the Dardanelles, from which they could project power into the eastern Mediterranean and the Middle East.

That night, Kennan and other top State Department officials met to consider whether America should come to Greece and Turkey's aid. It was a potentially historic shift. The eastern Mediterranean had been a traditional zone of British—not American—influence. (In their 1944 meeting in Moscow, where Churchill and Stalin agreed that the

United Kingdom would enjoy 90 percent control in postwar Greece, no American had even been in the room.) Moreover, defending Athens and Istanbul would cost far more than defending Tehran. Congress was in no hurry to send hundreds of millions of dollars to nondemocratic regimes whose fate had never troubled it before, and even some in Truman's own cabinet were dubious.

The State Department officials agreed that America should send the money. But Kennan shuddered when he heard the way Truman intended to justify it. "It must be the policy of the United States to support free peoples who are resisting attempted subjugation by armed minorities or by outside pressures," read the speech Truman was planning to give before Congress. This was just what Lippmann had warned against: a pledge to contain communism everywhere it reared its head. The Truman Doctrine "should not be a blank check to give economic and military aid to any area in the world where the communists show signs of being successful," Kennan pleaded in a memo to his superiors. The American people must be told that containment had limits. But Truman's political advisers said the universal rhetoric would help sway a skeptical Congress. Like the X article, it was a small, strategic breach of intellectual honesty, with consequences that turned out not to be small at all.

Once again containment worked. Congress appropriated the money, and by late 1949 the Greek communists (whom Stalin had never much cared for anyway) had laid down their arms and the Dardanelles was safe from Soviet threat. Truman's hard line boosted his approval ratings, as Democrats who had been pilloried as soft on communism in the disastrous midterm elections of 1946 regained their foreign policy edge. Things got even better in 1948, when the Soviets blocked land access to West Berlin and Truman responded with a massive airlift that forced them to back down. It was another step up the toughness ladder: Now the Americans weren't merely sending money to contain the U.S.S.R.; we were flying planes. And this brought yet more success. Abandoning Berlin to the Russians, warned one U.S. official, would have been "the Munich of 1948." Instead the Soviets caved, and Truman's reputation for firmly resisting communist aggression helped power him to an upset reelection victory that fall.

Containment was working beautifully; the Truman administration was flying high; and yet George Kennan—who had built the conceptual wings—was worried. "If I thought for a moment that the precedent of Greece and Turkey obliged us to try to do the same thing in China," he declared, "I would throw up my hands and say we had better have a whole new approach to the affairs of the world."

Almost everything Kennan knew about China he had learned from John Paton Davies, a man rather similar to himself. Although jollier than Kennan, Davies shared his talent for combining intimate knowledge of a foreign land with detached, almost clinical, judgments about its politics. Born to Baptist missionaries in the province of Sichuan, Davies had attended Yenching University alongside many of the people who would later populate China's political elite. After finishing his studies in the United States, he had returned to China as a Foreign Service officer, serving four diplomatic tours there over the course of twelve years. When Kennan became head of the Policy Planning staff, Davies was the first person he hired.

As Davies looked at China in 1948, he saw good news and bad news. The bad news was that there was no way on God's green earth that America could prevent Mao Zedong's Communist Party from taking power, since America's allies, the Nationalists of Chiang Kai-shek, whose corruption and incompetence Davies had witnessed up close, were in a state of terminal collapse. The good news was that Mao, who unlike the communist leaders in most of Eastern Europe had his own power base, was unlikely, in Davies' view, to remain Moscow's stooge for very long. If America played its cards right, in fact, Mao might even become America's ally in containing Soviet power. "A united communist China," suggested Kennan, reflecting Davies, "might threaten Russian security and Russian control of the communist movement."

Davies and Kennan proposed that America limit aid to Chiang in hopes of establishing decent relations with Mao once he took power. Forrestal, who had ascended to the newly created position of secretary of defense, objected. But Truman overruled him, declaring, with typical pool-hall eloquence, that sending money to Chiang was like "pouring sand in a rat hole." It was clear evidence that within the corridors of power, Kennan's modest version of containment remained alive and well. Forrestal notwithstanding, most Truman officials recognized that not every victory for communism was a victory for the U.S.S.R.; that containment applied in some parts of the world but not in others; and that given the limits of American money and power, sometimes the best thing the United States could do—even in the face of communist victories—was to leave well enough alone.

But that is not what they had told the public. From the X article to Truman's Greece and Turkey speech, the rhetoric of containment had soared higher than the reality, and Chiang's American supporters now demanded that the latter be brought in line. China, whose vast bounty of consumers and converts had long stirred the dreams of American business and the American church, occupied a special place in the national imagination. And as 1948 turned to 1949, a powerful China lobby, championed by *Time* magazine's founder, Henry Luce—the son of a Presbyterian missionary who had harvested souls in the city of Dengzhou—accused the administration of letting the communists win. Over administration objections, Congress appropriated billions in economic and military aid to Chiang.

Kennan urged his superiors to confront the "confusion and bewilderment in the public mind regarding our China policy," to say publicly what they were saying behind closed doors: that Chiang was a lost cause and Mao did not pose a grave threat. But Truman, like FDR before him, was more comfortable acknowledging the limits of American power privately than in public. His administration did not challenge the China lobby until it was too late, and by October 1949, when Mao hoisted a red flag in Tiananmen Square, the American China debate had turned rancid. The timing of Chiang's defeat was terrible: It came soon after news that Moscow had tested an atomic bomb, and amid of a wave of high-profile espionage trials, which fueled suspicions that Truman had let Mao take power because deep within the U.S. government, traitors were helping the communists win.

On February 9, 1950, with these flammable ingredients already in place, an obscure Wisconsin senator named Joseph McCarthy lit the match. He had been looking for an issue that would make his name. Friends suggested the St. Lawrence Seaway; then a national pension plan; then it hit him: treason! In a speech to the Women's Republican Club in Wheeling, West Virginia, he brandished a piece of paper allegedly listing 205 communists who worked for the State Department. The intersection of China's fall and McCarthy's lies, which he would hurl without scruple or shame for the next four years, played a crucial role in the growth of the hubris of toughness. China, after all, should have been a shining example of the limits of the Munich analogy. It should have convinced Americans of what Kennan and Davies knew: that contrary to the X article and contrary to what Truman had said in his Greece and Turkey speech, there was no earthly way America could man every anticommunist barricade across the globe. Had America genuinely tried to keep Mao from power—which would have required not merely epic sums of money, but U.S. casualties in numbers that would have made Okinawa look like afternoon tea—those limits would have become brutally clear. A serious effort at actually applying global containment in China would likely have brought the whole doctrine crashing down. But precisely because Truman hadn't done that, McCarthy and his allies could insist that holding China would have been easy: It would have required just a bit more aid, a few military advisers, and perhaps a little airpower. Fueled by McCarthyism, the legend grew that America had "lost" China not because it lacked the power to keep it, but because it refused to try.

Intellectually, the McCarthyite argument on China was preposterous. But its impact on American politics was immediate and profound. Within months of McCarthy's emergence and China's fall, Truman's job approval rating, which had been nearly 70 percent in January 1949, had been cut in half. Prominent Republicans were calling Secretary of State Dean Acheson a coward and his predecessor, George Marshall—the five-star general who had masterminded America's victory in World War II—a traitor. And the political claim that Truman officials were weak in their opposition to communism was often linked to a cultural claim, which had particular resonance in the toughness years: that they were weak as men. Acheson, sneered McCarthy, pranced around "with a lace handkerchief, a silk glove, and with a Harvard accent." GOP senator Everett Dirksen vowed to expose the administration's "lavender lads," and by 1953 more than four hundred State Department employees had been either fired or forced to resign for alleged homosexuality. Decades later, Lyndon Johnson, then a freshman senator from Texas, would remember that "Harry Truman and Dean Acheson had lost their effectiveness from the day that the Communists took over in China." When it was his turn in power, he would make sure nothing like that ever happened to him.

It was in this toxic political climate, on June 25, 1950, that the White House learned that communist North Korea had invaded its southern twin. Truman administration officials, consistent with Kennan's view that containment did not apply on the Asian mainland, had repeatedly implied, both publicly and privately, that they would not defend Seoul. But when war came, all that went out the window. Truman was in Independence, Missouri, when he learned of the invasion, and he quickly rushed back to Washington. "I had time to think aboard the plane," he later wrote. "In my generation, this was not the first occasion when the strong had attacked the weak. . . . Communism was acting in Korea just as Hitler, Mussolini, and the Japanese had acted ten, fifteen, and twenty years earlier." Later he added, "If . . . the threat to South Korea was met firmly and successfully, it would add to our successes in Iran, Berlin and Greece." Those were the choices: Munich or anti-Munich. Berlin and Greece had boosted Truman's approval ratings, while his failure to prevent Mao's takeover had sent them tumbling and fueled McCarthy's rise. The November midterm elections were only months away, and Republicans would surely cite another communist triumph as evidence that he was guilty of appeasement, if not treason. By 1950, Truman believed that acknowledging containment's limits was more than perilous. It was the fast lane to political death.

Within days, the United States was rushing troops to the battle. Inside the administration, the debate was no longer about whether America should fight; it was about what America was fighting for: the preservation of South Korea or the liberation of the communist North as well. When Kennan, who had been on leave at Princeton, returned to the State Department, Davies warned him that if U.S. forces tried to reunify the entire peninsula, China might enter the war. In response, Kennan proposed that Truman publicly pledge not to cross the 38th parallel, which divides the two Koreas. But in good toughness fashion, John Allison, director of the State Department's Office of Northeast Asian Affairs, accused him of espousing a "timid, half-hearted policy" of "appeasement." As the summer dragged on, Kennan felt the momentum swinging toward reunification. "The course upon which we are today moving," he wrote to Acheson in late August, "is one, as I see it, so little promising and so fraught with danger that I could not honestly urge you to continue to take responsibility for it."

But Acheson did take responsibility for it. In mid-September, General Douglas MacArthur pulled off a daring amphibious landing at Inchon, two hundred miles behind enemy lines, and began racing toward the 38th parallel. Halting him there would have required a heroic act of political will, since dealing with the egomaniacal general was, in Kennan's words, like "arranging the establishment of diplomatic relations with a hostile and suspicious foreign government." Besides, with the fall elections approaching, Truman needed a win, something that would silence the McCarthyite hordes. So far, every time he had taken another step up the toughness ladder, it had paid off.

So MacArthur crossed the 38th parallel into North Korea, and like Xerxes crossing the Bosporus, the gods took their revenge. In this case, the instruments of their vengeance were Chinese. In early October, with MacArthur's troops nearing the Yalu River, which separates North Korea from China, Stalin told Mao that if he threw five to seven divisions into the fight he could not only keep the Americans off his border, but deal them a crushing blow. After several sleepless nights, Mao decided—as Davies and Kennan had feared—to do just that. Within weeks, hundreds of thousands of Chinese troops were sneaking across the Yalu, carrying precooked meals so they wouldn't need to light fires visible to American planes. On October 26, U.S. forces took their first Chinese prisoner. Still, MacArthur pushed north, boasting that his men would be home by Christmas.

In late November, the Chinese launched an all-out attack, forcing U.S. troops into a panicked retreat. By spring, after months of ghastly fighting, American forces stabilized their positions roughly where the whole conflict began, at the 38th parallel, and State and Defense department officials began exploring negotiations to end the war. But MacArthur resisted, defying Truman's orders and demanding that America expand the conflict by blockading China's ports, bombing its military installations, and encouraging Chiang to launch an invasion from Taiwan. In April, after MacArthur's brazen insubordination became too much to bear, Truman fired him. But this just liberated the general to loudly accuse his former boss of being "blind to history's clear lesson" and practicing "a form of appeasement on the lines of the Munich conference in 1938." For his part, McCarthy called Truman's refusal to pursue total victory a "super-Munich," adding, "The son of a bitch ought to be impeached." The young senator from Texas, Lyndon Johnson, saw his office flooded with pro-MacArthur mail.

As MacArthur's supporters saw it, Truman—like Neville Chamberlain in 1938—was showing dangerous weakness in the face of aggression. But the charge just showed how unreal their definition of weakness had become. In the late 1930s, many Americans had denied that their nation's safety required the survival of France. Before 1947, few had believed American security had much to do with what happened in Turkey and Greece. In 1949, most observers assumed that the United States would abandon South Korea to its fate. Yet by 1951, Truman's opponents were handing him umbrellas because he refused to launch a war inside China to liberate all of Korea from communism. America's mounting global strength, as it filled the gap left by Europe's evaporating empires, combined with the terror of weakness inherited from World War II, had turned every foe into Hitler and every compromise into Munich. Politicians who once considered it no shame to let America's enemies have Paris now cried treason because we were letting them have Pyongyang.

MacArthur and his allies genuinely believed that the only reason containment kept expanding was that the communists kept confronting the United States with graver and graver threats. What they didn't grasp was that their definition of what constituted a threat had grown in parallel with America's growing confidence in its power. One big reason for that growing confidence was the U.S. economy. In the late 1940s, many Truman officials had seen it as fragile and thus unable to sustain too much defense spending. Their belief that the United States didn't need to stop communism everywhere in the world had been deeply bound up with their belief that America couldn't afford to. But by 1950, official views were beginning to change. In 1949, the new head of the Council of Economic Advisers, Leon Keyserling, began denying that deficits were as dangerous as most people assumed. Following the famed British economist John Maynard Keynes, he argued that the government spending that created deficits could also rev up the economy, thus producing higher economic growth—and higher tax revenues—and eventually bringing the budget back into balance. Keyserling cited World War II. After all, the massive military buildup—and resulting deficits—needed to fight the Nazis hadn't wrecked America's economy. To the contrary, they had ended the Depression. And in the years since, continued economic growth had helped balance the budget again. Recent experience, in other words, showed that the U.S. economy wasn't fragile at all. It could bear, and even benefit from, far more government spending than policymakers had previously imagined. The only real limits were in America's leaders' minds. As former undersecretary of state Robert Lovett told a committee tasked with putting America's containment strategy into written form, "there was practically nothing that the country could not do if it wanted to do it." His words might have served as an epigraph to the mounting hubris of the age.

In April 1950, that committee—led by Kennan's successor as head of Policy Planning, Paul Nitze—produced a sixty-six-page top-secret document called National Security Council Report 68, which declared that America could, and must, dramatically boost defense spending. NSC 68 was partly a product of Keyserling's economic optimism, an optimism nurtured by

America's burgeoning postwar boom. By 1955, Keyserling predicted, gross national product would grow to almost $300 billion, meaning that even at dramatically higher levels, military spending would not rise significantly as a percentage of GNP. Gradually, the Depression-induced caution of the 1940s was melting away. Truman officials were coming to see economic limits as a thing of the past.

In 1950, with NSC 68 claiming that the United States could afford to spend far more on defense, and Korea apparently proving that we needed to, a flurry of supplemental financing boosted military spending to more than $48 billion, an astounding increase over the roughly $13 billion in Truman's initial budget. Containment, which Kennan had conceived as primarily a political strategy, was now an unmistakably military one. And if NSC 68 eviscerated containment's financial restraints, it eviscerated its geographic ones as well. The report's authors admitted that not every communist victory represented a threat to the United States, but they insisted that America oppose communism's advance everywhere anyway because no matter how strategically insignificant a country might be, its loss would prove psychologically damaging. The *perception* of weakness, NSC 68 argued, would breed "doubt and recrimination," both among America's allies and within the United States itself. And with McCarthy becoming a national plague, no one needed to spell out just how debilitating that domestic "recrimination" could be.

NSC 68 highlighted a concept that would sit near the core of American foreign policy during the toughness years: "credibility," the belief that showing weakness in unimportant places would unnerve our allies and embolden the communists in important ones—which meant that unimportant places were important after all. America's challenge, therefore, was not merely to be strong, but to *appear* strong—too look "credible"— both to others and to ourselves. And in the funhouse mirror that was McCarthy-era Washington, where mighty America looked perpetually weak, creating the perception of strength required expanding containment to the far reaches of the earth.

Beneath the policy differences lay a deeper, philosophical difference between Kennan's version of containment and NSC 68's version. Kennan, like Lippmann, Morgenthau, and Niebuhr, believed that Wilson and the progressives had been naïve about human nature: There was darkness in the soul that made reason an insufficient guide to world affairs. "The fact of the matter," Kennan told students at the National War College, "is that there is a little bit of the totalitarian buried somewhere, way down deep, in each and every one of us." But for Kennan, that last clause was crucial: *Every one of us* was capable of evil, Americans included. That's why, although he believed America must sometimes use force to limit the evil done by its enemies, he also insisted that it limit itself, because it could do evil, too. For Kennan, containment was a partial doctrine because anticommunism was a partial truth—valuable in moderation, but prone, if stretched too far, to resemble the very fanaticism it was pitted against. NSC 68, by contrast, threw around words like *evil, irrational,* and *fanatical* liberally, but applied them only to the other side. Americans were uniformly described as moral, decent, and rational—so moral, decent, and rational, in fact, that they could take virtually unrestrained action to combat communism without ever imperiling their souls.

But by 1950, Kennan did believe that frenzied anticommunism was imperiling the American soul. There was madness in the air. Kennan's old patron, James Forrestal, whom Truman had recently replaced as secretary of defense, began claiming he had been fired because Soviet agents wanted to silence his warnings about Moscow's impending all-out attack. He also brooded endlessly over a Washington columnist's claim that he had shown cowardice when thieves came to steal his wife's jewelry. Within weeks of his dismissal, Forrestal was refusing to leave his house, talking in hushed tones for fear he was being bugged, keeping the blinds permanently drawn to guard against snipers. Friends convinced him to check into Bethesda Naval Hospital, where he was treated with sedatives, tranquilizers, and shots of insulin.

At first Forrestal's health seemed to improve. But a few weeks into his stay, he rose from his hospital bed in the middle of the night and began copying passages from Sophocles' poem "The Chorus from Ajax." The poem tells the story of the legendary Greek warrior Ajax, a man so confident in his martial skill that he spurns the goddess Athena when she offers her assistance on the battlefield of Troy. In retaliation for this and other acts of mortal hubris, she afflicts him with madness. Believing himself under attack by his enemies, Ajax pulls out his sword and slaughters his foes, only to find that he has murdered a herd of sheep. Shamed, he turns the sword on himself.

After finishing his transcription, Forrestal put the sheets of paper on the table beside his bed and walked across the corridor to an adjacent kitchen. There he opened a window and jumped. He landed facedown with a thud, thirteen floors below.

There was another man, even closer to Kennan, who by the early 1950s was checking for hidden microphones when he entered a room. And he, tragically, had reason to. In the summer of 1951, the Senate Subcommittee on Internal Security launched an investigation of John Paton Davies, charging that he had sabotaged Chiang, assisted Mao, and tried to infiltrate communists into the U.S. government. Kennan testified on Davies' behalf and even threatened to resign in protest. But given his opposition to global containment, Kennan was now a marginal, even suspect figure himself, and his protests had little effect. In 1953, Davies was removed from his post as assistant to the U.S. high commissioner in Germany and sent to stamp passports in Peru. When even that demotion didn't satiate McCarthy, the State Department launched yet another investigation into Davies' character, the ninth in five years—and then fired him outright. Rather than return to the United States, Davies remained in Peru. For the next decade, the American government's foremost China expert lived at the foot of the Andes, sustaining his family by building furniture.

In one of his last Davies-inspired memos, written in August 1950, just months before his friend was consumed by McCarthyite attack, Kennan focused on a little-noticed consequence of NSC 68: the decision to send military advisers to battle a communist-led revolt in the obscure French colony of Vietnam. "In Indochina, we are getting ourselves into the position of guaranteeing the French in an undertaking which neither they nor we, nor both of us together, can win," Kennan warned Acheson. But in a cover note he acknowledged that "like many of my thoughts, [these] will be too remote from general thinking in the Government to be of much practical use to you." He was right: his memo received no response at all.

CHAPTER SIX

THE PROBLEM WITH MEN

In an age of hysteria, Dwight Eisenhower was the political equivalent of Valium. He tacked to the soothing center; he spoke in apple-pie platitudes; he practiced his chip shot in the Oval Office. As a result, critics often accused him of passivity. Behold the Eisenhower doll, they quipped: Wind it up and it does nothing for eight years.

But an ice cube dropped into boiling water expends tremendous energy, even if the resulting liquid is merely lukewarm. In reality, Ike was not passive at all. Behind closed doors, he worked strategically, forcefully, even ruthlessly, to hold the price of containment down, in both dollars and American lives. And because he succeeded so well, the hubris of toughness, which had swelled massively between 1946 and 1950, deflated somewhat during his eight years in office. It was only once he exited the political scene that a new generation of leaders—emboldened by America's postwar success and lacking his hard-won knowledge of the limits of economic abundance and military might—flew the wings of containment into the sun.

When Eisenhower sought the presidency in 1952, some liberals feared that, because he had spent his life in the military, he might prove a warmonger. They could not have been more wrong. As a general, Ike had been the anti-MacArthur, a man known for his aw-shucks modesty, meticulous planning, and extreme caution. His mother was a pacifist, a philosophy he didn't share but never mocked. And his experience sending young men to their deaths bred in him an intimate hostility to war. Discussing dovish civilians in a letter to his brother in 1943, he wrote that "I doubt whether any of these people, with their academic or dogmatic hatred of war, detest it as much as I do. . . . [They] probably have not seen bodies rotting on the ground and smelled the stench of decaying human flesh." As the commander of Allied forces in March 1945, with Churchill pressuring him to race to Berlin before the Soviets got there, Ike refused; he would not take large casualties for a prize of little military value. Seven years later, he lashed out at those who criticized his decision, declaring that "none of these brave men of 1952 have yet offered to go out and pick the ten thousand American mothers whose sons would have made the sacrifice to capture a worthless objective."

If Ike learned wartime prudence the hard way, Americans in the early 1950s were learning it as well, in Korea. When the war began, only 14 percent of the public expected it to last more than a year. But by the time Eisenhower was elected, it had dragged on for thirty months, with no end in sight. For the last eighteen of those, little territory changed hands. It was World War I–style trench warfare: endless artillery barrages, impenetrable defensive lines, ghoulish conditions, dead boys piling up in the snow. To hold Pork Chop Hill, a place of no particular significance except that the military brass wanted it held, one army company lost 121 of 135 men. By the fall of 1952, a plurality of Americans considered the war a mistake. And with more than one hundred thousand Americans dead, wounded, or missing, the McCarthyite illusion that global containment was easy—that it required little more than cleansing Foggy Bottom of a few latter-day Benedict Arnolds—had become the least lamentable casualty of war.

Eisenhower's chief opponent for the Republican presidential nomination, Robert Taft, had proposed following MacArthur's advice and doubling down in Korea: taking the fight to Mao in an effort to quickly win the war. And once Ike got the nomination, some in his party pressured him to follow suit. But Eisenhower knew enough about war—and about MacArthur—to suspect that escalation was folly. And he had seen enough as a candidate to realize that the American people wanted peace more than they wanted victory. His running mate, a young, red-baiting California senator named Richard Nixon, and his chief foreign policy adviser, the starchy, self-righteous John Foster Dulles, excoriated the Democrats for appeasement and pledged to roll back communism in Korea and across the globe. But Ike, for the most part, promised not to win the war, but to end it. In October, as a new round of fighting sent U.S. casualties above one thousand per week, he called the war a "tragedy." American troops, he suggested, should cede much of the combat to South Koreans. And just eleven days before the election, in a nationally televised speech at Detroit's Masonic Temple, he promised that, if elected, he would go to Korea and make peace. "For all practical purposes," declared a political writer for the Associated Press, "the contest ended that night."

When Eisenhower arrived in Seoul in late November 1952, he found the South Korean president Syngman Rhee and America's top military commander, Mark Clark, itching for a major offensive aimed at unifying the peninsula once and for all. But Eisenhower politely ignored them, wrapping himself in a thick jacket, thermal boots, and a fur hat and trudging out to inspect the situation firsthand. After interrogating front-line officers, eating with troops, and scouting the terrain from the air, he concluded that his intuition had been right: Militarily, unification was a pipe dream. Instead, he passed word to the Soviets and Chinese that, if a truce wasn't reached soon, he might expand the war, and perhaps even use nuclear weapons. Having stoked his adversaries' fears of war, he then moved aggressively to make peace. He was aided in this effort by Stalin's death, since it was Stalin—more than Mao—who had been keen to pro-long the fighting. (After all, it was Chinese boys, not Russian ones, who were getting killed.) By summer, South Korean, North Korean, American and Chinese diplomats had negotiated an armistice, which left the border between North and South Korea almost exactly where it had been when the war began. Most Americans breathed a sigh of relief and vowed never to do anything like that again. For a time, Korea's bitter memory proved Ike's ally in keeping hubris in check—until its chastening influence was gradually buried under a new wave of success.

Eisenhower's behavior in Korea set the tone: no more endless expendi-tures of money and blood. The first constraint was almost as important as the second. Nothing in Ike's experience growing up in hardscrabble nineteenth-century Kansas predisposed him to believe that money was in-finite or debt a good thing. His economic views were shaped in an earlier, harsher era, when scarcity was the norm and free spending was a ticket to ruin. Since leaving the military, he had also become golfing buddies with a slew of prominent businessmen, who reinforced his antipathy to big government. Succeeding Truman, declared a congressman who shared Ike's economic views, was like "taking over a hussy who had spent all her husband's money and run up a lot of bills at the local department store."

Aided by the armistice in Korea, Eisenhower slashed military spending from 13 percent of GDP in 1954 to barely 9 percent when he left office. If America's generals had assumed that having one of their own in the White House would be good for business, they were bitterly disappointed. During Ike's presidency, two army chiefs of staff resigned and a third, Maxwell Taylor, retired to write a book attacking Eisenhower's policies on national defense.

If Eisenhower refused to pay for containment, he also refused to fight for it with American troops. There would be no more Koreas on his watch. This posed a problem, because, like the authors of NSC 68, he insisted that containment be global. He still wanted to contain communism everywhere; he just wanted to do so cheaply and bloodlessly (or, at least, cheaply and bloodlessly for the United States). And in that effort, two methods proved particularly useful. The first was covert action. The CIA—headed by John Foster Dulles's brother Allen—was busy during the Eisenhower years. It toppled leaders in Iran and Guatemala; attempted coups in Indonesia, the Dominican Republic, and the Congo; and discussed putting radioactive dust in Fidel Castro's shoes to make his hair fall out and rob him of the virile image that supposedly kept Latin leaders in power. Containing communism via assassinations and coups usually proved disastrous for the targeted nations, and for their long-term relations with the United States. But financially, if not morally, it was cheap. And few Americans got killed.

Eisenhower's second major method of containment didn't get Americans killed, either, although it did scare some of them half to death. That method was nuclear weapons. During his presidency, Ike wielded nukes to scare the Soviets and Chinese into settling the Korean War, to deter Moscow from invading West Berlin, and, in a kind of reductio ad absurdum, to stop the Chinese from taking Quemoy and Matsu, islands that weren't even vital to Taiwan's defense, let alone America's. If Moscow stirs up trouble, Ike informed congressional leaders in late 1954, the United States will "blow the hell out of them."

Eisenhower's discussion of the bomb—which he said America would employ "just exactly as you would use a bullet or anything else"—was chillingly nonchalant. And critics depicted him as a genial, vaguely senile, grandfatherly figure, mouthing banalities and cutting ribbons while subordinates prepared to blow the world to kingdom come. But as with many things Ike-related, appearances were deceiving. Eisenhower wasn't actually cavalier about nuclear weapons at all. To the contrary, he had seen enough of war to know that once unleashed, it could rarely be controlled. And as a result, he feared that if conventional fighting broke out between the superpowers or their proxies it might end in mushroom clouds. (One reason he favored covert action was its deniability, which allowed each side to pull back and save face before things got out of hand.) And so, in an audacious gambit, he loudly declared that, if conventional warfare broke out, he would initiate what he most feared: nuclear war. In so doing, he hoped not only to disabuse the Soviets of the belief that they could keep small wars small, but also to disabuse those in his own government. By insisting that any superpower conflict would descend the slippery slope to Armageddon, he hoped to prevent both sides from taking the first step.

So while Eisenhower's nuclear rhetoric was terrifying, his behavior was generally cautious. In 1954, with French troops besieged in the remote Vietnamese garrison of Dien Bien Phu, and Paris facing defeat at the hands of the communist Vietminh, Ike's air force and navy chiefs of staff, along with Vice President Nixon, urged American air strikes. But Ike, convinced that air strikes now would lead to ground troops or worse later, refused. On Quemoy and Matsu, although he threatened nuclear war, he resisted calls to initiate it, even after China shelled the islands. And in July 1958, although he sent troops to seize the Beirut airport in a show of support for Lebanon's beleaguered pro-Western regime, he overruled the joint chiefs, who wanted to occupy the entire country. By October, U.S. forces were gone.

Ike held down global containment's cost. Unfortunately, he never challenged the concept itself. He still hewed to the idea—enshrined in NSC 68—that if the United States permitted communism's advance anywhere, it would damage American credibility, thus emboldening communists everywhere. This idea, which eviscerated Kennan's effort to make distinctions between different regions of the globe and to see U.S. interests as something other than a function of communist threats, remained part of Eisenhower's conceptual arsenal. It was still lying there when he left office for his less skilled successors to pick up, like an unexploded grenade.

In this sense, for all his savvy, Ike paved the way for the hubris that followed. Because he reduced America's military resources without reducing the interests they were supposed to defend, he made it easy for younger men to attack him as passive. That was unfair: Eisenhower's strategy was not passive; it was downright daring. And for eight years at the height of the superpower standoff, it prevented war. "We kept the peace," he told a journalist. "People ask how it happened—by God, it didn't just happen." It was a remarkable achievement, but by the time Eisenhower left office, it no longer felt so remarkable. The Korean trauma was fading: Between the fall of 1952 and the fall of 1956 the percentage of Americans who considered the war a mistake fell by 15 points. Precisely because of Ike's success—because the communists made few significant gains and America avoided war—average Americans, and especially the foreign policy elite, came to believe that Korea had worked. It had been hard but necessary, a win. The conflict was being absorbed into the toughness catechism, the latest in a valiant litany of anti-Munichs that ran from Iran to Berlin to Greece. And increasingly a new generation of American leaders and thinkers was eager for more. They thirsted for new challenges—challenges that required danger, sacrifice, and strain. Eisenhower, who had seen enough danger and sacrifice to appreciate their absence, never grasped this yearning. He knew war too well to romanticize it, and he had nothing to prove. But he now belonged to a passing age.

In 1960, Ike's final year in office, a young intellectual named Daniel Bell published a book called *The End of Ideology*, in which he argued, among other things, that young intellectuals like himself were bored. They were bored because in America there were no more political crusades. In the Wilson years, intellectuals like Dewey and Beard had thrilled to progressivism's vision of a harmonious, rational world. In the 1930s, writers had flocked to communism's equally heady vision of inexorable scientific progress. But by the '40s, America's leading intellectuals—men like Niebuhr and the literary critic Lionel Trilling—had buried those grand visions and embraced tragic realism, not just in foreign policy but as a perspective on life.

It was a weary creed, befitting people who, as Bell noted, had seen horrors that made them old before their time. By the 1950s, few American intellectuals still believed—as had Dewey and the social gospel theologian Walter Rauschenbusch—that if government provided the right kind of education it could unlock the reason and love lodged in ordinary people's souls. Far from valorizing the common man, the intellectuals of the '50s breathed a sigh of relief that ordinary Americans didn't expect very much from politics. America's shopkeepers didn't harbor fascist dreams of national purity; they tried to keep taxes low. America's labor leaders didn't march for revolution; they tried to get their members a better dental plan. If the progressives had yearned to transcend humanity's petty, selfish concerns, the postwar realists embraced them. In cold war America, argued Bell, politics was about the "unheroic, day-to-day routine of living." After "living dangerously in the exciting land of either-or," declared Arthur Schlesinger, Jr., Americans had entered "the unromantic realm of more-or-less."

For older intellectuals like Niebuhr, who had seen the catastrophe that political crusades could bring, this realism represented an intellectual life raft. But for many of the younger thinkers who followed, it was more like a straitjacket. Since World War II generally marked the beginning of their political consciousness, not the end, they tended to take America's postwar success for granted. With the exception of Korea (and even that became a happier story in retrospect), they had seen no costly overseas traumas. At home they had experienced almost ceaseless boom. Yet instead of contentment, many experienced what the sociologist David Riesman called the "malaise of the privileged." There was something enervating about being told that 1950s America, with its private comfort and lack of public drama, was the best humanity could do. And the species of American that all this comfort produced—the conformist, sheeplike "Organization Man" that William Whyte portrayed in his widely read 1956 book—filled younger critics with disgust. "The young intellectual is unhappy because the 'middle way' is for the middle-aged, not for him," wrote Bell. "[There is] an underlying restlessness, a feeling of being cheated out of adventure, and a search for passion."

Among those who felt this restlessness was Schlesinger, a short, slight, balding, bespectacled, and bow-tied historian two years Bell's senior. Schlesinger was particularly worried about the impact that this lack of adventure was having on America's men. In 1958, in an essay for *Esquire* titled "The Crisis of American Masculinity," he warned that women were on the march, "an expanding, aggressive force, seizing new domains like a conquering army, while men, more and more on the defensive, are hardly able to hold their own and gratefully accept assignments from their new rulers." The signs of emasculation were everywhere: in the media's fascination with sex-change operations, in the rage for homosexuality, even in the movies. Why was it, Schlesinger asked, that Gary Cooper, Cary Grant, Clark Gable, and Spencer Tracy—each well into his fifties—kept playing romantic leads "opposite girls young enough to be their daughters?" Because none of the younger male stars looked man enough to pull it off.

The problem, he argued, stemmed from conformity and ease. Men drove in identical cars; worked in identical, air-conditioned offices; slept in identical, manicured suburbs. Nothing in their lives was unconventional; nothing involved danger. Postwar America had become homogenized, and it had gone soft. It no longer produced heroes; it produced drones. And a man without heroic dreams was not fully a man.

Satire, Schlesinger suggested, might be one antidote to this cultural malaise. Inspiring art was another. And a third tonic was politics. If a leader confronted the nation's affairs in a sharp and daring way, pushing Americans beyond their comfort zone, demanding they meet epic challenges without hesitation or fear, he could create a space in which men might become heroes. He could make public life "virile" again. Although Schlesinger didn't say so, he had a candidate in mind: the junior senator from Massachusetts, John Fitzgerald Kennedy.

It was an apt choice, since for Jack Kennedy, virility was virtually a philosophy of life. Like Theodore Roosevelt, who was in crucial ways his cultural and ideological ancestor, Kennedy's identity had been forged, in boyhood, by a struggle against physical weakness. Born asthmatic and with a deformed back, he spent much of his youth wheezing, on crutches, and wearing a brace. His mother described him as "elfin"; his grandfather offered him a dollar for every pound he gained. At two, he came so close to dying of scarlet fever that his father, Joseph Sr., promised God that he would donate half his money to charity if the boy lived. (God fulfilled his side of the bargain; Joseph Sr., it appears, did not.) At age fourteen, Jack suffered an attack of appendicitis so severe that a priest administered last rites. "At least one half of the days that he spent on this earth," said his younger brother Robert, "were days of intense physical pain."

It was not an easy family in which to be frail. Jack's father, who had clawed his way to great wealth only to find himself snubbed by Boston Brahmins who would not accept an Irish Catholic at any price, loathed weakness. And Jack's older brother, Joe Jr., a bigger, stronger boy with a brutal streak, punished it. At night, younger siblings lay awake listening to Joe smashing Jack's head against the wall; one particularly nasty confrontation landed Jack in the hospital with twenty-eight stitches. "Jack always had something to prove, physically," remarked his prep school friend Lem Billings, "to overcompensate and prove he was fit when he really wasn't."

It was hardly a surprise, therefore, that Jack Kennedy filled his adolescence with trials signifying manhood. Despite his size, he played football at Harvard. He racked up sexual conquests, first with prostitutes procured by his father and later with women he met on his own. (This particular show of physical vigor had the paradoxical effect of saddling him with yet another malady: the clap.) And he went to war. Rejected in 1941 by the army because of his bad back, he used his father's connections to find a spot in the navy, and then used them again to transfer from desk work into command of a PT boat in the South Pacific, one of the most glamorous jobs at sea. In July 1943, near the Solomon Islands, a Japanese destroyer sliced his boat in half. Two of his men were killed instantly; the other eleven were tossed into the ocean. With most of the surrounding territory controlled by the Japanese, Kennedy led his surviving men on a five-hour swim through crocodile- and shark-infested waters to a tiny deserted island. He towed one badly burned sailor by clenching a life-jacket strap in his mouth. Avoiding Japanese patrols, and despite repeatedly falling unconscious from exhaustion, he then led them on a second swim to a larger island two and a half miles away, where they survived on coconuts and water. There they were discovered by two Pacific islanders in a canoe, to whom Kennedy gave a coconut shell on which he scribbled a plea for help. The story of PT-109 would become legend. As president, Kennedy kept the coconut shell in a case on his desk.

Had Kennedy's quest to prove his manhood been a merely personal drama, it would hardly have mattered. But by World War II, his journey had taken on a distinctly ideological cast, one that mirrored the nation's as a whole. Joseph Sr., after all, was an appeaser. Like many in his generation, his views of foreign policy had been shaped by World War I, a war whose draft he had unapologetically dodged. Military service, he insisted, was a "sucker's game." Like Herbert Hoover, he considered war an inexcusable distraction from the real purpose of world affairs: making money.

In December 1937, FDR rewarded Joseph Sr. for the vast sums he had donated to his presidential campaigns by naming him ambassador to the Court of St. James (a position Kennedy wanted largely so he could lord it over his WASP tormenters back home). Arriving in London as Neville Chamberlain's appeasement policy was reaching its crescendo, Joe Sr. took an instant liking to the former Birmingham industrialist, a man who, like himself, valued commerce and loathed war. And he developed a sharp aversion to Chamberlain's rival, Winston Churchill, who struck Kennedy as not only militaristic, but perpetually drunk. When Chamberlain returned from Munich in September 1938, having purchased peace with Hitler at Czechoslovakia's expense, Joe Sr. called him the greatest statesmen of the age. A year later, with Hitler demanding part of Poland, Kennedy urged that London cede it to the Nazis as well. He even met secretly with the German ambassador, telling him that Britain lacked the stomach for war and accusing his own boss, Franklin Roosevelt, of succumbing to pressure from Jews.

By 1939, Joseph Kennedy, Sr., was a very controversial man. His declaration that "democracy in England is finished" and that fascism might prove a better alternative had evoked fury in both the British and American press, as had his prediction that if Britain did fight the Nazis, it would lose. The *New Statesmen* called him a "frightened rich man who thinks only in terms of money." When war broke out, Fleet Street pointedly noted that Kennedy had followed British women and children to the countryside, where fewer bombs fell, rather than stay in London with the nation's men.

None of this was lost on his college-age sons, Joe Jr. and Jack, who in the late 1930s split their time between London and Harvard. Joe Jr. eagerly embraced his father's isolationist line. In 1938, when Walter Lippmann criticized one of the ambassador's speeches, Joe Jr. accused the columnist of exhibiting "the natural Jewish reaction." The following year he toured civil war Spain, penning letters back home that championed the Fascist cause. Joe Sr., who thought becoming an author could boost a young man's career, hired an editor to turn his son's letters into a book. And when Joe Jr. returned to Cambridge in the fall of 1940, he helped found Harvard's Committee Against Military Intervention in Europe.

By 1940, however, with Paris under Nazi control and American opinion swinging hard against appeasement, publishers showed little appetite for a book praising Fascist Spain. Instead it was Jack who made his literary mark. In the fall of 1939, he had decided to write his Harvard thesis on British policy toward Hitler. Submitted in April 1940, "Appeasement at Munich" did not condemn Chamberlain's actions. To the contrary, it argued that he had faced little choice, since in a democracy like Britain's, public opinion naturally tended toward pacifism. Far from contradicting his father's view, Jack's paper subtly echoed it, implying that in a dangerous world, totalitarianism was the form of government most likely to survive.

When Joe Sr. suggested that his younger son now seek publication, Jack jumped at the chance to turn his thesis into a book. But he still didn't envision it as a brief against appeasement, telling his father that "I should like to get something in the conclusion about the best policy for America as learn't [*sic*] from a study of Britain's experience but of course don't want to take sides too much."

When Jack sent the manuscript to *New York Times* columnist and family friend Arthur Krock, however, that ambivalence quickly disappeared. Krock urged Kennedy to shift the blame from British democracy to Chamberlain himself. And he urged him to come down clearly in favor of U.S. intervention, to use Britain's example to warn Americans of the perils of lethargy in the face of the Nazi threat. Krock even suggested a title. Churchill had recently named a collection of his anti-appeasement speeches *While England Slept*. Kennedy should play off it, calling his book *Why England Slept*.

So it was that in July 1940, with his father and older brother still sympathetic to fascism, hostile to Jews, and inclined toward appeasement, John F. Kennedy published an interventionist manifesto, which lionized Churchill and demanded that America institute a draft. Hitting shelves only weeks after Paris fell, Jack's book became an instant bestseller (it didn't hurt that Joe Sr. bought copies in bulk) and identified its author with arguments about the inescapability of evil, the danger of aggression, and the necessity of force that would define his generation. By war's end, Joe Jr. was dead, having volunteered for a near-suicide mission over Germany that some believe he took in a desperate bid to clear his and his father's names. Joe Sr., having resigned his ambassadorship, was a political pariah, a symbol of policies that many Americans now associated with cowardice, if not treason. Jack, by contrast, was a famed interventionist and a war hero. It would be excellent training for a cold war climate in which toughness, both ideological and personal, became a national obsession. Schlesinger was right: Jack Kennedy had a gift for the politics of manliness. His mastery of it helped propel him to the White House. And his fidelity to it helped propel America into Vietnam.

What distinguished men like Kennedy and Schlesinger from their elders wasn't merely their fixation with Munich; it was their tendency to refract that experience through the lens of America's postwar success. As a result, they not only believed that communist expansion must be stopped (which Ike believed, too), but that America's capacity to stop it was nearly infinite (which Ike did not). For Kennedy and his supporters, in fact, it was precisely Eisenhower's unwillingness to confront communism with the full force of American ingenuity and might—even though he had acknowledged the danger—that made his presidency so dispiriting. The answer was to strip away Ike's self-imposed limits, to ask more of Americans, to get them fully into the game.

Had Sputnik not arrived, America's restless young intellectuals might have needed to invent it. On October 4, 1957, from out of the sky, came evidence that while Americans were sitting fat and happy on the couch, the other side was racing ahead. Sputnik was a beeping, two-foot-wide, 184-pound aluminum sphere, the first man-made satellite in space, and it was the Soviets who had launched it, not the United States. Two months earlier, in another shock, Moscow had launched the first intercontinental ballistic missile (ICBM). Suddenly America seemed to be falling behind in everything from missiles to science labs to math teachers.

Predictably, Ike greeted the furor with a yawn. When a committee led by Ford Foundation chairman H. Rowan Gaither used Sputnik to demand another round of NSC-68-style increases in defense spending, Eisenhower largely rejected the proposed buildup, blandly noting that it did not matter whether America had as many missiles as Moscow; we had enough to deter attack. (He also knew from secret U-2 spy planes that the Soviets had nowhere near as many ICBMs as panicked Americans assumed.) But this time, rather than tamping down national alarm, Eisenhower's low-key response inflamed it, further convincing his critics that he was lethargic and weak, the embodiment of what Schlesinger called "the politics of fatigue." One prominent journalist compared Eisenhower to Stanley Baldwin, the appeasement-minded British prime minister who had preceded Chamberlain. "Thank you, Mr. Sputnik," declared a commentator on the Mutual Broadcasting System. "You gave us a shock which hit many people as hard as Pearl Harbor. . . . You woke us up out of a long sleep. You made us realize a . . . nation, like a man, can grow soft and complacent." It was time for Jack Kennedy to make his move.

In his 1960 presidential bid, Kennedy made the post-Sputnik "missile gap" the symbol of an America that, like Britain in the 1930s, lay asleep as threats gathered. His opponent, Vice President Richard Nixon, was McCarthy in a better suit, a master at painting Democrats as pink. Running with Ike in 1952, he had labeled Democratic nominee Adlai Stevenson a "PhD grad of Dean Acheson's cowardly college of communist containment." But Kennedy was not Stevenson. In fact, he and his advisers saw the high-minded, self-consciously cerebral Illinois governor in much the same way Nixon did: as effete and soft. In Congress, Kennedy had joined McCarthy and Nixon in blasting Truman for losing China. Columnist Joseph Alsop dubbed Kennedy "Stevenson with balls," and Jack relished the line.

Kennedy beat Nixon at his own game. Picking up Schlesinger's theme, he linked Nixon to the soft, materialistic ethos that had supposedly enfeebled the nation. Referring to Nixon's famous "kitchen debate" with Soviet premier Nikita Khrushchev in a model house at the 1959 American National Exhibition in Moscow, Kennedy accused his opponent of having told the Soviet leader "that while we might be behind in space, we were certainly ahead in color television." The vice president "may be very experienced in kitchen debates," Kennedy jibed; "so are a great many other married men I know." But "I would rather take my television black and white and have the largest rockets in the world."

The virtues of a hard and strenuous life, and their importance to America's cold war struggle, were a central Kennedy theme. In December 1960, he penned an article for *Sports Illustrated* titled "The Soft American," in which he pointed to a recent study revealing that young Americans trailed Europeans in tests of flexibility and strength and warned that "our increasing lack of physical fitness is a menace to our security." In an era of "pre-cooked meals and prefab" houses, Kennedy wrote that same year, Americans were losing "that old fashioned Spartan devotion. . . . We stick to the orthodox, to the easy way and the organization man." Some wondered, he told an audience at Rice University, why America must beat the Soviets to the moon. "Why climb the highest mountain?" he responded, "[W]hy does Rice play Texas? . . . because that goal will serve to organize and measure the best of our energies and abilities."

Kennedy's mania for personal and ideological toughness represented, in a sense, Theodore Roosevelt's final, posthumous victory over Woodrow Wilson. In 1917, when Wilson launched his crusade to tame world affairs, Roosevelt's dissenting view—that seeing the world as a jungle would improve national character—had been a minority strain. But during World War II, cousin Franklin had embraced much of TR's vision, even as he spoke Wilsonian words. And now JFK, in his obsession with vigor, was echoing TR with uncanny precision. Roosevelt had ordered that all U.S. Marines be able to march fifty miles in under twenty hours; when Kennedy learned of this, he renewed the order and publicly extended it to the members of his cabinet. Roosevelt encouraged reporters to cover him hunting and hiking but refused to let them watch him play tennis, which he considered effete. For his part, Kennedy encouraged journalists to report that he played touch football (so long as they explained that in the Kennedy clan, touch football was quite savage), but he avoided being photographed playing golf, which was Ike's soft, grandfatherly game. On one occasion, Jack even screamed at a *New York Herald Tribune* reporter who had revealed that the reason he campaigned without an overcoat in New Hampshire and Wisconsin in winter was not that he possessed the kind of frontier grit that made him impervious to cold, but, rather, that he wore thermal underwear.

Vigor, that quintessentially Kennedy word, connoted not only toughness but youth. And when Kennedy spoke in generational terms, which was often, he emphasized that while he and his contemporaries were idealistic, they were not innocent. They were, he noted in his inaugural address, "tempered by war, disciplined by a hard and bitter peace." Unlike Wilson and the progressives, they knew that in the world, as in the heart, evil never died.

The Kennedy men, like the intellectuals and statesmen of the 1940s, saw themselves as coming of age in a world where innocence was no longer possible. (Kennedy's national security advisor, McGeorge Bundy, titled a 1963 foreign policy essay "From Innocence to Engagement.") In that way, toughness was central to the self-conceptions of both groups. But what they meant by it was different. In the '40s, moral toughness had meant recognizing that the world was inescapably tragic yet could not be escaped. It meant lowering one's expectations about what reason could accomplish without succumbing to despair. The ethos was defensive: The world must be engaged not so the children of light could transform it in their image, but to ensure that the children of darkness did not, either. For Kennedy, Bundy, and Schlesinger in the early '60s, by contrast, toughness connoted something more aggressive, something more like a crusade. It meant demanding more of oneself and one's nation, paying any price, bearing any burden, meeting any hardship, and thus achieving the impossible. As Schlesinger wrote about the early days of the Kennedy administration, "We thought for a moment that the world was plastic and the future unlimited."

For the older men, World War II was the last in a sequence of brutal blows. For the younger generation, it was the beginning of the nation's astonishing rise. In the '40s, toughness meant living with limits; by the '60s it meant transcending them. Kennan, Morgenthau, Lippmann, and Niebuhr were realists. The Kennedy men, by contrast, were what the radical sociologist C. Wright Mills called "crackpot realists" and David Halberstam called "ultra-realists." They started with the reasonable insight that evil could never be eradicated and that the currency of world affairs would always be power—and forgot that there were limits to these truths as well.

The men of Camelot were fond of comparing the 1950s to the '30s, with the implication that they too had come to power after a period of innocence, inaction, and mounting danger. Kennedy was obsessed with Churchill: He bestowed honorary American citizenship on the former British prime minister, claimed to have read all his books, and asked to hang one of his paintings in the White House. But Churchill and FDR had roused their nations for defensive action in a moment of genuine and frightening weakness. America in the '50s, on the other hand, was the opposite of weak. Even with Eisenhower's cuts, the NSC 68 buildup had left the Soviets—with their much smaller and less efficient economy—far behind. What the Kennedy men called appeasement was simply the cautious use of awesome strength. And what they called toughness in the face of aggression was something close to aggression itself.

"In the long history of the world," Kennedy declared in his inaugural address, "only a few generations have been granted the role of defending freedom in its hour of maximum danger. I do not shrink from this responsibility—I welcome it." But 1961, unlike 1938, was not freedom's hour of maximum danger. And claiming that it was smacked of generational envy. The Kennedy men, who had found the '50s boring and vaguely effeminate, wanted epic challenges. That's why they took fifty-mile hikes; it's why they wanted to go to the moon; it's why Rice played Texas. They thought such struggles would have a salutary effect on the national character, and the cold war, of course, was the greatest struggle of all. But Churchill and FDR had not waged war against Hitler because they feared their people were succumbing to a life of ease. It was ease that their Depression-wearied people desperately wanted, but came to accept that they could not have. That was what made the wartime generation's struggle both tragic and heroic. Kennedy's, by contrast, was a blend of tragedy and farce.

CHAPTER SEVEN

SAVING SARKHAN

In 1958, a novel about American foreign policy stunned the publishing world. It sat on bestseller lists for seventy-eight weeks and sold almost five million copies. Hollywood turned it into a movie starring Marlon Brando. Critics compared its impact to *Uncle Tom's Cabin* and *The Jungle*. By one estimate, it inspired twenty-one bills in Congress.

The novel was *The Ugly American*. It is set in the fictional Southeast Asian nation of Sarkhan, which American diplomats are trying to save from falling into communist hands. Unfortunately, most of the Americans in Sarkhan are "fat," "ignorant" organization men, more interested in gossiping at embassy cocktail parties than wading through remote swamps to battle an elusive and deadly foe. Sarkhan's only hope lies with the different breed of American embodied by Tex Wolchek, a "tall and muscular" army paratrooper intimately familiar with writings of Mao Zedong, and Colonel Edwin Hillandale, a charismatic, harmonica-playing expert on Southeast Asian languages and cultures who understands that the key to winning hearts and minds in Sarkhan is understanding that the natives are obsessed with astrology and palm-reading.

John F. Kennedy loved it. In 1959, he sent the novel to all his colleagues in the Senate. As president, he encouraged its co-author, William Lederer, to send him memos on counterinsurgency. He tried to appoint Major General Edward Lansdale—the man on whom Hillandale was based—as his ambassador to South Vietnam. When the State Department resisted, he made him an adviser on Cuba instead.

What appealed to Kennedy about *The Ugly American* was not just its theme—the intersection of personal toughness and anticommunism—but its setting: the third world. In 1951, as a thirty-four-year-old congressman, he had traveled with his brother Robert and sister Pat for seven weeks through India, Pakistan, Israel, Iran, Singapore, South Korea, Japan,

Thailand, and Vietnam. He came back moved by the poverty he saw, and by the depth of nationalist passion in countries long bridled by colonialism. "If one thing was bored into me as a result of my experience in the Middle as well as the Far East," he declared upon his return, "it is that Communism cannot be met effectively by merely the force of arms. The central core of our Middle Eastern policy is (or should be) not the export of arms or the show of armed might but the export of ideas, of techniques, and the rebirth of our traditional sympathy for and understanding of the desires of men to be free." Kennedy returned to this anti-imperialist theme repeatedly in his congressional career. In 1957, he gave a speech so critical of France's colonial war in Algeria that it elicited an angry visit from the French ambassador. But Kennedy stood his ground, keeping the ambassador waiting and purposely serving him an awful lunch.

The trip, and what he learned from it, spoke well of Kennedy. Like Eisenhower in 1952 trudging through the Korean snow, he had gone to see for himself. And in so doing he had partially freed himself from the intellectual shackles of global containment. At ground level, he had sensed the absurdity of trying to force Asia's left-wing, anticolonial movements onto Nazism's Procrustean bed. He had learned what Kennan learned in Moscow and Davies learned in Chongqing: that up close, a grand abstraction like "communism" explains far less than it appears to in Washington, D.C.

But Kennedy was also a child of Munich and McCarthyism. In his formative years, he had seen appeasement fail, containment succeed, and men destroyed for acknowledging that it couldn't succeed everywhere. So he tried to keep his heresies in check, to reconcile what he had seen in Asia with what his constituents read about in Sarkhan. In this regard, *The Ugly American* proved inspiring. Hillandale, after all, is no armchair anticommunist; he speaks the local language and cares about the local people. But all this knowledge and empathy is also a means to an end: crushing the communists. For Kennedy, Hillandale squared the circle between reality and ideology, between the need to understand and the need to be tough. Unfortunately, Hillandale wasn't a real person and Sarkhan wasn't a real place. Vietnam was.

In the late 1950s, with scores of Asian and African nations on the cusp of independence, JFK wasn't alone in his fascination with the developing world. It was widely assumed, among the restless toughness intellectuals, that their generation's greatest test—their World War II—would come in what Kennedy called, with a touch of romance, "the lands of the rising people." Politics may have grown boring in the West, argued Daniel Bell in *The End of Ideology*, but "the trajectory of enthusiasm has curved East, where, in the new ecstasies for economic utopia, the 'future' is all that counts." In economics and political science departments across the United States, modernization theorists began studying how to export capitalism to the poor world so the masses of Africa, Asia, and Latin America would see that Marxism was not the true path to prosperity. "The Communist bid to win Asia by demonstrating rapid industrialization is already launched," announced one such theorist, an ambitious MIT professor named Walt Rostow, in 1957. "It behooves the Free World and especially the United States to decide promptly whether it is to observe or participate in this struggle on which so much of our destiny hinges." Observing, of course, was what Ike did. Participating, shaping, mastering—that was the Kennedy way. When he became president, Kennedy made Rostow his deputy national security advisor.

As if to underscore Rostow's point, two weeks before Kennedy took office, Nikita Khrushchev gave a speech pledging Soviet support for "wars of liberation" across the third world. Kennedy opened his first National Security Council meeting by quoting the Soviet premier's words. The speech, he told a journalist, was "Khrushchev's *Mein Kampf.* He tells us what he's going to do." It was as if the Soviet leader had sounded an opening bell.

America's diplomatic corps and its regular army, Kennedy feared, were too slow, timid, and dumb for this epic contest; he needed Hillandales. He established mandatory counterinsurgency classes at the army war colleges and the State Department. High-level diplomats were summoned to a special "Interdepartmental Seminar" taught by Lansdale, Rostow, and others, where they read Mao and Che Guevara. Kennedy himself wrote the introduction to a book on counterinsurgency. And even volunteers for the newly created Peace Corps, whose work digging wells and building schools across the third world initially struck Kennedy as a bit soft, were imbued with the Hillandale spirit. Recruits were sent to jungle boot camps, where they awoke at 5 A.M. for calisthenics and a day of obstacle courses and grueling runs. At one Peace Corps camp in Puerto Rico, young do-gooders were taught to rappel down the face of a dam, and thrown, hands and feet bound, into a river to learn how not to drown.

Kennedy approved, but his real favorites—the men whom the Peace Corps civilians were trying to imitate—were the Special Forces. Fluent in obscure languages, familiar with the work of both Mao and Rostow, able to descend deep into the jungle for weeks at a time with neither food nor maps, they terrorized the regular army in mock battles. Generals would awake in the morning to find their bedsheets scrawled with the words "You are dead," and stickers notifying them that their jeeps had been blown up. Kennedy upped the Special Forces' budget fivefold and personally selected their equipment, which included sneakers with steel soles to protect against jungle booby traps. They were the only members of the U.S. military permitted to wear berets. In October 1961, Kennedy took the White House press corps on a field trip to Fort Bragg, North Carolina, where Special Forces troops staged ambushes and ate snake meat. Skydivers jumped from planes; frogmen swam from submarines. One soldier hoisted a rocket on his back, zoomed over a lake, and landed at the proud president's feet.

For Kennedy, the first round in this new cold war struggle came in Cuba, where on January 8, 1959, a thirty-two-year-old ex-pitcher named Fidel Castro—having turned to revolution after failing to make the Pittsburgh Pirates—had marched his guerrilla army into Havana, overthrowing the American-backed dictatorship of Fulgencio Batista. Initially, Castro's anti-imperialist rhetoric elicited Kennedy's sympathy. But when the Cuban leader's leftist proclivities became clear, that sympathy quickly dissipated. The NSC 68 rules—that America must stop any new scrap of territory from being painted red (especially one within spitting distance of Miami)—had been transgressed. And Kennedy, who had spent the campaign promising that in the fight against third-world communism he would get America off the couch and into the game, could not sit idly by.

The CIA, as it happened, had a plan to set things right. Buoyed by its success overthrowing leftists in Iran and Guatemala, the Agency had won Ike's permission to organize something similar in Cuba. Soon after Kennedy's inauguration, CIA deputy director Richard Bissell informed him that the Agency was training anti-Castro exiles in Guatemala for an invasion of Cuba's southern coast. Bissell and his boss, CIA director Allen Dulles, claimed that Castro's support was soft, both among the Cuban public and within the Cuban military. (The Agency's own Cuba analysts disagreed, but were ignored.) If Kennedy moved quickly, Dulles and Bissell promised, he could spark an uprising that, at worst, would enmesh Castro in a civil war. If he did nothing, by contrast, Moscow would help Castro consolidate control and foment Marxist revolution across the Americas.

The proposal pitted Kennedy's anti-imperialist instincts against his anticommunist ones. He was eager to launch the Alliance for Progress, an aid program aimed at putting the United States on the side of nationalist aspiration in the Americas, and he feared that invading Cuba would sabotage that effort. But while he had misgivings about the operation, he believed in the CIA. Smart, unconventional, tough—they were his kind of people. He particularly liked Bissell, a forceful, brainy alumnus of Groton and Yale who at a well-lubricated, late-night CIA dinner at Washington's Alibi Club had delighted the president by declaring himself a "man-eating shark." Bissell might not be Hillandale, but he reminded Kennedy of another favorite fictional character, James Bond, whose daring, elegance, anticommunism, and cost-free sexual promiscuity offered the president an idealized image of himself. (At a Georgetown dinner in 1960, Kennedy had asked Bond's creator, the British spy novelist Ian Fleming, how to handle Castro. Fleming suggested spreading rumors that Castro was impotent, which faintly echoed reality, since Schlesinger dubbed the CIA's invasion plan "Operation Castration.")

Schlesinger, to his credit, opposed the Cuba operation, as did some in the State Department; they feared the public relations fallout and didn't understand the rush. But Kennedy looked down on the bureaucrats at Foggy Bottom; they were organization men, cautious, unimaginative, and slow. "They're not queer at State," he told a friend, "but, well, they're sort of like Adlai [Stevenson] . . . [I]f I need material fast or an idea fast, CIA is the place I have to go." The difference between Eisenhower's foreign policy and Kennedy's, Rostow had explained, would be "a shift from defensive reaction to initiative." Now Kennedy had to choose between an audacious scheme run by an elite band or a policy of paper-pushing and wait-and-see. It was no contest.

The irony was that it would have required greater bureaucratic initiative, and greater political courage (a very different thing than cold war machismo), to rein containment in, as Eisenhower had done at Dien Bien Phu. But Kennedy was in a more vulnerable position than Ike: He didn't have his predecessor's experience running large, treacherous bureaucracies, and he didn't have the word *General* in front of his name. Near the end of a briefing on the invasion plan, Bissell looked at him and said, "Don't forget one thing. . . . If we take these men out of Guatemala, we will have to transfer them to the United States, and we can't have them wandering around the country telling everyone what they have been do-ing." The subtext was clear: Once Republicans learned that Kennedy had quashed a plan to overthrow Castro, a plan Eisenhower had put into mo-tion, he would face a firestorm on his right. (And the CIA, angry that its operation had been scrapped, and led by the brother of Ike's secretary of state, would discreetly fan the flames.) Dropping the invasion, Kennedy aide Kenny O'Donnell later explained, would have made the president look like an "appeaser of Castro."

That fear haunted Kennedy's advisers as well. At the State Depart-ment, Undersecretary Chester Bowles warned that the department's Latin America experts put the invasion's chance of success at only one in three, and he urged his boss, Secretary of State Dean Rusk, to try to dissuade Kennedy. But Rusk had spent his entire life ensuring that he was never caught on the wrong side of the appeasement charge. So poor as a child that he sometimes attended school without shoes, and so patriotic that he enlisted in ROTC not at the beginning of college, but at the beginning of *high school*, he had—rather miraculously—become a Rhodes scholar. And as a result, he had been in the audience in 1933 when the fashionable aristocrats of the Oxford Union declared that if war came to Europe again they would not fight for King and Country. It was an act that Rusk never forgot nor forgave. As Truman's assistant secretary of state for Far Eastern affairs, he had urged going north of the 38th parallel and helped organize the first American military aid to Vietnam. As a result, during the height of the red scare, he had come out of the Asia bureau unscathed, even as experts like John Paton Davies were destroyed. The reason was that Rusk wasn't really an Asia expert at all; he was a Munich expert.

Rusk's whole life, in other words, had taught him not to do what Bowles was urging. Privately, he did fear that the invasion was ill con-ceived. But he kept those worries to himself. And so did almost everyone else. "Nobody in the White House wanted to be soft," one aide later re-called. "Everybody wanted to show they were just as daring and bold as everybody else."

So Kennedy told the CIA to go ahead with the plan, so long as it kept America's fingerprints off it—wishful thinking, given that news of the exile training program had already shown up in the *New York Times*. On the evening of Saturday, April 14, fifteen hundred Cuban exiles loaded onto five freighters at a secret U.S. base in Puerto Cabezas, Nicaragua, bound for Cuba's southern coast. It would be, the historian Theodore Draper later wrote, "one of those rare events in history—a perfect failure." Castro, who knew an invasion was in the works, had arrested many of the people who might have supported one. And even if he hadn't, few Cubans were interested in joining an exile force composed of loyalists from the old regime. To make matters worse, the area the CIA had chosen for the exiles' landing, the Bay of Pigs, was a Castro stronghold, a place he often went to fish.

A preparatory strike aimed at disabling Cuba's small air force failed, so Castro's pilots sunk the boats carrying most of the exiles' ammunition and communications equipment before they even reached the shore. Once they did, the Cuban army was there in force to greet them. Bissell had assured Kennedy that even in the worst case scenario, the exiles could disappear into the nearby Escambray Mountains and launch a guerrilla war from there. But he had neglected to mention that between the beaches and the hills lay eighty miles of swamp.

For a day and a half, the rebels radioed frantically to their CIA trainers for air support. And in turn, the CIA and the military beseeched the president. But Kennedy—frightened of the international condemnation an all-out U.S. attack would bring—refused. He had, he was beginning to realize, been set up. Dulles and Bissell had never really believed that the Cubans could defeat Castro alone. They had simply wanted to get him on the hook. Once the fighting began, they assumed, he would have no choice but to invade.

That is certainly what the exiles believed. "Do not see any friendly air cover as you promised. Need jet support immediately," wired their commander at 1:45 P.M. on Tuesday, with his men facing imprisonment or death. And then, after his final plea was denied, this message: "And you, sir, are a son of a bitch." Back in Washington, Kennedy was sitting alone in his bedroom, in tears.

The Bay of Pigs had a paradoxical impact on Kennedy. Privately, he grew more wary of military action. He now understood the reason for the habitual optimism of the Joint Chiefs and the CIA: If things went awry, they would simply push him to escalate. He fired Dulles and Bissell and began relying more heavily on longtime aides, especially his brother Bobby. But publicly, he kept these anxieties to himself. Precisely because, like Truman on China, he had halted U.S. intervention before it cost significant American lives, he had not forced the foreign policy elite to acknowledge the limits of American power. Hawks could still say, as they did in 1949, that with airpower, military advisers, and guts, America could contain communism everywhere at modest cost.

Herein lay the tragedy of the toughness ethic: Even when presidents realized that global containment was impossible, they feared saying so publicly, and as a result, they perpetuated the political dynamic that held them captive. "Defeats" like Mao's takeover in China and the Bay of Pigs did not slow the climb up the ladder toward hubris, because politically they were processed not as battles America had lost but as battles America had chosen not to win. Since World War II, the only thing that had deflated the growing myth of American omnipotence and omniscience was agony, the agony of all those frostbitten American boys suffering and dying in Korea. But by 1961, that memory no longer stung.

So while Kennedy was privately chastened by the Cuban disaster, his political problem, as he perceived it, was not that he had been too much of a hawk, but too much of a dove. Within hours of the invasion's collapse, Bobby was demanding that the White House take steps to ensure that his brother did not look like a "paper tiger." Two days later, JFK himself warned that "our restraint is not inexhaustible," thus reassuring Americans that although he had quashed an all-out invasion, he might not be so sane the next time. Dulles and Bissell were replaced by men as hard-line as themselves. And the only other official to lose his job over the debacle was none other than Chester Bowles. At the White House, his warnings were seen not as evidence of good judgment but as confirmation of what everyone already suspected: that he was soft. Rusk, by contrast, soldiered on.

If Cuba was the first round, the second came in a country most Americans had never even heard of: the tiny, sleepy, landlocked nation of Laos. In early 1961, Vietnam was still an afterthought. (When he gave Kennedy a post-election foreign policy briefing, Eisenhower never even brought it up.) It was Laos, in Ike's words, that constituted the "cork in the bottle" in Southeast Asia. If it fell to communism, it might take the whole region down with it.

But for a country of such grave geopolitical import, the civil war in Laos had a distinctly Potemkin feel. The State Department wrote the speeches of the Laotian king. The Royal Lao Army, whose entire budget came from the Pentagon and CIA, showed little interest in fighting. Sometimes it canceled military maneuvers to pick flowers or swim. On one occasion, government and rebel troops skipped a planned battle and together participated in a local water festival. "My people," the king explained apologetically, "only know how to sing and make love." As a "military ally," quipped Kennedy's ambassador to India, John Kenneth Galbraith, "the entire Laos nation is clearly inferior to a battalion of conscientious objectors from World War I."

Despite these realities, or perhaps because of them, Rostow and the Joint Chiefs of Staff wanted to flood Laos with U.S. troops. When Kennedy asked Joint Chiefs chairman Lyman Lemnitzer for a backup plan in case the troops failed to stamp out communism, Lemnitzer exclaimed, "You start using atomic weapons!"

Kennedy was not going to be sent down a rabbit hole again. Instead he asked his European allies for advice. British prime minister Harold Macmillan and French president Charles de Gaulle suggested that he try cutting a deal with the Soviets. Moscow, it turned out, didn't particularly relish a communist victory in Laos, either, since the rebels were mostly allied with Beijing. So, in May 1961, U.S. Ambassador at Large Averell Harriman and Soviet foreign affairs minister Andrei Gromyko began negotiations that culminated in an agreement guaranteeing Laos's cold war neutrality; it could have military alliances with neither east nor west. With that, Laos quickly fell out of the news, returning to what Galbraith called "the obscurity that it richly deserves."

Once again Kennedy had acted wisely. And once again wisdom was politically perilous. Rostow was appalled by Kennedy's show of weakness. "The cease-fire in Laos came as a cold war defeat for the U.S.," declared Henry Luce's *Time* magazine. Because Kennedy could not say publicly what Kennan had—that a country like Laos did America little good as an ally and little harm as a foe—his good judgment there actually made it harder to show the same good judgment elsewhere. He was still climbing up the hubris ladder. Khrushchev, he told an aide, "must not misunderstand Laos and Cuba as an indication that the United States is in a yielding mood."

Unfortunately, that is exactly what Khrushchev seemed to believe. In June, Kennedy went to meet the Soviet premier in Vienna. Confident that he could win over fellow politicians with his intelligence and charm, Kennedy hoped to convince his Russian counterpart to avoid aggressive moves, especially in the third world. But if Kennedy wielded power like a stiletto, Khrushchev wielded it like a club. So uneducated that he could barely sign his name, Khrushchev had a primitive understanding of American politics. (He assumed that since New York governor Nelson Rockefeller hailed from a family of famous capitalists, he must be the real power in Washington.) Already inclined to suspect that a young, inexperienced man like

Kennedy was too weak to control the plutocrats who really ran American foreign policy, Khrushchev had felt his suspicions vindicated by the Bay of Pigs. By bullying the young leader, he hoped to win concessions, especially on Germany, which he cared about most.

So every time Kennedy said something gracious, Khrushchev said something brutal. Hour after hour, the Soviet leader lectured, hectored, and belittled, taking every Kennedy courtesy as an invitation for another verbal assault. Kennedy, who had expected a parlor discussion, found himself—unprepared—in a barroom brawl. And if the insults and bravado weren't enough, Khrushchev closed the summit with a threat: He would shut off the West's access to Berlin, the divided city located deep inside communist East Germany. If the United States resisted, there would be war.

In reality, Khrushchev's bluster stemmed more from weakness than strength. East Germany, the most economically advanced of the Soviet client states, supposedly a shining model of socialist development, was being slowly depopulated. In the summer of 1961 alone, more than twenty thousand East Berliners fled to the capitalist West, where per capita income was more than twice as high and consumers didn't have to wait two years to buy a refrigerator. If he did nothing, the Soviet leader feared, East Germany might collapse, thus creating the ultimate Russian nightmare: another unified, militarized Germany, this time backed by the United States. But the Munich paradigm predisposed America's leaders to see the Kremlin as strong, not weak, as zealously seeking new conquests rather than nervously trying to hold on to what it already had. And so for Kennedy, the obvious explanation for Khrushchev's belligerence was that after Cuba and Laos, the Soviet leader thought he was soft.

After their final meeting, Kennedy slumped on a couch in a darkened hotel room, a hat pulled over his eyes. "He savaged me," he told James Reston of the *New York Times*, who had scored an exclusive interview. "He thinks because of the Bay of Pigs that I'm inexperienced . . . thinks I'm stupid. Maybe most important, he thinks that I had no guts." Reston thought the president was in shock. In London the next day, en route back to the United States, an emotional Kennedy told the hawkish columnist Joseph Alsop, "I just want you to know, Joe, I don't care what happens. I won't give way. I won't give up, and I'll do whatever's necessary. I will never back down, never, never, never." Back in Washington, Rostow was composing a memo to the president. The spring and summer of 1961, he wrote, were analogous to 1942, when the Allies suffered setback after setback. But just as Churchill and Roosevelt had turned the tide, so would they. The place to start was Vietnam.

Kennedy had some experience with Vietnam. He had gone there in 1951, as part of his trek with Bobby and Pat across Asia. And he had found a country at war. France, which had lost its Southeast Asian colonies to Japan during World War II, had managed to regain them afterward, thus boosting its war-wounded pride. But French control was more apparent than real. Under Japanese occupation, Ho Chi Minh's Vietminh had developed into a powerful force, and for a brief period after Tokyo's defeat even governed the country. When Paris tried to reassert colonial rule, Ho's troops launched a guerrilla war. By the time Kennedy arrived, they controlled two-thirds of the countryside. With France buckling under the war's economic strain, the United States—which saw the conflict less as an anticolonial struggle than as an anticommunist one—was increasingly paying the bills.

French colonialism left a poor impression on the young congressman. He found French officials obstinate and crudely racist, far inferior, from his Anglophile perspective, to their British counterparts in India, who had granted independence before the real ugliness began. In keeping with his generally anticolonial bent, Kennedy in 1953 introduced Senate legislation conditioning U.S. aid for the French effort on Paris's willingness to grant its Southeast Asian colonies independence. Unless Vietnam developed a sovereign, anticommunist "native army" free from French control, he insisted, Indochina was lost.

Then, in 1954, one seemed to magically materialize. Rocked by their devastating loss to the Vietminh at Dien Bien Phu, the French agreed to grant Vietnam independence. But to deny Ho the full fruits of victory, Vietnam was to be temporarily partitioned. North of the 17th parallel, where the Vietminh enjoyed overwhelming control, would be a communist state. But in the South, where the French enjoyed greater influence, a pro-Western government would be established. In 1956, internationally supervised elections would unite the country once again.

For the young John Kennedy, Vietnam's partition was an intellectual trap. He badly wanted to believe that America could fight Vietnamese communism without becoming colonialism's heir. And on paper, the newly independent nation of South Vietnam enabled him to believe exactly that. It had an army, a flag, and a devoutly Catholic, avowedly anticommunist leader named Ngo Dinh Diem. Kennedy liked what he saw and in 1956 helped found the American Friends of Vietnam. South Vietnam, he declared, summoning all his best cold war metaphors, "represents the cornerstone of the Free World in Southeast Asia, the keystone in the arch, the finger in the dike."

But it was a mirage: There was no South Vietnam. The Vietminh, by far the most powerful force in the country, had only accepted temporary partition because they were promised elections to reunify the entire nation, elections they (and everyone else) assumed Ho would win. South Vietnam, by contrast, was led by men who had either sat out their country's anti-imperial struggle or opposed it. In Washington, their titles and crisply pressed uniforms appeared impressive. But back home they were just the same colonial flunkies serving a new foreign master. "Like water turning into ice," David Halberstam would later write, "the illusion crystallized and became a reality, not because that which existed in South Vietnam was real, but because it became real in powerful men's minds." And John F. Kennedy, who should have known better, was one of them.

In the fall of 1961, with the Kennedy administration hungry for a win after Cuba, Laos, and Vienna, Vietnam forced its way onto the White House agenda. Two years earlier, having abandoned hope that national elections would ever be held, and after watching Diem ruthlessly suppress Vietminh supporters in the South, Ho's government in Hanoi began aiding the Vietcong, a southern insurgency dedicated to reunifying the country by force. The southern countryside soon erupted in violence, and Diem responded by forcing peasants from their guerrilla-infested villages, which alienated them further. By September 1961, the Vietcong were launching ever more brazen attacks, even briefly seizing a provincial capital fifty-five miles from the southern capital, Saigon. Alarmed, the Joint Chiefs proposed dispatching a large contingent of U.S. troops. But Kennedy distrusted their advice. So he dispatched Rostow and his special military advisor, General Maxwell Taylor, to travel to South Vietnam to investigate.

The choice of fact-finders predetermined the facts they found. If Kennedy, Rusk, and National Security Advisor McGeorge Bundy thought victory in Vietnam was important, Rostow went further: He believed it was virtually preordained by history itself. Rostow, the son of left-wing Jewish immigrants who in homage to their adopted land had named their three sons Walt Whitman, Ralph Waldo, and Eugene Victor (for Socialist leader Eugene Victor Debs), was socially insecure; colleagues in Cambridge and Washington often thought he was trying just a little too hard to fit in. (This was easy for people who graduated from Groton to say about someone who graduated from New Haven's Hillhouse High.) But what Rostow lacked in social self-confidence he more than made up for in intellectual self-confidence. He had entered Yale at fifteen, had won a Rhodes scholarship four years later, and by twenty-nine had become the youngest person ever offered a full professorship at Harvard. In his masterwork, *The Stages of Economic Growth: A Non-Communist Manifesto*, he set out to disprove Marx. The great philosopher was correct to see history as moving inexorably in one direction, Rostow conceded, but he had gotten the direction wrong. It was actually moving toward American-style capitalism. For older intellectuals like Niebuhr, accepting that capitalist America was the best humanity could do had been a sobering, even tragic realization, since it fell so far short of their earlier utopian dreams. But for Rostow, America *was* utopia. We had solved the big problems— we had reached the end of ideology—and our answers were universal. Since American society was "now within sight of solutions to the range of issues which have dominated its political life since 1865," explained Rostow, domestic policy was "becoming a bore." But helping third-world countries stuck far down on history's ladder was exciting as hell. Like Walter Lippmann and John Dewey in 1917, Rostow believed that he had cracked history's code, which made him serenely confident that—with his guidance—America could win in Vietnam.

Rostow's approach to foreign policy was the mirror image of Kennan's. Kennan had started with an intimate knowledge of a particular country at a particular time and then crafted a theory around it. Rostow, by contrast, started with a theory that was applicable always and everywhere and then grafted it onto a specific time and place. As a result, he was better at seeing forests than trees. Returning once from a trip to Latin America, he astonished colleagues by declaring that the key to understanding Hispanics was to realize that they were really Asians. "Oh, for God's sake, Walt," someone replied, "why are you talking about something you know nothing about?"

Maxwell Taylor also had too much invested in South Vietnam to see it for what it really was (or was not). Even by Kennedy administration standards, he cut an imposing figure. Movie-star handsome, he spoke French, Spanish, Chinese, German, and Japanese, and was given to quoting Thucydides in the original Greek. The *New York Times* called him a cross "between Virgil and Clausewitz." During World War II, he had sneaked behind enemy lines to scout out locations for a potential American attack. Since the war, he had run the U.S. Military Academy at West Point and then, in a dazzling cultural shift, the Lincoln Center for the Performing Arts.

He had come to Kennedy's attention after leaving the Eisenhower administration and then slamming its overdependence on nuclear weapons, in a book called *The Uncertain Trumpet*. Ike had threatened nuclear war because he believed that small wars, once unleashed, could not be effectively controlled. But Taylor believed they could: *The Uncertain Trumpet* was a primer on how to fight them, and win. "We can trigger near total destruction," declared its book jacket. "But can we defend Berlin-South Korea-Vietnam-Iran-Thailand-America?" Now Taylor was Kennedy's Special Military Advisor, and his job as he saw it was to answer that question yes.

In their report, Rostow and Taylor conceded that South Vietnam was suffering a "deep and pervasive crisis of confidence." But rather than linking that crisis to the artificiality of the state itself, they attributed it to the Vietcong's recent gains and to America's neutrality deal in Laos. The implication was clear: Kennedy's soft line in Laos had destabilized the region, and only a hard line next door could stanch the bleeding. Rostow and Taylor recommended sending eight thousand U.S. combat troops. Officially, their mission would be to help repair the devastation from a recent flood in the Mekong Delta. But in reality they would be there to buck up the South Vietnamese and scare the North into stopping the war.

Kennedy hesitated. He suspected that if he sent the troops and the North's aggression did not cease, he would soon face pressure to send more. "It's like taking a drink," he told Schlesinger. "The effect wears off, and you have to take another." But he was determined to hold South Vietnam. It had become, said Bundy, "a sort of touchstone of our will." So as he had on Cuba, Kennedy split the difference, refusing to send combat troops but substantially boosting the number of U.S. advisers and the amount of aid. As a model, he cited Truman's aid to Greece, which had halted communism's advance with U.S. money but without U.S. soldiers. It was a bad analogy: Greece was a real country; South Vietnam was not. But it was one of a number of comforting analogies that the Kennedy administration had inherited from almost two decades of foreign policy success. And neither Rostow nor Taylor nor Kennedy himself knew enough about Vietnam to grasp how poorly it applied.

By 1962, Kennedy was growing more confident. On Berlin, Khrushchev had backed down, abandoning his threat to block Western access to the city, and instead building a wall halfway through it, to keep the East Germans in. And if Kennedy's Berlin victory helped erase the humiliation at Vienna, he soon won an even bigger victory, which helped erase the humiliation of the Bay of Pigs.

That aborted invasion hadn't spelled the end of U.S. efforts to topple Castro. Kennedy, who had himself attacked Nixon for "losing" Cuba to the communists, knew that if it stayed lost he could be vulnerable, too. So in late 1961 he put Edward Lansdale, of *Ugly American* fame, in charge of a secret effort to overthrow the Cuban leader. The effort was dubbed Operation Mongoose, since Lansdale, who owned two of the catlike carnivores, insisted that they excelled at killing dangerous snakes.

Once again Castro was a step ahead. He asked the Soviets for help deterring another U.S. attack, and Khrushchev, angry that America had installed nuclear missiles near his border in Turkey (and worried that Castro might shift his allegiance to Beijing), decided to return the favor by putting nukes near America's shores. At first, evidence of the missile deployments in Cuba was murky, and Kennedy downplayed the rumored buildup. But as the summer of 1962 turned to fall, his GOP opponents—led by New York Senator Kenneth Keating—seized on the rumors and denounced the White House as weak for standing idly by. Then, at 8:45 A.M. on Tuesday, October 16, McGeorge Bundy walked into the president's bedroom and handed the bathrobe-clad commander in chief photos proving that the buildup was real. "Ken Keating," Kennedy muttered, "will probably be the next President of the United States."

At first Kennedy was not sure why the Cuban missiles posed such a grave threat. ("What difference does it make?" he mused. "They've got enough to blow us up now anyway.") But when he met with his advisers, a consensus quickly formed that the missiles were intolerable. Whether or not they *actually* threatened America, Kennedy later explained, they "would have appeared to, and appearances contribute to reality." Whether this hoary old chestnut, enshrined by NSC 68, was really true—whether concessions in one part of the world really did embolden the communists elsewhere—was far from obvious. (In fact, a mountain of political science literature would later cast doubt on the claim.) But to men who had seen Hitler exploit every British and French concession, it made intuitive sense. And besides, even if weakness didn't embolden the communists; it would certainly embolden the GOP. Allowing Castro not merely to survive but to obtain missiles able to incinerate much of America, Bobby declared, would get his brother impeached.

Several advisers proposed a sneak attack while America still enjoyed the element of surprise. But Kennedy, rightly terrified that an attack might spark World War III, decided to start with an embargo instead. "The 1930s taught us a clear lesson," he told the nation on Monday, October 22: "aggressive conduct, if allowed to go unchecked and unchallenged, ultimately leads to war." The embargo worked. Ships carrying Soviet arms were met before reaching Cuba by American vessels (including a 2,200-ton destroyer named for Joe Kennedy, Jr.), and rather than risk Armageddon, the Soviet ships turned around. But on the island, construction continued. For all Kennedy knew, the Soviets already had everything they needed to install and arm the missiles, even if another ship never got through.

By Sunday, October 28, the Kennedy administration was thinking seriously again about air strikes, to be launched before dawn on Tuesday morning. White House officials were handed envelopes marked "To Be Opened in Emergency," which explained that in the event of nuclear war they would be ferried by helicopter to shelters in Maryland's Catoctin Mountains. (Each official was allowed to save one secretary.) Privately, Kennedy put the chance of war at "somewhere between one and three and even." Then, that morning, Khrushchev cut a deal, agreeing to dismantle the missiles in return for a U.S. promise not to invade Cuba and a secret pledge to withdraw the missiles in Turkey.

On the surface, the outcome confirmed the toughness paradigm. "If we have learned anything from this experience," declared the minutes of an October 29 National Security Council meeting, "it is that weakness, even only apparent weakness, invites Soviet transgression. At the same time, firmness in the last analysis will force the Soviets to back away from rash initiatives." As many of Kennedy's advisers saw it, the missile crisis was his equivalent of Truman's aid to Greece or the airlift in Berlin, his very own anti-Munich. And like Truman, he reaped the rewards. By the end of October, Kennedy's approval rating was eleven points higher than it had been in August, when Republicans were slamming him as weak. That fall, Democrats broke the historic pattern for a president's first midterm elections and gained seats in the Senate.

But Kennedy knew—even more vividly than before—that the toughness ethic that he publicly upheld was at odds with reality. In truth, it wasn't only Khrushchev who had blinked. By promising not to invade Cuba and to withdraw U.S. missiles from Turkey, Kennedy had made two key concessions of his own. He just hadn't forthrightly explained them to the American people. Soon after the crisis, Kennedy brought in Schlesinger, who served as the White House's court historian, and told him that commentators were drawing dangerous conclusions from the missile crisis. Khrushchev, the president insisted, had backed down partly because the United States had met him halfway and partly because he didn't consider the Cuban nukes crucial to Soviet survival. "It was because of these factors that our policy worked," wrote Schlesinger in his notes, "not just because we were tough. . . . He [Kennedy] worried that people would take the wrong lessons away from the crisis."

Had he known the truth, which was that Castro furiously opposed Khrushchev's concession because he believed the nukes were crucial to *his* survival, Kennedy might have taken his private heresies even further. It was one thing to expect Moscow to back down in showdowns far from home. It was another to expect indigenous communist movements to do so when they were fighting on their own soil. A generation earlier, Kennan had insisted on exactly this distinction—between Soviet interests and the interests of local communist movements—but by the early 1960s he had been in intellectual exile for a decade. And so when Kennedy began venturing, tentatively, away from the logic of global containment, he found few influential guides and few well-lit conceptual paths. When it came to the cold war, his top advisers, for all their surface brilliance, were intellectually conventional and timid. And they knew very little about the third world. Secretary of Defense Robert McNamara, in fact, was already thinking about how to apply the lessons of the Cuban Missile Crisis—which were, of course, the same old lessons of Munich—to another place where he believed communist aggression needed a firm American reply: Vietnam.

Conditions there were going from bad to worse. The new American aid, which Kennedy dispatched after the Rostow-Taylor mission, was supposed to come with strings. U.S. officials were unhappy with Diem, a short, squat man who liked to hear himself talk—he could answer a single question with an hour-long reply—and couldn't be bothered to listen. His government either neglected or abused South Vietnam's rural villagers, the overwhelming bulk of the population. He closed down newspapers and sent political opponents to "reeducation centers." And his main criterion for choosing political advisers was blood. At one point the president's three brothers comprised half the South Vietnamese cabinet.

But when American officials tried to pressure Diem to make reforms, he either ignored them or flew into a rage, and Washington—unable to credibly threaten to abandon Saigon—backed down. Rather than grow more representative, Diem's regime grew more insular, with power increasingly confined to the president; his drug-addict brother, Ngo Dinh Nhu; and his poisonous wife, Madame Nhu, popularly nicknamed the Dragon Lady. Censorship increased, and the regime began requiring official permission for all public gatherings, including weddings and funerals.

Soon this repression blew up in the government's face. On May 8, 1963, in violation of government orders, revelers in the city of Hue flew religious flags in a mass celebration of the Buddha's birth. Diem's troops responded by firing into the crowd, killing two adults and six children. In an overwhelmingly Buddhist country ruled by a small, repressive Catholic clique, that was like throwing a lit match on dry grass. Buddhist priests launched hunger strikes and mass protests. Then, on June 11, as cameras rolled, an elderly monk burned himself alive.

Madame Nhu called the immolation a "barbecue" and offered gasoline and matches to anyone who wanted to follow suit. For his part, Diem sent troops to arrest monks by the thousands and ransack their pagodas. He had declared war on the most revered figures in Vietnamese society, and soon, in Halberstam's words, "a form of madness seemed to take over in Saigon." University students walked out of their classes in protest, followed by high school students, and then elementary school students. Back in Washington, Kennedy's whiz-kid advisers strained to understand the crisis. They kept referring to the leader of the Buddhist rebellion as Mr. Bonze, not realizing that *bonze* was not his actual name; it was the Vietnamese word for "monk."

In August, South Vietnamese generals asked U.S. ambassador Henry Cabot Lodge, Jr., grandson of the man who had battled Woodrow Wilson over the League, whether Washington would oppose a coup. At first the Kennedy administration was noncommittal. But spooked by rumors that Diem was secretly negotiating a deal with Hanoi to reunify Vietnam and expel the Americans, the White House eventually encouraged the plotters. That even America's handpicked ally was prepared to liquidate South Vietnam should have been a sign that it was not the place to draw an anticommunist line in the sand. But instead, Diem's overtures helped convince U.S. officials that he needed to go.

In the early afternoon of November 1, renegade troops occupied military bases and communications centers across Saigon and demanded that Diem resign. Within hours they had virtually encircled the government palace. By morning, Diem and his brother had fled through a secret un-

derground tunnel and taken refuge in a Catholic church in the district of
Cholon. There they gave confession and took communion before making
a deal with the coup leaders for safe passage out of the country. But the
deal didn't hold. After being picked up by an armored personnel carrier,
they were shot and hacked to death by rebel troops, their bodies deposited
in an unmarked grave. Informed of the news during a White House meeting, Kennedy turned pale and fled silently from the room.

By 1963, John F. Kennedy had become a mildly seditious man. The missile crisis had shaken him, and his apparent victory over Khrushchev had
made him less vulnerable to attack from his right. In May, he asked a few
White House aides to begin drafting what they called "the peace speech,"
and not to show it to anyone in Rusk's State Department or McNamara's
Pentagon until the last minute.

On June 10, Kennedy took the podium on commencement day at
American University, and gently challenged some of the key assumptions
that had governed U.S. foreign policy since NSC 68. First, he described
the U.S.S.R. less as bloodthirsty and aggressive than as haunted and
afraid. "No nation in the history of battle ever suffered more than the
Soviet Union in the Second World War," he declared. "A third of the
nation's territory, including two thirds of its industrial base, was turned
into a wasteland." This was a far cry from equating the Soviet Union
with Nazi Germany in 1938. Kennedy was finally describing Soviet leaders as Kennan had—as leaders of a particular country, guided in large
measure by their own historical experiences—rather than as mere vessels
for a demonic ideology. Yes, Kennedy acknowledged, Soviet communism
sometimes appeared crazed and frightening, but Americans should take
care "not to fall into the same trap . . . not to see only a distorted and desperate view of the other side." Anticommunism, in other words, was only
a partial truth; toughness could be taken too far. It may have been only
a coincidence, but in the months before the speech, Kennedy had grown
close to a new member of the National Security Council staff, thirty-five-
year-old Michael Forrestal, whose father's own distorted and desperate
anticommunism had taken his life fourteen years before.

In August, Kennedy and Khrushchev signed a treaty partially banning
nuclear tests; it was the first arms control deal of the cold war. And while
some Republicans denounced the treaty as appeasement, it easily passed
the Senate. When Kennedy mentioned the agreement on a western tour
in late September, in fact, he was surprised to find that it elicited loud

applause, even in Billings, Montana, and Salt Lake City, Utah. Perhaps the country was changing; perhaps ordinary Americans were no longer stuck in 1950. Perhaps after a big win in 1964, Kennedy might voice even greater heresies.

Perhaps America didn't need to be in Vietnam. "If I tried to pull out completely now from Vietnam, we would have another Joe McCarthy red scare on our hands, but I can do it after I'm reelected," Kennedy told Senate Majority Leader Mike Mansfield. "We don't have a prayer of staying in Vietnam," he insisted to his old friend Charlie Bartlett over drinks. "Those people hate us. They are going to throw our asses out of there at almost any point. But I can't give up a piece of territory like that to the Communists and then get the American people to reelect me."

Was this just talk? It was Kennedy's gift, and perhaps also his curse, to see better than his advisers that toughness had become a conceptual prison. But he was smarter than he was brave. Forced to choose between the risk of losing in Vietnam and the risk of really trying to win, his first choice, as it had been all along, was not to choose. In September, he once again suggested that they send Lansdale to Saigon; maybe he could square the circle. A few weeks later, as the president's motorcade approached the Texas School Book Depository in Dallas, he was shot and killed. The nation grieved, and the hubris of toughness continued to swell.

CHAPTER EIGHT

THINGS ARE IN THE SADDLE

John F. Kennedy, said French president Charles de Gaulle, was America's mask. Lyndon Johnson was its face. With Kennedy, power was rarely crude. He hid it with elegance, irony, wit. Throughout his political career, he had left the unsightly work to others. Joe Sr. handled the bribes; Bobby leveled the threats. Jack was seldom in the room.

Raw politics—horse-trading, patronage, intimidation—bored and embarrassed him. His grandfathers had played that game on the ghetto streets of Irish Boston. He had been sent to Choate and Harvard to transcend it. And although he had seen more combat than advisers like Bundy, McNamara, and Rostow, he was squeamish about raw force as well. When he thought of those Cuban exiles lying dead on the beach, or of Diem's mutilated body, he became almost physically ill. The icon was Bond: force veiled by style. The bad guys were dispatched in dramatic fashion, but you never actually saw them die.

With Lyndon Johnson, by contrast, power was naked, corporeal, animal. "You really felt as if a St. Bernard had licked your face for an hour, had pawed you all over," commented the *Washington Post*'s Ben Bradlee. "He never just shook hands with you. One hand was shaking your hand; the other hand was always someplace else, exploring you, examining you." As a twenty-three-year-old congressional assistant, Johnson began his lifelong habit of demanding that subordinates accompany him not only into the bathroom, but into the stall. By the time he reached the Senate, he was urinating openly in his office washbasin during meetings. In mixed company, sometimes he pulled down his pants to scratch his rear end. Fellow members of Congress described encountering him in bathrooms and watching with discomfort as he shook his penis in their direction while commenting on its size. In the words of Robert Caro, "None of the body parts customarily referred to as private were private when the parts were Lyndon Johnson's."

Unlike Kennedy, he had never had the luxury of being above the fray. Lacking money and family connections, he had early on developed an ability to make men serve his interests through sheer force of will. With those more powerful than himself, he employed brazen, epic flattery. With those less powerful, he relied on humiliation. Kennedy outcompeted the people around him; Johnson emasculated them. He would brutalize an aide, berate him in public, reduce him to a whimpering wreck, and then turn around and give him a car or some other lavish gift. As president, he threatened cabinet members that if they resigned he would sic the FBI and IRS on them. Kennedy did not enjoy seeing men abase themselves; it made him uncomfortable. For Johnson, by contrast, it was the beginning of a good working relationship. "How loyal is that man?" he asked a White House aide about a potential hire. "Well, he seems quite loyal, Mr. President," the aide replied. To which Johnson erupted, "I don't want loyalty. I want *loyalty*. I want him to kiss my ass in Macy's window at high noon and tell me it smells like roses. I want his pecker in my pocket."

There was nothing subtle about the way Lyndon Johnson practiced the politics of toughness. He saw doves as little boys or, worse, as women. Informed that one official was growing squeamish about Vietnam, he shot back, "Hell, he has to squat to piss." Munich, of course, was at the heart of it. "Lyndon's ideas were set in concrete by World War II," wrote a former Senate colleague after he became president. "Every big action he takes will be determined primarily on the basis [of] wether [*sic*] he thinks any other action will look like a Munich appeasement." In the summer of 1941, he had been a thirty-two-year-old congressmen, deputized by FDR, and by his mentor, Speaker of the House Sam Rayburn, to make sure that legislation instituting a military draft passed. In a country still unreconciled to war, resistance had been fierce. One congressional aide said that "in forty years on the Hill he had never seen such fear of a bill." When the tally hit 203–202 in favor, Rayburn halted the voting, in blatant violation of House rules. For his young protégé, the spectacle was not disturbing; it was inspiring. This was how real men wielded power, how evil was countered, and how democracy survived.

In the late 1940s, the lessons Johnson learned in the anti-Nazi struggle transferred seamlessly to the cold war. "We have fought two world wars because of our failure to take a position in time," he declared in 1947, in endorsing Truman's aid to Turkey and Greece. "We must apply to Dictator Stalin the same doctrine that we applied to the Kaiser and to Hitler." In anticommunism, Johnson's instincts and his interests aligned. As military budgets skyrocketed, defense became a big industry, especially in Texas. And Johnson, who secured a seat on the Senate Armed Services Committee just in time to support the NSC 68 buildup, became the industry's champion, and its beneficiary.

If historical analogy and home-state pork weren't reason enough to back global containment, there was, of course, political fear. Like Kennedy, Johnson saw McCarthyism from both ends: as hunter and prey. In 1948, he shamelessly linked his Democratic Senate opponent, Coke Stevenson, to labor unions allegedly under communist control. But soon after winning, he watched his support for Truman become a frightening liability in a state going wild for MacArthur. Johnson never fully got over the McCarthy years. As president, he sometimes told younger aides that they didn't know what it was like, didn't realize the value Congress placed on Asia, didn't understand the hell that would break loose if a Democrat let the communists paint another Asian ally red. That fear lodged deep within him, where it merged with terrors of an even more visceral sort. "Every night when I fell asleep," he told a biographer about his years as president, "I would see myself tied to the ground in the middle of a long, open space. In the distance, I could hear the voices of thousands of people. They were all shouting at me and running toward me: 'Coward! Traitor! Weakling.' They kept coming closer. They began throwing stones. At exactly that moment I would generally wake up . . . terribly shaken."

Taking the helm of an anguished nation in late November 1963, Johnson decided that the best way to win its affection was to be more faithful to Kennedy's legacy than Kennedy would have been himself. He tried to retain Kennedy's foreign policy team, since they represented continuity. (Insecure around all those eggheads, he nonetheless tried to create a little camaraderie. A few weeks after the assassination, he invited several staffers to join him for a swim in the White House pool: naked, of course. But one egghead was so nervous that he dove in without taking off his glasses. And they spent much of the rest of the time trying to recover them.) The key Kennedy holdovers—Bundy, McNamara, Rusk—assured LBJ that continuity meant holding South Vietnam. It was the awful irony of the Kennedy-Johnson succession. In his final months, Kennedy had sensed that the path he was following in Vietnam was leading to disaster. (Whether he would have gotten off it is another question.) But Johnson insisted on following that path, at least initially, because it had been Jack's.

For most of 1964, Johnson tried to keep Vietnam out of the newspapers. He told the eggheads to manage the problem quietly so he could focus on winning the presidential election that fall—the election that would make him a real president, not just a surrogate. But subtly, American policy was beginning to shift. Kennedy's decision to send advisers in 1961 had been meant to deter the North from supporting the Vietcong. His 1963 decision to ditch Diem had been meant to create a government in the South capable of rallying the country. By 1964, however, both efforts had failed. Diem's overthrow produced not a legitimate government, but virtually *no* government, as a dizzying array of factions jockeyed for control. In January, a group of young officers led by General Nguyen Khanh took control. But Saigon remained consumed by demonstrations, strikes, and rumors of additional coups. Khanh took up residence in a houseboat on the Saigon River in case he needed to make a fast getaway. Meanwhile, the South Vietnamese public was becoming increasingly anti-American and increasingly pro-reunification, and the flow of weapons and fighters heading south from Hanoi grew and grew.

Watching with mounting frustration, the Johnson administration decided to make the North pay for its meddling. The real problem, of course, was the South, where the government and indeed the entire state enjoyed little popular allegiance. But the Munich and Korea analogies distorted Washington's view, inclining U.S. officials to see Vietnam as yet another case of cross-border aggression. And no one knew how to solve the South's internal problems anyway. So in a classic example of policymakers having a hammer and deciding that their problem was a nail, Johnson's advisers decided to take the fight to Hanoi.

It was this decision that led, in the summer of 1964, to something called OPLAN 34A: South Vietnamese commando raids against North Vietnam. On August 2, soon after one of those raids, the destroyer U.S.S. *Maddox* journeyed north of the border to conduct electronic espionage off the North Vietnamese coast, in a place called the Tonkin Gulf. Perhaps thinking that the *Maddox* had been part of the commando raids, North Vietnamese torpedo boats opened fire, and after a brief skirmish, were warded off. Johnson was furious. Moscow and Beijing think "we're yellow and don't mean what we say," he told an aide. So he ordered the *Maddox* back into North Vietnamese waters, where on the night of August 4, sixty miles off the coast, it and another ship, the *Turner Joy,* radioed that they were under attack. In response, Johnson ordered air strikes against North Vietnamese ships and oil storage facilities. More important, he asked

Congress for a resolution authorizing him to take "all necessary measures to repel any armed attacks against the forces of the United States."

It was a blank check for war. (Johnson compared the resolution to "grandma's nightshirt . . . it covered everything.") Yet Congress quickly complied. Presidential power had grown massively in the quarter century since Franklin Roosevelt repealed the Neutrality Acts. Truman had waged war in Korea without asking Congress first. In 1955, Ike had demanded—and received—authority for military action over Quemoy and Matsu. Kennedy had kept Congress in the dark about the Bay of Pigs. For many in the bipartisan foreign policy elite born under FDR, the lesson of the 1930s was that average Americans were too parochial to respond proactively to foreign threats. And the lesson of the McCarthyite 1950s was that Congress was filled with wild men who, if allowed too much oversight of the executive branch, would wreak havoc with their demagogic crusades. Occasionally lying to Congress and the public, in other words, was part of protecting the country. And who could argue with the results? As the "imperial presidency" grew, so did American prosperity and power. Sometimes, it seemed, the American people and their representatives in Congress didn't even want to know all the things being done to keep them safe. In 1958, according to a University of Michigan study, 71 percent of Americans trusted their leaders to do the right thing all or most of the time. As late as 1966, asked why Congress didn't more carefully scrutinize the activities of the CIA, Massachusetts Senator Leverett Saltonstall explained that there was "information and knowledge on subjects which I personally, as a Member of Congress and as a citizen, would rather not have."

Lyndon Johnson, who had been manipulating other officeholders since he stole the student council elections at Southwest Texas State Teachers' College, lustily embraced this tradition of "benign" duplicity. ("How do you know when Lyndon Johnson is telling the truth?" went a joke of the time. "When he strokes his chin, pulls his earlobe, he's telling the truth. When he begins to move his lips, you know he's lying.") Administration officials never mentioned the South Vietnamese raids that had preceded the August 2 attack, thus leaving the impression that it was entirely unprovoked. And they never mentioned their suspicions that the August 4 attack had not actually occurred at all. The Senate debated the Tonkin Gulf Resolution for less than ten hours, during which the chamber was mostly empty; then passed it, 88–2. The House took forty minutes, and its vote was unanimous.

* * *

For Johnson, it was a masterstroke. He invoked the now-familiar litany, comparing his actions in the Tonkin Gulf to Truman's in Greece, Turkey, and Berlin, and to Kennedy's in the Cuban Missile Crisis. And as it had for them, toughness paid political dividends, with the crisis boosting his popularity thirty points almost overnight. His Republican presidential opponent, Barry Goldwater, a man who seemed altogether too comfortable contemplating nuclear war, was marginalized on the far right. And having kept both the communists and the Republicans at bay, and with a big lead in the polls, Johnson seemed to have the country in the palm of his hand. "I never had it so good," he exclaimed at a rally in Detroit, where he was endorsed by both auto titan Henry Ford and the head of the United Auto Workers, Walter Reuther. It was the perfect end-of-ideology image: the inclusive, centrist leader flanked by the enlightened capitalist and the anti-revolutionary worker. "The times were good now," wrote Halberstam about that fall of 1964, "and there were better ones, golden ones, ahead."

The country was booming: Unemployment was 4.5 percent; inflation was almost nonexistent; there hadn't been a recession in nearly five years. Militarily, Johnson declared, the United States was stronger "than at any other time in our peacetime history." African-Americans, having watched Johnson muscle through the Civil Rights Act, seemed grateful. ("Those Negroes," LBJ declared, "cling to my hands like I was Jesus Christ.") A cover story in *Time* commented on the wholesome, deferential quality of America's youth, noting that "the classic conflict between parents and children is letting up." The country appeared almost devoid of worry. Its capacity to achieve seemed limited only by its capacity to dream—and in Lyndon Johnson it had a president who dreamed big. "I'm sick of all the people who talk about the things we can't do," he boomed. "Hell, we're the richest country in the world, the most powerful. We can do it all."

His election victory was massive: 61 percent, the largest in American history. In the Senate, Democrats now claimed sixty-eight seats. In his inaugural address, Johnson called America a nation of "miracles" and announced that he would do FDR one better: He would end poverty for good. Aides thought he had his eye on Mount Rushmore, or higher. "I understand you were born in a log cabin," commented West German chancellor Ludwig Erhard on a visit to the LBJ Ranch. "No, Mr. Chancellor," Johnson replied. "You have me confused with Abe Lincoln. I was born in a manger."

* * *

Johnson was on top of the world, and to stay there all he had to do was defeat one "raggedy-assed little fourth-rate country." In November, administration officials drew up bombing plans aimed—depending on whom you asked—at either reducing Hanoi's support for the Vietcong, halting it altogether, or buoying morale in the South. But Johnson hesitated. It was his war now, and although less attuned than Kennedy to the power of anti-imperial aspiration, he knew enough to know that Vietnam was a bad place to pick a fight. "What the hell is Vietnam worth to me?" he screamed at McGeorge Bundy. "I don't think it's worth fighting for." Standing on the precipice, like Agamemnon before the purple robe, he paused, even trembled, searching for a way to avoid the fateful step. But a trembling man can still commit hubris. Johnson raged against the prison of his own assumptions—assumptions both about America's enemies and about America's people—but they imprisoned him nonetheless. "If you start running from the Communists they just chase you right into the kitchen," he told Bundy, moments after declaring that Vietnam was not worth fighting for. And then, in a conversation the same day with his old Senate mentor, Richard Russell: "They'd impeach a President who'd run out, wouldn't they?"

Seeking an alternative to both escalation and defeat, or perhaps just seeking to delay the reckoning, Johnson sent Bundy to South Vietnam to investigate. But Bundy was as much a prisoner of the toughness ethic as he. A legend at Yale, where he had arrived as the first student in school history with three perfect scores on his college entrance exams, Bundy in 1940 had broken with the prevailing isolationism on campus. Just weeks after Jack Kennedy published *Why England Slept*, he had issued his own interventionist manifesto in a book of student essays titled *Zero Hour: Summons to the Free*. When he returned to academia after the war as a preternaturally young government professor at Harvard, his Munich lecture became famous; students crowded into the lecture hall to hear his voice crack as he described the betrayal of little Czechoslovakia. "It is certainly a peculiar statistical coincidence," remarked Bundy's brother William, who himself served as Johnson's assistant secretary of state for the Far East, "that of the decision-making group [on Vietnam], Johnson, Rusk, McNamara, my brother, Walt Rostow, myself, all of those of us who were of age in 1940–41 were interventionists in the isolationist-interventionist debate." In making Bundy his envoy, Johnson was sending yet another

American who knew too much about Central Europe in 1940 and too little about South Vietnam in 1965.

During Bundy's trip, the Vietcong attacked a U.S. Army barracks in the town of Pleiku, killing eight and wounding 126. Within hours, Bundy sent word back to Washington: The bombing must begin immediately; America must stand firm. The next day, Johnson spoke from the East Room of the White House in his best Kennedyese: "We love peace," he declared. "We shall do all we can in order to preserve it for ourselves and all mankind. But we love liberty the more and we shall take up any challenge, we shall answer any threat. We shall pay any price to make certain that freedom shall not perish from this earth." Letting South Vietnam go communist was now the equivalent of letting freedom perish from the earth. And in this climate of rhetorical absurdity, in which America's leaders could not distinguish marginal threats from mortal ones, the air war against North Vietnam began.

Eleven years earlier, when the leaders of the navy and air force proposed air strikes to save the French at Dien Bien Phu, U.S. Army Chief of Staff Matthew Ridgway had warned them not to be naïve: The air strikes would fail, and American ground troops would surely follow. Back then, Korea was still an open wound; Lyndon Johnson himself had counseled caution. By 1965, however, the only caution that Korea induced was tactical: If America went to war again in Asia, it must not draw China in. Memories of Korea had brightened over the intervening decade. Polls showed that almost two-thirds of Americans now considered the war worthwhile, close to twice the number in 1952. Korea was no longer considered a bloody stalemate; it was considered a win. In 1954, having fought one world war and one large regional war in less than a decade, the public was tired. But by 1965, success had made Americans malleable. They weren't crying out for war in Vietnam, but where their leaders pointed, they dutifully followed. It had worked so far.

So the bombing commenced and ground troops quickly followed. In late February, General William Westmoreland, head of the Military Assistance Command in Saigon, requested two battalions—roughly 3,500 troops—to protect the U.S. air base at Da Nang. To fight an air war, after all, you needed ground troops to guard the planes.

By March, it was clear the bombing wasn't working. (The CIA's own reports had predicted it would not work, but Johnson rarely read even the summaries.) In South Vietnam, which had just endured yet another coup, morale was as low as ever, and the North was gearing up for a major new offensive. Westmoreland now requested two entire army divisions, not merely to guard bases but to fight the enemy on the ground. Johnson agreed to send forty thousand, but limited their mission: They were only to fight in "enclaves" within fifty miles of a U.S. base. And they would stay only long enough for the South Vietnamese to build up their forces and for the bombing to take its toll. Johnson still believed he was standing on the precipice, that he could pull back whenever he wanted. But it was an illusion; he was already over the edge.

Things were moving quickly now. By summer, the South Vietnamese army was visibly crumbling; desertion rates among draftees approached 50 percent. The North was no longer just sending supplies and advisers to the Vietcong; it was sending full regiments, who decimated the South Vietnamese in battle. Westmoreland asked for 150,000 troops and more intense bombing. Bundy, McNamara, and Rusk all concurred.

But one top official, Undersecretary of State George Ball, began to furiously resist. He was a bit older than Bundy, McNamara, and Rostow, and he wasn't awed by them. For all their supposed brilliance, they struck him as parochial. Unlike them, he knew the French experience in Vietnam well, having been the French government's attorney in the United States for much of the 1950s. While other administration officials invoked America's history in World War II, Greece, Berlin, Korea, and Cuba, the familiar histories of toughness and triumph, only Ball talked about France's history in Vietnam, a history of defeat. French president Charles de Gaulle, Ball noted, believed America would suffer the same agony in Vietnam as his country had. And that view was shared by most of America's European and Asian allies, as well as by the United Nations itself, which had endorsed the war in Korea but refused to endorse the war in Vietnam. Urged to heed the warnings of America's allies, Johnson replied, according to notes, "that he did not pay the foreigners at the UN to advise him on foreign policy, but that he did pay [Dean] Rusk."

As spring turned to summer in 1965, four graying intellectuals echoed Ball's fears: Walter Lippmann, George Kennan, Reinhold Niebuhr, and Hans Morgenthau, the very men who had done so much to define the toughness creed a generation before. In February, Lippmann called the administration's escalation "supreme folly. While the war hawks would rejoice when it began, the people would weep before it ended." Johnson and

Bundy responded by repeatedly inviting Lippmann to the White House, where they lathered him with flattery and tried to deceive him into thinking they shared his views. But Lippmann—who had been schmoozing with presidents for half a century—was neither dazzled nor intimidated, and kept pounding away in print. So the White House decided to play rough. Two teams of administration aides launched the "Lippmann Project," an investigation of everything he had written since the 1920s, aimed at exposing him as a fool. Johnson began telling anyone who would listen that Lippmann, now in his late seventies, was cowardly or senile or both. The campaign took a toll: Some of Lippmann's longtime friends began to snub him; government officials stopped answering his calls. But the old man did not buckle. "The root of [Johnson's] troubles has been his pride, a stubborn refusal to recognize the country's limitations or his own," wrote Lippmann in one particularly acid column. "Such pride goeth before destruction and an haughty spirit before a fall."

"I am dreadfully worried about what our people are doing in Southeast Asia," George Kennan confided to a friend in March. He tried to speak to Bundy but got nowhere. Finally he agreed to testify before the Senate. "I think that no episode, perhaps, in modern history has been more misleading than that of the Munich conference," he declared. "It has given to many people the idea that never must one attempt to make any sort of political accommodation in any circumstances. This is, of course, a fatally unfortunate conclusion."

"The policy of restraining Asian communism by sheer military might is fantastic," added Reinhold Niebuhr in September. When Hubert Humphrey journeyed to New York to give the keynote address at a dinner in his honor, the seventy-three-year-old theologian was appalled to find the vice president "claiming my anti-Nazi stance of the thirties [as justification] for the present war."

In May, Hans Morgenthau took the stage at the first National Teach-In against the war. Yes, he conceded, South Vietnam would likely go communist if U.S. forces withdrew. But to claim, as the Johnson administration did, that it would therefore become a client of Beijing was just plain ignorant, given the historic animosity between the two nations. (In fact, communist Vietnam would go to war with communist China four years after the United States withdrew.) In June, Morgenthau accepted a CBS News invitation to debate Bundy on the war. But the younger man outmaneuvered him, selectively quoting from Morgenthau's articles to make it appear as if the aging professor were contradicting himself. Lacking

Bundy's debating skills, or his instinct for the jugular, Morgenthau left the exchange embittered, later admitting to a friend that he felt a "general discouragement, sometimes bordering on despair."

Outside government, the Daedaluses of the toughness ethic—the men who had helped craft the conceptual wings that Johnson was now flying into the sun—saw disaster ahead. And inside the government, George Ball saw it, too. On June 18, 1965, he sent Johnson a memo that began with a poem by Emerson:

> *The horseman serves the horse,*
> *The neat-herd serves the neat,*
> *The merchant serves the purse,*
> *The eater serves his meat;*
> *'Tis the day of the chattel,*
> *Web to weave, and corn to grind,*
> *Things are in the saddle,*
> *And ride mankind.*

Things—or as the Greeks would have put it, gods—were in the saddle, not men. Vietnam was not a place that the United States, for all its might, could control. America's leaders needed to lower their sights, appreciate their limits, husband their strength, maybe even save their souls. None of the men who mattered agreed, or even really understood. And Lyndon Johnson sent another hundred thousand troops to South Vietnam.

CHAPTER NINE

LIBERATION

Understanding how the hubris of toughness fell—and what replaced it—requires a half step back in time.

On Wednesday, April 19, 1961, McGeorge Bundy slipped out of his office on the third floor of the Old Executive Office Building and into a conference room across the hall, where the National Security Council staff sat waiting. Two days earlier, Cuban exiles had waded ashore at the Bay of Pigs. By Wednesday, those who remained alive were being hunted down by Castro's men in the Zapata swamp. The Kennedy administration's first major foreign policy initiative had proved a catastrophe. Yet Bundy's tone was self-assured, even cocky. The former Harvard dean compared the Cuban exiles, with their blind faith that Washington would come to their rescue, to assistant professors who naïvely insisted they would gain tenure, until the fateful day arrived. Some of the men in the audience—which included Walt Rostow, Arthur Schlesinger, and a few others—chuckled. Others picked at a large bowl of fruit in the center of the table. But near the back of the ornate conference room, one junior staff member silently seethed. The gilded surroundings, the haughty repartee, the fruit juice dripping from powerful men's lips as they joked about weaker men's fate—it reminded him of ancient Rome. "I guess [Castro's ally] Che [Guevara] learned more from Guatemala than we did," quipped Bundy, referring to the 1954 coup on which the Cuba invasion was partly modeled. For Marcus Raskin, that was the final straw. "It's interesting that Che learned from Guatemala," he interjected, "but what have we learned?" Bundy didn't respond; he just glared. Someone else scolded Raskin for indulging in recriminations. That afternoon, the twenty-six-year-old received a call from Bundy's assistant. His presence at National Security Council meetings would no longer be required.

In an administration of self-styled Hillandales, Raskin was that oddest of creatures: a man uninterested in appearing tough. The son of a Jewish immigrant plumber from Milwaukee, he had been, in his youth, a musical prodigy. He mastered the first-year piano instruction manual in a week; he was playing concertos by age twelve. But at his core, the disheveled young pianist was less a musician than a moralist. As a teenager, he walked with a classmate named Jerry Silberman through the hardscrabble neighborhood near his childhood home, burning with outrage at the squalor. Silberman later moved to Hollywood and changed his name to Gene Wilder. Raskin moved to Washington, where he began causing trouble.

Several left-leaning Democratic congressmen had banded together to form something called the Liberal Project, and in 1959 they hired Raskin as the group's secretary. Several days into the job, he sent his bosses a memo. "Americans are bored. They are apathetic about politics. They are afraid," he wrote. "They have withdrawn from the awesome complexity and almost hopeless dread, which is the general social and political scene." On the surface, it was a familiar attack on the malaise of the late Eisenhower years, not much different from what Schlesinger and Bell were writing at the time. But beneath the surface lay the seeds of an ideological revolt. As Schlesinger saw it, Americans were bored because their government was not fighting the cold war with vigor and nerve. Once it did, public life would become exciting, and men could again become men. For Raskin, by contrast, the problem was not that America's leaders were waging the cold war lethargically; it was that they were waging the cold war at all. The militarization of America's foreign policy, he argued, was sucking the life out of America's people. Their creative, hopeful instincts were being suffocated by fear. It was a critique that, propelled by Vietnam, would move from the margins of American politics to the center during the 1960s, shattering the hubris of toughness and wrecking the careers of many of the smug young men who sat alongside Marcus Raskin that April day, eating fruit.

Bundy hired Raskin knowing that he was, by Kennedy standards, on the extreme left. But Bundy liked to be intellectually challenged. Silencing dissenting views was too easy; it was more satisfying to allow radicals a hearing, and then beat them in a fair fight. Bundy, a champion debater from his days at Groton, envisioned Raskin as a kind of sparring partner, someone who would keep him sharp and then politely retire once the decisions were made.

But Raskin did not play by the rules. Not content with mere intellectual banter, he crossed into open moralizing—something the hard-nosed Camelot men could not abide. A few months after the Cuba meeting, Bundy asked an aide to draw up plans for a nuclear first strike if the Soviets invaded West Berlin. When Raskin found out, he said something indelicate: He called his colleagues war criminals. "How," he asked, "does this make us any better than those who measured the gas ovens or the engineers who built the tracks for the death trains in Nazi Germany?" It was the ultimate taboo, turning the Munich analogy on its head and seeing America's leaders, not the communists, as Hitler's heirs. For hours Raskin and the aide screamed at each other, until both broke down in tears. Things were getting out of hand. "I can't take Marc anywhere," Bundy muttered, "without worrying that he won't pee on the floor."

In the spring of 1962, Raskin's former bosses in Congress published a book, *The Liberal Papers*, which thanked him by name. American foreign policy, declared its opening essay, was warped by "fear about masculinity . . . distorted into a need to feel tough." Later chapters proposed that America cut its nuclear stockpiles, slash military aid to anticommunist dictatorships, and support China's admission into the UN—in short, that America try to end the cold war. Republicans quickly pounced. Senate Minority Leader Everett Dirksen suggested that the book be renamed "Our American Munich." With the GOP vowing to make Raskin an issue in the fall midterm elections, he was reassigned to domestic affairs. By year's end he was gone.

In his rage at the cold war, however, and at the tough-guy politics that sustained it, Raskin was not alone. He represented the tip of a large and growing iceberg floating beneath the surface of American public life. In 1964, Stanley Kubrick produced *Dr. Strangelove*, which depicted America's generals as nuclear-obsessed madmen convinced that communists were trying to contaminate their bodily fluids. The *New York Times'* fifty-nine-year-old film reviewer condemned the movie as disrespectful, since it depicted "virtually everybody" in power as "stupid or insane—or what is worse, psychopathic." But a younger critic wrote in to object, praising the film for exposing how "after some 20 years of living under the constant threat of the great bomb we are intellectually and emotionally spastic." In 1959, a naval officer turned historian named William Appleman Williams published *The Tragedy of American Diplomacy*, which argued that it was America's insatiable search for overseas markets, not Soviet aggression, that had sparked the cold war. Williams was soon hauled before the House Un-American Activities Committee. But over the course of the 1960s, his academic disciples began hacking away at the idea that the Soviets represented evil and America embodied good. And in June 1962, fifty-nine students gathered at a United Auto Workers camp in Port Huron, Michigan, where they formed Students for a Democratic Society and issued a generational manifesto that would sell sixty thousand copies. Its message was similar to Raskin's: that the cold war was not a geopolitical necessity, forced upon America by a slave superpower bent on world domination. It was an elite strategy, used to keep Americans in a state of emotional terror and political lockdown so they would stand mute while their leaders pursued policies of violence and exploitation from Harlem and Mississippi to the far corners of the earth.

In the mainstream press, these cracks in the political ice went largely unnoticed. But they were signs of a deeper change. The cold war was beginning to thaw. By the late 1950s, Joseph McCarthy had drunk himself to death, and with anticommunist hysteria no longer at fever pitch, peace groups began emerging from the forced hibernation of the red scare years to demand a halt to the testing—and ultimately the production—of nuclear bombs. In Moscow, Stalin was dead, too, replaced by less fearsome men. In Beijing, Mao was growing increasingly anti-Soviet, giving lie to the specter of a unified communist menace. And, most important of all, in the United States, a generation with no memory of Munich was coming of age. They flooded America's universities: College enrollment nearly quadrupled between 1946 and 1970. Most of the students were apolitical, but a vocal few, like those who gathered at Port Huron, began contrasting the catechisms of the cold war with the new realities of their time. The younger generation, Lyndon Johnson later remarked in disgust, wouldn't "know a Communist if they tripped over one." But that was precisely the point. For the twenty-six-year-old Raskin, and the even younger activists who trailed behind, Munich was not a template; it was a cliché. Soviet communism—those tired old bureaucrats in Moscow mouthing slogans even they seemed to no longer believe—did not strike them as particularly menacing. What was menacing was *anti*communism. America's "democratic institutions and habits have shriveled in almost direct proportion to the growth of her armaments," declared the Port Huron Statement. For a whole generation of younger activists, the cold war was not a struggle for democracy; it was the antithesis of democracy. It was not something to wage, but something to overcome.

The struggle against the cold war was, at its core, a struggle against the ethic of toughness that marked the Camelot age. It was a revolt, in the eyes of those who instigated it, against the violence and oppression that characterized American policies at home and abroad. And that revolt linked the anti–cold warriors to another group of Americans struggling against violence and oppression: the activists of the burgeoning civil rights movement. How intimately the two movements were intertwined can be glimpsed in the unusual career of one Albert Bigelow. Bigelow had been a naval captain in the Pacific during World War II. But ten years after the war, his life changed forever when two horribly disfigured victims of Hiroshima who had come to the United States for plastic surgery stayed at his house. (Again the World War II analogy was being flipped on its head: For the anti–cold warriors the good war wasn't quite so good.) Haunted by what his government had done, Bigelow decided that the next time America detonated a nuclear weapon, he would be there. So in March 1958 he and three other pacifists set sail in a thirty-foot sailboat, *The Golden Rule,* for the Marshall Islands, where the United States was set to test a hydrogen bomb. Arrested near Honolulu, Bigelow reluctantly concluded that placing himself in the path of nuclear explosions was impractical. But he still wanted to lay down his body for justice and peace. So in May 1961 he boarded a bus in Washington along with twelve other members of the Congress of Racial Equality (CORE), determined to integrate interstate transport across the South.

CORE's founder, James Farmer, had cut his teeth working for the pacifist Fellowship for Reconciliation, which was formed in 1915 to oppose World War I. Bayard Rustin, the Fellowship's other African-American staffer, went on to organize the 1963 March on Washington. And the other two major civil rights organizations—the Southern Christian Leadership Conference and the Student Nonviolent Coordinating Committee (SNCC)—were founded on pacifist principles as well. So the movement to end the cold war helped to incubate the movement for civil rights—and the pollination also went the other way. It was a summer spent working for SNCC in 1961 that helped convince the University of Michigan's Tom Hayden to cofound Students for a Democratic Society the following year.

"I am going," wrote Bigelow before setting sail for the Marshall Islands, "because however mistaken, unrighteous, and unrepentant governments may seem, I still believe all men are really good at heart." Philosophically, this was the great divide, separating the practitioners of cold war toughness from the anti–cold war activists who revolted against them. For Hans Morgenthau, Reinhold Niebuhr, and the realist intellectuals of the 1940s, the beginning of wisdom was recognizing that people were not necessarily good at heart, that evil could never be eradicated from the world because it could never be eradicated from the soul. This premise had profound implications for foreign policy: It meant that conflict was endemic, that evil means might be required to ensure that greater evil did not prevail. That had been the lesson of Munich: that sometimes the horror of war was the only way to prevent greater horror. And it was this lesson that "ultra-realists" like Bundy, Rusk, and Lyndon Johnson not only learned but overlearned, and that helped lead America into Vietnam.

Bigelow didn't see the world that way, because he didn't see human beings that way. And neither did many others on the New Left. It is the "theological gospel according to Reinhold Niebuhr," wrote Marcus Raskin, "namely, that we live in an immoral world that offers the statesmen no option but to choose evil, albeit a lesser one," that has "made sonsofbitches of us all." From the struggle to end the cold war to the struggle for civil rights, the activists of the early 1960s were generally more hopeful; they refused to believe that when you stripped humanity naked you found the snake of original sin. (Even Martin Luther King, Jr., who admired Niebuhr, found his view of human nature too bleak.) It was no accident that America's leading pacifist journal, which published white anti–cold warriors along with civil rights leaders like King and Rustin, was called *Liberation*. The assumption was that while society was filled with violence, fear, and hate, people were not born that way. If you liberated them from oppressive institutions like segregation and militarism, their inner goodness could break free.

If this optimism about human nature underpinned the struggle for racial justice and world peace, it was even clearer in the third leg of the 1960s protest triad: the counterculture. The closest thing hippies had to a philosopher was a rather square classics professor named Norman O. Brown. Brown was a sunny Freudian, which Freud himself might have considered an oxymoron. In Freud's view, the human subconscious was a dark and dangerous place. People had an instinct for sex, which if not constrained by cultural taboo would destroy civilization. And they had an instinct for death, which could easily explode into violence. So repression was tragic because it denied people their deepest desires, but it was also necessary because those desires, if unleashed, could spawn barbarism and chaos. Brown disagreed. He denied that people had an instinct for death, and he denied they should repress their instinct for sex. In his view, people's truest selves were not frightening; they were glorious. Brown urged Americans to unlock their inner Dionysus, the Greek god of wine and ecstasy, which "breaks down the boundaries; releases the prisoners; abolishes repression." Liberate people from social constraint, he insisted, and they will become virtual geysers of love.

By the late 1960s, thousands of young hippies—clad in tie-dye and tripping on LSD—were attempting exactly that. Like the overall-wearing civil rights activists of SNCC and the coffeehouse peace activists of SDS, they believed they were championing love against hate, nonviolence against violence, liberation against repression. All three movements believed that the men who ran America, and kept the nation in a state of conformity and fear, were morally bankrupt and needed to be peaceably overthrown. All three believed that once they were, the love and compassion locked inside the American heart would break free. All three believed that ultimately the ethic of toughness was an ethic of death. And all three were radicalized and empowered when that ethic blew itself to bits in Vietnam.

In the summer of 1965, George Ball had warned Lyndon Johnson that once he began escalating in Vietnam, he would not be able to stop: The number of U.S. troops might eventually reach half a million. Robert McNamara called that prediction "outrageous." But within eighteen months it had come true. The problem was simple: Every time America upped the ante, so did Hanoi. In July, when Johnson made the fateful decision to throw a hundred thousand U.S. troops into the war, there were two North Vietnamese regiments fighting in the South. By November, there were as many as nine. America escalated in the air as well. By 1967, the United States had dropped more bombs on Vietnam than it did in all of World War II. But Hanoi built tunnels, thirty thousand miles of them. And China and the Soviet Union, which had once supported a negotiated settlement, began vying to see who could aid Hanoi more. Between 1965 and 1968, by one estimate, the bombing cost America almost ten dollars for every one dollar of damage it did to North Vietnam.

Johnson's actions didn't only spark a counter-escalation in Vietnam; they also sparked a counter-escalation at home. In the early 1960s, anti–cold war activists had mostly focused on nuclear disarmament, not Southeast Asia. In five thousand words, the Port Huron Statement only mentioned Vietnam once. But with American planes and then American GIs pouring into Vietnam, the New Left's priorities began to shift. On Easter Sunday 1965, SDS held the first major rally against the war. In May, the

University of Michigan held the first antiwar "teach-in," and by summer this exercise in guerrilla education had spread to 120 campuses. In November, thirty thousand people protested in Washington; the following March, fifty thousand turned out in New York. Over the course of 1965, the membership of SDS quadrupled.

With large public majorities backing Johnson on the war, it was easy to dismiss the protesters as a lunatic fringe. But soon mainstream figures began rebelling as well. In March 1967, with roughly forty thousand Americans wounded or dead in Vietnam, Robert Kennedy called the war a "horror" and urged Johnson to halt the bombing. A month later Martin Luther King publicly denounced it, too, describing America as "the greatest purveyor of violence in the world today." And with the antiwar movement growing bigger and angrier, and key members of the political establishment defecting to its side, the Johnson administration began to feel like an institution under siege. In November, when Dean Rusk traveled to New York for a speech, five thousand people turned out to protest, and members of SDS threw bottles and bags of blood. When Bundy went to defend the war at Harvard, a campus that a few years earlier had held him in awe, demonstrators carried signs asking, "When Will Bundy Pay for His War Crimes?" His old academic colleagues increasingly shunned him; his brother William's office at MIT was firebombed. Stephen Bundy, a Harvard undergraduate, walked into the offices of the *Harvard Crimson* newspaper one day to find its editors throwing darts at a picture of his father mounted on the wall.

McNamara, whose own son was becoming a radical peace activist, began to break under the strain. In November 1965, a thirty-two-year-old Quaker man sat down at the river entrance to the Pentagon, a mere fifty yards from McNamara's office, poured kerosene over his body, and set himself and his eighteen-month-old daughter on fire. (The baby lived; her father did not.) It was not an easy image to forget. At a New Year's Eve party a few weeks later, McNamara began talking with a woman who opposed the war, and then suddenly began to sob. Over the next two years the scene repeated itself several times. The man with the slicked-back hair and the clipped, machine-gun voice, who once epitomized the ultratough Camelot style, now seemed hard-pressed, when talking about the war he had helped launch, to keep from bursting into tears.

The heady optimism of 1964, when America seemed invincible and Lyndon Johnson held it in the palm of his hand, was a distant memory now. In 1966, George Ball quietly resigned. When McNamara suggested that the war could not be won, Johnson accused him of disloyalty and froze him out. He turned on Bundy, too, convinced that he was secretly colluding against him with Bobby Kennedy. By the end of 1967, Bundy and McNamara were both gone.

Replacing Bundy as national security advisor was Rostow, whom one aide described as "Rasputin to a tsar under siege." Rostow was a jovial figure, far gentler than the caustic Bundy. But on the subject of Vietnam, his historical determinism made him the truest of true believers, "a fanatic in sheep's clothing," in one detractor's words. He carefully edited intelligence reports before Johnson saw them, weeding out bad news and trumpeting signs of progress, however trivial. To more pessimistic colleagues, his insistence that America was winning in Vietnam seemed increasingly divorced from reality. After one dispiriting report on the war in 1967, he turned to the briefer in a slightly desperate tone and demanded, "You do admit that it'll all be over in six months?" "Oh," the man replied with a smirk. "I think we can hold out longer than that."

But Rostow possessed the quality that his boss prized most: absolute loyalty. As former aides and allies turned against him on the war, Johnson, an insecure, unlovable man in the best of times, crawled into a dark space inside himself. He was, he insisted, the victim of a conspiracy. Communists had infiltrated the "highest counsels" of government; antiwar senators were taking orders from the Kremlin. No longer able to travel without being besieged by demonstrators, he became a virtual prisoner inside the White House, which was itself ringed by protesters chanting, "LBJ, how many kids have you killed today?" and "Lee Harvey Oswald, where are you now?" Increasingly, Johnson found it difficult to sleep; his doctor worried that his health might give out. Robert Kennedy privately described the president as "very unstable." At a private meeting in 1967, when reporters repeatedly badgered him about why America was in Vietnam, Johnson finally unzipped his pants, pulled out his penis, and screamed, "This is why!"

"The only difference between the Kennedy assassination and mine," Johnson moaned, "is that I am alive and it has been more torturous." The awful irony was that he didn't care very much about Vietnam. His real love, he insisted, was the Great Society, his massive effort to alleviate poverty, not "that bitch of a war." But he remained captive to the toughness ethic, convinced, as he later told a biographer, that if he let South Vietnam fall, people would say "I was a coward. An unmanly man. A man without a spine." So in November 1967 he summoned the commander of U.S. forces in Vietnam, William Westmoreland, to Washington to tell Americans that victory was near. "We are making real progress," the square-jawed general told the National Press Club. "We have reached an important point where the end begins to come into view."

Westmoreland's timing was very bad. Even as he spoke, North Vietnamese and Vietcong units were quietly slipping into cities and towns across the South. A diversionary attack on a Marine base near the border with Laos drew U.S. forces out of urban areas. Many South Vietnamese troops had left their bases as well, having returned to their home villages to celebrate the lunar New Year, which the Vietnamese called Tet. And then, on January 30, 1968, all hell broke loose. At 2:45 A.M., nineteen Vietcong troops blew a hole in the wall surrounding the American embassy in Saigon. For six hours they crouched behind flowerpots in the embassy garden and traded fire with U.S. military police, until they were finally subdued. Over the next twenty-four hours, Hanoi attacked thirty-six of South Vietnam's forty-four provincial capitals, as well as the airport, the South Vietnamese military headquarters, and the presidential palace in Saigon. Communist forces seized control of the picturesque city of Hue, ancient seat of the Vietnamese empire. By the time U.S. Marines retook it a month later, it was "a shattered, stinking hulk, its streets choked with rubble and rotting bodies."

Back home, commentators reacted with shock. "What the hell is going on?" demanded CBS newscaster Walter Cronkite, who according to polls was the most trusted man in the nation. "I thought we were winning the war!" In purely military terms, Tet was actually an American victory. The Vietcong, who excelled at guerrilla warfare, were no match for U.S. forces in a conventional fight, and they sustained massive losses in the attack. But for ordinary Americans, who had repeatedly been told that victory was near, that was cold comfort. If the enemy could make it to the gates of the U.S. Embassy itself, the war was clearly far from over. In February, *Washington Post* columnist Art Buchwald conducted a mock interview with General Custer at the Battle of Little Big Horn. "We have the Sioux on the run," Custer insisted, "the Redskins are hurting badly and it will only be a matter of time before they give in."

Westmoreland proposed upping the ante yet again: mobilizing the reserves and calling up another two hundred thousand troops. But politically, something had snapped. To replace McNamara, Johnson had chosen establishment superlawyer Clark Clifford, a man with doubts about the war and an ego so large that not even Lyndon Johnson could subordinate it to his own. (Clifford-watchers often told the story of the company president who called to ask him how to handle a legal problem. Clifford told the man to say and do nothing, and charged him ten thousand dollars. When the man called back to complain about the bill and to ask why he had to remain mum, Clifford replied, "Because I told you to," and billed him another five thousand dollars.) Since his days teaching elementary school in Texas, Johnson had always made sure to establish dominion over his aides, to make it brutally clear who was beholden to whom. But he was weak now; he needed Clifford more than Clifford needed him. The new defense secretary quashed Westmoreland's request for massive reinforcements. And he gathered a group of seasoned cold warriors—men like Dean Acheson and Council on Foreign Relations president John McCloy—to tell Johnson the ugly truth. There could be no more escalations. The question was no longer how to win; it was how to get out.

In important ways, America had become a different country. Underpinning the hubris of toughness had been several assumptions, all born from the nation's extraordinary post–World War II success. The first assumption was that when America put its shoulder to the wheel, no military force on earth could long stand in its way. That faith had shattered at Tet. The second assumption was that the American people would bear whatever burdens and pay whatever price their leaders demanded to stop communism. By the late winter of 1968, that too looked dubious. Tet had sent public opinion reeling. Peace protests were drawing hundreds of thousands into the streets, and in March, for the first time, Gallup showed that a plurality of Americans opposed the war.

Thirdly, the hubris of toughness had presumed boundless resources. Kennedy and Johnson had basically embraced NSC 68's assumption that when it came to foreign policy, the government had a blank check. But by 1968, resources no longer looked so boundless. Johnson's spending on the war and the Great Society—and his refusal to raise taxes to pay for them until it was too late—had sparked inflation and a gaping budget deficit. With the U.S. economy suddenly weak, foreign investors began to convert their dollars into gold, shaking the world's faith in the greenback. Johnson arranged a Band-Aid fix and the momentary crisis passed, but as Acheson acknowledged, "The gold crisis has dampened expansionist ideas." Just a few years earlier, men like Rusk, Bundy, and McNamara had taken it as a virtual given that America had the money, the firepower, and the will to stop communist expansion anywhere on earth. By the spring of 1968, that confidence seemed like the fantasy of a bygone age.

On March 26, Acheson, McCloy, and the other establishment graybeards went to the White House. Unlike Kennan, Lippmann, Morgenthau, and Niebuhr, most of them had nodded approvingly as the Camelot generation stretched containment far beyond its original size and shape. They had found it flattering that younger men wished to replicate in the third world what they had achieved in Europe. And they hadn't known enough about the world's hotter, darker regions to see that the analogy was horribly flawed. But by 1968, these foreign policy elders were less afraid of Ho Chi Minh than of their own grandchildren, who, radicalized by the war, seemed to be turning against them and everything they had built. Losing in Vietnam, they had come to believe, would endanger the United States less than staying there. Johnson, who had spent his entire career looking over his right shoulder, making sure that men like these never saw him as weak, was aghast. "The establishment bastards," he fumed, "have bailed out."

Five days later, Johnson did, too. Politically he had become radioactive. On March 12, an obscure antiwar senator named Eugene McCarthy had stunned the political world by coming within four thousand votes of beating him in the New Hampshire primary. Polls suggested that the outcome in Wisconsin, the next state to vote, would be even worse. Johnson, interestingly enough, had been thinking a lot about Woodrow Wilson, the last president destroyed by hubris in war. As LBJ later told a biographer, he was haunted by images of Wilson's final eighteen months, when the twenty-eighth president was "stretched out upstairs in the White House, powerless to move, with the machinery of the American government in disarray around him." Johnson would not let that happen to him: He would end the war and leave office, thus saving his political reputation and perhaps his life itself. And so at 9 P.M. on March 31, he went on television to announce that America was prepared to curb the bombing and negotiate a peace agreement with Hanoi—and that he would not seek reelection as president.

In a sense, LBJ's gambit worked. He returned to Texas, grew his hair long, ate like a pig, smoked like a chimney, terrorized the workers on his ranch, and lived for another five years. But he had only saved his body. Close observers suspected that his soul was already dead. "My daddy committed political suicide for that war in Vietnam," said Johnson's daughter Luci. "And since politics was his life, it was like committing actual suicide."

The war, and the hatreds it spawned, haunted Johnson to the grave. He was dissuaded from attending the 1968 Democratic convention for fear that his presence might fuel violence. He wasn't even invited to the Democratic convention in 1972, and his portrait did not hang alongside past presidents and current candidates in the party's gallery of honor. When he died, Johnson's *New York Times* obituary declared that "his vision of the Great Society [had] dissolved in the morass of war in Vietnam." The first southerner to occupy the White House since Wilson, he too had seen war turn rancid, and died with its taste on his lips.

Johnson wasn't the only one thinking about World War I. In myriad ways, the New Left also reached back in time, to the days before American foreign policy was swallowed by the Munich analogy, to the peace crusaders of interwar years. Randolph Bourne, who had languished in obscurity since World War II, suddenly became hot. In 1964, Harper & Row published an anthology of his essays; in 1966, E. P. Dutton put out another. The New Left historian Christopher Lasch devoted an admiring chapter to him in a book that took swipes at Niebuhr. In 1967, a radical young linguist named Noam Chomsky quoted Bourne at length in an essay on intellectuals and war.

If Vietnam revived Bourne's reputation, it resuscitated his two mentors as well: John Dewey and Charles Beard. Marcus Raskin called Dewey's bid to outlaw war a model for post-Vietnam foreign policy. And while writing the Port Huron Statement, Tom Hayden was influenced by Dewey's 1927 book, *The Public and Its Problems*, in which the philosopher argued, contra Walter Lippmann, that with the right education, ordinary people could effectively govern themselves. For Hayden and his intellectual mentor, C. Wright Mills—who wanted a "participatory democracy" and loathed the postwar belief that elites should keep the irrational masses far from the levers of power—Dewey was an inspiration. They even set out to reclaim the word *utopian,* which in the toughness years had rarely been uttered without a sneer.

While New Left sociologists and philosophers were resurrecting Dewey, New Left historians—led by William Appleman Williams—were resurrecting Beard, hailing his warnings about presidential power and the economic motives for war. "The war in Indochina . . . has led many liberals and radicals to look again at the criticism offered by such men as Charles Beard," wrote a young, militant historian named Ronald Radosh. Even Herbert Hoover's foreign policy received a second, more positive look.

Invoking Beard, Dewey, and Bourne was a way of summoning an earlier time, before American soldiers patrolled the global beat, before American children cowered under their desks during nuclear drills, before America's leaders lathered themselves with machismo to cover their fear. Not everyone in the anti-Vietnam movement was a Deweyian pacifist, of course. Most protesters didn't oppose all wars, just this one. Still, as the fighting ground on, killing sixteen thousand Americans and hundreds of thousands of Vietnamese by the beginning of 1968, even many nonpacifists began to feel, as Raskin and SDS and the civil rights marchers had from the beginning, that Vietnam was a symptom of a larger disease: a penchant for violence that had become endemic to American foreign policy and American life.

If intellectuals and activists were creating post-toughness politics—a politics aimed at liberating American from the Cold War—in the late 1950s that political ethic moved from the streets into the Democratic Party. The first presidential candidate to give it voice was Eugene Mc-Carthy, the man whose shocking success in the 1968 New Hampshire primary had chased Lyndon Johnson back to Texas. McCarthy was an unlikely trailblazer. A conventional cold warrior for most of his career, he had spent the late 1950s slamming Eisenhower as an appeaser. In 1964, he told Johnson that he would make a better running mate than Hubert Humphrey because Humphrey was apt to go soft on Vietnam. As late as 1965, with U.S. troops pouring into Indochina, McCarthy still held his tongue, hoping that LBJ might appoint him ambassador to the UN.

But by year's end, as it became clear that he wouldn't get a glamorous new job, McCarthy grew bitter and bored. An introvert who as a young man had toyed with becoming a Catholic monk, he didn't really like the Senate. He often skipped committee meetings to read theology or write poetry. And as time passed, his personal alienation morphed into a broader ideological discontent. For one thing, administration officials kept lying to him. One evening in February 1965, he went with a group of senators to the White House, where Johnson lavished praise on South Vietnamese leader Nguyen Khanh, insisting that he was popular and in control. The next day Khanh was ousted in a coup. The following year, in a Senate Foreign Relations Committee hearing, McCarthy asked Dean Rusk whether he believed MacArthur's old adage that America could not win a land war in Asia. Rusk replied with a blizzard of obfuscation, to which McCarthy responded, "I don't think that quite answers my question." Rusk looked at him smugly and declared, "I know it didn't, sir."

By 1967, McCarthy had become that most dangerous of Washington creatures: a politician with little to lose. He had just turned fifty; his job annoyed him, his marriage had turned sour, and he had the gnawing sense that life was passing him by. Young peace activists were scouring the Senate, looking for someone to run against Johnson in the primaries. McCarthy's children, two of them peace activists themselves, urged him to do it. An old monk told him to seize the day. He saw the chance to inject some excitement and purpose into his life, maybe even to become a hero.

As it turned out, he was not a very good hero. Although honest and intelligent, he was also lazy, arrogant, mean-spirited, reclusive, and opaque. Instead of prepping for an important debate, he passed the time singing Irish ballads. He made an influential *New York Times* columnist wait outside his office while he finished writing a poem about wolverines. One afternoon in Los Angeles, arriving early for an event, he simply walked to the podium and gave his prepared speech, even though the crowd had not yet arrived.

After New Hampshire, he was joined in the race by Robert Kennedy, who immediately seized the hero's mantle. In this competition with Camelot, with the fairy-tale story of the shy, tortured younger sibling who rises to avenge his brother's death and save the nation in its moment of peril, McCarthy's deficiencies became downright painful. At his events, listeners nodded politely, or nodded off. At Kennedy's, girls screamed; fans ripped at his clothes; grown men cried. Emotionally McCarthy was a dud. But ideologically he was the truly momentous figure: the vehicle through which post-toughness politics entered the political mainstream.

Robert Kennedy was too wedded to the past. Earlier in his career, he had taken the Kennedy cult of toughness to truly lunatic extremes. As a Senate investigator, he once challenged Teamster leader Jimmy Hoffa to a push-up contest. On a trip down the Amazon in 1965, he jumped into a river filled with piranhas and dared his boatmates to join him. The following year, on safari in Africa, he dismounted from a jeep, walked to within fifteen feet of a rhinoceros, and stared at the beast until it trotted away.

On foreign policy, he had been a rigid, brutal, hawk. In the 1950s, he hadn't merely worked for Joseph McCarthy; he had made the drunken demagogue godfather to his first child. A maniac for Edwin Hillandale and James Bond, he had kept a green beret on his Senate desk and organized such fierce mock counterinsurgency battles at the Kennedy vacation compound that relatives warned he was scaring the kids. In his brother's administration he was known for advocating the hardest of hard-line stances, and then calling his colleagues wimps when they pointed out the risks.

By 1968, his views had evolved, and he called for deescalating in Vietnam. But he never confronted the underlying assumptions that made the war possible. Attacking global containment itself—linking the horrors of the late 1960s to the hubris of the early 1960s—would have meant defiling his brother's memory, the one thing that he would not, under any circumstances, do. And so his critique of the war was passionate yet incoherent. No longer a cold warrior, and yet not an anti–cold warrior, his true foreign policy ideology, if it can be called that, was simply Kennedyism: the belief that through the alchemy of Camelot, America could be both hawkish and dovish, tough and humane, crusading and restrained, that it could bear any burden in its long twilight struggle against communism yet escape burdens like Vietnam.

McCarthy was more lucid and more radical. Vietnam, he declared, "originated in the containment doctrines of the 1950s." It was the bitter fruit of NSC 68, of the belief that every time communist guerrillas advanced across some godforsaken jungle outpost it was 1938 all over again. For McCarthy, enough was enough. He called not merely for deescalating in Vietnam but for recognizing Cuba, ending weapons sales to the third world, and slowing the nuclear arms race. It was the same agenda that Raskin had been fired for supporting in 1962. And McCarthy's philosophical assumptions also mirrored those of the New Left. He accused America's leaders of being too quick to see evil lurking around every corner, too unwilling to believe that people would respond to love, not force. We must place our "hope and trust in our fellow men," McCarthy told antiwar activists. "We do not have . . . to be afraid."

It was no surprise that many of those activists joined his campaign, believing—for the first time in their lives—that a mainstream presidential candidate saw the world as they did. In the late 1960s, the hubris of toughness, which had been building like a wave for two decades, finally broke. In the '70s, post-toughness politics would take over first the Democratic Party and then the government itself.

CHAPTER TEN

THE SCOLD

"The tragedy of Vietnam is the tragedy of the catastrophic overextension and misapplication of valid principles." With those words, published in 1967, Arthur Schlesinger traced not only the rise of the hubris of toughness, but the arc of his own career. In the 1940s he had been the junior member of a group of realist thinkers—led by his mentor Reinhold Niebuhr—who told Americans that the struggle against evil was forever, that reason would never govern the world. In the early 1960s he had watched contemporaries like Bundy and Rostow stretch that toughness ethic past the breaking point, turning the struggle against evil into a global crusade. And by decade's end he had become a senior member of a new group of intellectuals, many of them too young to remember Munich, who rejected toughness politics and began imagining a post–cold war world.

Eugene McCarthy lost the Democratic nomination in 1968, first to Robert Kennedy—who had taken a decisive lead by the time he was gunned down—and then to his old rival Hubert Humphrey, who was even less inclined to challenge cold war orthodoxy. But McCarthy's ideas won. After 1968, the antiwar movement began taking over the Democratic Party. That year, for the first time since the cold war began, northern Democrats in Congress voted en masse against a major weapons system: the Sentinel antiballistic missile. In 1970, peace activists in Washington State rewrote the state party platform to demand a moratorium on the production of nuclear weapons. In 1971, Senate Majority Leader Mike Mansfield introduced legislation to cut the number of American troops in Europe in half.

Underlying these heresies was the belief—once confined to a scruffy New Left fringe—that, as Schlesinger put it, "the military hang-up of our foreign policy is an exercise in futility." In Vietnam, America was using more firepower than any nation in history. (The United States would ultimately drop 14 million tons of explosives, more than five times as many as it dropped during World War II.) Yet it was losing to the Vietcong, a ragtag guerrilla force, backed by North Vietnam, a country that used one-fifth as much electricity as was generated annually by the Arlington, Virginia, branch of the Potomac Electric Power Company. "Perhaps the principal lesson of the past decade," wrote Leslie Gelb and Paul Warnke, two rising stars in the post-Vietnam foreign policy establishment, "is that military force is a singularly inept instrument of foreign policy."

American foreign policy, argued the post–cold war thinkers of the early 1970s—again echoing the New Left radicals of the early 1960s— had been too governed by fear. The Soviet Union, they insisted, was not Nazi Germany. It was a tired, defensive gerontocracy, eager, above all, to be left alone. To be sure, communist movements might be gaining steam in certain pockets of the third world. But they were no more controlled by Moscow than they could be defeated by the United States. The key was to stop worrying about these obscure conflicts, to see them as local affairs that could not harm the United States unless the United States used them as an occasion to harm itself. What truly threatened America, in other words, was not the Soviet Union, but America's excessive fear of the Soviet Union, which led it to do self-destructive things. "If we find ourselves in a hostile world," declared Warnke, "it's going to mean our foreign policy has been idiotic."

Cold warriors often accused the new intellectuals of isolationism. But that was mostly wrong. The post-toughness Democrats were well aware that America could not escape the world. In fact, they popularized the term *interdependence* to describe the myriad new ways in which America was entangled in it. But it was precisely this entanglement, they argued, that required that America change the *nature* of its international involvement from conflict to cooperation. "Perhaps the greatest challenge to American foreign policy makers in the next generation will be to find constructive ways in which to cooperate with other nations in 'managing interdependence,'" wrote Anthony Lake, a talented post–cold war thinker who, like Gelb, was born too late to remember Munich. The language was calmer and dryer than that of the New Left. It was the language of the seminar room, or even the boardroom, not the language of the soapbox and the street, but many of the same ideas were there: If you liberated people from their fear you could unlock their decency; if America's leaders got over their obsession with appearing tough, they could build a more cooperative, more humane world.

Although their criticism of military force and their talk of global community sometimes made them sound like John Dewey, the post–cold war thinkers were neither pacifists nor utopians. Dewey had opposed *all* U.S. military commitments, even after Hitler took power. The post–cold war thinkers of the 1970s, by contrast, only suggested that America limit the scope of its military commitments and stop viewing communism as a unified menace—which made sense, given that Beijing and Moscow threatened each other more than either threatened the United States. In the quarter century between Pearl Harbor and Pleiku, America's leaders had gradually forgotten that there were limits to American power, and as a result, that there must be limits to American fear. Now Vietnam had shown that America was not omnipotent, and so men like Lake, Warnke, and Gelb usefully pointed out that America did not need to be.

With their greater optimism about international cooperation, and about human nature itself, the post–cold war intellectuals were temperamentally different from realists like Kennan and Morgenthau. But when it came to global containment, the views of the two groups tended to coincide. The reason was that the post–cold war intellectuals were not rebelling against Kennan's and Morgenthau's ideas; they were rebelling against "catastrophic overextension and misapplication" of Kennan and Morgenthau's ideas. They were not rebelling against the realism of the 1940s; they were rebelling against *ultra-realism* of the 1960s. In suggesting that Lyndon Johnson and Dean Rusk's worldview was overly dark, men like Gelb, Warnke, and Lake were actually doing something similar to what Kennan and Morgenthau had done when they called John Dewey's worldview overly rosy: They were deflating an ideological tradition that had swelled too far. In insisting that the truths of realism, like the truths of progressivism, had limits, the post–cold war thinkers were offering the beginnings of an answer to the hubris of toughness. All they needed was a leader who could make Americans listen.

That was the hard part, as Democrats learned in 1972, when they nominated George McGovern for president. Unlike Eugene McCarthy, McGovern was no latecomer to peace politics; it was in his blood. He hailed from South Dakota, two of whose three congressmen had opposed World War I. As a young man in the 1940s, he had joined the United World Federalists, which denounced the UN as too weak. In the presidential election of 1948, he had supported Henry Wallace over Harry Truman, and thus cast his lot with the last prominent Democrat to oppose the nascent cold war. McGovern never repudiated those early views. To the contrary, in his autobiography he insisted that "the peace of the world would have been better served by the hopeful and compassionate views of Wallace than the get tough policy of the Truman administration." If Marcus Raskin and others in the antiwar movement yearned to revive the spirit of peace progressivism, they finally had their man.

When McGovern called Wallace's worldview "hopeful," he was referring to something specific: the hope that the United States and the Soviet Union, having cooperated to vanquish fascism, could make their continued cooperation the foundation of a peaceful world. McGovern never abandoned that hope. Throughout his career, he worked to demilitarize, and ultimately end, the cold war. At the height of McCarthyism, when many politicians were baying for war not merely with Pyongyang, but with Beijing as well, he opposed U.S. intervention in Korea. A decade later, in his first Senate speech, he proposed recognizing Castro's Cuba. By the time he ran for president in 1972, he was demanding that the Pentagon's budget be slashed by more than a third.

Even George McGovern was no John Dewey or Herbert Hoover; he wasn't calling for an end to America's core military commitments in Western Europe and Japan. But he was the most radical critic of global containment ever nominated by a major party. "The war against communism is over," he told the journalist Theodore White. "Somehow we have to settle down and live with them."

Like Warnke, who became his campaign's top foreign policy adviser, McGovern did not believe that America was fated to live in a hostile world. In his acceptance speech at the Democratic convention, he urged Americans to stop being "so absorbed with fear and danger from abroad"; on the campaign trail, he conjured "a future that is not based on outdated stereotypes of military confrontation and power politics." It was a revealing critique. Ever since the 1940s, America's most influential foreign policy thinkers had insisted that power politics was not a choice but a fact. It was the inevitable expression of the lust for power that resided in the human heart. But McGovern, like the New Left, rejected that tragic view. McGovern's Christianity was not Niebuhr's. In fact, his hero was the progressive theologian whom Niebuhr had rebelled against: Walter Rauschenbusch, father of the Social Gospel. Christianity's task, wrote Rauschenbusch, is "transforming the life on earth into the harmony of heaven." For Niebuhr this was utopian nonsense. But Rauschenbusch's words so inspired McGovern that decades after reading them in college, he quoted them in his autobiography.

Paradoxically, it was this more hopeful view of human nature, and of foreign policy, that made George McGovern such an angry man. What Raskin introduced into the Kennedy White House, McGovern introduced into the 1972 presidential campaign: the language of moral fury. For Bundy, who considered himself a student of Niebuhr, who believed that evil was forever present in world affairs, and that "grey is the color of truth," sermonizing about foreign policy was like "pee[ing] on the floor." But McGovern, like Raskin, believed that foreign policy was not a choice between lesser evils but between evil and good. So he said what he believed, which was that Vietnam was more than a mistake. It was a crime.

"This chamber," McGovern told his Senate colleagues in September 1970, in perhaps the harshest attack on the toughness ethic ever by a national politician, "reeks of blood. Every senator here is partly responsible for that human wreckage at Walter Reed and Bethesda Naval and all across our land—young men without legs, or arms, or genitals, or faces, or hopes. There are not very many of these blasted and broken boys who think this war is a glorious adventure. Do not talk to them about bugging out, or national honor, or courage. It does not take any courage at all for a congressman, or a senator, or a president to wrap himself in the flag and say we are staying in Vietnam, because it is not our blood that is being shed. But we are responsible for those young men and their lives and their hopes. And if we do not end this damnable war, those young men will some day curse us."

George McGovern knew something about shedding blood. In early 1943 he had been a twenty-two-year-old college student with a pilot's license, a pregnant wife, and an absolute dread of flying a plane. Nonetheless, he piloted thirty-five bombing raids over Germany, sometimes amid enemy fire that forced him into terrifying emergency landings. On one mission his right tire blew out upon takeoff, meaning that his plane would likely crash when it tried to land. Rather than bailing out, McGovern and his crew completed their nine-hour bombing run, all the while knowing that they might die upon their return to base. On another flight, McGovern's plane was struck by more than a hundred pieces of shrapnel during withering German antiaircraft fire. One of his men was hit in the leg. The plane lost oxygen, and the heated suits that insulated the crew from the subzero weather gave out. Still, McGovern found a makeshift airstrip and successfully landed the plane.

One of McGovern's men, a streetwise, hard-drinking engineer from Bridgeport, Connecticut, was shattered by these harrowing ordeals. After the war he entered a mental asylum, and then stepped in front of a truck, in what may have been a suicide. McGovern's navigator, a somber Milwaukeean with dreams of entering the ministry, was flying with another pilot when he was forced to bail out over German soil just before his plane blew up. McGovern spent the remainder of the war sleeping next to the man's empty bunk, gazing at his clothes and photos of his family, praying that he would return. He never did. One of the deep ironies of the 1972 campaign, a campaign in which many Americans came to see McGovern as a pacifist, if not a coward, was that he was one of the greatest war heroes ever to run for president. He had been awarded the Distinguished Flying Cross for Bravery four times. But in his speech accepting the Democratic nomination, he excised mentions of his war record. "It seemed hypocritical," said one of his aides, to boast that "'I killed more people than any of you guys.'"

George McGovern would not play the politics of toughness. He attacked the idea that "tough talk and big Pentagon budgets are somehow synonymous with national manhood" and said he would, if necessary, go on his hands and knees to Hanoi to retrieve American prisoners of war. He refused to boast about his wartime heroism, perhaps because he feared dishonoring the dead. He was a very unusual cold war politician: one with no need, or desire, to prove himself a man.

It was all quite admirable. There was just one problem: He lost forty-nine states.

Politically, rejecting the hubris of toughness was treacherous: Americans didn't want to go on bended knee to Hanoi. But sustaining it was also treacherous, since Americans didn't want to keep sending their sons to fight Hanoi, either. In 1968 and 1972, as the Democratic Party was abandoning the politics of machismo, America elected and reelected a man who said "combat is the essence of politics" and boasted about eating sheep's testicles. No one worked harder than Richard Nixon to maintain America's image of ferocity, and his own. Yet Nixon's bravado couldn't conceal the fact that fewer and fewer of his citizens were willing to pay the price to contain communism around the world. And so his foreign policy often resembled a man who hangs a "beware of dog" sign on his house while trying to rouse an exhausted and indifferent mutt. Nixon repeatedly bluffed foreigners about what the United States might do and lied to Americans about the commitments he had made in their name. Those efforts were sometimes ingenious but they involved too much secrecy, dishonesty, and immorality for a public that no longer gave its presidents the benefit of the doubt. Ultimately Nixon's efforts to practice the politics of toughness in an era of diminished power failed. And in 1977, Jimmy Carter did what McGovern could not: smuggle post-toughness foreign policy into the White House.

It was hardly surprising that Richard Nixon considered fear a more powerful force than love. His own life proved it. At an early age he discovered that people didn't much like him. In college his brooding manner and harsh tone repelled girls. After law school he was rejected by the fancy Manhattan firms. But he overcame this personal rejection through brass-knuckle determination. In his first political campaign, he defeated an entrenched congressman by insinuating that he espoused the "communist line." Once in Congress, Nixon exposed establishment golden boy Alger Hiss as a spy, thus making himself a hated figure among right-thinking liberals but helping win himself a spot as Dwight Eisenhower's running mate. Even that triumph, however, didn't end the social rejection. In eight years as president, Ike never once asked his grasping understudy to join him in the White House private quarters. "What starts the process really are laughs and slights and snubs when you are a kid," Nixon once told a friend. "But if you are reasonably intelligent and if your anger is deep enough and strong enough you learn that you can change those attitudes by excellence, [and] personal gut performance, while those who have everything are sitting on their fat butts."

If Nixon saw life as a pitiless struggle in which grit—not charm—usually won out, he reinforced that worldview with his choice of national security advisor. Growing up in the Bavarian town of Furth in the 1930s, young Heinz Kissinger watched his meek, kindly father destroyed by repeated experiences of degradation and defeat. He developed such a terror of non-Jews that even after he immigrated to the United States as a teenager and changed his name to Henry, he still instinctively crossed the street when gentile boys approached. He abandoned his own, deep religious faith after the Holocaust claimed thirteen of his relatives. And in its place he substituted a view of life that would have made Hobbes shiver. Explaining the Holocaust to an American acquaintance after the war, Kissinger wrote that "the intellectuals, the idealists, the men of high morals had no chance. . . . Having once made up one's mind to survive, it was a necessity to follow through with a singleness of purpose inconceivable to you sheltered people in the States. Such singleness of purpose broached no stopping in front of accepted sets of values; it had to disregard ordinary standards of morality. One could only survive through lies, tricks and by somehow acquiring food to fill one's belly. The weak, the old had no chance."

As an adult, Kissinger spoke little about the harrowing anti-Semitism of his youth. ("Any people who have been persecuted for two thousand years," he joked, "must be doing something wrong.") But its impact was palpable. "Life is suffering. Birth involves death," he wrote in his undergraduate thesis, a 388-page behemoth titled "The Meaning of History"— the longest in Harvard's history—which spurred the university to limit the length of all future submissions. If McGeorge Bundy and Dean Rusk had in their youth witnessed the Western democracies' failure to resist Hitler, Kissinger had witnessed something even more chilling: the failure of Germany's own citizens—his family's Christian neighbors—to resist the destruction of their fragile democracy from within. As a result, even more than native-born Americans of his generation, he tended to see democracies as weak and ordinary citizens as irresponsible and apathetic. Only strong leaders, acting independently of their people's whims, could restrain the chaos and evil lurking in international affairs, and within the hearts of women and men.

Kissinger, his colleagues were astonished to learn, feared that he might become the victim of a vicious anti-Semitic backlash if he presided over defeat in Vietnam. (It didn't help that Nixon himself often baited him with anti-Semitic remarks.) Asked once about White House chief of staff H. R. Haldeman, Kissinger replied that "Haldeman was the kind of guy who would send you to the showers." It was hard to imagine a worldview more at odds with the one emerging inside the Democratic Party, which proposed that by ending the cold war, America's leaders could foster international harmony and unlock the decency buried inside the nation's soul.

Given their dread of weakness, it was no surprise that Nixon and Kissinger considered post-Vietnam America—and especially the post-Vietnam foreign policy elite—to be dangerously soft. And they saw their challenge as ensuring that the United States did not pay dearly for the infirmity of its ruling class. In part that meant giving America's enemies a reason not to exploit its vulnerability. Nixon opened diplomatic relations with China and he pursued détente with the Soviet Union. And by playing the two communist giants off against each other, and holding out the prospect of economic and diplomatic favors if they eschewed aggression, he tried to achieve the goals of containment without its price. Instead of containing America's foes, he gave them incentives to contain themselves.

Nixon's overtures to Moscow and Beijing helped slow the arms race and reduce the threat of nuclear war (at least to the United States; in the late 1960s, Moscow and Beijing came alarmingly close to war with each other). Thawing the cold war also promoted democratic change in the Soviet bloc, since the 1975 Helsinki Accords, a centerpiece of détente, helped spur dissident movements in Eastern Europe. That was the good news. The bad news, from Nixon and Kissinger's perspective, was that Moscow would not—or could not—contain communism's spread in Vietnam and other corners of the third world. For post-toughness liberals like Eugene McCarthy and George McGovern, that wasn't a big problem. After all, they argued, America's rapprochement with China proved Kennan's old argument that not every communist regime was America's enemy or Moscow's friend. For McCarthy and McGovern, it hardly mattered what form of government third-world countries chose: They couldn't bloody the United States unless we used them as an excuse to bloody ourselves. If America reached out its hand in friendship, in fact, it could cooperate with left-wing regimes to build a more just and peaceful world.

But Nixon and Kissinger had little interest in a foreign policy of kumbaya. What McCarthy and McGovern saw as the hand of friendship looked, through Nixon and Kissinger's darker lenses, like the white flag of surrender. Individually, third-world communist regimes might not threaten the United States, they conceded. But the perception of weakness did. The more blood there was in the water, the more ferocious the sharks would grow. For Nixon and Kissinger, it was easy to imagine just how merciless America's enemies would become once they realized Uncle Sam could be pushed around. All Nixon and Kissinger had to do was envision how merciless *they* would be. "We'll get them on the ground where we want them. And we'll stick our heels in, step on them hard, and twist," Nixon told aides one day in 1971, in explaining how to handle a wounded opponent. "Get them on the floor and step on them, crush them, show them no mercy." No wonder he didn't want to be the one lying on the ground.

If McCarthy and McGovern took America's post-Vietnam weakness as a license to stop viewing foreign policy as a contest of military strength, for Nixon and Kissinger the lesson was the opposite. The weaker America grew, the more crucial the old NSC-68 idea of "credibility"—the belief that America would stand up for its allies and against its adversaries—became. Nixon called it "the madman theory." The more irrational it became for America to use massive force, the more America's enemies needed to believe that its leaders were just crazy enough to do it.

In their efforts to appear insane, Nixon and Kissinger occasionally convinced each other. Without his calming presence, Kissinger told aides, "that drunken lunatic" would "blow up the world." He later informed them that "our peerless leader has flipped out." For his part, Nixon mused that Kissinger required psychiatric care. But the Dr. Strangelove act made less of an impact on its intended audience: America's third-world foes. Soon after taking office in 1969, Nixon tried to bolster his reputation for bloodcurdling toughness by secretly bombing North Vietnamese sanctuaries in Cambodia, something Johnson had refused to do. But Hanoi was not intimidated and kept trying to topple the government of South Vietnam. The following year—furious that Vice President Spiro Agnew had accused him of "pussyfooting," and after considerable drinking and several screenings of *Patton*, a movie that begins with the legendary World War II general standing in front of a vast American flag and declaring, "We're not just going to shoot the bastards, we're going to cut out their living guts and use them to grease the treads of our tanks"—Nixon startled aides by ordering that U.S. troops invade Cambodia outright. "If, when the chips are down, the world's most powerful nation, the United States of America, acts like a pitiful, helpless giant," Nixon told the nation, "the forces of totalitarianism and anarchy will threaten free nations and free institutions throughout the world." Two years later, the president—now sporting a pipe in homage to his new favorite tough guy, General Douglas MacArthur—responded to a communist offensive by blockading North Vietnam's ports and mining its harbors. "The bastards," he vowed, "have never been bombed like they're going to be bombed this time."

In opening relations with China and initiating détente with the U.S.S.R., Nixon and Kissinger broke from cold war orthodoxy. But in their obsession with "credibility" they were still ultra-realists, guided by the axioms of NSC 68. For realists like Kennan and Morgenthau, credibility had meant convincingly pledging to defend America's core interests. Such pledges should be made sparingly, they believed, because making them credible required convincing the American people that the country in question was worth incurring huge costs to defend. Nixon and Kissinger, by contrast, suggested that American credibility was on the line wherever communists and anticommunists squared off. Like his old boss, Dwight Eisenhower, Nixon was trying to limit containment's price without curtailing its scope, which created a mismatch between means and ends. Eisenhower had found that mismatch easier to conceal, because with much of the third world still under colonial rule, global containment wasn't yet truly global, and because he could deploy the CIA without Congress looking over his shoulder. For Nixon and Kissinger, however, the discrepancy was harder to mask. No amount of bombing or bluster could obscure the fact that the American people were no longer willing to sacrifice their sons for Saigon.

In January 1973, the United States and North Vietnam signed a peace deal that on paper kept South Vietnam alive. Afraid that American credibility would be damaged by Saigon's rapid military defeat, Nixon secretly promised its leaders that the United States would "respond with full force should the settlement be violated by North Vietnam." But it was a bluff. Within months Congress passed legislation banning any further U.S. military operations in Vietnam. By April 1975, with Saigon on the verge of defeat, Gerald Ford, who had assumed the presidency after Nixon's resignation, made a last-ditch appeal to Congress for aid to save the crumbling regime: "U.S. unwillingness to provide adequate assistance to our allies fighting for their lives would seriously affect our credibility throughout the world," Ford told a joint session of Congress in a speech written by Kissinger. Not a single member of Congress clapped, and two Democrats walked out. Congress rejected the aid and South Vietnam fell.

Vietnam was only the most graphic evidence that global containment was becoming the foreign policy equivalent of a bounced check. Nixon's resignation prompted a Democratic landslide in the midterm elections of 1974, which pushed Congress firmly into the post-toughness camp. "Scratch a new House member," went the expression, "and you're likely to find an old peace activist." Despite Nixon and Ford's efforts, the percentage of GDP spent on defense dropped from more than 8 percent in 1970 to less than 5 percent by 1977. Nixon and Kissinger helped get Chile's leftist president overthrown in 1973, just as Eisenhower had done in Guatemala in 1954. But Congress no longer tolerated that sort of thing. And when news of America's role in the coup became public, the Senate held investigations so grueling that they led to the conviction of Nixon's former CIA director for perjury. When civil war broke out in Angola, Ford funneled arms to anticommunist forces and Kissinger warned that "if Moscow gets away with this one, it will try again soon in some other area." But Congress soon banned the aid and Ford meekly abandoned the policy, leading Kissinger to denounce his boss to reporters.

What only a few New Left radicals had argued in the early 1960s— that American democracy was threatened less by communists overseas than by what America's government did to fight them—was becoming conventional wisdom. In 1974, Nixon resigned as a result of Watergate, a vast exercise in White House–authorized criminality set in motion by his fury over newspaper leaks about Vietnam. And Watergate was only the most famous of the crimes that the U.S. government was found to have committed in the cold war's name. A few months after Nixon left office, the *New York Times* reported that the CIA had conducted a secret audit of its illegal behavior over the preceding decades, which insiders dubbed "the family jewels." Pressed by Congress, Director of Central Intelligence William Colby tacitly admitted that the CIA had conspired to kill a bevy of foreign leaders. It had secretly opened and photographed hundreds of thousands of pieces of domestic mail. And it had administered LSD and other narcotics to unsuspecting Americans in an effort to develop tactics for mind control. In New York and San Francisco, the CIA had actually run brothels in which government-employed prostitutes lured men to secretly monitored rooms and then plied them with drugs to see how they would react. The Agency called it Operation Midnight Climax.

American foreign policy had passed through the looking glass. John Paton Davies was now back from Peru, with his government clearance restored. The policy that he had been persecuted for espousing—diplomatic relations with communist China—was becoming a reality. A *New York Times* columnist proposed that he be named America's first ambassador to the People's Republic. In 1971, the Senate Foreign Relations Committee— once a symbol of congressional timidity, now a symbol of congressional power—invited him to a kind of coming-out party for global containment's early foes. As cameras flashed, Missouri's Stuart Symington called Davies a "great American." Arkansas' William Fulbright added, "It is a very strange turn of fate that you gentlemen who reported honestly about conditions [in China] were so persecuted because you were honest about it. This is a strange thing to occur in what is called a civilized country."

Meanwhile, in 1975, the Senate committee created to investigate the CIA hauled in Edward Lansdale, the inspiration for Edwin Hillandale of *Ugly American* fame, to answer questions about U.S. efforts to assassinate Fidel Castro. In incredulity and disgust, senators asked whether it was really true that Lansdale had proposed spreading a rumor among Cubans that Christ was about to return to take vengeance upon the godless Castro. (A U.S. submarine was then supposed to shoot flares into the night sky to convince Cubans that the Second Coming had begun.) Exploits that once seemed exciting and noble now looked immoral and a tad pathetic. The hubris of toughness, which Nixon and Kissinger had been keeping on life support, appeared dead. And in 1976, America elected the man who many assumed would be its undertaker.

Unlike George McGovern, James Earl Carter had not spent his life opposing the cold war. In fact he hadn't thought about it much at all. His campaign autobiography, published in 1975, barely mentioned foreign policy, and he almost canceled a trip to Japan that same year when it turned out that none of his aides owned passports.

But the revelations of America's overseas crimes touched something deep within him. Jimmy Carter was, at his core, a scold. Growing up in a family marred by alcoholism, infidelity, and depression, he had been the straight arrow. When his father sent him to gather boll weevils that had fallen off the cotton shrubs near their Georgia home, young Jimmy rebuked other children for picking them directly from the plants. For his presidential inaugural address, he proposed a quote from Chronicles— "If my people . . . turn from their wicked ways; then I will hear from heaven, and will forgive their sin, and will heal their land"—until staffers convinced him it was too self-righteous. Well into his presidency, he still corrected aides' grammatical mistakes.

Carter's moralism was also stoked by his experience with race. While only recently alerted to America's sins abroad, he was intimately familiar with the sins for which it needed to repent at home. And as a white Georgian who supported civil rights, he saw himself as something of an agent for that repentance. Carter's father was a staunch segregationist; he may even have taken part in a lynching. But Jimmy, the dutiful son in every other respect, bucked his father's views—once even taking a date to a black church. As an adult, he courageously refused to join the racist White Citizens' Council, even though some members boycotted his business in response. In his inaugural address as Georgia's governor in 1971, he dramatically denounced racism in the state, thus gaining national attention. So while Carter entered presidential politics without firm foreign policy views, he was by temperament and experience conditioned to see world affairs less as a contest for power than as a quest for justice, less as a question of us versus them than right versus wrong. If the civil rights movement had overthrown a system of violence, fear, and hate, thus liberating the decency buried within his beloved South, Carter hoped to extend that struggle to America's role in the world. At home, building Martin Luther King's "beloved community" had required overcoming segregation. Around the world, building a "global community"—one of Carter's favorite phrases—would require overcoming the cold war.

In 1976, human rights were also shrewd politics. Liberals were outraged by the abuses of the CIA and by America's support for anticommunist tyrants overseas. Conservatives were furious over détente, which had led Nixon and Ford to downplay repression inside the U.S.S.R. During the campaign, Carter skillfully courted both sides, using human rights to attack Ford from both left and right. Unlike McGovern, Carter didn't challenge the toughness ethic head-on. In fact, his attacks on détente won him the support of many of the unreconstructed cold war Democrats, or "neoconservatives," who had abandoned McGovern four years before.

But when Carter entered the White House, his balancing act collapsed. The reason was that when liberals and conservatives talked about human rights, they meant fundamentally different things. For conservatives, the essence of world affairs remained conflict. America was locked in an existential battle against an implacable foe, and human rights were a useful way of justifying and waging that battle. Since the cold war was a struggle between moral and immoral ideologies, anything that helped America win it was, in the long run, moral. For post-toughness liberals, by contrast—as for the early New Left—human rights were not a weapon in the cold war, but its antithesis. A moral foreign policy could only be built on love, not fear. And that meant seeing the world not as warring camps, but as a moral community, where all nations, regardless of ideology, cooperated to improve humanity's common estate.

If Carter flirted with both views during the campaign, as president he tilted clearly to the liberal side. His administration was not devoid of cold warriors: National Security Advisor Zbigniew Brzezinski was a hawk, although, sensing Carter's proclivities, he did not fully bare his talons at first. But the bulk of Carter's appointments were more dovish. Key post–cold war intellectuals such as Gelb, Warnke, and Lake all got important jobs. Carter's ambassador to the United Nations, Andrew Young, and his assistant secretary of state for human rights, Patricia Derian, both came out of the civil rights movement. His first choice to head the CIA, Theodore Sorensen—an outspoken liberal with no experience in intelligence work—so frightened America's spooks that they derailed his nomination.

Carter criticized human rights abuses in the Soviet Union but emphatically rejected the right's effort to use them to scuttle détente. To the contrary, he argued that it was obsessive anticommunism that had damaged human rights around the world. "We are now free of that inordinate fear of communism which once led us to embrace any dictator who joined us in that fear," he declared a few months after taking office. The implication was clear: The world was less dangerous than the cold warriors had claimed. And if Americans stopped projecting their nightmares onto it, they could make it a more decent place.

In part that meant acknowledging the diminished significance of military force. Carter's first budget slashed defense spending by 5 percent, and he halted production of big-ticket weapons systems like the B-1 bomber and the neutron bomb. And he wasn't only less willing to build up America's armaments; he was less willing to deploy them. Six days into office, Carter pledged to withdraw U.S. troops from South Korea, whose authoritarian regime he had repeatedly condemned. "We have an aversion to military involvement in foreign countries," he explained. "We are suffering, or benefiting, from the experience that we had in Vietnam." And there was more. Carter not only abandoned Lyndon Johnson's policy of sending U.S. troops to fight third-world communism; he also moved away from Richard Nixon's policy of selling weapons to help third-world allies resist communism themselves. If Nixon and Kissinger had tried to practice global containment on the cheap, Carter moved away from the doctrine altogether. He announced sharp cuts in U.S. arms sales to the third world and he halted, or slashed, aid to a throng of anticommunist dictatorships. Carter was not always consistent: Some pro-American tyrannies continued to get aid. But compared to every president since NSC 68, his policies marked a genuine shift. "We have witnessed perhaps, the end of a phase in our own foreign policy, shaped largely since 1945," declared Brzezinski. "Preoccupation particularly with the cold war as a dominant [concern] of U.S. foreign policy, no longer seems warranted."

If post-toughness foreign policy meant reducing military support for pro-Western tyrants, it also meant reaching out to the kind of leftist regimes that the United States had once tried to overthrow. In 1977, Carter began to do what Raskin had proposed a decade and a half earlier: normalize relations with Cuba. He removed U.S. travel restrictions to the island, and the two nations established low-level diplomatic ties. In an interview with ABC, Castro told Barbara Walters that Carter was the first American president he actually liked. Carter also moved to normalize relations with communist governments in Angola and a newly unified Vietnam.

The Carter administration pointed to its Angola policy with particular pride. When Marxists seized power there, and Cuban troops began patrolling the country, Kissinger and others had predicted disaster unless the United States intervened. Instead Congress cut off aid to anticommunist forces, and Carter refused to reinstate it. And by 1977, the regime in Luanda was seeking closer ties to Washington and the Cuban troops were guarding the refineries where American companies drilled for oil. For Carter officials, the lesson was clear: "If we maintain calm, keep our doors open, we are likely, in most cases, to find common interests with new governments," explained Undersecretary of State David Newsom.

That was too rosy. Not every leftist regime would abandon its anti-Americanism for a price. Carter sometimes seemed to forget that if nations had interests in common, they also had interests in conflict. But if Carter failed to establish a global community that cut across the cold war divide, neither did his pullback from global containment imperil American security. Some third-world countries went communist on his watch, but given their lack of military and economic power, they added little, if anything, to Soviet strength. Communism brought misery to people in countries like Angola, but the anticommunist alternatives were usually little better, and by refusing to arm regimes and rebel movements that brutalized their people, Carter won the United States some modest goodwill. By the middle of his term, the burying of the hubris of toughness appeared well under way. "In the early period of the Carter administration," wrote Marcus Raskin, "attempts were made . . . to organize a foreign policy that was not totally dependent on military power."

And then, in the blink of an eye, events intruded, and the post-toughness moment was gone.

CHAPTER ELEVEN

FIGHTING WITH RABBITS

On April 20, 1979, Jimmy Carter got into a fight with a rabbit. He was fishing on a pond at his Georgia farm, and had just cast his rod into the water when he noticed the creature paddling toward him. It seemed angry. Its nostrils were flaring, its teeth were bared, and it was hissing in a menacing way. When it tried to board his boat, Carter clubbed it with his oar.

Upon returning to the White House, he recounted the story over lemonade to several aides. They were dubious. Rabbits, they pointed out, are not known either for swimming or for attacking people. And if the animal was so threatening, why had the Secret Service agents onshore not come to the president's aid? Carter did not take kindly to having his honesty questioned. He had, after all, campaigned for president by pledging never to tell a lie. And so, propelled by two of his signature personality traits—self-righteousness and micromanagement—he went searching for a photo to corroborate his account. Eventually he found one: a White House photographer had captured the furry predator speeding away from his boat. Triumphant, the president showed the evidence to his staff.

That should have been the end of it. But in August, Press Secretary Jody Powell, perhaps looking to humanize his boss, passed on the anecdote to the Associated Press. Big mistake. The media, desperate for copy in a slow news month, leaped on the story. "Bunny Goes Bugs: Rabbit Attacks President," screamed a front-page article in the *Washington Post*. In homage to the 1975 blockbuster about a killer shark, the paper dubbed Carter's assailant "Paws."

All three nightly news programs featured the incident. Carter was questioned about it for a week, and when journalists learned that there was a photo floating around the White House, they filed a Freedom of Information Act request for its release. Conservative columnists Robert Novak and George Will alleged that Carter's weak response to his amphibious attacker mirrored his weak response to the Soviet threat. (It didn't help that rather than admitting that he had clubbed the rabbit, Carter, fearful of offending animal rights activists and in violation of his no-lying pledge, told the press that he had merely splashed water on it with his oar.) Evangelical leader Jerry Falwell explained that in the book of Revelation, swamp rabbits are associated with Satan. Carter, he declared, should have throttled the beast.

Journalists loved the rabbit story for the same reason they loved stories about Gerald Ford falling down stairs: It confirmed their stereotypes. By the summer of 1979, the press no longer viewed Carter as an idealistic outsider bringing down-home decency to Washington. They viewed him as a wimp. Post-toughness foreign policy, with its vision of a world defined more by cooperation than conflict, more by love than fear, had appeared politically ascendant in the mid-1970s, with the public sick of war and comforted by détente. But by the time rabbit-gate hit the papers, the world looked nastier, and for Jimmy Carter, Washington was getting nastier, too.

Carter had been blunt: There would be no more Vietnams. In 1978, when Soviet arms and Cuban troops poured into Ethiopia to help its Marxist president retake a stretch of desert called the Ogaden from neighboring Somalia, Carter rebuffed Brzezinski's suggestion that America send a naval task force to the Indian Ocean in response. His caution made sense, given that Somalia had committed the initial aggression, and that the Soviets and Cubans were actually restraining Ethiopia from launching a wider war. Still, among hawkish foreign policy watchers, the events in Africa's horn rankled. Détente had been sold to Americans, in part, as a way of curbing Soviet meddling in the third world. (It was one of Kissinger's creative untruths: Moscow had never promised to do any such thing.) Yet while U.S. diplomats were backslapping their Russian counterparts at chummy superpower summits, communists had taken Vietnam, then Angola, and now Ethiopia. Brzezinski warned that the mood in Washington was shifting. American losses in the third world, he told his boss, "are creating the conditions for domestic reaction."

He didn't know how right he was. The Ogaden may not have mattered very much to the United States, but Iran did. For a quarter century it had been a prize ally. In 1973, with the U.S. economy reeling from an Arab oil embargo, Iranian crude had been an economic lifeline. And in Vietnam's wake, with the United States no longer willing to contain third-world communism with American boys, the Nixon administration had turned the Shah's kingdom into a virtual U.S. armory, selling it some $20 billion of weapons designed to keep the Soviets from the oil fields of the Persian Gulf. By 1976, there were seventy thousand Americans living in Iran, mostly tending to its oil industry and burgeoning military machine. And while Carter privately chided the Shah for his dismal human rights record, even he considered Tehran too valuable to alienate because of moral qualms. On December 31, 1977, the Carters went to celebrate New Year's Eve at the Shah's royal palace, which one journalist compared to Versailles in the days of Louis XIV. As the minutes counted down to 1978, and waiters distributed caviar, Dom Perignon, flaming ice cream, and diced partridge, the president called Iran "an island of stability in one of the most troubled areas of the world." He noted the "respect and the admiration and the love" that Iranians felt for their leader. And he said he felt the same way: "There is no leader with whom I have a deeper sense of personal gratitude and personal friendship."

Eight months later, the island of stability was underwater. Riots gripped the country; the Shah spent the following New Year's Eve under siege; by mid-January 1979 he was in exile, replaced by the Ayatollah Ruhollah Khomeini, whose relationship to the United States was not exactly defined by gratitude and friendship. He called America the "Great Satan."

Then, in June, Nicaragua's pro-American dictator, Anastasio Somoza, fell to the leftist Sandinistas, giving Moscow its first new potential beachhead in the Western Hemisphere since the coup in Chile in 1973. In reality, these setbacks weren't as disastrous as Carter's critics claimed. Iran's new theocratic regime, while bad news for the United States, was bad news for the U.S.S.R., too, since it helped inspire Muslim opposition to communist rule across the border in Afghanistan and within the Soviet Union itself. Communist governments in Angola and Ethiopia offered the Soviets few benefits and lots of headaches. By decade's end, Moscow's new African clients were costing so much money and proving such poor Marxists that some Kremlin officials suggested pulling the plug. And, most important, even as communism made gains in the developing world, it was starting to rot inside the U.S.S.R. In 1975, a massive agricultural shortfall forced the Soviet Union to begin importing grain, something it would do every subsequent year until it ceased to exist. Despite high oil prices, which kept the Soviet economy afloat, Russians by 1979 were loudly complaining about shortages of medicine, soap, toothpaste, and thread. The Soviet Union was like a boxer who kept landing jabs but whose coronary arteries were contracting, slowly shutting off blood to the heart.

But in the late Carter years, Americans weren't particularly focused on the Soviet Union's weakness; they were too preoccupied with their own. A president skilled in the politics of manhood might have found ways of assuaging those fears at limited cost. But as a Democrat with little foreign policy experience, Carter was not well positioned to do so. And he didn't even fully grasp the problem. By 1979, the ethic of toughness—which had appeared dead—was rising from the grave. It would not be laid to rest until Carter was out of office and Ronald Reagan had taken his place.

The story of the resurgence of the toughness ethic in the 1970s is also the story of neoconservatism. And the story of neoconservatism is, to a large degree, the story of an impish, pugilistic garment worker's son named Irving Kristol. Even as a child, Kristol had the instincts of a conservative, which is to say, he revered authority and tradition. He experienced Orthodox Judaism in its crudest and dumbest form: The rabbi at his Brooklyn yeshiva told him to spit when he passed a church, slapped him in the mouth when he misbehaved, and made him deliver his bar mitzvah speech in Yiddish, a language Kristol did not understand. Yet unlike many of his contemporaries, Kristol venerated traditional Judaism nonetheless, even if he could never quite explain why.

In the late 1930s, he enrolled at City College of New York and became a Trotskyist, ostensibly committed to world revolution. Yet underneath he remained a conservative. He began reading Reinhold Niebuhr and other Christian theologians, which was an odd thing for a Jewish Trotskyist to do, and found himself attracted to their emphasis on original sin. He venerated Lionel Trilling, the great Columbia literary critic, who, like Niebuhr, emphasized the tragic element of the human condition. Then, after graduating, Kristol was drafted into World War II, where he fought in a unit "heavily populated by thugs or near-thugs from places like Cicero, [Illinois] (Al Capone's home base)." The experience left him even more skeptical of human nature. His fellow enlisted men, Kristol discovered, were "inclined to loot, to rape, and to shoot prisoners of war. Only army vigilance kept them in check." And so for Kristol, the vigilance of people in power became very important. "My wartime experience," he later wrote, "did have the effect of dispelling any remnants of antiauthority sentiments (always weak, I now think) that were cluttering up my mind."

By the 1950s, Kristol was no longer a Troskyist; he was a cold war liberal. Calling himself a conservative barely occurred to him, since, as far as he could tell, there were no conservative intellectuals in 1950s America, and if there were, they hated the New Deal, which he did not. But it was the conservative elements of '50s liberalism—its anticommunism, its anti-utopianism, its belief that ordinary people were no better than the societies in which they lived, and quite possibly worse—that Kristol liked best. In the '50s, when you could be a liberal and a believer in the toughness ethic at the same time, Irving Kristol was both.

But by the early 1970s, that was no longer possible. The Democratic Party was repudiating the toughness ethic, and so Irving Kristol decided that to remain an old-fashioned toughness liberal, he had to become, oddly enough, a "conservative." Many other "neoconservatives" clung to the term *liberal*. They accused the McGovern Democrats of linguistic theft. But for his part, Kristol embraced the new term. He explained that when you're named Irving, you learn not to worry about labels.

What made Irving Kristol a neoconservative was not his support for Vietnam. In fact, like many neocons in the late 1960s and early '70s, he was ambivalent about the war. What Kristol and his allies really cared about was not the war but the movement against it: the New Left. And what they hated about the New Left was its revolt against authority, its belief that by liberating Americans from their existing institutions it could build far better ones in their place. For Kristol, who believed that people liberated from traditional constraints were little more than animals, this was utopianism: "the main source of all evil in the world." And if liberals were no longer willing to do what the army had done to those thugs from Cicero, Illinois—put down their convulsions by force—then he would ally himself with whoever would.

The first institutions that the neoconservatives set out to defend against the utopian threat were universities, which held a special place in their heart. At City College in the 1930s, Kristol and his friends had been too poor to buy a sandwich. Yet they used that ramshackle institution as a springboard from the ghettos of Brooklyn and the Bronx into a world of intellectual exploration and middle-class comfort. It was quite a shock, therefore, to watch the "rich college fucks" (in Daniel Patrick Moynihan's words) of SDS denounce America's great universities as tools of cold war repression. Amazement turned to horror as the pampered radicals employed rhetorical and sometimes actual violence to achieve their demands. And horror turned to fury as campus administrators buckled before the student mob.

By the late 1960s, SDS, radicalized by Vietnam and by the brutality with which the police often greeted antiwar protest, had abandoned nonviolence in favor of what some called "armed love." ("We must learn to fight as well as seek love," explained a New York–based SDS faction called the Motherfuckers, which blended hippie values with antiwar fury. "We must take up the gun as well as the joint.") Kristol and the neoconservatives were not surprised: In their mind, liberation from existing authority usually ended in blood. In 1968, the SDS leader at Columbia, a suburban Jewish kid from New Jersey named Mark Rudd—whose mother once brought him a home-cooked meal in the middle of a sit-in—sent a letter to the university's president. "There is only one thing left to say," he wrote. "It may sound nihilistic to you, since it is the opening shot in a war of liberation. . . . 'Up against the wall, motherfucker, this is a stickup.'" Two years later, a garment worker's son turned sociologist named Nathan Glazer, who had graduated from City College in 1944, announced that he had seen enough. "How does a radical" like himself, he asked, "end up by early 1970s a conservative?" His answer: by watching nihilists like Rudd, who sought the "destruction of authority—any authority." He had "learned," Glazer explained, "in quite strictly conservative fashion, to develop a certain respect for what was."

If that respect applied to universities, it also applied to cities. Neoconservatives loved cities; urban America was the only America most of them knew. And in the late 1960s, they came to believe that the New Left and its soft liberal enablers were destroying city life. In 1965, Kristol, Glazer, and Daniel Bell helped found the *Public Interest*, which cast a skeptical eye on liberal efforts to radically improve the estate of the urban poor. The writers for the *Public Interest* had no ideological problem with government; unlike traditional conservatives, they did not worship the free market. But they believed in authority. And they decided that in its efforts to liberate the poor from the conditions that oppressed them, the New Left was undermining the traditional forms of authority—be they family, police, ward boss, church, or neighborhood school—that made ghetto life tolerable at all. When reformers bused kids across town, or increased welfare payments to single mothers, or tore down old neighborhoods to build new ones in their place, they may have believed they were merely removing the obstacles that denied people opportunity and happiness. But they were actually meddling with entrenched patterns of life, patterns that reflected human nature as it really was, not as the New Left utopians wished it to be. Trying to change those patterns, warned Kristol and his colleagues, would create more misery and violence, not less. The more dramatic the effort at progress, the more dramatic the failure that would result.

What was true for universities and cities, the neocons argued, was also true for families. Patriarchal authority might be arbitrary and unjust; it might involve force. But it was rooted in human nature. Efforts to liberate women and children from the traditional family, insisted *Commentary* editor Norman Podhoretz—the second most important neoconservative polemicist, after Kristol—would only make things worse.

Neocons made a similar argument about the Democratic Party, an institution to which, in the 1970s, many of them still felt attached. Between 1968 and 1972, a commission headed by—who else?—George McGovern had instituted a flurry of reforms aimed at fumigating the party's smoke-filled rooms. But a little-known neoconservative political scientist named Jeane Kirkpatrick insisted that the reforms had backfired. Rather than making the party more democratic, they had merely shifted power to a new class of political consultants who were even more selfish and unrepresentative than the party bosses they replaced. It was no surprise that Kirkpatrick, like Kristol, admired Niebuhr. Like Niebuhr, she insisted that the language of reason often concealed the reality of interest, and that the pursuit of political perfection was a dangerous game. "In party reform as in life," Kirkpatrick wrote, "good intentions are not enough, and wishing does not make it so."

If neoconservatives loved universities, cities, patriarchy, and the Democratic Party, they also loved American power. Since many of them were second-generation American Jews, who had entered the middle class just a few years after their European brethren entered Dachau and Buchenwald, they felt with particular intensity that only American strength kept barbarism at bay. And since most of them knew little about Russia, but a lot about communism, having witnessed Stalinist thuggishness and duplicity from the other side of the City College cafeteria, they generally interpreted Soviet actions as the result of a fanatical ideological drive. Stalinism and Nazism had taught them that evil was real and that America—even in its worst moments—was not it. In the late 1960s they watched in disgust as the New Left compared Lyndon Johnson to Adolf Hitler and Fidel Castro to Jesus Christ. And in the 1970s that disgust only grew as they saw America retreat, its enemies swarm, and post-toughness liberals declare that harmony was breaking out across the globe.

The neoconservatives generally insisted that when it came to foreign policy they held the same beliefs as John F. Kennedy and Lyndon Johnson; it was the post-toughness liberals who had changed. "In order to become a neoconservative," Kristol quipped, "all you had to do was stand in place." There was a lot of truth in that. Like Kennedy and Johnson, the neocons saw Hitlers everywhere, and like Kennedy and Johnson, they saw American power as constrained not by finite resources but merely by an irresolute will. What the neocons did not acknowledge was that in carrying on the ultra-realist tradition of the Best and the Brightest they were actually abandoning the older realist tradition of Niebuhr, Morgenthau, Lippmann, and Kennan, which had recognized that American power and wisdom were limited, and had seen anticommunism as a vital but partial truth. While Kristol and Kirkpatrick paid homage to the tragic toughness of the 1940s, they were actually heirs to the crusading toughness of the 1960s. In their obsession with anti-utopianism, they forgot that anti-utopianism, when pushed too far, can become hubristic, too.

During the 1976 campaign, the neoconservatives had held out some hope for Jimmy Carter. They were still mostly Democrats, and he had attacked Ford from the right as well as the left. But when Carter stocked his administration with post-toughness liberals like Warnke, Lake, Gelb, and Young, the neocons were enraged. Carter offered only one of the neocons a job: as special envoy to Micronesia. ("Not for Polynesia. Not for Macronesia," joked Podhoretz's stepson-in-law, Elliott Abrams. "But Micronesia.") So they decided to tear him apart.

If the Camelot gang had often compared the 1950s to the '30s, the neoconservatives now called the '70s America's decade of drift and peril. "The parallels with England in 1937 are here, and this revival of the culture of appeasement ought to be troubling our sleep," wrote Podhoretz in 1977. (If that was too subtle, the essay was accompanied by a drawing of Carter carrying an umbrella.) Kennedy had cited the Gaither Commission as evidence that Moscow was pulling ahead while America slept. Now the neocons pointed to Team B, a group of cold warriors— led by Paul Nitze, the crusty author of NSC 68; Richard Pipes, a right-leaning historian of Russia; and a young arms control wonk named Paul Wolfowitz—who claimed that while America was slashing its defense budget Moscow had taken a decisive edge. And like the men of Camelot, the neocons often claimed that America's "culture of appeasement" was a culture of effeminacy; the missile gap stemmed, once again, from a manliness gap. Podhoretz, echoing Schlesinger in 1958, announced that homosexuality was sapping America's martial spirit, just as it had sapped Britain's after World War I. Carter, a writer for the *Wall Street Journal* declared, "wouldn't twist arms. He didn't like to threaten or rebuke."

He donned sweaters. "He even kissed Brezhnev!" He was America's first female president.

The neocon critique of Carter was even harsher than Kennedy's and Schlesinger's attack on Ike. For the restless cold war intellectuals of the late 1950s, Eisenhower was simply tired and complacent. For the neocons, by contrast, Carter's problem went deeper. He did not merely lack the energy to wield force; he inhabited a fantasy world in which force no longer mattered. He was like the university presidents who spoke of understanding student radicals because they lacked the guts to expel them or the mayors who talked about the root causes of crime because they feared locking up rioters. Like other authority figures unwilling to defend their domains, Carter was using the rhetoric of love to mask a failure of will.

The result, from the neocon perspective, was predictable. Because Carter would not use force to maintain order, the world's "juvenile delinquents"—a phrase Kristol applied to anti-American regimes in the third world—were running wild. And as in America's cities and college campuses, the result was not harmony and love, but barbarism and chaos. It was true, Kristol conceded, that American "power may indeed corrupt." But "we are now learning that, in the world of nations as it exists, [American] powerlessness can be even more corrupting and demoralizing."

For Jeane Kirkpatrick, the Oklahoma-born Baptist and mother of three who would become an unlikely icon in a movement dominated by Jewish men, it was déjà vu. The same post-toughness naïveté that had undermined traditional authority in the Democratic Party was now undermining American authority around the world. Once again a romantic view of human nature, and a lack of respect for institutions as they actually were, was producing unintended consequences of the ugliest kind. Kirkpatrick was very upset about Nicaragua and Iran. In her view, Somoza and the Shah were a bit like the bosses who had run the old Democratic Party. Their authority did not stem from democratic principle; it stemmed from tradition, and it relied on force. And so when antigovernment militants in Nicaragua and Iran rebelled, demanding liberation from authoritarian rule, Carter sympathized. He told Somoza and the Shah that America, with its new fidelity to human rights, would not support a military crackdown. Instead America's longtime clients would have to find a more democratic basis on which to continue their rule. In Kirkpatrick's view, this was naïveté on stilts. Authoritarian regimes like Nicaragua's and Iran's were not thin crusts, which could be torn off to reveal a glimmering democracy underneath. They reflected the culture of their societies, societies no more capable of sustaining liberal democracy than of flying a man to the moon. In a 1979 *Commentary* essay titled "Dictatorships and Double Standards," Kirkpatrick wrote that "democratic governments have come into being slowly, after extended prior experience with more limited forms of participation." The process usually took "decades, if not centuries." Believing Iran and Nicaragua could democratize overnight was like imagining that you could eliminate poverty by bulldozing a slum. Carter, she argued, "is, *par excellence*, the kind of liberal most likely to confound revolution with idealism, change with progress, optimism with virtue."

For Kirkpatrick, it was the same, awful story: Believing people were fundamentally better than the institutions that governed their lives, and would create a freer, more loving society if liberated from external constraint, was as naïve in Iran and Nicaragua as in Harlem. It was, she believed, high time that Americans accepted the world for what it was: a nasty place in which people were generally no better than their governments, professions of goodwill usually weren't worth the paper they were written on, and moral progress—when it came at all—was grindingly slow. Foreign policy was a long, hard, often ugly battle between a lesser evil called the United States and a network of greater evils headquartered in Red Square. To think it was anything else, or could become anything else anytime soon, would only make things worse. "We are daily surrounded by assertions that force plays no role in the world," she declared. "Unfortunately it does."

In the final months of 1979, those assertions suddenly stopped. On November 4, Iranian militants scaled the walls of the U.S. Embassy in Tehran; then flung open the gates for their armed colleagues as local police stood by. By nightfall, seventy-six Americans sat blindfolded, tied to chairs or beds, in the ambassador's house. Nothing could have more graphically illustrated the neocons' claim that America was weak in a dangerous world.

The hostage crisis transfixed the nation. American television reporters set up shop outside the embassy gate, where they filmed shackled and blindfolded hostages being paraded before jeering crowds. Local news stations hunted down hostage relatives from their city or state, staking out their houses and broadcasting their anguished pleas. On the *CBS Evening News,* Walter Cronkite signed off every night by announcing the number of days the Americans had been held captive. If more than a decade earlier his on-air revulsion during the Tet offensive had marked the end of the toughness era, now his on-air revulsion at the hostage crisis seemed to mark the end of the post-toughness era as well.

Carter tried diplomacy and imposed sanctions. But the big question—the neocon question—was force. Secretary of State Cyrus Vance argued no. He had been a junior member of the cold war establishment: Yale and Yale Law, a gunnery officer during World War II, a distinguished corporate lawyer, McNamara's deputy secretary of defense. But Vietnam had changed him; he was now a senior member of the post-toughness foreign policy class, a patron to men like Gelb, Warnke, and Lake. He was less of a moralist than McGovern and Carter but he shared their belief that military force was usually useless or worse. Diplomacy and accommodation, while hardly glamorous, were the best tools America had. "Most Americans," he declared after the hostages were seized, "now recognize that we alone cannot dictate events. This recognition is not a sign of America's decline. It is a sign of growing American maturity in a complex world."

His chief bureaucratic rival, National Security Advisor Zbigniew Brzezinski, did not believe that for a minute. Born in Poland and educated in Canada and at Harvard, Brzezinski was naturally aggressive. His relentless style and angular features led Carter's chief of staff to dub him "Woody Woodpecker." Playing soccer once at a White House picnic, he charged the ball with such ferocity that colleagues ran for cover.

In his early career, he had been an unabashed hawk. During the Cuban Missile Crisis, he called on the Kennedy administration to bomb. In 1965, he debated alongside McGeorge Bundy, and against Hans Morgenthau, in favor of American intervention in Vietnam. As late as 1968, he was still urging America to stay there, if necessary for thirty years. We must show "we have the staying power," he pleaded. America must not "chicken out."

But by the 1970s, Brzezinski could see which way the wind was blowing. And he was not one to tilt at windmills. So at Columbia, where he taught, he changed the name of his center from the Research Institute on Communist Affairs to the Research Institute on International Change and shifted his focus from anticommunism to interdependence. By the time he entered the Carter administration he was, like his boss, heralding the emergence of a "global community" where "elements of cooperation prevail over competition."

Critics suspected that this was all an act, that in his heart, Zbig still considered the world a harsh and unforgiving place. As early as 1978, he had urged a show of force in response to the Soviet intervention in the

Horn of Africa. And in 1979, when Somoza and the Shah fell, he sloughed off his prior talk of global cooperation like an outer skin. On Iran he urged military action, even though it might endanger the hostages, so America's enemies would learn that molesting the United States still carried a price. He mocked Vance as "the last Vietnam casualty," declaring that his colleague was "so worried after the disastrous use of force in Vietnam that he lacks the will to use force again." When Carter fretted about the loss of life that an attack on Iran might bring, Brzezinski bluntly informed him that international affairs were "not a kindergarten."

It was a sign of how far the pendulum had swung—of how eager Carter now was to use force—that he authorized a rescue effort that stood little chance of success. (The plan's proponents actually called that one of its virtues: The mission was so implausible, they argued, that it would take the Iranians by surprise.) On April 24, 1980, eight U.S. helicopters flew to a secret location in the Iranian desert. There they were meant to rendezvous with cargo planes that would help them refuel before flying hundreds more miles to another secret base, nearer Tehran. From there the rescuers would transfer to trucks, drive to the U.S. Embassy, overwhelm the captors, and bundle the hostages onto helicopters, which would fly to another secret location, where planes would whisk them out of Iran. That was the plan. In reality, as the helicopters reached their first desert outpost, massive clouds of dust caused several of them to malfunction, leading Carter to abort the mission. To make matters worse, a helicopter and a plane crashed into one another as they were trying to leave, killing eight American troops. At 1 A.M. on April 25, Carter broke the news to the American people. It was yet another humiliation. Vance had already resigned in disgust.

Analytically, it made little sense to blame the disasters in Iran on post-toughness foreign policy. The Islamic revolution, after all, was the bitter fruit of America's decades-long support for that country's tyranny. The real presidential villains were not Carter—who just happened to be in office when the dam broke—but Eisenhower, who had overthrown a democratically elected Iranian prime minister in 1953, and Nixon, whose endless weapons sales had made the Shah think he was invincible. The problem wasn't that America hadn't seen Iran through a cold war prism; it was that America had *only* seen Iran through a cold war prism. And had Carter followed Kirkpatrick's advice and taken forceful military action to save the Shah, he just would have made Iranians hate the United States, and its puppet regime, even more. As for the hostages, Carter did try military intervention. The failure of his rescue effort actually illustrated one of post-toughness foreign policy's main axioms: Military force often doesn't work.

But for Carter politically, the events in Iran were a catastrophe. And, incredibly, things got worse. In the middle of the hostage nightmare, eighty-five thousand Soviet troops entered Afghanistan to prop up a communist regime threatened by Islamist rebels (rebels who, ironically, drew inspiration from Iran). Never before had Moscow invaded a country it had not conquered during World War II. The neocons brandished their Munich analogies. Brzezinski warned that Afghanistan was the first step in a Soviet bid for the oil fields of the Persian Gulf, an area he called "the arc of crisis."

In truth, as Soviet archives would later reveal, Kremlin officials had no such designs; they merely hoped to avoid a fundamentalist Afghan regime that might stir up rebellion among Muslims inside the U.S.S.R. At one point, Soviet premier Leonid Brezhnev even asked an aide where this "arc of crisis" that the Americans kept talking about actually was. Far from being a launching pad for further conquests, the Afghan invasion left the Soviet Union dangerously overstretched. To hear the neocons tell it, the U.S.S.R. in the 1970s was like Germany in the 1930s, growing ever more powerful as the Western democracies buried their heads in the sand. But, geopolitically, the 1970s actually proved a net loss for the Soviets, since both Egypt, once Moscow's most important ally in the Middle East, and China, once Moscow's most important ally in Asia, shifted to America's side. The Afghan invasion proved the biggest net loss of all, as Moscow threw vast sums of money and vast numbers of lives into a savage, unnecessary, unwinnable war that hastened the demise of the Soviet Union itself.

But to the Carter administration, already reeling from a string of setbacks and humiliations, the Afghan invasion looked downright apocalyptic. Aides offered Carter a range of retaliatory measures. He chose them all. He embargoed grain exports to the U.S.S.R., expelled Soviet diplomats, cut cultural ties, restricted Aeroflot flights into the United States, and announced that America would boycott the 1980 Olympic Games in Moscow. And that was only the beginning. If he had backed away from post-toughness foreign policy when the hostages were seized, he now ripped it to shreds. Human rights became an afterthought, as America rushed aid to autocratic allies like Pakistan. The CIA was unleashed. ("We need to remove unwarranted restraints on America's ability to collect intelligence," declared Carter, failing to mention that he was one of the people who had imposed those restraints in the first place.) Arms control was now dead, as Carter told the Senate not even to bother voting on the SALT II nuclear agreement, which his administration had painstakingly negotiated. U.S. forces flooded into the Middle East as the Pentagon scrambled to build military installations in the dictatorships of Kenya, Somalia, and Oman. Americans, Carter declared, must abandon their dream of a "simple world" of "universal goodwill"—who could have encouraged that?—and face the "sometimes dangerous world that really exists."

It was like a time warp; all the old analogies came flooding back. Brzezinski compared Carter's response to the Afghan invasion to Truman's response to communist aggression in Turkey and Greece. The White House unveiled a "Carter Doctrine," deliberately meant to echo Truman's in 1947, under which America would forcibly repel any threat to the Persian Gulf. "We must pay whatever price"—where had Americans heard that before?—"to remain the strongest nation in the world," Carter declared. "Aggression unopposed," he warned, "becomes a contagious disease."

That spring, Vice President Walter Mondale addressed the U.S. Olympic Committee, urging them to comply with Carter's call for a boycott of the 1980 games. Going to Moscow, he declared, would be like going to the Berlin Olympics of 1936. "The story of Hitler's rise," he declared, was "a chronicle of the free world's failure: of opportunities not seized, aggression not opposed, appeasement not condemned." Now America faced the same choice and "history holds its breath."

It was a resurrection story, or so it seemed. Born from the shame of Munich, toughness foreign policy had grown to excess in the golden decades of the early cold war, died in the rice paddies of Vietnam, and was now being reborn in the snowcapped peaks of the Hindu Kush. Except that it was not; it was all a mirage. The hubris of toughness was not being revived, any more than Franklin Roosevelt had revived Woodrow Wilson's hubris of reason during World War II. It had simply never received a proper burial. No post-Vietnam president had successfully reconciled the politics of manhood with the reality of limits. Until Ronald Reagan, a man who considered himself FDR's heir, and in a strange sort of way, actually was.

CHAPTER TWELVE

IF THERE IS A BEAR?

One summer day, sometime in the 1980s, an army colonel named Samuel Trautman went to see a Green Beret who had served under his command in Vietnam. The man had fallen on hard times. Upon his return to the United States, hippies had thrown garbage on him and called him "baby killer." Brutal flashbacks from his days as a prisoner of war had left him unable to hold down a job. Now he was a prisoner again, arrested for rampaging through a northwestern town where he had gone to find the last remaining member of his unit, only to learn that the man had died from Agent Orange.

Trautman came with an offer. The Green Beret could go free on one condition: He must return to Vietnam, on a top-secret mission to find the POWs still under Hanoi's control. In the courtyard of a prison labor camp, as inmates broke rocks under a sweltering sun, John Rambo looked at his former commander and asked, "Do we get to win this time?"

In 1985, *Rambo* became the smash hit of the summer by answering that question with an earsplitting, kick-ass "Yes!" In the film, Rambo returns to Vietnam, kills communists by the dozen, frees his imprisoned comrades, and is awarded the Medal of Honor by America's patriotic new president, Ronald Reagan.

Art, if you can call it that, was imitating life. In the 1980s, Reagan—a career Hollywood actor whose wife sometimes greeted him when he left the Oval Office by yelling "cut"—often seemed to be starring in a movie of his own, which followed the *Rambo* script. Less than a month after taking office, he publicly awarded the Medal of Honor to a Green Beret who had saved the lives of eight men while enduring thirty-seven bullet and bayonet wounds in eastern Laos, but had been denied the award because of a technicality. America's Vietnam veterans, Reagan told a Pentagon crowd, "came home without victory not because they'd been defeated but because they'd been denied permission to win."

In *Reagan: The Movie*, America, after a long string of indignities and defeats, finally remembers how to win. It returns to Vietnam, slays its demons, and recaptures the confidence of a bygone age. It was as if the country had turned back the clock to 1964, relegating the entire post-toughness era to the status of a bad dream. (The other big movie of 1985 was actually titled *Back to the Future*.) In 1982, Hasbro Toys resumed producing the G.I. Joe action figure, which it had discontinued the year after Saigon fell. By 1985, it was America's bestselling toy. That same year in New York, Vietnam vets got the ticker-tape parade they had long been denied. "This country has really needed to flex its muscles," declared the man who played Rambo, Sylvester Stallone. "The other little nations were pulling at us, saying, 'You're bullying. Don't tread on us.' So we pulled back. . . . And what happened, as usual, is people took kindness for weakness, and America lost its esteem. Right now, it's just flexing. You might say America has gone back to the gym."

Reagan: The Movie delighted audiences. But it was a fantasy. When it came to foreign policy, the real Reagan didn't turn back the clock to 1964. He never seriously considered enforcing global containment with U.S. troops. Instead of refighting Vietnam, he created Potemkin Vietnams where America won because it could not possibly lose. He served up victories on the cheap, triumphs without risk. Reagan's critics often accused him of reviving the chest-thumping spirit that had led to Vietnam. But they missed the point. For Reagan, chest-thumping was in large measure a substitute for Vietnams, a way of accommodating to the new restraints on U.S. power while still helping Americans feel strong and proud. Reagan didn't revive the hubris of toughness. He did what Carter had tried but failed to do: He performed the last rites.

The key was his relationship to evil. Or, more precisely, his lack of relationship. For a half century, evil had lain at the heart of the toughness ethic. American intellectuals had discovered it in the 1930s and '40s as the Depression and Stalinism and Nazism shattered the progressive dream of reason. By the mid-1960s, when America entered Vietnam, the discovery had become an obsession. Since evil was terrifying, so were the Vietcong, and since evil was permanent, it was always 1938. The New Left, and for a time, Jimmy Carter, had rebelled against that, insisting that despite their differences, nations—like individuals—were capable of cooperation, harmony, even love. But Carter's timing and his political instincts were both bad, and in 1979 post-toughness foreign policy collapsed.

For neoconservatives like Kristol and Kirkpatrick, Carter's failures reaffirmed the toughness ethic. They showed that in foreign policy, as in urban policy, dramatic progress was usually a mirage; people just weren't built that way. The key was to embrace the lesser evil and defend it with force. When Reagan's ideological allies talked about overcoming the Vietnam syndrome and getting back to fighting the cold war, that's what they meant. It's what they believed Ronald Reagan had been elected to do. But the joke was on them because Reagan didn't believe in evil, at least not as a powerful or permanent thing. He believed what the New Left believed: that beneath all bad authority lay good people, waiting to be set free. Communism, the nuclear standoff, the cold war—he saw them as ugly crusts concealing the glorious humanity concealed beneath. "We have it in our power to begin the world again," Reagan liked to say, quoting Thomas Paine. It was, the columnist George Will complained, "the least-conservative sentiment conceivable," a hint of the transgressions to come.

Reagan didn't fret about evil because he didn't see misfortune. He was America's Mr. Magoo: No matter how many buildings collapsed around him, he kept marching to the merry beat of his internal drummer. (In this regard, he was the temperamental opposite of Richard Nixon, who seethed with bitterness no matter how high he climbed.) Reagan's childhood can fairly be described as harrowing—yet he refused to be harrowed. His father was a drunk, a man who kept destroying the family's screen door because rather than opening it he would walk right on through. One day when Ron was eleven, he found his old man passed out, spread-eagle in the snow, and had to drag him by the armpits to bed. Not surprisingly, the family was poor. No longer able to afford chicken on Sunday, they began eating liver, which the local butcher usually sold as pet food. Yet in his autobiography, Reagan called his hometown "heaven" and his childhood "one of those rare Huck Finn–Tom Sawyer idylls." About the Depression, which hit just as he went looking for his first full-time job, and which sent armies of vagrants scavenging for food, Reagan remembered "that there was a spirit of helpfulness and yes kindliness abroad in the land." Among friends, his optimism became a running joke. In Hollywood, when rain threatened to cancel a shoot, he would tell the director, "Don't worry. It's going to clear up any time." He couldn't play villains because he couldn't get in touch with his dark side. And he hated attending funerals. During the recession of 1982, economic advisers showed him a litany of grim statistics in a bid to convince him to change his policies. To their amazement, Reagan focused on the only positive one.

This almost surreal optimism may have grown out of Reagan's childhood efforts to cope with his father's alcoholism. The children of alcoholics sometimes take refuge in dream worlds, and Reagan, remarked an early girlfriend, "had an inability to distinguish between fact and fantasy." He also inherited his cheeriness from his mother, a deeply religious woman who raised him in the Church of Disciples, a spinoff from Unitarianism that downplayed the idea of sin. Reagan spent his youth marinating in this theological tradition. He spent every Wednesday night and virtually all of Sunday in church, dated the local minister's daughter, and attended Eureka College, a Disciples' institution. And from the Disciples he acquired a view of human nature right out of the Social Gospel. "My personal belief," Reagan wrote to a friend in 1951, "is that God couldn't have created evil so the desires he planted in us are good." In a 1981 speech, Kirkpatrick, whom Reagan appointed ambassador to the United Nations, called her boss's election a reproach to "an age and to a society like ours, which regularly deny the existence of evil." But she had it all wrong. In a fundamental sense, Reagan did deny the existence of evil: He saw it as merely the transitory and artificial absence of good.

To be sure, Reagan loathed communism. His animosity was ancient and intimate; it dated from his fights with communists in Hollywood in the 1940s. But while Reagan shared the right's hatred of communism, he didn't share its fear. Communism didn't scare him because he saw it as an aberration, not as the manifestation of something terrible lodged in the soul. So while his passionate anticommunism set him apart from SDS, and from post-toughness liberals such as McGovern and Carter, his confidence that communism could be transcended, thus unlocking a harmonious world, created a strange ideological kinship with the New Left. The New Left believed it could end the cold war, and so did Reagan. He thought he could end it by overcoming communism; they thought they could end it by overcoming anticommunism. But intellectually, the connective tissue was a belief that evil—however defined—could be cast off, revealing something far better underneath.

If Carter's optimism about cooperating with the Soviets led him to initially ease up on the cold war, Reagan's optimism about defeating the Soviets led him to initially escalate it, in the hope that Moscow's empire might crack. To this end he dramatically boosted defense spending and aided anticommunist regimes and rebels in the third world. But even in those early years, when his cold war policies were most aggressive, Reagan still placed careful limits on the force he was willing to employ and the risks he was willing to take. In 1981, when Poland's communist government cracked down on the independent labor movement, Solidarity, Jeane Kirkpatrick, Irving Kristol, Norman Podhoretz, and George Will all urged Reagan to declare Warsaw in default of its overseas loans and thereby economically destabilize the Polish regime. But Reagan deferred to America's Western European allies, who feared that calling in the loans would destabilize the entire European banking system. Reagan, noted George Shultz, who in 1982 took over as his secretary of state, was "a cautious though decisive man, sensitive about being viewed as too pugnacious." He wanted to halt, and even roll back, Moscow's advance, but not at too high a cost. He didn't ask Americans to "pay any price, bear any burden" to keep countries from going communist; he didn't think he had to, since time was on our side. He didn't see America's back as perpetually against the wall.

In part that is because *his* back wasn't perpetually against the wall. Since, unlike Truman, Kennedy, Johnson, and Carter, he hailed from the political right, Reagan didn't have to worry as much about looking soft. In fact, he recognized, in a way his conservative supporters rarely did, that after Vietnam, looking like too much of a hawk was almost as bad as looking like too much of a dove. Aided by shrewd political advisers like James Baker, Michael Deaver, and Richard Wirthlin, Reagan grasped the public's curious relationship to Vietnam. On the one hand, they wanted to avenge it, to show the bad guys once and for all who was boss. When Carter rubbed salt into the wound by allowing America to be humiliated in the third world yet again, they turned on him. On the other hand, most Americans had internalized Vietnam's basic lesson: America should never again put its boys on the line trying to keep communists from power in the third world. The public loved *Rambo* because in the movies, America won and no Americans died. Reagan knew that if he could gin up a war like that, Americans would cheer. If he couldn't, it was better to let the bad guys win.

Although neoconservatives sometimes called Reagan Kennedy's heir, the two presidents reflected their very different times. In the early 1960s, sacrifice had been a political winner. Almost two decades of nearly unbroken success had convinced the country, and particularly its foreign policy elite, that it was capable of anything, as long as it got off the couch. In the early '80s, by contrast, Americans didn't want to be told to sacrifice; between Vietnam and stagflation they had been sacrificing more than enough. (In 1984, when Walter Mondale pledged that if elected he would raise taxes, that call for sacrifice helped ensure his landslide defeat. For his part, Reagan told Americans that they could have it all: a big defense increase, popular domestic programs, and lower taxes.) Americans no longer wanted their president to find mountains for them to climb, because they were no longer sure mountains could be climbed. They wanted him to find a molehill and call it Mount Everest. It was like the 1984 Olympics, which the Soviets boycotted. Did Americans cheer any less loudly because their best opponents weren't on the field? Hell no. These weren't the Camelot years, when Kennedy compared the cold war to a game between Texas and Rice. Playing the best, he had said back then, brings out your best. By the '80s, America was happy to play Grenada. We needed the win.

Most of Reagan's intellectual allies didn't understand this. They were still ultra-realists, too afraid of Moscow to endorse post-Vietnam limits on America's cold war fight. And since they didn't have to get elected themselves, they could afford to ignore the public's post-Vietnam terror of war. The result was a series of skirmishes, which pitted Reagan's conservative supporters and a few hard-line aides against his more pragmatic advisers, and most importantly, against Reagan himself. Time and again, the result was the same. When looking tough was risk-free, Reagan played the part for all it was worth. But when the ante went up, he cashed in his chips.

Take Central America. To Reagan's liberal critics, it was Vietnam all over again. They watched in horror as he invoked the dreaded domino theory, warning that if the communists in Nicaragua consolidated their rule, and the pro-American regime in El Salvador fell, Mexico might be next. "Here we go again," warned a dovish senator. "El Salvador," read a bumper sticker of the time, "is Spanish for Vietnam."

Had Reagan wanted to refight Vietnam, El Salvador would have been a logical place. Its smaller size and greater proximity to the United States would have made the logistics easier. The American military had a better grasp of its language, culture, and terrain. Its leftist rebels were less unified and battle-hardened than the Vietcong. And if you saw those rebels as agents of Moscow, as Reagan did, the Monroe Doctrine offered a rationale for intervention that dated back to the nineteenth century.

But Reagan never even considered it. At a meeting early in 1981, his first secretary of state, Alexander Haig—the senior official most interested in picking a direct fight south of the border—told his colleagues that if Cuba didn't stop arming El Salvador's communist rebels, "I'll make that island a fucking parking lot." Among conservative intellectuals, Haig's vision of a 1962-style blockade to prevent Castro from exporting arms caught on. William F. Buckley, the longtime editor of *National Review*, endorsed it explicitly, adding that it should be accompanied by declarations of war against both Cuba and Nicaragua. Norman Podhoretz wrote that "the United States must therefore do whatever may be required, up to and including the dispatch of American troops, to stop and then to reverse the totalitarian drift in Central America." Irving Kristol added that, "The world should know, and the American people must be told, that under no circumstances will we permit El Salvador to go the way of Cuba and Nicaragua, and that we will use all means to prevent this." But among Reagan's key advisers, Haig's talk of direct military action was greeted with a chorus of anxious references to Vietnam. Assured that Haig had been merely speaking for effect, Michael Deaver, the White House aide closest to the president, replied, "It certainly had a good effect on me. It scared the shit out of me." From that meeting until Haig resigned more than a year later, Cuba wasn't blockaded; Haig was. Deaver ensured that Haig never saw the president alone. Chief of Staff James Baker kept him off television. And they were almost certainly reflecting the wishes of their boss. As Deaver later noted, Haig's comments "scared the shit out of Ronald Reagan," too.

Haig's problem was that his tough-guy act wasn't an act. He couldn't turn off the projector. "You get a band of brothers from CIA, Defense, and the White House and you put together a strategy for toppling Castro. And in the process we're going to eliminate this lodgment in Nicaragua from the mainland," he instructed his assistant Robert McFarlane. But when McFarlane went looking for that band, he found the Pentagon in a distinctly unfraternal mood. Reagan's defense secretary, Caspar Weinberger, was a hoarder: He loved buying weapons; he just didn't like using them. He considered Vietnam "the crime of the century" because America's leaders had tried to fight it without enduring popular support. (In 1984, in a speech that left a lasting impression on his young military aide, Colin Powell, he would lay out a series of tests for military intervention—including sustained public backing and clearly defined goals—that El Salvador, like Vietnam, plainly flunked.) When McFarlane reported that the Defense Department had little interest in another Bay of Pigs, Haig flew into a rage, denouncing the Pentagon's caution as "limp-wristed, traditional cookie-pushing bullshit." But Weinberger was living in reality, as manifested not only by opinion within the military, but in the nation at large.

Three-quarters of Americans, according to a 1985 survey, opposed invading Nicaragua. A 1983 ABC poll found that Americans opposed sending troops to El Salvador by a margin of almost six to one, even if that meant allowing a communist takeover. That same year, when Reagan reassured a joint session of Congress that "to those who invoke the memory of Vietnam: there is no thought of sending American combat troops to Central America," members of both parties erupted in cheers.

Congress was a big reason for Reagan's restraint. An essential condition of the hubris of toughness had been the explosion of executive authority in the years between FDR and LBJ. In the 1980s, however, Reagan enjoyed no such free rein. Americans generally liked him, but after Vietnam they no longer associated unchecked presidential power with national success. Even his efforts to aid the Nicaraguan Contras provoked a brawl on Capitol Hill. In the 1950s and '60s, the CIA had overthrown third-world governments whenever presidents felt like it; Congress read about it in the newspapers, if at all. But in 1984, when the *Wall Street Journal* revealed that the Agency had been mining Nicaragua's harbors, the Senate denounced the move, 84–12. That same year, Congress outlawed covert aid to the Contras, and when news broke that the Reagan administration had been aiding them anyway—selling arms to Tehran and funneling the profits to the Contras in what was dubbed the Iran-Contra Affair—the political fallout was devastating. Reagan's chief of staff was forced to resign. Nine administration officials were indicted, and six convicted. Reagan's advisers feared he might be impeached.

The president's conservative backers were forever chanting "let Reagan be Reagan," implying that a pantywaist State Department, a gun-shy military, an insolent Congress, and a slothful public were squelching his hawkish convictions. But for Reagan, convictions were often a malleable thing; he molded them to political reality without ever believing he had changed his mind. He certainly wanted to keep Central America from falling to communism, and his efforts in that regard helped get a vast number of Nicaraguans and Salvadorans killed. (As a percentage of its population, Nicaragua lost more people in the Contra-Sandinista war than the United States did in the Civil War, World War I, World War II, Korea, and Vietnam combined.) But Reagan knew he couldn't get any of his *own* citizens killed. So when hawks proposed sending U.S. troops to Central America, he pointed out that Latins didn't like "the colossus of the north sending in the Marines." As hawks like Buckley, Podhoretz, and Kristol recognized, this anti-interventionism undermined Reagan's anticommunism, since the Contras were unlikely to topple Nicaragua's government by themselves. But Reagan refused to see the contradiction, because, like the public, he wanted to fight the cold war without sacrifice or risk. After Iran-Contra, when it became clear that Congress would no longer permit military aid to the Contras, Reagan did not dwell on his failure to overthrow a Soviet ally less than a thousand miles from the Rio Grande. He saw the bright side: He had prevented another Vietnam. Whereas he had once vowed to stare down Moscow and Havana, he now congratulated himself for staring down his own right-wing base. "Those sons of bitches won't be happy until we have 25,000 troops in Managua," he told his chief of staff triumphantly in 1988, "and I'm not going to do it."

Reagan knew how to do what Truman and Eisenhower had done in Korea but Kennedy and Johnson never managed in Vietnam: cut his losses. The best example came late in his first term, as he was turning his eye toward reelection. Over Weinberger's objection, and without thinking through the details (which for Reagan was not uncommon), he had approved the deployment of U.S. peacekeepers to Lebanon as part of a deal to end fighting between Israel and its Palestinian and Syrian foes. Originally the deployment was scheduled to last only a month. But U.S. Marines were sucked into the quicksand of Lebanese politics and eventually took up arms to assist a Christian, pro-Western government at war with segments of its own society. In the early morning of October 23, 1983, those segments took revenge. A terrorist with a thick mustache and a wide grin drove a Mercedes truck past Marine guards and around a series of concrete and barbed-wire obstacles before detonating twelve thousand pounds of TNT at the mouth of the four-story building where the 1st Battalion, 8th Marine Regiment slept. The building was shoved off its foundation and collapsed, killing 241 Marines. For more than six hours, rescuers waded through the rubble searching for wounded, guided by screams. Reagan called it the "saddest day of my presidency, perhaps the saddest day of my life."

Democratic House Speaker Tip O'Neill demanded a U.S. withdrawal. But Reagan refused and accused him of appeasement. "He may be ready to surrender," Reagan declared, doing his best Rambo imitation, "but I'm not." Unfortunately for Reagan, the American people were. Immediately his approval ratings began to sink. Nearly 60 percent of the public disapproved of his policy. Congressional Republicans were growing restless.

And so five days after calling O'Neill Neville Chamberlain for wanting to pull U.S. troops out of Beirut, Reagan decided to do exactly that. Midge Decter, Podhoretz's wife and a noted neoconservative in her own right, declared herself "disgusted." Baker and Deaver were thrilled: Their boss's poll numbers soon ticked back up.

But the Lebanon retreat only illustrated part of Reagan's political brilliance. He buried the hubris of toughness not only by avoiding unpopular wars (after all, even Jimmy Carter did that), but by staging popular ones and thus expunging the stench of defeat. The day the Marine barracks were destroyed, Reagan was confronted by a different foreign crisis. In the tiny Caribbean island of Grenada, known for its exports of nutmeg (if for anything at all), coup plotters had murdered the prime minister. Anarchy reigned and U.S. officials claimed that the eight hundred American students at the St. George's Medical School might be in danger. Reagan also feared that the coup would help Castro turn Grenada into a communist base. As with many of Reagan's pronouncements, the conviction was sincerely held but not entirely logical, since Castro's ties had been to the murdered prime minister, and the Cuban leader denounced the coup. Reagan's insistence that Cuba was building an airfield in Grenada for Soviet planes did not pass muster, either, since the planned runway was too small.

But Reagan was not a stickler for detail. (He was the same man, after all, who the year before had toasted the people of Bolivia at a state dinner in Brazil, and then explained away the mistake by claiming he was going to Bolivia next, which bewildered the government of Colombia, where he was due the following day.) So at 5:36 A.M. on October 25, just two days after the Lebanon attack, four hundred Marines landed on Grenada's western shore. A little after 6 A.M., U.S. Army Rangers parachuted into the southern part of the island. Thirty hours later, the war was won and American medical students were kissing the ground in Charleston, South Carolina. More than 70 percent of Americans backed the Grenada invasion. Mail to the White House ran more than ten to one in favor.

According to his foremost biographer, Lou Cannon, Reagan didn't hatch the invasion to divert attention from Lebanon. But he was "shameless" in using it to turn the public mood from humiliation to pride. The Grenada invasion spawned more medals per soldier than any war in U.S. history. Reagan hosted the St. George's medical students at the White House, where they waved little American flags. Three years later, he staged another Potemkin war, dropping more than ninety 2,000-pound bombs on Libya in retaliation for its involvement in a Berlin terrorist attack that killed a U.S. soldier. Like Grenada, the Libya bombing garnered more than 70 percent support at home. "Every nickel-and-dime fanatic and dictator knows that if he chooses to tangle with the United States of America, he'll have to pay a price," declared Reagan, morphing into Rambo again. The last phrase was key: It was our enemies who would pay the price, not us. Americans could thump their chests, swell with pride, and never get up from the couch.

"There's a bear in the woods." Viewers heard a menacing, pounding beat. Against a bleak landscape, the mammoth creature prowled across America's television screens in the fall of 1984. "For some people, the bear is easy to see. Others don't see it at all. Some people say the bear is tame. Others say it's vicious and dangerous. Since no one can be really sure who's right, isn't it smart to be as strong as bear? If there is a bear?" The screen went blank and then Reagan's picture appeared, above the words PREPARED FOR PEACE.

At first glance, it was everything the neoconservatives believed, in visual form. The world was a wilderness, devoid of civilization's rules. America shared it with a predator, deaf to morality and reason. And America was alone. In the commercial's final frame, you see the grainy outline of a man, standing only a few paces from the beast on a barren hillside. Does he have a gun? You sure as hell hope so.

But what about the last line: "If there is a bear?" You could almost see Jeane Kirkpatrick and Irving Kristol furrowing their brows. Of course there is. The bear is eternal. You can call him Hitler or Brezhnev or Beelzebub, but he's always there. And you had better not let down your guard.

The last line, however, was classic Reagan. For the moment, sure, there's a bear. But if you show you're strong he may turn into a teddy, in which case you can throw away your gun and have a picnic.

Kirkpatrick had made her reputation explaining why that couldn't happen. In her famous essay, "Dictatorships and Double Standards," she argued that authoritarian regimes like those of Somoza and the Shah took decades or centuries to democratize, and that totalitarian regimes like Moscow's would never democratize at all. Her neoconservative allies generally agreed. There is not "the slightest possibility that even the most minimal degree of civil or political liberty will ever be allowed under Communism," declared Norman Podhoretz in 1976. "All hopes for 'liberalization' in the Soviet Union are so chimerical," added Irving Kristol in 1984. "There may be occasional spasms in this direction, but they do not and cannot last."

That's what Kirkpatrick thought her boss believed as well. "The Reagan administration . . . is less optimistic than recent American governments about the evolution of Soviet society," she explained in 1981. But that was wrong: Reagan was actually *more* optimistic about the evolution of Soviet society than his predecessors. Although his right-wing allies generally described the U.S.S.R. as a fearsome juggernaut, he thought—correctly—that it was running out of gas. In part that stemmed from his deep conviction that communism defied the laws of economics. But more fundamentally it flowed from his belief that evil systems could not forever keep good people down. Communism, he wrote in 1975, "is a form of insanity—a temporary aberration which will one day disappear from the earth because it is contrary to human nature." Seven years later, in a speech to the British parliament, he said the Soviet Union was experiencing a "revolutionary crisis" because it "runs against the tide of history by denying human freedom and human dignity to its citizens."

This was a far cry from the toughness ethic. For intellectuals like Niebuhr and Schlesinger, who in the 1930s had seen democracy fall in the most advanced societies on earth, and had watched the best minds of their generation enlist willingly in totalitarianism's armies of the night, the idea that an inexorable "tide of history" lifted humanity toward freedom was a progressive fantasy. Kirkpatrick and Kristol, as children of the toughness age, were heirs to that harsh view. "Americans," demanded Kristol in 1982, should stop "thinking that time is on our side. . . . For the Soviet leaders are utterly convinced that time—i.e. history—is on their side. And so far there is precious little evidence to prove them wrong."

This skepticism about progress was intimately related to skepticism about human nature. Among the cold war liberals of the 1950s, it had been widely accepted that people were generally no better than the political systems that governed them, and that when they sought revolutionary change, things usually got worse. For Kristol and Kirkpatrick, both former cold war liberals, the actions of the New Left graphically confirmed this view. And their new allies on the right, traditional conservatives like Buckley and George Will, took an ever darker view of humanity's propensity to evil. In a gurgling infant, they saw original sin. In a Siberian gulag, Reagan saw the untapped goodness of man.

If cold war conservatives and neoconservatives took a sober view of humanity in general, many took a particularly dim view of Russians. For Kirkpatrick, culture mattered; it was the soil in which abstractions like democracy either sprouted or withered. And Russia, she suspected, was arctic tundra. Like Kennan, she saw Soviet communism not merely as an ideological veneer but also as the latest expression of a tradition of "oriental despotism" that stretched back to the czars. The historian Richard Pipes, who had helped paint a menacing picture of Soviet strength on Team B and served as the first Soviet expert on Reagan's National Security Council, called Russian communism the triumph of the "*muzhik*, the Russian peasant. And the muzhik had been taught by long historical experience that cunning and coercion alone ensured survival. . . . Marxism . . . has merely served to reinforce these ingrained convictions."

Reagan didn't know from *muzhiks*. When he thought about Russians, he imagined "Ivan and Anya," decent folks who, but for an accident of birth, could have appeared on *Father Knows Best*. "Just suppose with me for a moment that an Ivan and an Anya could find themselves, say, in a waiting room, or sharing a shelter from the rain or a storm with a Jim and Sally, and that there was no language barrier to keep them from getting acquainted," he told the nation in 1984, in a passage he wrote by hand. "Before they parted company they would probably have touched on ambitions and hobbies and what they wanted for their children and the problems of making ends meet. And as they went their separate ways, maybe Anya would say to Ivan, 'wasn't she nice, she also teaches music.' Maybe Jim would be telling Sally what Ivan did or didn't like about his boss. They might even have decided that they were all going to get together for dinner some evening soon. Above all, they would have proven that people don't make wars."

There was a parallel with Reagan's view of the United States. Commentators often noted that he saw America as one giant Hallmark card, obscured by an overgrown and dastardly federal government. But because of his reputation as a screeching hawk, they mostly failed to notice that he saw the U.S.S.R. the same way: as a land of good-hearted Ivans and Anyas laboring under an oppressive but artificial regime. In particular, critics were misled by his phrase "evil empire," a term that drove the Kremlin and many liberals batty but that Reagan used exactly once, in 1983. It was the handicraft of a speechwriter named Anthony Dolan, a Buckley protégé. And the original text read: "Now and forever, the Soviet Union is an evil empire." Reagan cut out the first three words. Asked four years later, when he and Mikhail Gorbachev were fast friends, why he no longer used the phrase, Reagan replied, "I was talking about another time and another era." The bear was no longer a bear.

Because he believed the Soviet Union could radically change, Reagan also believed something else alien to the toughness tradition: that America could eventually disarm. In the mid-1940s, before he became a conservative, he, like McGovern, had been a member of the United World Federalists, a group that sought to outlaw nuclear weapons. And even as he moved right, Reagan clung to his dreams of nuclear abolition, although both his allies and his adversaries largely overlooked them, since they didn't fit his image as a cold war hawk. As was often the case with Reagan, movies played a big role. Reagan loved science fiction. And Colin Powell, who eventually became Reagan's national security advisor, believed his boss had been particularly affected by a 1951 film called *The Day the Earth Stood Still*. In it, earth is visited by an ambassador from a higher civilization, which has abolished violence by handing over military power to an army of intergalactic robots, who are programmed to destroy all planets that still resort to war. The aliens have been watching events on earth with dismay. They fear the earthlings will learn how to travel through space and thus infect the solar system with their bloodlust. Before departing in his flying saucer, the alien visitor warns that unless the people of earth abandon war, the robots will descend and kill them all.

For Reagan, this was powerful stuff. Again and again during his presidency, he cited a potential alien invasion as one reason America and the Soviets must overcome their differences. In 1985, he mentioned the idea to Gorbachev, who gave him a strange look and changed the subject. He frequently discussed the alien threat with aides, who worked feverishly to keep it out of his speeches. Whenever Powell got word that Reagan was ruminating about an attack from outer space, he would roll his eyes and say, "Here come the little green men again."

When it came to ensuring peace and harmony between the superpowers, Reagan saw nuclear abolition as the long-term goal. His short-term strategy was "Star Wars," a missile shield that would protect America until the Soviets disarmed (and presumably protect against space aliens as well). To most scientific experts, seeing Star Wars as a quick fix was downright bizarre, since they didn't believe it would work *ever*, let alone anytime soon. And even the hawks who favored Star Wars generally did so because they considered it an alternative to disarmament treaties, a way of reducing the nuclear threat without having to rely on the Soviets to keep their word. But in the Disney movie that played inside Reagan's head, Star Wars and disarmament complemented each other. America would build a missile shield and share it with Moscow, thus rendering both sides invulnerable to attack. Then they would eliminate their nuclear weapons, eventually leaving them with no swords, only massive shields. Reagan—who believed in reincarnation—told Gorbachev that in a prior life he had probably been the inventor of the shield.

Star Wars sparked excitement among Reagan's backers and ridicule from his critics. But during his first term, his desire to abolish nuclear weapons attracted much less notice. One reason is that Reagan, like many conservatives, believed that in the 1970s Moscow had taken the nuclear lead, and he didn't want to negotiate seriously on arms control until America had caught up. For many of those conservatives, however, the belief that America was too vulnerable to let down its guard was perpetual. It did not reflect their assessment of the superpower balance at one moment in time; it reflected their eternal view of world affairs. In 1984, the hawkish Committee on the Present Danger warned that the "the gap between U.S. and Soviet military capabilities continues to grow." But Reagan disagreed. By the end of his first term, after several years of massive military expenditure, he believed that America had caught up.

In 1983, several events also triggered Reagan's latent terror of nuclear war. It began, predictably, with movies. In June, he took in *WarGames,* which starred Matthew Broderick as a hacker who breaks into the North American Aerospace Defense Command (NORAD) and almost starts a nuclear war. Reagan—who, as his view of Star Wars revealed, tended to believe that pretty much anything was technologically possible—found the film disturbing. Two days later, at a meeting with members of Congress, he asked if anyone had seen the movie. None had, so Reagan began describing the premise. "I don't understand these computers very well," he explained, "but this young man obviously did. He had tied into NORAD!" A few months later, the president was even more deeply affected by *The Day After,* a made-for-TV movie that depicts Lawrence, Kansas, in the aftermath of a nuclear war. "It is very effective and left me greatly depressed," he wrote in his diary. "[W]e have to do all we can . . . to see that there is never a nuclear war."

Reagan didn't draw a sharp contrast between reality and celluloid. (In the most bizarre example, he told Israeli prime minister Yitzhak Shamir that he had visited Nazi concentration camps at the end of World War II, an odd claim, given that he had spent the entire war inside the United States. In truth, he had seen footage of the camps in a film.) And so just a few days after viewing *The Day After,* when he attended a briefing on U.S. military procedure in the event of a Soviet nuclear attack, Reagan saw the movies he had watched playing out before him in real life. His dread only grew when he learned that his nuclear buildup and anti-Soviet rhetoric had so terrified the Kremlin that they interpreted a NATO "war game" (there was that phrase again) called Able Archer as preparation for a real attack, and put their military on high alert.

In his mounting fear of war, Reagan once again reflected the public mood. After several years of epic military spending, cold war vitriol, and no superpower summits, the percentage of Americans favoring arms control had shot through the roof. In 1983, the House even passed a resolution endorsing a mutual freeze on the production of nuclear arms. Reagan's political advisers urged him to make an overture to the Kremlin so Democrats could not portray him as a warmonger in his 1984 reelection campaign.

Shultz, a beefy ex-academic with a tattoo of a Princeton tiger on his behind, was also eager to get disarmament talks going. And so in January 1984, more than a year before Gorbachev took power, Reagan began a dramatic rhetorical shift. Declaring that "nuclear arsenals are far too high," he told the nation that "my dream is to see the day when nuclear weapons will be banished from the face of the earth." Then, in September at the UN, he quoted Gandhi as saying that "I have been convinced more than ever that human nature is much the same, no matter under what clime it flourishes, and that if you approached people with trust and affection, you would have tenfold trust and thousand-fold affection returned to you." It was the kind of thing that could make a neocon's hair stand on end.

Even after Reagan won reelection, the antinuclear talk continued. "I just happen to believe that we cannot go into another generation of the world living under the threat of those weapons," he said in November. Irving Kristol was starting to get anxious. "To be conciliatory on the arms-control issues during the election campaign might have been politically expedient," he wrote in December. "But why take it so seriously now? Nothing of any significance is going to happen, and it would be the height of naivete to think otherwise." By toughness standards, however, Reagan was indeed naïve, and growing more so. Nothing in the Soviet Union had yet changed, but he was insisting, with mounting vehemence, that cooperation, not conflict, must define relations between the United States and the U.S.S.R. He was increasingly repudiating the toughness ethic. For conservatives, he was becoming a dangerous man.

Of course, none of this would have mattered nearly as much had the Politburo in March 1985 not chosen Mikhail Sergeyevich Gorbachev as general secretary of the Soviet Communist Party. The fifty-four-year-old Gorbachev represented a younger generation that had come of age during Nikita Khrushchev's post-Stalin thaw, and had spent the subsequent decades quietly seething at the geriatric, mulish, often inebriated commissars who were driving Russia's economy into the ground. Gorbachev didn't want to abandon communism; he wanted to save it. But to revitalize Soviet industry and agriculture he had to slash defense spending, which by the mid-1980s was consuming a jaw-dropping 40 percent of his government's budget. And he knew that the Politburo wouldn't permit those cuts unless superpower relations improved and America looked like less of a threat.

Reagan made that possible. Within hours of Gorbachev's selection, and without knowing anything about the radical reforms upon which the Soviet leader would later embark, Reagan invited him to a summit meeting, without preconditions. Two days later, Gorbachev met with Vice President George H. W. Bush, who was in Moscow to attend the funeral of Gorbachev's predecessor. Bush's talking points would have made Jeane Kirkpatrick twitch. (As it happened, she had left her post several months earlier, declaring her "disappointment" that so "many of the people in the administration" did not share her "foreign policy objectives.") "We should strive to eliminate nuclear weapons from the face of the earth," read Bush's notes. "We should seek to rid the world of the threat or use of force in international relations."

By now Reagan was hungry to meet Gorbachev and begin dismantling the nuclear weapons he had spent so much money to build. Although for years he had denounced the never-ratified SALT II treaty on the grounds that it enshrined America's nuclear inferiority, in 1985 he overruled administration hard-liners and quietly scrapped some older submarines so that America would not exceed the treaty's limits and upset the Kremlin. In March, when Soviet troops in East Germany killed a U.S. soldier, giving Reagan a perfect excuse to avoid scheduling a superpower summit, he instead told journalists that such incidents just made him want to meet his Soviet counterpart more. And once the summit was set for November, he did something that astonished his aides: study. (This was, after all, the same man who in 1983 had refused to prepare for a meeting with world leaders because, as he told James Baker, "the *Sound of Music* was on last night.") Whereas in the past Reagan's advisers had been forced to prepare one-page "mini-memos"—or actually commission short documentary films—to get him to digest information, he now devoured twenty-five briefing papers on Soviet politics and culture. It was like that other classic Stallone character, Rocky, rousing himself from his lethargy before the big fight. At their meeting in Geneva, Reagan and Gorbachev were scheduled to meet one-on-one for only fifteen minutes. Instead they talked for almost five hours. At one point Reagan leaned over to his Soviet counterpart and whispered, "I bet the hard-liners in both our countries are bleeding when we shake hands."

They were bleeding soon enough. No treaties came out of Geneva, but the atmosphere was chummy and the two leaders agreed to meet again the following year, with both sides confident an arms control deal was just a matter of time. Podhoretz was beside himself. Gorbachev, he insisted, was the shrewdest, most sinister Soviet leader yet. He was duping the West into believing that Russia no longer posed a threat, and thus getting the United States to pay to revitalize the collapsing Soviet economy. But since the arms control deals that Washington and Moscow were contemplating were (of course) ridiculously skewed in the Kremlin's favor, Gorbachev was dismantling none of his massive military machine. With the West's guard down, and the military balance shifted further to Moscow's advantage, the Soviet leader would then follow up his charm offensive with a military one, ripping off his smiley mask and revealing himself as the Hitler-Stalin that he really was. As Reagan and Gorbachev edged closer to a disarmament treaty, Podhoretz published a stream of articles with titles like "How Reagan Succeeds as a Carter Clone" and "What If Reagan Were President?" In 1986, when Reagan refused to use Moscow's imprisonment of an American journalist as pretext to cancel his second summit with Gorbachev, Podhoretz accused him of having "shamed himself and the country" in his "craven eagerness" to give away the nuclear store. George Will said the administration had crumpled "like a punctured balloon." What particularly galled Podhoretz and Will was that Reagan, because of his tough-guy reputation, was burying the toughness ethic more successfully than Carter could have dreamed of. If Gorbachev was a wolf dressed as a sheep, Reagan was a sheep dressed as a wolf, and as a result, America was being eaten alive.

At first most conservatives avoided attacking Reagan personally. Throughout 1985 and 1986, they largely blamed his advisers for sabotaging his allegedly hawkish views. But the truth was closer to the reverse. In January 1986, Gorbachev suggested eliminating all nuclear weapons by the year 2000. Administration hard-liners were appalled, but Reagan was intrigued, reminding aides that "I have a dream of a world without nuclear weapons."

That fall, when the two leaders met in Reykjavik, Iceland, Gorbachev again proposed complete abolition, this time by 1996. It was a mind-bending proposal. Nuclear weapons were crucial to America's alliance with Western Europe, not to mention its power in the world. America's European allies were shocked, but Reagan was typically uninterested in the details. This was the happy ending that he believed every drama should have. He told Gorbachev that they should return to Iceland in ten years, each carrying their country's last nuclear missile. Then they would publicly demolish them and "give a tremendous party for the whole world."

It never happened, since Gorbachev conditioned his offer on a U.S. promise not to deploy Star Wars, a promise Reagan would not make. Still, the following December they agreed to eliminate all Intermediate-Range Nuclear Forces (INF). This went far beyond anything Nixon or Carter had done. SALT I and II had merely limited how many new nuclear weapons the superpowers could build. Under Reagan, for the first time, the two sides began dismantling them.

The INF deal was sharply slanted in America's favor. Desperate to cut military spending, Gorbachev had given in on almost everything: agreeing to dismantle Soviet missiles in Asia as well as Europe, long a Kremlin red line, while allowing Britain and France to keep theirs. But for many on the American right, details weren't the point. Reagan was accepting that the Soviet Union was no longer really evil, that America no longer lived in a 1930s world. It was as if Churchill had invited Hitler over for tea and begun dismantling the Royal Air Force.

No longer could conservatives pretend that the president had been hoodwinked by his liberal advisers. In his final year in office, with Reagan himself almost giddy about his relationship with Gorbachev, the ultra-realists attacked him in exactly the same way they had long attacked the post-toughness left. Will accused Reagan of "elevating wishful thinking to the status of political philosophy." The *Wall Street Journal* called him "a utopian." Kirkpatrick insisted that Gorbachev "is not the architect of Soviet retreat." A Republican congressman named Dick Cheney called *glasnost* a fraud.

To fight the INF deal, Podhoretz called for reviving the Committee on the Present Danger, originally founded in 1950 to lobby for NSC 68 and then reestablished in 1976 to oppose détente. It had grown moribund after Reagan's election, but now that Reagan had proved even worse than Carter, Podhoretz proposed reconstituting it. At a Ramada Inn in Tysons Corner, Virginia, another group of conservatives formed the Anti-Appeasement Alliance, which aimed to defeat the treaty in the Senate. In conservative newspapers across the country, they ran full-page ads declaring, "Appeasement is as unwise in 1988 as in 1938." Underneath the text were pictures of Ronald Reagan, Mikhail Gorbachev, Neville Chamberlain, and Adolf Hitler. It was the reductio ad absurdum, the comic ending to a story that was turning out magically well. If Munich had been a genuine tragedy, and in Vietnam the Munich paradigm had produced a blend of tragedy and farce, this was pure farce. Invited to a wedding, a few die-hards insisted that it was really a funeral and donned sackcloth and ashes, which made the whole affair merrier still.

With only months left in his presidency, Reagan traveled to Moscow, where Ivans and Anyas mobbed him everywhere he went. "Systems may be brutish, bureaucrats may fail. But men can sometimes transcend all that," he told students at Moscow State University, sounding a little like the Port Huron Statement. "The accumulated spiritual energies of a long silence yearn to break free."

A year later, Eastern Europe did break free, and then the captive nations of the soon-to-be-former U.S.S.R. Reagan's vision of a world liberated from the cold war and of a Soviet Union liberated from communism had come true. In future years, many on the right would try to reverse the causality of these earthshaking events, arguing that it was Reagan's merciless pressure that brought the Soviet regime to its knees. Had that truly been the case, there would have been no reason for their anguish, no reason to call him a dupe and a fool. Conservatives called Reagan those things for good reason: because he had betrayed them, and the toughness tradition itself. He had refused to spill American blood fighting communism in the third world. And he had turned away from nuclear confrontation and toward nuclear peace before Gorbachev even took power, let alone deposited communism in history's dustbin. "If Reagan had stuck to his hard-line policies in 1985 and 1986," wrote Anatoly Dobrynin, longtime Soviet ambassador to the United States, "Gorbachev would have been accused by the rest of the Politburo of giving everything away to a fellow who does not want to negotiate. We would have been forced to tighten our belts and spend even more on defense." The cold war ended, and Soviet communism collapsed, not because Reagan made America more frightening but because he made it less so; not because he learned the lessons of Munich but because he unlearned them, and helped the American people unlearn them, without surrendering their pride.

The claim that America had overthrown the Soviet empire by threat of force was a toxic seed that would grow in future decades into a hubris of its own, nourished by the rich soil of post–cold war success. But no one could foresee that in the years of wonder that brought the cold war, and the hubris of toughness, to a close. George Kennan was an old man now, the sole surviving member of that extraordinary realist circle—including Niebuhr, Morgenthau, and Lippmann—that had helped bury the hubris of reason a half century before. The others had all died in political despair. Lippmann had paid dearly for opposing Vietnam, eventually leaving Washington, where Lyndon Johnson and his henchmen had made him a figure of ridicule. It is a "much less pleasant world to live in," he told an interviewer in his final days. "I think it's going to be a minor Dark Age." Niebuhr also found Vietnam a brutal blow. "I am scared by my own lack of patriotism," he confided near the end. "For the first time I fear I am ashamed of our beloved nation." Morgenthau admitted to "a general discouragement, sometimes bordering on despair" in his declining years. "I am no longer 'my usual combative self,'" he told a friend in 1972. "I have become convinced, however reluctantly, of the hopelessness of fighting. . . . Perhaps to save one's soul is all that is left."

By temperament, Kennan was the darkest of them all. Like the others, he had preached the gospel of toughness in the 1940s, when it was still an insurgent idea. He had told Americans to see world politics as an arena of struggle, not accord, a sphere governed by power, not reason or love, where progress was often a delusion and evil never died. But his ideas, so precious in moderation, had grown monstrous in the hands of the ultra-realists. He was containment's Dr. Frankenstein, and by the 1980s he had spent so many years raging at his creation to no avail that he too fell prey to national self-loathing and moral despair. "Poor old West," he wrote, "succumbing feebly, day by day, to its own decadence." Maybe it wasn't even worth defending. Maybe the greatest evil was within.

Reagan, he initially assumed, was rock bottom. His administration was "piling weapon upon weapon, missile upon missile . . . like the victims of some sort of hypnosis, like men in a dream, like lemmings headed for the sea." Nearing eighty, Kennan turned back to the disaster that in his view had started it all: World War I. He embarked on a three-volume history of the origins of the conflict, all the while suspecting that it was mere prelude to the far greater catastrophe toward which America now blindly marched: a nuclear World War III "from which there can be no recovery and no return." Perhaps, he mused, it was just recompense for Hiroshima and Nagasaki. The chickens were coming home to roost.

All this cultural gloom made him an easy mark. The *New Republic* declared him senile and suggested that he be "put out to pasture." "The great container," it jeered, "springs a leak."

His opinions marginalized, and his faculties derided, Kennan in 1981 suggested scrapping SALT II as insufficiently bold and instead cutting all nuclear stockpiles in half, en route to the eventual abolition of all nuclear weapons. It was a bizarre turn of events: the architect of the toughness ethic trying to stop his own runaway tradition by embracing the most utopian of post-toughness ideas. He never thought his proposal would gain a serious hearing. It was a cry in the dark, the desperate offering of a sad old man to a president he assumed held his views in contempt.

Instead, five years later, Reagan offered the same proposal himself. For Kennan it was something of a mystery. He was suspicious of Reagan. He distrusted people who lacked an appreciation for tragedy, who insisted that history was forever moving in the right direction. But it was Reagan's peculiar sunniness, his insistence that the world need not be a dark and dangerous place, that helped tame Kennan's Frankenstein monster. The ethic of toughness, which had begun with the recognition that the optimism of the progressive era had its limits, was now ending with the recognition that the pessimism of the cold war had its limits, too. It was the unlikeliest of outcomes, led by the unlikeliest of peacemakers. But it was glorious nonetheless. The end of the cold war, admitted Kennan, "is a fit occasion for satisfaction." For him, that was a scream of joy.

PART III

THE
HUBRIS OF
DOMINANCE

CHAPTER THIRTEEN

NOTHING IS CONSUMMATED

In April 1989, with the Soviet empire teetering on the brink of collapse, the Senate Foreign Relations Committee called eighty-five-year-old George Kennan to explain what was happening to the world. It was a very unusual hearing. For one thing, Kennan was the sole witness. For two hours, he alone held the floor, his voice occasionally quivering, and dropping to little more than a whisper. For another, the senators didn't bloviate, at least not as much as usual. They listened like attentive schoolchildren, prefacing their questions with little tributes, telling him what a privilege this was, offering the thanks of a grateful nation. Since the United States had won the cold war as much with ideas as with armies—and since containment was the most famous idea of all—the little old man across the dais was the closest thing America had to a conquering general. (If the senators knew that Kennan had renounced containment almost four decades earlier, and watched in horror as it grew into a global crusade, they didn't let on.) When the hearing closed, everyone rose to applaud: the senators, their staff, even the committee stenographer. Of course she did, wrote the *Washington Post*'s Mary McGrory, "Considering the guff she usually has to take, the self-serving, pettifogging, jargon-ridden, sometimes semiliterate chest-pounding, Kennan's profound and luminous comments must have sounded like a Mozart oboe concerto."

There was only one discordant note. It came from Senator Daniel Patrick Moynihan, no intellectual piker himself, a former Harvard government professor and ambassador to the UN and India, a man, quipped George Will, who had written more books than most of his colleagues had read. Moynihan had a score to settle on behalf of Woodrow Wilson. For half a century, the melancholy Kennan had been shadowboxing with America's twenty-eighth president, landing blow after blow against Wilson's righteous optimism, his faith that reason and democracy would eventually rule the world. For a long time now, Kennan had had his fellow Princetonian on the ropes. The Somme, Auschwitz, Hiroshima, the Gulag, Cambodia, Iran: History itself had sent Wilson down for the count. In the battle between the rationalizer and the misanthrope, Kennan had a mountain of skulls on his side.

But now, in the virtual blink of an eye, history had turned. From Asia to Latin America to Eastern Europe, from the mud villages of the third world to the citadels of totalitarian power, freedom was washing across the world like a mighty wave. It was time, Moynihan believed, to give Wilson his due.

"I know you have been skeptical about Wilson," Moynihan ventured, tiptoeing toward a rebuke. But has not the world "accommodated [itself] to Wilson's visions?" And there, in front of everyone, after a lifetime of intellectual combat, the conquering general raised the white flag. "You are right. I was long skeptical about Wilson's vision," Kennan replied. "But I begin today in the light of just what has happened in the last few years to think that Wilson was way ahead of his time."

Nineteen eighty-nine was a year that blew men's minds. Six months after Kennan's testimony, Yoshihiro Francis Fukuyama, deputy director of the State Department's Office of Policy Planning, traveled to the south of France. Like Jeane Kirkpatrick and Richard Pipes, he was a Sovietologist and a conservative, not a combination that traditionally inclined one toward optimism about world affairs. But he had been watching in wonder as Mikhail Gorbachev and his top advisers abandoned the language of Marxism and began speaking a tongue that sounded remarkably like America's own. Pipes was still comparing Gorbachev to Brezhnev. Norman Podhoretz was still comparing him to Lenin. But Fukuyama believed that something big was happening; he could feel the earth moving beneath his feet.

When he arrived at the French Riviera for a NATO meeting, his European counterparts told him to calm down. Forget about it, declared the West German representative: Germany won't be unified in my lifetime. When Fukuyama went to Germany to see for himself, he heard more of the same. In East Berlin he snapped photos of demonstrators in the streets. Forget about it, said U.S. Embassy officials: East German communism is here to stay.

Two weeks later, the Berlin Wall fell.

An old mentor of Fukuyama's, a University of Chicago professor and pied piper of the intellectual right named Allan Bloom, was hosting a lecture series titled "Decline of the West?" It was the same old conservative gloom: America and its fellow democracies were culturally weak, morally weak, ideologically weak, militarily weak. It was still 1938. But Fukuyama had said good-bye to all that, and when Bloom asked him to take part, he gave a talk that was not gloomy at all. It was, in fact, the most breathtakingly optimistic statement by an American intellectual since Walter Lippmann and John Dewey glimpsed utopia in the rubble of World War I. Fukuyama's title: "Are We Approaching the End of History?"

"The twentieth century," he explained, "has made us all into deep historical pessimists." He recounted the litany of horrors: "two destructive world wars, the rise of totalitarian ideologies, and the turning of science against man in the form of nuclear weapons and environmental damage." Thoughtful people like Richard Pipes and Allan Bloom had witnessed so much despair over such a long time that they had come to believe it was the normal order of things. They were so used to tragedy that they could not recognize triumph, even when it stared them in the face.

But there it was, the triumph of democracy—of Wilson's dream—bursting out across the globe. In Poland, Lech Walesa was quoting Jefferson; in China, students hoisted a Goddess of Democracy in Tiananmen Square. Things had come full circle. In 1918, with liberal revolution seemingly cascading across Europe, the *New Republic* had declared, in a fit of Wilsonian ecstasy, that "at this instant of history, democracy is supreme." Now, more than seventy years later, after decades upon decades of blood and tears, Fukuyama was saying so again. History with a capital *H*, history as the epic struggle over how humanity should be ruled, was coming to an end. The fascist idea had died; the communist idea was dying; the democratic idea stood alone. Democracy was history's ultimate victor, the final ideological destination of man.

The battle of ideas was over. Intellectually, there was nothing left to contain. At the core of the toughness creed had been the belief that America's enemies were fearsome not merely militarily, but ideologically as well. The fascists and communists were terrifying because they tapped into something deep in human nature. They were part of us, the dark side of the soul. Ronald Reagan had never believed that: For him, there was no dark side of the soul. And now Fukuyama was turning Reagan's insurgent optimism into a new creed, a Wilsonianism for a one-superpower age. The ethic of toughness, which had emerged to explain and combat the great totalitarian ideologies of the twentieth century, was being laid to rest alongside them. In Social Sciences 122, a lecture hall at the University of Chicago, the ethic of dominance was being born.

Fukuyama's lecture became an article, and then a book, and made him a star. *The end of history* became a synonym for intellectual audacity, for the biggest of big-think. But as a foreign policy doctrine, the ethic of dominance actually began life small. Like the ethic of reason and the ethic of toughness before it, it started off as a limited creed and then grew with success. Fukuyama in 1989, like Kennan in 1946, was Daedalus. He built wings, then watched his intellectual progeny fly them into the sun.

For Fukuyama it was democracy that was dominant, not necessarily the United States. America, in fact, was ending the cold war a bit like it ended World War II: triumphant, but exhausted and insecure. As one top U.S. official put it, we "crossed the finish line out of breath."

America's totalitarian competitor may have been crumbling, but the United States wasn't the only democracy on the world stage, and in the late 1980s it wasn't the most vigorous. America's share of world GDP had been dropping for as long as anyone could remember. Its economic growth was sluggish; its education and health-care systems were embarrassing; its inner cities were terrifying; and everyone in the country seemed to be in debt, the government included. During the Reagan years, America had gone from the world's largest creditor to its largest debtor. And on October 19, 1987, roughly two years before the Soviet empire fell, the United States had suffered a day of reckoning of its own. The stock market plunged 22 percent, more than the worst day of 1929.

So while Fukuyama was dubbing the end of the cold war "the end of history," he shared intellectual center stage with another academic super-star, Yale historian Paul Kennedy, who gave it a less triumphal name: "the end of the American era." The United States and the Soviet Union, Kennedy claimed, were partners in decline, twin victims of a historical dis-ease called "imperial overstretch," in which great powers assume overseas obligations they cannot afford. In winning the cold war, in other words, America had actually lost. Germany and Japan, nations that could build cars and VCRs as opposed to missiles—products people actually wanted to buy—were eating our lunch. In a miracle of good timing, Random House published Kennedy's book, *The Rise and Fall of the Great Powers*, only months after the stock market crash. Initially it printed nine thou-sand copies; within a year it had sold 225,000. For thirty-four weeks, Kennedy's book graced the *New York Times*' bestseller list. He had delivered his message, and the intellectual class heard him loud and clear: Before getting too self-congratulatory about the demise of our slow, dumb, overmuscled foe on the other side of the earth, Americans should take a hard look in the mirror. If the Soviets were geopolitical mastodons, we were *Tyrannosaurus rex*.

So 1989 was a heady time, but not a hubristic one. The American idea—democracy—was triumphant, but America itself was not. Economically the nation was anxious, and just as in the early days of containment, economic anxieties made America's leaders wary of projecting military power. One of the most popular phrases in early post–cold war Washington was *peace dividend*. With no threat on the horizon, and America's military expenditures bleeding it dry, politicians looked forward to cutting back foreign deployments and cutting the Pentagon down to size. By scaling back America's ambitions overseas, they hoped to bind its festering wounds at home.

There was another inhibition as well. It wasn't just that America lacked the money to run its vast military machine; many influential Americans doubted whether that military machine was much use anyway. The conventional wisdom, as the cold war wound to a close, was that big armies were like Bentleys: nice to look at but unlikely to take you very far. In Afghanistan the Soviets had driven theirs into a ditch, and the Japanese and Germans were zipping ahead on civilian power alone. For its part, America hadn't fought a real war since it left Saigon with its tail between its legs in 1975, and Reagan, for all his Rambo bluster, had never dared send American troops in large numbers into harm's way. If the American economy stood in the shadow of a stock market crash and a prophecy of decline as the cold war closed, the American military, Reagan notwithstanding, still stood in the long shadow of Vietnam.

Between 1989, when the cold war ended, and 2003, when America invaded Iraq, all this changed. Militarily America won and won, until America's leaders found it hard to imagine America could lose. Economic decline turned to economic mastery, and soon resources seemed almost infinite. Success was like helium, and the ethic of dominance grew. Fukuyama's belief that history was moving democracy's way became George W. Bush's belief that within every dictatorship lay a democracy ready to burst free. And Fukuyama's belief that democratic ideals were supreme became Bush's belief that American power was supreme, that the United States could dominate—ideologically, economically, and militarily—every important region on earth. If the cold warriors had inflated the lessons of 1938 until they believed that virtually every foe could and must be stopped, the post–cold warriors inflated the lessons of 1989 until they believed that virtually every foe could and must be smashed. By 2003 the ethic of dominance had become the equivalent of the ethic of reason in 1917 and the ethic of toughness in 1965: a good idea pushed too far, a theory that explained too much, a species of hubris, a very dangerous thing.

It started with a war Americans barely even remember, and a hidden struggle that presaged much that was to come. In 1988, with Ronald Reagan on his way out the door, a bureaucratic fight broke out within the bowels of his administration. It pitted two men who were almost caricatures of their breed. On one side was Admiral William Crowe, chairman of the Joint Chiefs of Staff, a big, bald, sixty-three-year-old good ol' boy whose lazy Oklahoma drawl made Ivy League civilians forget that he had a master's from Stanford and a doctorate from Princeton. On the other was Assistant Secretary of State Elliott Abrams, the quintessential neocon: a Jewish, New York–born ex-liberal who had attended high school with the children of Julius and Ethel Rosenberg, lived in Nathan Glazer's attic during law school, and married Norman Podhoretz's stepdaughter. The subject was Panama, where the United States and its longtime client, Manuel Noriega, were in the midst of a nasty divorce.

Given his pedigree, observers often assumed that Abrams was an ideological clone of Jeane Kirkpatrick, the Reagan administration's most famous neocon. But there was a generational difference. Kirkpatrick was fifty-five when Reagan took the oath of office; Abrams was only thirty-two. As a result, Abrams, like his generational contemporary Francis Fukuyama, was more deeply influenced by the wave of democratization that began sweeping the world in the 1980s. In Reagan's first term he served as assistant secretary of state for human rights and humanitarian affairs, a job in which, critics assumed, he would follow the Kirkpatrick line: soft on pro-American dictators. But that line was premised on pessimism that third-world autocracies could democratize anytime soon (and a corresponding belief that if America pushed them hard to do so, the result would be not democracy, but communism). In 1983, Irving Kristol was still writing that "the traditions—political, religious, cultural—that shape Latin American thinking and behavior are such as to make it exceedingly difficult for the countries of Southern America to proceed along the [democratic] lines followed by Northern America and Western Europe." And yet, between 1980 and 1986, ten former autocracies embraced democracy in Latin America alone. So Abrams grew more optimistic about democracy's chances, more willing to push for human rights even in places like Chile, Paraguay, South Korea, and the Philippines, where the thugs were on our side.

In the summer of 1985, Abrams took over the State Department's Latin America bureau. A few months later, Noriega—then receiving a $200,000 annual salary from the CIA (the same amount the U.S. government paid Ronald Reagan)—ordered his henchmen to abduct Hugo Spadafora, a prominent critic of the regime. Over the course of seven hours, Noriega's men repeatedly rammed a stick up Spadafora's rectum, pounded his genitals until they were grotesquely swollen, pushed sharp objects under his fingernails, slashed the muscles in his thighs, broke several of his ribs, and then beheaded him. After obtaining photos of his brother's corpse, Winston Spadafora traveled to Washington, where he showed the pictures to anyone with the stomach to look at them. Even Jesse Helms was appalled.

Quietly, Abrams began pushing for a tougher line against Noriega. By 1987 he was denouncing him publicly. And by 1988 he was proposing military force. It helped that with the superpower standoff winding down, Noriega's role as a conduit for aid to the Nicaraguan Contras no longer seemed as vital. During the cold war, the ethic of toughness had required embracing lesser evils like Noriega to contain the greater evil, communism. Now, in a world where America was becoming ideologically dominant, America could set its sights higher. Rather than merely containing evil, it could impose good.

Abrams's desire to topple Noriega wasn't purely humanitarian, of course. In the late 1980s, Americans were in a frenzy over drugs. According to a 1988 Gallup poll, they rated drugs their number-one concern. Noriega was up to his eyeballs in the stuff, and in February of that year two Florida grand juries indicted him for helping to smuggle marijuana and cocaine into the United States. (In the past, U.S. officials had leaned on prosecutors not to charge Noriega, because he was providing useful cold war intelligence. But by 1988 he was no longer important enough to protect.) So Abrams's argument for armed regime change wasn't premised solely on spreading democracy. But in his mind, as in Woodrow Wilson's, spreading democracy and safeguarding American security were interlinked: Regimes that did bad things at home generally did bad things abroad. It was a crucial intellectual shift. Cold war conservatives like Jeane

Kirkpatrick had generally warned against trying to rapidly export democracy, both because they believed it wouldn't grow in barren third-world soil and because they feared the effort would imperil American security, which they believed pro-American thugs were serving rather well. Abrams, by contrast—like many other young, post–cold war conservatives—believed that democracy could quickly take root in foreign lands; they had seen it happen before their eyes. And they feared that if it didn't, continued dictatorship might put American security at risk. For Kirkpatrick, toughness meant recognizing that in a world where evil never dies, trying to impose American values can imperil American interests. For Abrams, dominance meant believing that evil could be vanquished, and that by imposing its values, America could further its interests at the same time.

Unlike Abrams, William Crowe wasn't particularly ideological. And that was precisely the point. In war games, he noticed that civilians were generally quicker than military men to resort to force, and he suspected that it was because for them, war *was* a kind of game. They lived in the world of theory; war didn't turn their stomach because they had never smelled it up close. Abrams, Crowe often noted, had procured an educational deferment to avoid Vietnam. His views "were both naïve in their formulation and reckless in their casual commitment of our military men and women. This latter phenomenon is not unknown among young political appointees who have never served in uniform. However, Mr. Abrams raised it to an art form."

Throughout 1988, Abrams and Crowe battled. Abrams wanted to establish a democratic government in exile on Panamanian soil to pressure Noriega; Crowe didn't consider Panama's opposition very effective or very democratic. Abrams said the United States could overwhelm Noriega's Panamanian Defense Forces with a small force; Crowe made invading Panama sound like D-day. Abrams said the Panamanians would welcome U.S. troops as liberators; Crowe asked why, if the Panamanians loved democracy so much, they weren't fighting for it themselves. Abrams believed that Crowe couldn't get over Vietnam; Crowe believed that if Abrams had managed to get himself over *to* Vietnam in the 1960s, he wouldn't be demanding a repeat performance south of the border now.

In 1988, the Crowe-Abrams feud was a fairly one-sided affair. Ronald Reagan—that dove in hawk's feathers—had sent ground troops into battle only twice: in Grenada, against an opposing army of six hundred men, and in Lebanon, a disaster that still haunted his sleep. (According to his press secretary, Reagan's final words in the Oval Office were "The

only regret I have after eight years is sending those troops to Lebanon.")
Ideologically, Reagan was a transitional figure: more convinced than
most of his conservative contemporaries that democracy would triumph
across the globe, but more skeptical than many younger conservatives
that U.S. troops could help bring that triumph about. So to Abrams's
dismay, the president sided with Crowe, opposing any action that might
require "counting up the bodies." Instead Reagan tried unsuccessfully
to cut a deal with Noriega: Prosecutors would drop the drug charges if
he gave up power. America, insisted Deputy Secretary of State Lawrence
Eagleburger after Reagan's decision, "will *never* invade Panama."
The Pentagon gleefully sent around a mock invitation to Noriega's
farewell party as Panamanian leader. Across the front it read "Due to
circumstances beyond Elliott Abrams's control the farewell party has
been indefinitely postponed."

But Abrams had the last laugh. When Reagan's vice president, George
H. W. Bush, ran to succeed him in 1988, Reagan's offer to drop the drug
charges became a millstone around his neck. It didn't help that as director of the CIA in the 1970s, Bush had met Noriega and kept him on
the U.S. payroll. Democratic Party bumper stickers read "Bush-Noriega
'88—You know they can work together." And in response, Bush was
forced to promise that if he won the White House, there would be no
deals.

Once elected, Bush tried fomenting a coup in Panama. But the effort
ended in disaster as Noriega's men foiled the plot and tortured its ringleader to death. In May, Panamanians went to the polls and overwhelmingly voted Noriega's allies out of office. But the dictator refused to cede
power. Outraged, demonstrators flooded into the streets chanting, "Down
with the pineapple" (a reference to Noriega's acne-covered face). They
hung pineapples from telephone polls in disgust. (The fruit-based protest
may also have been a subtle reference to Noriega's flexible sexuality, since
it was an open secret that he was a bisexual who periodically dressed in hot
yellow, lathered himself in perfume, and cavorted with a pilot boyfriend.)
But Noriega foiled the protests by buying up the entire country's pineapple supply. And when Guillermo Endara, the man who had won Panama's
presidential election, led a protest rally, government goons smashed him
in the face with an iron bar.

Finally, with the United States demanding that Noriega step down,
Panamanian forces detained a U.S. Marine at a roadblock and then

shot him as he tried to drive away. An American navy lieutenant who had the misfortune to witness the murder was blindfolded and repeatedly kicked in the head and groin. His wife was threatened with rape and forced to stand with her hands above her head until she collapsed.

Reagan might still have resisted military action; no one would have dared call him a wimp. But George H. W. Bush had a high-pitched, squeaky voice, goofy-preppy diction, and people did call him a wimp, often. In his native milieu, the fading world of Brahmin New England, where bravado was for the nouveau riche and the first-person singular was a dirty word, Poppy Bush had actually been something of a stud. He'd been tapped for his first secret society at fourteen, joined the navy as its youngest flier four years later, captained the baseball team at Yale, and made his first million by thirty. But the media did not care that he made his motorcade stop at red lights so as not to inconvenience other drivers or that he played a mean game of horseshoes. He might have been a masterful president before television, or perhaps radio. But he proved dismally unable to cultivate his public image in the CNN age. And because, unlike Reagan, he did not look like a wartime leader, he had to become one.

In late December 1989, wearing a pair of bright red socks—one reading "Merry" and the other reading "Christmas"—Bush authorized U.S. troops to invade Panama, taking the first, small step on the ladder from inhibition to confidence to hubris that would culminate with his son's invasion of Iraq. As Americans were just beginning to learn, war was easier in the post–cold war age: Noriega, unlike Ho Chi Minh, had no rival superpowers to funnel him arms. Militarily the invasion was a rout. Noriega, who was visiting a brothel when the U.S. attack began, took refuge at the house of the papal nuncio, the Vatican's man in Panama. For two days and nights, U.S. troops surrounding the building blasted rock music (which Noriega loathed), including such topical choices as "You're No Good," "I Fought the Law," and "Voodoo Child," a reference to the discovery in Noriega's apartment of tamales in which he had written the names of his political enemies. Eventually Noriega gave himself up and was flown to a Florida jail. In the entire operation, twenty-three U.S. soldiers died.

Before the war, according to polls, only a third of Americans had wanted to invade. But once the invasion proved a success, support jumped to 80 percent, and Bush's approval ratings hit 76 percent, the highest for any U.S. president since Vietnam. It was the first sign that in the post–cold war age, starting a war could prove a political boon.

For a small war, the Panama invasion boasted an astounding number of code names: Nimrod Dancer, Sand Fleas, Ma Bell, Post Time, Klondike Key, Stumbling Block, Lima Bean, Purple Storm, Blind Logic, Prayer Book, Gabel Adder, Silver Bullet, Acid Gambit, Nifty Package, Robin Quart, Pole Tax, Krystal Ball, Just Cause, and Promote Liberty. But the last one proved the most apt. In the run-up to war, critics had called the idea of invading to impose democracy absurd. "Surely it is a contradiction in terms and a violation of America's best ideals to impose democracy by the barrel of a gun," declared Massachusetts Senator Edward Kennedy. Jimmy Carter warned that "any sort of military involvement there would immediately alienate the Panamanian people."

But on the ground, to the astonishment of many foreign observers, U.S. troops were in fact greeted as liberators. A CBS poll found that the war was even more popular in Panama than in the United States. When Vice President Dan Quayle arrived in Panama City soon after the invasion, he was greeted with signs reading "Viva Quayle" and "Gringos Don't Go Home. Clean Panama First." (Only later, when the Bush administration reneged on its promise of $1 billion to "clean up" post-Noriega Panama, did the pro-gringo sentiment fade.) And confounding pessimists like Jeane Kirkpatrick and Irving Kristol, Panama didn't revert to dictatorship. Endara was sworn in as president, and then peacefully voted out four years later as Panama joined the swelling ranks of Latin American democracies.

How was all this possible in a country with weak democratic traditions and a history of nationalist resentment toward the United States? In a sermon several weeks after the war, a Panamanian priest offered an answer. The real Panama, he insisted, had not been invaded by the United States. The real Panama—which offered its people freedom and a better life—had actually been invaded years earlier by Noriega and his goons. It was this imposter nation, this tyrannical "anti-country," whose sovereignty the United States had trampled, not the real Panama's. And because the United States had done so, "the true country now starts to take its first difficult and costly steps toward its reconquest: a free country with justice and liberty." Authoritarianism, the priest was saying—repudiating Kirkpatrick's argument in "Dictatorships and Double Standards"—was not organic; it was artificial, unreal. When America peeled away Panama's ugly, tyrannical veneer, it found a democracy trapped underneath. Fukuyama had said history was on democracy's side; now America was learning that through war it could speed history up. Years later, Bush's defense secretary Dick Cheney would call the invasion "good practice," "a trial run."

On November 27, 1990, a little less than a year after Noriega fell, William Crowe, now retired, went to visit the man holding his old job, Chairman of the Joint Chiefs Colin Powell. That August, Saddam Hussein had invaded Kuwait. He had done so for simple, primal reasons, the same reasons a hungry predator devours farm-fattened prey. Iraq was a big, powerful country and it was starving. Its eight-year war with Iran had left it bankrupt, and without money to lubricate his rule, Saddam feared losing power. His regime owed $30 billion to its smaller, richer neighbor to the south, and the Kuwaitis—disregarding Saddam's suggestion that they forgive his debts because he had defended the Arab world against Persian marauders—were demanding that he pay up. The Kuwaitis also refused to cut oil production and thus hike prices, which Saddam hoped would raise some much-needed cash. And they refused to stop the "slant drilling" that Saddam claimed was stealing Iraqi oil. In principle, Kuwait's ruling Al-Sabah family had every right to spurn Saddam's pleas: Iraq's desperation was his own fault. But there was something brazen about their unwillingness to strike a deal. It was like a rich man walking past a street tough and waving his gold watch.

The CIA thought Saddam would merely dip his toe across the border and snatch a few oil wells. Instead he swallowed Kuwait whole. Suddenly, Saddam controlled 20 percent of the world's proven oil reserves, and within the Bush administration, there was unanimity on one thing: He could not have Saudi Arabia, which would give him more than 45 percent. Within days, America was dispatching troops to Saudi soil.

But when it came to Kuwait itself, the Panama divisions broke open again. In theory, everyone was for sanctions, which were meant to push Saddam back across the border nonviolently. In reality, however, some people were more for them than others. Many of Crowe's military buddies wanted to wait a year or more for the embargo to take its toll, in hopes of avoiding war. Top Pentagon civilians, by contrast, led by Cheney and his undersecretary for policy, Paul Wolfowitz, believed the United States should wait only a few months, and then draw its sword. On the surface, it was merely a debate about means: would economic pressure or military force best restore Kuwait? But under the surface, it was also emphatically a debate about ends. The doves wanted to restore the status quo ante; the hawks wanted to change it. The doves just wanted Saddam out of Kuwait. The hawks wanted him not only out of Kuwait, but so emasculated that he could never again threaten his neighbors and hopefully, so emasculated that he lost power.

During the cold war, the hawkish option would probably not even have been on the table. From the late 1950s until the collapse of Soviet power, Iraq had been Moscow's ally, and the Kremlin would not have permitted the United States to beat Iraq to a pulp. Had Saddam invaded in 1980 instead of 1990, America would have been lucky to get him out of Kuwait at all, thus restoring the prewar balance of power. But now, in a one-superpower world, Cheney and Wolfowitz realized that they could do more than hold the line; they could extend it. They could do more than contain Saddam; they could cripple him. Still squinting in the light of a changed world, they were coming to see that with the cold war over, no great power could prevent America from expanding its dominance. In the Persian Gulf, the door to U.S. hegemony was wide open. The American military simply needed to walk through.

Crowe, like many in uniform, dreaded the thought. He worried that Cheney and Wolfowitz—who like Abrams had not served in Vietnam—were unlearning its lessons, and that they didn't have blood in the fight this time, either. (He did. His son Blake was a Marine captain sweltering in the Saudi desert.) "War is not neat, it's not tidy. It's a mess," Crowe told a Senate panel; the Pentagon had already ordered sixteen thousand body bags. For him the cold war was still the template, and the cold war proved the virtue of patience. Attacking Iraq before sanctions had time to work, he argued, would be like trying to militarily roll back the U.S.S.R.

Sitting in his old Pentagon office, now redecorated by Powell, Crowe tried to nudge his successor into doing what he wished his predecessors had done in Vietnam: lay his career on the line to prevent an ill-conceived war. The country isn't eager to fight, Crowe insisted; Powell nodded in agreement. War can be legitimate, Crowe went on, but only after you've been attacked. Powell nodded again. They were sitting at an antique table, as a steward in a bright yellow jacket served lunch. Finally, Crowe asked Powell straight out: "Where are you on the Gulf deployment?" Powell gestured toward the window, which offered a gorgeous view of the Potomac River, and Washington, D.C., on the other side. "I've been for a containment strategy," he replied, "but it hasn't been selling around here or over there."

For Colin Luther Powell, the army was more than a career; it was his way of claiming ownership of his country. His Jamaican immigrant parents had not felt fully American; they lived in a West Indian cultural bubble, dancing to calypso and revering the queen. His own generation knew America better, but for young Caribbean-Americans growing up in the shadow of Jim Crow, to know America was not necessarily to love it. Colin Powell was not the only son of West Indian immigrants to figure prominently in the politics of late-twentieth-century America, after all. The other three were Stokely Carmichael, Louis Farrakhan, and Malcolm X.

As a young man, Powell had seen the army as both a refuge from the actually existing United States and a vision of what it might one day be. By the time he entered the service in the late 1950s, the military was fully integrated, but America itself was not. As a young officer stationed in the South, leaving base meant searching in vain for a restaurant that would serve him a hamburger or let him use the bathroom. When he and his young wife went looking for housing near Fort Bragg, North Carolina, in 1962, a real estate agent showed them an empty shack in the middle of a garbage dump. The army, by contrast, offered camaraderie, shared risk, and common purpose, across the color line. While hardly devoid of racism, it rewarded his talent and his willingness to play the game. "I don't shove it in their face," he said years later, in explaining his success. "I don't bring any stereotypes or threatening visage to their presence," and "I perform well." That he did, in both senses of the word.

He had risen within the army by fiercely protecting its interests and risen in Washington by skillfully serving the interests of the civilians for whom he worked. But now, with Iraq's invasion of Kuwait, those imperatives collided. Crowe had done only one tour in Vietnam; Powell had done two and thus the war was even more seared into his consciousness. He had seen it corrupt and degrade his beloved army; he had watched "recently healthy young American boys, now stacked like cordwood" and sat on helicopter rides alongside the bodies as they began the long journey home. Lebanon had left its mark, too. As Caspar Weinberger's military aide in 1983, he had taken the late-night call from Beirut informing him that hundreds of Marines had been blown to bits. Weinberger had responded by setting the bar for military action so high that it virtually guaranteed Reagan would not use force again, and that suited Powell just fine.

When Bush administration officials discussed how to respond to Iraq's invasion of Kuwait, Powell felt the ghosts of Vietnam and Lebanon in the room. Civilians kept asking him about military options for liberating Kuwait; he kept asking whether Kuwait was worth dying for. Claims that Iraq could be defeated from the air terrified him; that's what Rostow and company had claimed in Vietnam. Bush said Iraq didn't look that tough; it hadn't even managed to beat Iran. "But Iran paid in manpower," Powell replied. It lost close to half a million men.

Powell's objections didn't help his relationship with Secretary of Defense Dick Cheney. Powell wasn't crazy about his boss. Only half in jest, he called Cheney a "right wing nut," and as Crowe had done with Abrams, Powell often noted Cheney's deferments during Vietnam. It was harder to know what Cheney thought about Powell, or about most things, for that matter. That's why people called him "the Sphinx." But he had a reputation for humiliating generals. "You're chairman of the Joint Chiefs," he bellowed after a meeting where Powell raised questions about the wisdom of war. "You're not secretary of state. You're not the national security advisor. And you're not secretary of defense. So stick to military matters."

Powell didn't give up. If he was "the skunk at the picnic," he said later, they could all "take a deep smell." That fall, with U.S. troops streaming into the Gulf and war drawing nearer, he mounted another protest. First he tried to convince Cheney that they should give sanctions more time to work, but Cheney replied that Bush had already made up his mind. Then he went to Secretary of State James Baker, who he suspected wanted a diplomatic way out, and received mild encouragement. Then he talked to Brent Scowcroft, Bush's national security advisor, who seemed annoyed that Powell wouldn't let the matter lie. Finally he got his meeting with Bush himself. It was a Friday afternoon in early October, and the president was in one of his light and goofy moods, which made Powell even more nervous. Powell wanted to make a stand, to do what he— like Crowe—wished his predecessors had done in Vietnam. He tried to make the case for containment, which he called "strangulation" because it sounded more aggressive. He acknowledged that his strategy could take as many as two years, but said he was convinced that the embargo would eventually work, at least in restoring the prewar status quo, which was all he thought America needed to do. What he didn't say—it was just too difficult in that atmosphere; it might have elicited even harsher rebukes— was that two years was sure as hell an acceptable wait if it meant sparing the thousands of young Americans about to be sent into the meat grinder, young Americans who were a lot more real to him than to his country-club bosses. But Powell didn't say that. He had built his career on giving people in power what they wanted and he knew even better than they did what they didn't want: an angry, self-righteous black man "peeing on the floor." He could live, he declared, with whatever decision Bush made. And Bush—who liked Powell, liked what he represented about America, sometimes almost forgot he was black—seemed pleased.

A decade later, when America's military confidence had swelled, and the Gulf War looked in retrospect like an easy and obvious thing, people often forgot how close the United States came to not fighting the war at all. The military wanted to give sanctions more time to work. Baker was itching to cut a diplomatic deal. Scowcroft was trying to anticipate Bush, and Bush—although he grew more certain over time—was tentative early on.

For its part, the public was hardly gung ho. Polls showed that Americans supported military action, but only when told it would be cheap and quick and that the blood would be spilled by the other side—exactly what retired military leaders were telling them not to think. In Congress, virtually the entire Democratic leadership, and most of the rank and file, voted no; it was the closest vote on authorizing military force since 1812. Zbigniew Brzezinski warned that the war could produce twenty thousand U.S. casualties and "potentially devastating economic consequences." Aging veterans of Camelot such as Robert McNamara and George Ball came out of the woodwork in opposition. In testimony opposing the war, Arthur Schlesinger urged senators to "take a few quiet moments to revisit the Vietnam monument" before casting their votes. "Vietnam hangs in the collective subconscious like a bad dream, a psychic wound," observed *Newsweek*, "It hovers over politicians and policymakers, the past that will not die."

But the past did begin to die. Analogies don't live forever, and just as the "lessons" of Munich had superseded the "lessons" of World War I, America's victories in the cold war and the Gulf began to undermine the "lessons" of Vietnam. Before the war, Saddam had warned that "the United States relies on the Air Force and the Air Force has never been the decisive factor in a battle in the history of wars," which was pretty much Colin Powell's view. But they were both proved wrong. During the post-Vietnam years, while army men like Powell had been cursing airpower and the mirage-like promise that it would make warfare cheap and clean, airpower had undergone a technological revolution. During Nixon's Christmas bombings of Hanoi in 1972, the average B-52 missed its target by 2,700 feet. In the Gulf War, by contrast, planes using laser-guided smart bombs regularly came within a foot or two. The air force could now choose not merely which building to hit, but which floor. At one point a stealth fighter guided a bomb through an air shaft in the roof of the Iraqi air command.

The desert, it turned out, was an easier battlefield than the jungle; and Iraq, a more urbanized, industrialized country than North Vietnam, was more vulnerable to attacks from the air. War, Crowe had insisted, is never neat and tidy. But for America in the Gulf, the air war came close. Leslie Gelb called it "immaculate destruction."

By the time American troops went in on the ground, the Iraqi military had been decapitated. Its various units were like limbs without a brain, unable to communicate with one another or do much of anything except flail. In one massive battle, Iraq lost eight hundred tanks; America didn't lose a single one. One hundred hours after the ground fighting began, it was over. Terrified Iraqi soldiers were streaming back along Highway 80 into Iraq. It was a terrible, savage war, but only for one side. Perhaps one hundred thousand Iraqis died. For the United States, the figure was 146.

War, American policymakers were coming to believe, was easier in a one-superpower world. Saddam, like Noriega, had no geopolitical big brother to come to his aid. The flip side was that because America didn't have to worry about holding off communists in other parts of the globe, it could focus its firepower on the Gulf. And because communism was dead, and the Bush administration had rallied most of the world to America's side, other countries paid the bills. Germany and Japan both ponied up; the Saudis alone pledged $15 billion. The Gulf War didn't land the U.S. government deeper in debt, as Brzezinski and others feared. By some accounts, America actually turned a profit on the war.

"By God," exclaimed President Bush, "we've kicked the Vietnam syndrome once and for all." In New York City, Colin Powell rode in a white Buick convertible, with Cheney in the car behind, while a million yellow ribbons fluttered, six tons of confetti fell, and close to five million people cheered. A radio station in Raleigh, North Carolina, began opening its programs with the words "broadcasting from the most powerful nation in the world." House Minority Whip Newt Gingrich and spy novelist Tom

Clancy consulted with an expert in Latin and coined the phrase *Venimus, Vidimus, Icimos Gluteos*: "We came, We saw, We kicked ass."

But for all the chest-thumping, the victory felt incomplete. It "is like *coitus interruptus*," declared Norman Podhoretz. "Nothing is consummated." Precisely because America's military had proved so awesome, critics began to wonder why it hadn't removed Saddam from power. As in Korea forty years earlier, after MacArthur's stunning maneuver at Inchon, the thrill of success bred the temptation to reach for more. Polling showed that while only a third of Americans supported overthrowing Saddam before the war began, by a week into the air campaign the figure had more than doubled.

Bush also wanted Saddam gone. He had assumed that when Saddam's military rivals saw him thrashed they would oust him from power. The problem was that Saddam had no military rivals; he had murdered everyone with talent or an independent power base. As the war drew to a close, Bush encouraged Iraqis to "take matters into their own hands, to force Saddam the dictator to step aside," but Bush's intended audience—Saddam's fellow generals—didn't respond. Instead, it was Iraq's Kurds and Shia, whom Saddam and his Sunni Arab clique had brutally repressed, who rose up against his rule.

No senior Bush administration official argued that America should march to Baghdad and depose Saddam itself. That idea, which many hawks would come to support in retrospect, was still virtually unthinkable in 1991. Assumptions about what the U.S. military could achieve, how much the U.S. economy could afford, and how much the American people would bear remained far too low, even after America's success in expelling Saddam from Kuwait. The hubris bubble had not yet fully swelled. The Vietnam syndrome was dying but not yet dead.

There was a narrower debate, however, which was revealing in its own right. Bush did not want to help the Kurds or Shia at all. On Powell's recommendation, he would not even ground Saddam's helicopters, which were slaughtering the Shia rebels who had risen up on what they thought was Bush's command. Partly that was because he feared getting sucked into an Iraqi civil war. But his reluctance ran deeper than that. Bush worried that if he helped the Shia, the Sunni Arab regimes that had backed the war, but deeply feared a Shia, pro-Iranian regime in Baghdad, would denounce him, shattering the international coalition that had prosecuted the war. That mattered because in 1991—as opposed to 2003—America's leaders didn't believe they could fight Saddam on their own dime. Poverty bred multilateralism and multilateralism induced restraint. In an era of economic scarcity, America needed rich allies like Saudi Arabia to foot the bill.

What's more, Bush agreed with his Sunni Arab allies. Like them he saw the Shia rebellion not as a democratic uprising but as a sectarian one, likely to replace one species of tyranny with another, this one quite possibly fundamentalist. For Bush, the Shia rebellion in Iraq's south evoked less Poland in 1989 than Iran in 1979, and in this respect he was closer to Jeane Kirkpatrick than to Francis Fukuyama, closer to William Crowe than to Elliott Abrams. He didn't believe that within the cocoon of Saddam's dictatorship lay an Iraqi democracy ready to sprout wings. He had hoped for a coup that would produce a milder, pro-American version of Saddam, a Sunni general who would be our bastard, not his own. But when he saw that this was not possible, he stuck with Saddam as the lesser evil.

For Bush, lesser evils were about the best one could expect. The last American president to serve in World War II, he had come of age when democracy remained a rather exotic and beleaguered form of government. And his realist instincts were reinforced by Scowcroft, a Kissinger disciple with a gentle disposition but an extremely harsh view of the world. Bush simply did not believe that trying to promote American ideals would promote American interests; like Irving Kristol, he feared the unintended consequences of good intentions. He had not been particularly happy about the breakup of the Soviet Union, preferring autocratic stability to democratic chaos, preferring the devil he knew to the one not yet born. For him the Panama invasion had been about drugs, domestic politics, and American honor. Democracy hadn't played much of a role. And now he would not lift a finger to give democracy a chance in Iraq. His signature word—it became a running joke on *Saturday Night Live*—was *prudence,* which Merriam-Webster defines as "caution or circumspection as to danger or risk." He was like Eisenhower in this way. His foreign policy was Hippocratic: First, do no harm.

Cheney and Wolfowitz were different: more optimistic and more aggressive. Cheney had been slow to recognize the changes in the U.S.S.R., but once he did, he wanted to dismember America's old foe. He wasn't trying to promote democracy necessarily; he wasn't a terribly idealistic sort. But he wanted to expand American power. To be truly safe, he believed, America must extend its dominance, and in that way he exemplified the spirit of the emerging post–cold war age. Bush was comfortable limiting America's power and his own; limits protected you against bad things.

He didn't want to restore the imperial presidency; he believed Johnson had only hurt himself by hoodwinking Congress on Vietnam. So he let Congress vote on the Gulf War, thus giving himself cover if it went bad. Cheney had opposed that. For him, letting Congress in the game wasn't cover; it was a straitjacket. Cheney hadn't wanted to run the war through the UN, either. Why give anyone else a say? And when the war ended, he was less inclined than Bush to defer to America's Arab allies. He didn't want to get drawn into an Iraqi civil war any more than Bush did, but he supported shooting down Saddam's helicopters. He didn't necessarily believe that a Shia government would be democratic, but he did think it might be more pliant. He was hungrier than Bush to extend America's reach and more willing to take risks to bring that about.

Like Cheney, Wolfowitz also believed in extending American dominance. But he saw the spread of democracy and the spread of American power as more closely intertwined. Like Abrams and Fukuyama, who was his close friend, he had been in his thirties in the early Reagan years. And like them he had been shaped by the great tide of democratization that washed across the developing world. If Abrams's crucible had been Panama, Wolfowitz's had been the Philippines. At first, the Reagan administration—in keeping with the Kirkpatrick line—had embraced Manila's pro-American dictator, Ferdinand Marcos. But as Marcos weakened and a communist insurgency grew, Wolfowitz, as assistant secretary of state for East Asia, began pushing for reform. By late 1985, Kirkpatrick, now out of government, was accusing her former colleagues of the same democratic "purism" that had led to the downfalls of Somoza and the Shah. "Anyone who thinks it [the Marcos regime] will be succeeded by an authentic parliamentary government with sensible economic policies," added Irving Kristol, "has been swallowing too many happiness pills." But Wolfowitz kept swallowing them. In 1986, when Marcos rigged an election and Filipinos poured into the streets in protest, Wolfowitz helped convince Secretary of State George Shultz that America must force Marcos out, and Shultz in turn convinced Reagan. A democratic, pro-American government (with sensible economic policies) took power, and the communist insurgency faded. Wolfowitz later called it "the high point of my career."

Not surprisingly, then, Wolfowitz saw the Shia rebels differently than Bush and Scowcroft did: not as fundamentalists, but as democrats. Of all the top Bush officials, he was the most adamant that America come to their aid. He kept urging that Saddam's helicopters be shot down, even after the bureaucratic fight was lost.

In so doing, he slammed up against Powell. Like Abrams and Crowe before them, Wolfowitz and Powell were oil and water. Powell loathed disorganization; in Wolfowitz's office, the papers were stacked up so high you could barely see. Wolfowitz was a nerd: He spoke six languages and kept ten ballpoint pens inside his suit pocket. Powell distrusted intellectuals. "There are some people at that school you hang out at," he later told a Harvard professor, "who contribute nothing else to the national GDP except to create these great schemes."

In late March, with Wolfowitz still kicking up a fuss about the helicopters, Powell screamed at him that the question was closed. Inside the administration, cautious realism still held the upper hand. But things were changing. Before the war, critics had mostly called Bush reckless for going to war too early. It was a sign of the shifting political mood that they now called him timid for ending it too soon.

By any reasonable standard, the Gulf War was a success. Facing a public, a Congress, and a military still deeply scarred by Vietnam, Bush had taken America to war and won it at a price—in money and lives—that Americans were willing to pay. Yes, Saddam was still in power, but he was weaker. His military had been ravaged and as the price of defeat he was forced to accept a humiliating regime of sanctions and inspections designed to ensure that it was not rebuilt. The Kurds and Shia paid a terrible price for America's decision to end the war early, but the Bush administration eventually established a safe haven in Iraq's north that gave many Kurds some respite from Saddam's horrors. And in both north and south, American planes patrolled the skies to make sure Saddam did not menace his neighbors again. The hawks who had wanted not merely to liberate Kuwait but to cripple Iraq had achieved their goal.

The war also dramatically expanded American dominance of the Middle East. When Reagan took office, Central Command (CENTCOM), which oversaw U.S. military operations from Kenya to Kazakhstan, and ran the war in the Gulf, had not even existed. Back then, America had jostled for influence in the Middle East with the U.S.S.R. Now Syria, one of Moscow's former clients, had joined a U.S.-led war against one of its other former clients, Iraq, while the Soviets watched from the sidelines. In the 1980s, America's military presence on Middle Eastern soil had been limited to a navy supply base in tiny Bahrain and a few hundred peacekeepers in the Sinai desert. Now, after the Gulf War, U.S. ground troops occupied six Middle Eastern countries and U.S. planes patrolled the skies

above much of Iraq. In 1985 the United States had accounted for 15 percent of arms sales to the Middle East. A decade later it was 72 percent.

Yet despite all this, as George Bush began his 1992 reelection campaign
he found himself attacked by both conservatives and leading Democrats
for having left Saddam in power. In April 1991, Tennessee Senator Al
Gore, who only months earlier had agonized about whether to support the
war at all, demanded that Bush develop a strategy of democratic regime
change so that "we advance the day not just when Saddam Hussein is out
of power, but when there exists in Iraq a government that grants reasonable consideration to all its peoples." At the 1992 Democratic convention,
Georgia governor Zell Miller said that President Bush "talks like Dirty
Harry but acts like Barney Fife." It was the same fate that had befallen
Harry Truman when he pulled back from MacArthur's bid to liberate the
Korean peninsula from communism. American power had expanded dramatically, just not as dramatically as American confidence. It was a telltale
sign that hubris was starting to build. The higher America flew, the more
fervently critics demanded that it fly higher still.

CHAPTER FOURTEEN

FUKUYAMA'S ESCALATOR

For Colin Powell, Bill Clinton's presidency constituted an irony, maybe even a cruel joke. He had been sad to see the Bushies go. Sure, Cheney was a bit icy. Wolfowitz's ideas were a little grand. But Powell liked Republicans. Like him, they believed in authority, discipline, order. With Republicans, meetings started on time; decisions got made; you knew who was in charge. He particularly liked Reagan and Bush. They were courtly and dignified, traditional in a way he appreciated. They looked the way he thought presidents should look. Powell considered many in his own age cohort, and among the slightly younger baby boomers, to be self-indulgent, whiny, unkempt. The military was a museum of old-fashioned virtues, and when it came to civilians, Powell found them best expressed in men and women a generation older than him.

It was a challenge, therefore, to serve a commander in chief who, on at least one occasion, sat in the Oval Office in his sweatpants, munching a banana smeared with peanut butter. Powell, who still had eight months left as chairman of the Joint Chiefs when Clinton took office, was a fanatic about punctuality: At his morning staff meetings, everyone took their seats at 8:30; he walked in at the stroke of 8:31. The new White House, by contrast, ran on what reporters called "Clinton Standard Time": anywhere from a few minutes to a few hours late. (It was an irony of the Clinton-Powell relationship. Powell had developed his mania for promptness in part to combat racist stereotypes of blacks. Clinton, a son of the redneck South, conformed to those stereotypes and partly as a result was dubbed "the first black President.") Powell rose early; Clinton stayed up late, often working—and expecting his staff to work—until 2 A.M. Powell was stoic: When injured for the second time in Vietnam, he asked the army not to inform his wife. Clinton loved to emote. At a touchy-feely, get-to-know-you session for cabinet members and White House staff, he described how painful it was as a young boy to be fat.

Finally, Powell prized proper appearance. When meeting someone for the first time, he instinctively looked to see if their shoes were shined. Clinton, by contrast, often took reporters' questions in his tracksuit, while sweating profusely from his morning run. He was also a binge eater, whose intake increased with stress. And some of his advisers were worse. Powell's new boss at the Pentagon was Les Aspin, a disheveled, doughy man known to lunch on potato chips doused in mayonnaise. At a meeting with Jordan's King Hussein, Powell watched in horror as Aspin ate thirteen hors d'oeuvres.

There should have been at least one note of consolation: At least these graying beatniks wouldn't start many wars. Powell distrusted foreign policy makers who hadn't served in uniform, but he actually preferred those who had avoided Vietnam because they opposed it to couch-potato militarists like Abrams, Wolfowitz, and Cheney. In the culture war, he and the Clintonites were often on opposite sides, but when it came to real war, Powell hoped that the children of Woodstock and the children of Khe Sanh would find common ground in their resolve to never do anything like Vietnam again.

That's why the Clinton years were a cruel joke, because the hippies had changed. Vietnam no longer traumatized them; they were starting to appreciate war. It wasn't apparent right away, but by the mid-1990s, after some humiliating false starts, the Clintonites began learning many of the same lessons about the post–cold war age that the Bushies had learned: that military force worked, that America's political and economic model was universal, and that when America pushed hard in one direction, its allies fell in line. There were partisan differences, to be sure, differences that would become clearer when Cheney, Wolfowitz, and Powell returned to office under a new President Bush. But by Clinton's second term, the hubris bubble was again starting to swell. The ethic of dominance, it turned out, was ideologically ambidextrous. In the Clinton years, American hegemony learned to hit from the left side.

The Clintonites, to be fair, had never really been hippies. During Vietnam they had formed the right edge of the antiwar movement, the students who wanted to save the system, not smash it, and the left edge of the establishment, the people nudging it to adapt to changing times. Many of them—like Sandy Berger and Clinton himself—had worked in the McGovern campaign. Some—like Anthony Lake and Richard Holbrooke—had served as discontented young diplomats in Vietnam. Virtually none had joined SDS.

But Powell was not wrong to assume that he could dissuade the Clintonites from using force. Compared to Cheney, who ate generals for breakfast, the Clintonites were more deferential to the military, or maybe just more afraid of it. And the military certainly wasn't very deferential to them. Two months into his presidency, when Clinton visited the U.S.S. *Theodore Roosevelt*, its crew mocked him behind his back. "What did Clinton do when a protester threw a beer at him?" went one joke making the rounds of the ship. "It was a draft beer so he dodged it." Two months after that, an air force general accused Clinton of "pot-smoking, draft-dodging [and] womanizing." Clinton had taken office promising to lift the ban on gays in the military, but after encountering fierce resistance from Powell and the other Joint Chiefs, he backed down. After that, many in the military considered him an easy mark. Presumably he would back down even faster if they opposed going to war.

Clinton had talked tough on foreign policy during the campaign, slamming Bush for leaving Saddam in power, tolerating "ethnic cleansing" in the former Yugoslavia, and coddling the dictators in Beijing. But less important than what he said was how little he said it. In his four-thousand-word convention acceptance speech, foreign policy received only 141 words. He had been elected, Clinton told ABC's Ted Koppel, to "focus like a laser beam on the economy." Foreign policy aides—except for those who dealt with international economic issues like trade—were warned not to take too much of his time. When a deranged man crashed a small plane onto the White House lawn in 1994, insiders joked that it was Clinton's CIA director trying to get a meeting.

Powell suspected that as long as Clinton didn't devote much personal attention to foreign policy, his foreign policy wouldn't be particularly aggressive. And he was right. In Clinton's first two years as president, his personal hesitancy on international affairs—combined with America's ongoing economic troubles and the fading, but not yet dead, legacy of Vietnam—left him hesitant to use military force. The early Clinton years were to the rising ethic of dominance what the Eisenhower years had been to the rising ethic of toughness: an interregnum, a pause before the march toward hubris resumed.

A big reason for that pause was the U.S. economy, which in 1993 still looked every bit as fragile as it had five years earlier when Paul Kennedy predicted America's decline. Echoing Kennedy, top Clinton officials privately warned that America couldn't afford costly new foreign commitments. "We simply don't have the leverage, we don't have the influence, [or] the inclination to use military force," conceded Undersecretary of State Peter Tarnoff in May. "We don't have the money to bring positive results any time soon."

Western Europe, by contrast, was feeling its geopolitical oats. No longer dependent on America for protection against the Soviet Union, many Europeans believed they were finally becoming masters of their own house. And in 1993 they proudly inaugurated the European Union, which was meant to help the continent speak with one loud voice on the world stage. The Gulf War may have left America dominant in the Middle East, but in Europe, American power appeared to be in retreat. As EU chairman Jacques Poos exulted, "The Age of Europe has dawned."

If European problems were now meant to have European solutions, the test case was Yugoslavia, which in the early 1990s had ruptured in a gush of blood. From the beginning the Europeans called the tune. In 1991, when first Slovenia and then Croatia seceded from the Serb-dominated Yugoslav federation, the Bush administration warned against recognizing their independence. But Germany—recently unified, historically allied with Croatia, and basking in its newfound geopolitical power—brushed Washington aside. Bonn recognized the two breakaway nations; the rest of Western Europe quickly followed suit, and American officials watched with a combination of passivity and indifference. As Secretary of State James Baker liked to say, "We don't have a dog in this fight."

Soon all hell broke loose. Slovenia contained barely any Serbs and thus made a fairly clean getaway. But in Croatia, Serbs comprised more than 10 percent of the population, and many of them wanted no part of an independent Croatian state led by ultranationalist, anti-Serb bigots. In May 1991, Serbs in the small eastern Croatian town of Borovo Selo raised the old Yugoslav flag. Croatian policemen arrived to take it down. Suddenly the Serbs began shooting. They not only killed a dozen policemen but also gouged their eyes out. It was a small taste of things to come.

A year later, in the spring of 1992, heavily Muslim Bosnia seceded as well. But with Orthodox Christian Serbs comprising roughly a third of its population, and Catholic Croats another fifth, it was the most vulnerable of all. On April 6, the Europeans recognized Bosnia's independence; the next day America did, too. Two weeks later, local Serbs, backed by the Yugoslav army, laid siege to the Bosnian capital, Sarajevo, shelling it night after night. By midsummer, the city that had hosted the 1984 Winter Olympics was a horror show. With its cemeteries overflowing, people began burying bodies in public gardens, even backyards. There was barely any water, food, or fuel. "What is the difference between Sarajevo and Auschwitz?" went a local joke. "In Auschwitz at least they had gas."

It got worse. In the Bosnian hinterland, the Serbs were building concentration camps. In the town of Banja Luka, they dispossessed Bosnian Muslims of their homes and possessions, shaved their heads, herded them onto cattle trains, and dumped them at a place called Omarska, an old iron mine now encased by barbed wire. Inside, inmates were made to endure—and commit—unspeakable crimes. Men were forced at gunpoint to castrate their sons and rape their daughters. Lacking adequate supplies of food and water, some inmates were reduced to drinking their own urine; others tried to survive by eating grass. The goal was to "cleanse" Bosnia of Muslims: Some would be killed, others deported, and the rest would be too psychologically shattered to ever assert their independence again.

Europe's response was to push through a UN arms embargo, which, given that Belgrade controlled virtually all of the Yugoslav army's heavy weaponry, merely prevented the Bosnians from defending themselves. Britain, France, and several other European nations also sent troops as part of a UN "peacekeeping" force. But in Bosnia there was no peace to keep. Allowed under UN guidelines to defend their aid convoys, but not allowed to defend Bosnians, the peacekeepers proved worse than bystanders. They became potential hostages, whom the Serbs could seize if the West threatened military action. Useless at preventing Bosnia's destruction, the UN troops proved effective only at preventing the world from doing anything about it.

The Clintonites were appalled. Vice President Al Gore, National Security Advisor Anthony Lake, Ambassador to the United Nations Madeleine Albright, and Ambassador to Germany Richard Holbrooke began pushing a policy called "lift and strike," aimed at saving Croatia and Bosnia by lifting the UN arms embargo and striking Serb positions from the air. But implementing "lift and strike" required overruling the Europeans, who still saw the former Yugoslavia as their show. And before the Clintonites could overrule the Europeans, they first had to overrule Colin Powell.

That proved impossible. On national security, Powell was Clinton's most powerful adviser. He had more foreign policy experience than the new president and many of his top aides, and was vastly more popular on Capitol Hill. Within the military, he sometimes seemed to be the only person preventing outright insubordination. Clinton was also terrified that Powell would resign and run against him for reelection in 1996. In their first clash, on gays in the military, it was the president—not the general—who had caved. "Powell simply overwhelmed the administration," noted Holbrooke. "He regarded the new team as children. And the new team in turn regarded him with awe."

The Gulf War had made Powell a national hero, but it hadn't altered his reluctance to use force. Bosnia struck him as a cross between Lebanon and Vietnam. As in Lebanon, America would be putting itself in the middle of a Byzantine, brutal civil war that had little direct bearing on its security. And as in Vietnam, the former Yugoslavia's terrain was a forbidding patchwork of forests and mountains, nothing like the open desert of Kuwait. Civilians sometimes called the Serbs a lackluster force, undisciplined and frequently drunk, but Powell likened them to the Vietcong: tenacious defenders of what they considered their home soil. Powell grew particularly agitated when Clintonites suggested that America could defeat the Serbs from the air. "When I hear someone tell me what airpower can do," he declared, "I head for the bunker." When administration officials asked him what it would take to save Bosnia, Powell replied two hundred thousand ground troops. That was meant to be a conversation-stopper, and usually was.

Despite Powell's opposition, by May 1993 Clinton officials thought they had reached a consensus on "lift and strike." Secretary of State Warren Christopher was dispatched to consult with the Europeans. But the key word was *consult*. The Christopher trip was multilateralism in its truest form. America was not presenting its allies with a fait accompli; it was genuinely seeking their opinion. It was a sign that when it came to Europe, the balance of power had changed: America would ask, not tell.

The Europeans, in turn, told Christopher to go to hell. British and French leaders refused to endanger their peacekeepers by taking steps sure to enrage the Serbs. "We do not interfere in American affairs," declared one top European official. "We hope they will have enough respect not to interfere in ours."

When Christopher reached Bonn, he got a call from Les Aspin telling him he might as well come home. Clinton had changed his mind; he no longer supported "lift and strike." Powell had given him a copy of Robert Kaplan's travelogue, *Balkan Ghosts*, which Clinton read as suggesting that the peoples of the region had been slaughtering each other for five hundred years. Soon Christopher was saying the same thing. In testimony before Congress soon after his return, he emphasized that the killing went both ways. State Department underlings were dispatched to find examples of atrocities committed by Bosnians against Serbs.

Christopher's trip had been "an exchange all right," snarled former Reagan official Richard Perle. "Warren Christopher went to Europe with an American policy and he came back with a European one." But it was more complicated than that. The Europeans had rolled Christopher in large measure because Powell had again rolled Clinton. As usual there was a close association between presidential power and American assertiveness. It was no surprise that at the same moment *Time* magazine dubbed Clinton "The Incredible Shrinking President," many in Europe were talking about the incredible shrinking United States.

Eventually Clinton's fortunes would turn around, and America would resume the upward climb that had begun with Panama and would end with Iraq. But in 1993 that was still several years away. Before the Clinton administration could ascend the ladder of dominance, it would taste even more humiliation first.

If the administration's inability to end the war in Bosnia was embarrassing, equally embarrassing was its inability to win the war three thousand miles to the south, in Somalia. In Bush's final months, in a move that struck some observers as exceedingly noble and others as exceedingly cynical, America had sent thirty thousand troops to Somalia, a cold war pawn turned failed state whose people were starving because warlords prevented the distribution of food. Throughout 1992, top Bush officials—and particularly Powell—had opposed military intervention, before quickly reversing themselves after Clinton's election in November. In trying to explain the about-face, some observers cited the mounting horror of the situation, captured on CNN: the skeletal children too weak to swat away flies. Cynics, who included some inside the Bush administration itself, offered a harsher explanation: Bush and Powell were going to war in East Africa so Clinton wouldn't be able to in the Balkans. In Somalia, admitted Scowcroft, it was "a lot cheaper" than in Bosnia "to demonstrate that we had a heart."

American troops were supposed to be gone by the time Bush left office, but the deployment lingered on in a fit of absentmindedness. In the early Clinton administration, where foreign policy was often an afterthought, Africa was less than an afterthought, and anything closely associated with George Bush was suspect, Somalia received a fraction of the high-level attention devoted to the controversy at the White House travel office. No top Clinton official even visited Somalia. Aspin had wanted to go, but his less than robust constitution was so unsettled by the required inoculations that he canceled his trip.

But someone was paying attention: Boutros Boutros-Ghali, the imperious secretary-general of the United Nations. As an Egyptian, he had strong feelings about the hellhole nearby, and one of those feelings was intense hostility toward its strongest warlord, Mohammed Farrah Aidid. Over the course of 1993, Boutros-Ghali stood at the intersection of two dangerous trends. Militarily the peacekeeping force in Somalia was growing weaker as third-world troops gradually replaced their more intimidating U.S. counterparts. But politically the operation was growing more ambitious as peacekeepers expanded their mandate from merely safeguarding the distribution of food to building a functional Somali government, which required sidelining Aidid.

One day in June, Pakistani peacekeepers announced that they were planning to confiscate some of Aidid's weapons. Aidid's men responded by killing twenty-four of them and mutilating their bodies. The American general in charge of UN forces then upped the ante by putting a bounty on Aidid's head.

By now, Powell wanted U.S. forces removed. So did Aspin, but he was shouting into a void. (During his yearlong tenure as defense secretary, he had a grand total of two meetings with the president.) Powell might have gotten Clinton's attention, but he was weeks away from retirement and had largely checked out. So the White House sleepwalked toward a showdown with a warlord it knew little about in a war it thought it could ignore.

On October 3, U.S. Special Forces tried to capture two of Aidid's top lieutenants at a hotel in downtown Mogadishu. Suddenly one of their Black Hawk helicopters was shot down, leaving wounded American soldiers trapped in the middle of a Somali slum. Aidid's men burned tires to alert their comrades, and throngs of AK-47-wielding fighters descended on the besieged Americans. By day's end, eighteen Americans—along with roughly one thousand Somalis—were dead. CNN showed a Somali mob dragging one U.S. soldier's disfigured corpse through the streets.

Holbrooke called it "Vietmalia": Vietnam plus Somalia. Powell had retired, but the Powell Doctrine—that America should go to war in only the most extreme and auspicious of situations (and then with overwhelming force)—was now more alive than ever. For one thing, after Mogadishu, presidential power over foreign policy declined even further. Four days after the massacre, West Virginia Senator Robert Byrd, chairman of the powerful appropriations committee, literally screamed at Clinton that unless he withdrew U.S. troops immediately, Congress would cut off funding. A few months later, they were gone. When Clinton awarded a posthumous Medal of Honor to one of the soldiers killed in Somalia, the man's father told him he was unfit to be commander in chief.

If Bosnia had proved America weak in Europe, suddenly it looked weak almost everywhere. In Haiti, the UN had brokered a deal with the junta that overthrew elected president Jean-Bertrand Aristide. Two hundred U.S. troops were to begin training the Haitian military as part of a plan to return the island to democracy. But when the U.S.S. *Harlan County* steamed into Port-au-Prince a week after the killings in Mogadishu, it encountered a mob on the dock chanting "Somalia! Somalia!" For a day the ship idled offshore as Clinton officials tried to decide what to do. Then it turned around and headed back home. Critics called it Somalia Two.

Rock bottom came the following year in the tiny central African nation of Rwanda. In January 1994, a Canadian major general named Roméo Dallaire, who was heading the small UN peacekeeping mission there, received chilling news. An informant told him that Hutu militiamen, trained by Rwanda's overwhelmingly Hutu military, were planning to exterminate the country's Tutsi minority. Everything was in place. They were simply awaiting the signal to begin.

Dallaire faxed UN headquarters in New York. The informant was willing to show him where the militias stashed their weapons, and Dallaire wanted permission to seize them. His UN superiors refused. He was expanding the scope of his mission. It sounded like Somalia. The Clinton administration would never go along.

Three months later, a plane carrying Rwanda's president was shot down: It was the signal to begin. Teams of Hutu extremists began hunting down Tutsis and moderate Hutus village by village, neighborhood by neighborhood, house by house. In the city they used guns; in the countryside, where guns were scarcer, they used machetes, clubs studded with nails, hammers, screwdrivers, even the handlebars of bicycles. At a school where two thousand Tutsis had taken refuge among a few peacekeepers, the Tutsis pleaded with the UN troops to shoot them rather than leave. Death by machine gun was preferable to being hacked to death.

Dallaire appealed for reinforcements, but the Pentagon resisted fiercely.

Even though the peacekeeping mission didn't include Americans, Clinton officials feared that the United States might get dragged in if the peacekeepers got in trouble. The best way to ensure that Rwanda didn't become Somalia was to withdraw the UN altogether. For its part, the U.S. Embassy staff was already on its way out. The American ambassador's chief steward, a Tutsi, begged to be taken along. "We're in terrible danger," he pleaded. "Please come and get us." The ambassador refused. The steward's family and more than thirty other Rwandans employed by the United States were subsequently killed.

Spooked by Somalia, the Clinton administration did not hold a single high-level meeting to discuss intervention. Government spokesmen tied themselves in rhetorical knots to avoid calling the killings "genocide," for fear that if they did, America would be obligated to try to stop it. By the time the carnage was over, one hundred days later, eight hundred thousand Rwandans were dead in what one journalist called "the most efficient killing since the atomic bombing of Hiroshima and Nagasaki." Back home in his native Quebec, Dallaire swallowed a fistful of pills, washed them down with a bottle of Scotch, lay down on a park bench, and tried to die.

Four years later, on a tour through Africa, Bill Clinton arrived in Rwanda's capital and apologized. He met survivors and handed the country's new president a plaque commemorating the dead. Back home, critics were scathing: Clinton apologized too often and too well. He had done so twice already on the same trip.

But the apology mattered. It mattered that when asked to name the greatest regret of his presidency, Clinton did not mention Somalia, as Reagan had mentioned Lebanon, an intervention gone bad. He mentioned Rwanda, an intervention never carried out. It mattered because in death the Rwandans came to matter more to American foreign policy than they had in life. Rwanda became the anti-Vietnam: a parable about the horror of *not* going to war. Along with Bosnia, it became the occasion for an historic reconciliation between the American left and American power.

Understanding this reconciliation requires understanding that intellectually, Elliott Abrams and Paul Wolfowitz were not the only people who embraced Fukuyama's vision. In the 1990s, many liberals also came to believe that they were living at history's end. Among liberals, the idea that all people hungered for democracy and human rights was not new. Carter had said as much in the 1970s when he made human rights the rhetorical centerpiece of his foreign policy. But during the cold war, there had been a problem: If all people wanted human rights, why did so few governments—especially in the third world—respect them? Mainstream post-toughness liberals (unlike some on the New Left) did not admire dictators like Fidel Castro and Ho Chi Minh. Yet many liberals suspected that for all their nastiness, Castro and Ho were giving their people things that they genuinely wanted: the dignity that comes from casting off imperial rule, the rough-hewn social justice that comes from breaking up feudal estates and giving long-suffering peasants some land. Why else had the Vietnamese been more willing to fight for Hanoi than for Saigon and had the Cubans refused to rise up against Castro, despite the CIA's best efforts? Most liberals in the '70s still considered political freedom a universal desire. But many accepted that, in some places at least, economic development and economic justice might have to come first. In a nod to third-world autocracies, Carter even defined human rights as comprising not just freedom of speech and the secret ballot but also the right to food, shelter, health care, and education. "A lot of third world countries . . . insisted that 'human rights begins at breakfast, and you cannot expect us to worry about frills like civil and political liberties until we can feed our people,'" noted Jessica Tuchman, the human rights expert on Carter's National Security Council, "and there was a lot of sensitivity to that in the State Department."

It was the liberal equivalent of Jeane Kirkpatrick's argument for why America should tolerate anticommunist third-world thugs. Kirkpatrick and other cold war conservatives said dictators like Somoza and Marcos were necessary evils. We should support them because at least they were better than communism. Many post-Vietnam liberals, by contrast, saw communists like Castro, Ho, and the Sandinistas as the necessary evils. We should tolerate them because at least they were better than imperialism. In the long, long term, both liberals and conservatives thought greater freedom was possible. Kirkpatrick believed that over the course of decades, or perhaps centuries, pro-American autocracies could evolve toward democracy. Liberals hoped that after fulfilling their people's basic needs, leftist tyrants would loosen up. But in the near term, both sides accepted that oppression was sometimes inevitable. Demanding immediate change might just make things worse.

The democratic revolutions of the 1980s—culminating in the euphoria of 1989—changed that. They not only shattered the right's cultural pessimism; they shattered the left's cultural relativism as well. From Asia to Africa to Latin America to Eastern Europe, dozens of poor countries embraced democracy, giving the lie to the claim that freedom must wait until everyone ate a good breakfast. What's more, it became painfully obvious that left-wing tyrannies hadn't been providing good breakfasts anyway; in former communist bastions like China and Vietnam, ex-revolutionaries began scrambling to lure investment from the capitalist West.

The impact on liberals was profound. The more human rights spread in the poor world, the easier it was to unequivocally condemn those regimes that still denied them. If South Africa was embracing liberal democracy, what was Rwanda's excuse? If democracy was breaking out in Budapest and Bucharest, what justification could there be for Belgrade's crimes? Communism's collapse also meant that where dictatorships remained, they often lost their leftist coloration. With the end of the cold war it became easy to see that Serbia's murderous leader, Slobodan Milosevic, was less a Marxist than a racist hypernationalist. The Bosnian genocide, many liberals insisted, was being perpetrated not by the far left, but by the far right: by "fascists." Among liberals, fighting European fascism—as George Orwell had done during the Spanish Civil War—had a very good pedigree. And the fact that the Serbs were ravaging Bosnia, which was diverse, secular, and tolerant—a liberal dream—made the cause particularly compelling. "Bosnia isn't Vietnam," declared Todd Gitlin, a former SDS president who shed his pacifism during the Balkan wars, "it's Spain."

Liberals didn't always acknowledge that they were viewing Bosnia and Rwanda through Fukuyama's prism. In fact, some suggested that these spasms of evil proved that he had been too optimistic about history's course. But Fukuyama had not claimed that political evil would disappear with history's end; he had claimed that political evil would no longer be intellectually respectable. Bad leaders in Serbia and Rwanda might still do bad things, but unlike the fascists or communists of old, their rationales would no longer capture the imagination of people across the world. This was a big change. In 1949, in the *Vital Center*, Schlesinger had argued that what made totalitarianism so frightening was its genuine appeal, even to the people fighting it. It had seduced some of the greatest minds in the Western world. It wasn't just an awesome foe militarily; it was an awesome foe intellectually, even spiritually.

The Serbs and Hutus, by contrast, didn't seduce anyone. They were not just militarily puny; they were intellectually puny. They offered no sweeping alternative vision for how to organize society. They had no sympathetic intellectuals at Berkeley and the Sorbonne. Unlike Nazi and Soviet totalitarianism, which had provoked deep philosophical inquiry, the ideologies of Milosevic and the Hutu *genocidaires* were interesting only as anthropology. They represented not a competing vision of humanity's future, but an ugly vestige of its tribal past. When liberals remarked in horror that genocide was occurring in *1994*, they were assuming that 1994 should be better than 1944. They were buying into Fukuyama's belief that history climbs upward, that progress is the normal course of things. And they were doing so because with the cold war's end they had seen so much progress themselves. In this regard, they were, like Fukuyama himself, heirs of Woodrow Wilson and the optimistic early-twentieth-century progressives. And they were unlike the chastened toughness intellectuals who emerged in their wake. Arthur Schlesinger and Reinhold Niebuhr had not assumed that 1944 would naturally be better than 1904. They had not assumed that history only marched one way.

So intellectually, Clinton-era liberals were more confident than their cold war predecessors that human rights were achievable everywhere, soon. And militarily, they were more confident that America could defend those rights at the point of a gun. Comparing the liberal reaction to Bosnia and Rwanda to the liberal reaction to another genocide—two decades earlier—illustrates the point. In 1975, with the Khmer Rouge on the verge of taking power in Cambodia, President Gerald Ford had warned that their victory would spark a "bloodbath," and urged continued military aid to prop up their anticommunist opponents. His warning was prophetic: In their three and a half years in power, the Khmer Rouge would kill or starve two million of Cambodia's seven million people. But in the mid-1970s, Americans were far too exhausted by Vietnam to seriously contemplate military action. And liberals, in particular, found it almost impossible to imagine—after all the destruction the United States had wreaked in Southeast Asia—that further American military action could serve the cause of human rights. "It is argued that we must give military aid because if we do not there will be a bloodbath," noted New York Democratic congresswoman Bella Abzug, but "there is no greater bloodbath than that which is taking place presently and can only take place with our military assistance."

If liberals in the 1970s found it hard to believe that an American military intervention could have humanitarian effects, they found it equally hard to believe that such an intervention could have a humanitarian intent.

Ford's moral pleas, many assumed, were bogus; he just did not want another cold war domino to fall. But by the 1990s, when genocide broke out in Bosnia and Rwanda, the world was no longer a superpower chessboard, and liberals had to squint to see sinister geopolitical motivations behind the call for humanitarian war. To the contrary, what struck many as sinister was America's palpable indifference to atrocities in countries where it had few economic interests. If America could save Kuwait's oil, many liberals suggested, surely it should save Bosnia's people. Leslie Gelb, the young post-toughness intellectual who in 1971 had written that "military force is a singularly inept instrument of foreign policy," was by 1993 a columnist for the *New York Times*. "Diplomacy without force is farce," he thundered in a column on Bosnia; NATO should begin bombing immediately.

By the time Clinton took office, liberals were also shedding another inhibition against American military force: They were losing faith in the United Nations. In the Bush years, with the cold war over, the UN had gotten a new lease on life. Since Washington and Moscow no longer lined up on opposite sides of every third-world squabble, the Security Council finally seemed able to resolve them. It authorized the Gulf War, and in the early 1990s blue-helmeted UN peacekeepers fanned out across the world, including to Somalia, Bosnia, and Rwanda. During the 1992 campaign, Clinton even mused about a standing UN army. Perhaps, many liberals hoped, it would be UN—not U.S.—muscle that saved poor Bosnia from destruction.

But by 1994, the UN dream was dying. In Somalia, even congressional Democrats were furious that Boutros-Ghali had led America by the nose into the Mogadishu massacre. In Rwanda, Dallaire's peacekeeping force had proved impotent. In truth, these disasters were as much America's fault as the UN's: In both cases the organization was largely doing America's bidding. But the UN, and especially its prickly secretary-general, were convenient scapegoats. Even among liberals, the vision of a UN-led order was losing its luster.

Bosnia killed it. The Europeans had run the show there from the beginning, and they had operated through the UN, which in theory was what multilateral-minded American liberals wanted. The problem was that in practice the UN operation was a gruesome farce. As Blue Helmets watched, the Serbs swallowed much of Bosnia. By the fall of 1994, those Bosnian Muslims who were not exiled or dead had mostly clustered in a few "safe havens," supposedly protected by UN troops. Except that they weren't really protected at all. In October the Serbs attacked the "safe haven" of Bihac, shelling it with cluster bombs and napalm. (It was a symbol of how the moral symbolism had flipped. In Vietnam we used napalm. In Bosnia our enemies did.) When NATO warplanes bombed Serb positions in retaliation, the Serbs took UN peacekeepers hostage and threatened to kill them unless NATO stopped. So NATO stopped and the rape of Bihac went on.

By 1995, the UN in Bosnia had become a symbol of moral equivalence between murderers and the murdered, of pitiful weakness in the face of genocide. The post–cold war liberal faith in international institutions had collided with the post–cold war liberal faith in human rights. Eight years later, many liberal hawks would see the Iraq War in similar terms: as a choice between form and substance, process and outcome; between solidarity with a set of procedures and solidarity with flesh-and-blood human beings crying out in muffled agony from across the globe. For a whole generation of liberal intellectuals and policymakers, Bosnia preconditioned the answer. After Bosnia, full-throated multilateralism never regained its allure.

The disillusionment with the UN also reflected disillusionment with Europe. On the Security Council, the Russians protected their fellow Orthodox Christians, the Serbs. The French, who had been Belgrade's allies between World Wars I and II, sympathized with the Serbs as well, especially under French president François Mitterrand. The Germans looked out for their World War II allies, the Croats. Few European governments seemed to care much about the Bosnian Muslims, who they sometimes implied were not fully European because they were not Christian.

For American liberals, especially Jewish-American liberals, who were thickly represented among the Bosnia hawks, it all sounded hideously familiar. "Fifty years ago, there was no room for Jews in Europe. Now there is no room for Muslims in Europe," wrote the *New Republic*'s Leon Wieseltier, whose parents both survived the Holocaust. "There is no such place as 'the West.' There is Europe and there is America and they are distinct chapters in the history of decency." The implication was clear. The world's natural champion of human rights, the true executor of Fukuyama's vision of a postdictatorial world, was not the UN or even "the West"; it was the United States. With their post–cold war confidence that human rights were universal, their post–Gulf War confidence in military force, and their Bosnia-induced disgust with Europe and the UN, many American liberals were turning, without apology, to American power.

In Washington, this intellectual ferment mattered. Bosnia didn't stir passions in Peoria, but Bill Clinton closely followed elite, as well as mass, opinion. (He was so frequently influenced by op-eds that Vice President Gore began writing rebuttals to ones he felt were leading the president astray.) And by 1995 the anger in elite circles over Bosnia was boiling over. The *New York Times*, which some suspected was trying to expiate its guilt for its inadequate coverage of the Holocaust, was relentless. ("What the fuck would they have me do?" Clinton exploded after reading one *Times* column.) Low-level State Department officials resigned in protest. Twenty-seven liberal human rights, religious, and professional groups—many with histories of near pacifism—jointly demanded military action. (In Bosnia, even the Quakers were for war.) At the dedication of the newly opened Holocaust Memorial Museum on the National Mall, famed Holocaust survivor Elie Wiesel demanded that "something, anything, must be done." As Clinton shifted uncomfortably in his seat, the audience broke into applause.

In June 1995, the House of Representatives voted to unilaterally defy the UN arms embargo. It was a declaration of independence: America would no longer practice multilateralism at Bosnia's expense. The Europeans were furious; if America stopped enforcing the embargo, the UN's Bosnia mission would collapse. But it was collapsing anyway. In July, the Serbs went after the "safe haven" of Srebrenica, where forty thousand Bosnian Muslims had taken refuge. First the Serbs captured thirty of the Dutch peacekeepers supposedly keeping Srebrenica safe, dispossessing them of their weapons, vehicles, and berets. Since the UN had to approve any NATO military response, and such a response might get the Dutch hostages killed, that gave the Serbs an open door. While UN peacekeepers watched, the Serbs put Srebrenica's women, children, and elderly on buses, which eventually dumped them at another "safe haven" a few hours away. Along the route, Serb soldiers periodically stopped the buses to select attractive women for roadside rape. Srebrenica's adult men were blindfolded, stripped to their underwear, taken into the countryside, and, by the thousands, shot. For their part, the Dutch peacekeepers left Srebrenica for the Croatian capital of Zagreb, where they celebrated their freedom by drinking and dancing into the night. The Dutch commander informed reporters that "the parties in Bosnia cannot be divided into 'the good guys' and 'the bad guys.'"

After Srebrenica, the UN in Bosnia was effectively done. Congress, led by Senate Majority Leader Robert Dole, who was preparing to run against Clinton in 1996, voted again to lift the arms embargo. Britain and France reminded the White House that since their peacekeepers were members of NATO, the United States would be obligated to help airlift them out. It was the worst of both worlds. America would have to intervene militarily and Bosnia would still die.

For Bill Clinton, Srebrenica culminated a miserable first two years in office. The public had massively repudiated him in the 1994 midterm elections, and he was so overshadowed by the new Republican Congress that he was reduced to reminding reporters in April 1995 that "the president is relevant." His foreign policy team was a shambles. Secretary of Defense Les Aspin had been fired in 1993. Secretary of State Warren Christopher had tried to resign in late 1994. National Security Advisor Anthony Lake had seriously considered doing the same. CIA director James Woolsey had left in early 1995 and would soon endorse Dole's presidential bid. After a trip to Washington, French president Jacques Chirac said the position of leader of the free world was "vacant."

For liberal Bosnia hawks, Clinton was a wretched disappointment. At the very moment that American liberals were reconciling themselves to American power, a liberal president seemed unable to wield it. But Bill Clinton was a bad person to count out, because his entire career was a story of improbable last-second triumphs, of victories snatched from the jaws of defeat. He had lost his first race for reelection as Arkansas' governor before coming to politically dominate the state. Rumors of adultery had almost blown up his presidential candidacy early in the primaries, leading him to dub himself "the comeback kid" when a strong showing in New Hampshire saved his campaign. He had begun the general election in third place, behind Bush and Ross Perot. At 4:30 A.M. on inauguration morning, he was still writing his inaugural address.

If Clinton had a rendezvous with destiny, he usually arrived late. He often seemed unable to focus until his own indecisiveness or irresponsibility had brought events to the brink of crisis. But the worse things got, the better he performed. "Do you know who I am?" he once told Republican House leader Newt Gingrich. "I'm the big rubber clown doll you had as a kid. . . . [T]he harder you hit me, the faster I come back up."

Now, after Srebrenica's fall, it was five minutes to midnight. Congress and the Europeans were forcing Clinton's hand, and Bosnia was even starting to hurt politically. Clinton's political Svengali, Dick Morris, who had previously opposed military intervention, now decided that America's failure in Bosnia was reinforcing the public's perception of his boss as weak. All of a sudden, the famously tentative Clinton administration grew focused, aggressive, even domineering. Lake was dispatched to inform the Europeans that the next time the Serbs did something awful, America would bomb the hell out of them. The contrast with Christopher's 1993 trip was clear: America was no longer asking; it was telling. To make such bombing possible, the Clintonites bluntly informed Boutros-Ghali that NATO would no longer ask his permission to use military force. On Bosnia, the UN was being cut out of the game. Inside the administration, the military was also being shunted aside. Powell's successor as chairman of the Joint Chiefs, John Shalikashvili, was a capable, well-liked soldier with a made-for-Hollywood life story. Born in Poland to refugees from Soviet Georgia, he had come to America at age sixteen and learned English from John Wayne movies. But he was not a national icon able to awe the Clintonites into submission, and unlike Powell, he saw himself as part of the Clinton team. He deferred to them rather than the other way around.

Almost as important to the military's new willingness to use force was Shalikashvili's director of strategic policy and plans, a cerebral, tightly coiled lieutenant general named Wesley Clark. Within the military, Clark wasn't particularly popular. He didn't backslap, and he liked showing people how smart he was. He was the sort of person, one associate remarked, who would finish a three-hour exam in two hours and then tell you how easy it was.

Some of Clark's uniformed colleagues considered him an apostate, a guy who cared more about helping the White House than sticking up for the military. The fact that he, like Clinton, was a Rhodes scholar from Arkansas fueled suspicions that the two men were close. That wasn't true, but the anti-interventionists in the Pentagon had reason to worry nonetheless. The reason was that Clark—unlike Crowe on Panama or Powell on the Gulf or the Balkans—was becoming emotionally invested in Bosnia's fate. On the morning of August 19, 1995, he was driving with a group of U.S. officials toward Sarajevo along a notoriously dangerous mountain road that they had been forced to traverse because Milosevic refused to offer them a safer route through Serb territory. Suddenly one of the American vehicles veered off the cliff and hurtled down the side of the mountain. Clark rushed to where it had landed and found the bodies of two of his colleagues. From then on, observers noted, he spoke about Milosevic

with a particular intensity. Powell's voice had only welled up with emotion when discussing the army grunts in Vietnam or the dead Marines in
Beirut. Clark began speaking that way about the people Milosevic killed.

On August 28, the Serbs did something awful: They shelled a Sarajevo
market, killing thirty-seven Bosnian civilians. NATO responded with a
massive bombing campaign. Planes from the U.S.S. *Theodore Roosevelt*,
the same aircraft carrier whose sailors had ridiculed Clinton two years earlier, blasted Serb positions. The bombing coincided with a ground push by
the Croats, who had evaded the arms embargo with clandestine American
help. Together these attacks forced the Serbs—for the first time since the
fighting began—into a headlong retreat. And with the war suddenly turning against them, they showed a new willingness to discuss peace.

The man charged with discussing peace with them was Richard Holbrooke, who came to symbolize the newly aggressive Clinton style. On
Bosnia, Holbrooke was white-hot. He had traveled there before Clinton
even took office, meeting with concentration camp survivors and staying
in a Sarajevo Holiday Inn whose rooms were stained with blood. He had
also been on that road with Clark when their colleagues plunged to their
death. But other U.S. officials were passionate, too. What distinguished
Holbrooke was his combination of moral passion and comfort with raw
power. He was a steamroller, far more interested in results than in proce-
dural niceties. When you gave Holbrooke a mission, heads butted, china
broke, feathers were often deeply ruffled, but the job got done. In this
sense he was the temperamental opposite of Warren Christopher, whose
1993 trip to Europe had been characterized by excellent form, unfailing
courtesy, and utter failure. Holbrooke embodied the new liberalism that
in 1995 was being born: idealistic about ends, but somewhat brutal about
means, committed not merely to containing evil but to vanquishing it,
and confident that vanquishing evil and extending American dominance
were usually one and the same.

Holbrooke summoned the Balkan leaders to an air force base in Dayton,
Ohio. It was a symbol that America—not Europe—was now in control.
And after twenty days of extremely muscular diplomacy, he brokered a
deal, which U.S. peacekeepers would help enforce. For Clinton, Dayton
was part of a larger political comeback. That winter he bested the Gingrich
Republicans in a standoff over the budget and the following year cruised to
reelection. And for American foreign policy, Dayton was part of a comeback
story as well: After two years of hesitation and failure, American confidence

was again starting to build. A year earlier, the Clinton administration had finally forced out the Haitian junta that humiliated the U.S.S. *Harlan County*. If the Bushies had overthrown Noriega, the Clintonites had now also deposed a dictatorship that posed no military threat to the United States, and installed an elected leader in its place.

If the Gulf War had shown liberals that not every war need be Vietnam, Bosnia underscored the point. In the entire NATO bombing campaign, not a single American died. And for a whole generation of liberal foreign policy makers and commentators, Bosnia taught lessons directly at odds with the lessons of Vietnam: First, that military force could serve moral ends. Second, that the Pentagon's reluctant warriors did not always know best; their prophecies of disaster should be taken with spoonfuls of salt. Third, that the UN should be relegated to the backseat because it lacked the moral clarity and military muscle for humanitarian war. And fourth, that the Europeans were useful allies, but not equal partners, good to have along for the ride but utterly unable to steer the car. The Clintonites were still more multilateral than the emerging post–cold war conservatives. They weren't ideologically hostile to international institutions. But they did want America firmly in command. As Clark exclaimed during Anthony Lake's trip to Europe, after U.S. officials told America's allies that Bosnia policy was about to change, "The big dog barked today." It would bark even more loudly before the Clinton administration was through.

In 1996, the White House helped dump Boutros Boutros-Ghali as head of the UN, making him the first secretary-general ever denied a second term. The message was clear: Secretaries-general shouldn't get too haughty or independent. In managing the post–cold war world, Turtle Bay would be subordinate to Pennsylvania Avenue, not the other way around.

If the age of the UN was now over, so was the age of Europe. In the early 1990s, many on both sides of the Atlantic had predicted that with no Soviet Union to contain, NATO would wither, and with it America's political and military hold over Western Europe. Instead the European Union would assume control, in partnership with the UN and perhaps the reconstituted Organization for Security and Co-operation in Europe. The former Yugoslavia was to be the showcase for this new institutional order.

But NATO saved Bosnia and Bosnia in turn saved NATO, extending it—and American power—deep into the former Soviet bloc. Holbrooke, whose position as assistant secretary of state for European affairs put him in charge of NATO expansion, steamrolled a skeptical Pentagon, and America steamrolled Russia, which was desperately opposed. Even America's NATO allies, some of whom opposed expanding the alliance, and some of whom wanted to admit a different slate of members, were shunted aside. "Washington was riding roughshod over its allies," lamented Germany's representative to NATO. "When exactly did the Americans go from leadership to hegemony?" wailed a top French official. The Russians ranted and raved but were plainly impotent. In the end they merely pleaded for NATO to stay off former Soviet soil.

America's appetite, however, grew with the eating. The success of NATO in incorporating countries like Poland, Hungary, and the Czech Republic seemed to illustrate what academics called "democratic peace" theory—the theory that democracies did not war with one another. Fukuyama had argued that all nations were moving toward democracy; now democratic peace theory suggested that the move toward democracy would also be a move toward peace. So NATO did eventually push onto former Soviet soil. There was nothing for Moscow to do but scream.

The Bosnia intervention was to Eastern Europe what the Gulf War had been to the Middle East: the gateway for a massive extension of American power. And Kosovo was to Bosnia what the Korean War had been to Truman's 1947 decision to aid Greece and Turkey: the same principle, stretched further. If the ethic of toughness had grown in the late Truman years because it brought success, the ethic of dominance grew in the late Clinton years for the same basic reason.

Unlike Bosnia, Kosovo was not an independent country. It was a province of Serbia whose overwhelmingly Albanian Muslim population had enjoyed autonomy until Milosevic withdrew it as part of his effort to Serbianize everything. As the regime in Belgrade grew more brutal, a Kosovar guerrilla movement emerged, and its presence just made the Serb authorities crack down harder. Finally, in January 1999, the Serbs executed forty-five Kosovars—men, women, and children—and left their bodies in the snow.

In Bosnia, it had taken years of this sort of thing, and worse, to overcome the Clinton administration's inhibition about military force. But the Clintonites were not inhibited anymore. For one thing, a different foreign policy team was now in charge. Wesley Clark had become Supreme Allied Commander in Europe, and while others in the Defense Department still feared military action, his position as NATO's top military man meant that the Pentagon's efforts at obstructing military action were virtually doomed from the start.

Running things on the civilian side was Clinton's second-term secretary of state, Madeleine Albright, who in her own way embodied the administration's newly muscular style as much as Holbrooke did. If Holbrooke's comfort with American power was a function of personality, Albright's was largely a function of genealogy. Her father had been a high-ranking Czech diplomat who fled the Nazis and returned home after World War II only to flee again when Moscow took Czechoslovakia in its grip. Czechoslovakia's betrayal at Munich, Albright often noted, was her formative foreign policy experience, even though it happened when she was only one year old. She experienced it through her parents, and as frequently happens in such circumstances, personal and political history fused, giving the latter a special intensity. Albright's relationship to Vietnam, by contrast, was unusually distant. She had spent the Vietnam years as a wife, mother, and student, not a frustrated young diplomat like Anthony Lake or an earnest young protester like Clinton or Sandy Berger. Because her life had been touched by the crimes of America's enemies but not by America's own, and because America was her refuge, not merely her home, she combined a deep idealism about the United States with a very personal hostility to its dictatorial foes. And in combination, that made her the top Clinton official with the fewest moral or practical qualms about American force.

With Albright in charge, America no longer negotiated with Milosevic; it gave him orders at gunpoint. In February 1999, at a château in Rambouillet, France, he was handed a take-it-or-leave-it offer: Grant Kosovo autonomy, withdraw most Serbian troops from the province, and see them replaced with NATO peacekeepers, or else be bombed. It was an audacious proposal. Kosovo, unlike Bosnia, was not an independent country. America was now dictating internal Serbian affairs, telling Belgrade to remove soldiers from part of its own country.

Milosevic tried stalling for time. At Rambouillet, the Serb delegates did little serious negotiating and much serious drinking, downing almost four hundred bottles of wine in the first five days alone. (This in turn spawned a great deal of singing, as woozy Serbs crooned patriotic tunes late into the night.) The Kosovars, by contrast, stayed sober and took America's deal. As a result, six days after the conference closed, the bombing began.

The air war started slowly and inefficiently, as the Clinton administration's fear of U.S. casualties limited its scope. But gradually, NATO airpower (which was to say, American airpower, since U.S. planes flew 80 percent of the sorties) proved ferociously effective—more effective, even, than during the Gulf War. In 1991, only 9 percent of the bombs dropped on Iraq had been precision-guided; by 1999 in Kosovo, it was over 60 percent. Bombarded by NATO from the air and by Kosovar rebels on the ground, Serbian troops began to desert in mounting numbers. NATO also targeted the houses and businesses of Serbia's notoriously corrupt political elite, making them pay a highly personal price for keeping Kosovo in their clutches. Finally, Milosevic caved and Serbian forces withdrew. The United States and its allies had threatened to send in ground troops, but unlike in the Gulf, they never needed to. As the military historian John Keegan declared, "There are certain dates in the history of warfare that mark real turning points. . . . Now there is a new turning point to fix on the calendar: June 3, 1999, when the capitulation of President Milosevic proved that a war can be won by airpower alone." In thirty-four thousand sorties, only two NATO planes had been shot down. Not a single American died in combat.

If the 1990s were a story of rising American military and ideological dominance, Kosovo was the apex. First, it represented yet another case of war made easy. From Panama to the Gulf to Bosnia and now to Kosovo, almost everything the U.S. military touched had turned to gold. The sole exception was Somalia, and even there the losses had been comparatively small: only eighteen dead, nothing like Vietnam, Korea, or even Beirut. By 1999, Somalia's cautionary tale had been forgotten amid the broader narrative of Clinton-era military success.

Second, Kosovo illustrated the increasing marginalization of the United Nations. Knowing that Russia would veto military action at the Security Council, America had circumvented it, using NATO instead. But once Milosevic was vanquished, the UN retroactively blessed the war and took over Kosovo's political administration. It was a lesson many liberal hawks would remember four years later during the debate over Iraq: Not only could the United States win wars without the help of Turtle Bay, but once it had, the UN would help tidy up the mess.

Third, Kosovo underscored America's growing willingness to violate its enemies' sovereignty. In Panama, in the "no-fly" zones in northern and southern Iraq, in Haiti, and now in Kosovo, America had used force to prevent regimes from committing abuses on their own soil. In each case, U.S. officials had claimed those abuses threatened the United States or its allies, but that was a stretch; the threats were hardly imminent or grave. In reality, as America expanded its power, it was expanding its definition of what constituted a threat, just as it had during the early cold war, when the last hubris bubble grew. The more confident America's leaders became in the hammer of military force, the more closely they looked for nails.

Fourth, presidential power over foreign policy was again on the rise. Even as Clinton battled for his political life against Republican efforts at impeachment over the Monica Lewinsky affair, he waged war in Kosovo and bombed Iraq without congressional authorization. In late 1994, in a move of dubious legality, he also bypassed Congress to bail out Mexico, which was on the verge of financial default. In the 1970s and '80s, the left had championed restraints on presidential power. But now many liberals were beginning to feel about executive authority the way they had felt during the McCarthy years. The congressional Republicans of the late 1990s, like the congressional Republicans of the early 1950s, seemed so primitive and reckless that many liberals hoped the presidency would become a bit more imperial. When it came to foreign policy, they didn't want Congress in the game.

Finally, there was something else that made Kosovo a harbinger of things to come: It was a "preemptive" war. In Kosovo, the United States did not merely respond to ethnic cleansing faster than it had in Bosnia; it responded before most of the ethnic cleansing even began. When NATO started bombing, the Serbs had not yet forced large numbers of Kosovars out of their homeland (although significant numbers had been internally displaced from their villages). In other words, the war was justified less by what Milosevic had already done in Kosovo than by what Americans believed he would do there in the future, judging by his behavior elsewhere. For many human rights–minded liberals, this was what made Kosovo so exciting. It suggested that the United States and its allies could act before the slaughter and expulsions truly began. In Kosovo, the argument for preemption was strong: Given Milosevic's record in Bosnia, and what was already starting to happen on the ground, there was good reason to believe that terrible crimes were about to occur. But in America, where leaders did not usually justify war on preemptive grounds, Kosovo nudged open an intellectual door, a door George W. Bush would fling wide open four years later, when he cited "preemption" to justify his invasion of Iraq.

*   *   *

It was a giddy time, in some ways even more intoxicating than 1989. As the new millennium dawned, America dominated the world not only ideologically and militarily, but now economically as well. Germany and Japan, which a decade earlier had seemed poised to leave America in the dust, were by Clinton's second term growing at less than half the rate of the United States. Between 1992 and 1996, the U.S. stock market doubled in value. By 1997, unemployment was at its lowest rate in twenty-four years and economic growth was rising at a phenomenal 8 percent per year. In Washington, it began raining revenue, and by 1998 the budget deficit that Paul Kennedy had cited as evidence of America's impending economic doom was virtually gone. Entrenched cultural maladies like crime, welfare dependency, and teen pregnancy, which had vexed intellectuals and policymakers for decades, also began to improve dramatically.

The *New York Times* called the erasing of the deficit "the fiscal equivalent of the fall of the Berlin Wall," and the psychological effect was comparable. By 1999, Americans had seen their country repeatedly conquer enemies and problems that supposedly could not be conquered. Saying America couldn't accomplish something looked like a fool's bet. Ideologically, the spread of democracy had erased the cultural pessimism and relativism that flourished on both right and left in the 1970s and '80s. Militarily, America's triumphs in Panama, the Gulf, and the Balkans had largely erased the specter of Vietnam. And economically, these wars hadn't bankrupted the United States. To the contrary, the further America extended its military power, the richer it grew.

Paul Kennedy offered a mea culpa. "Nothing has ever existed like this disparity of power, nothing," he wrote in evident wonder. America was spending more on defense than the next nine nations combined, yet fiscally it wasn't even breaking a sweat. "Being Number One at great cost is one thing," he observed. "Being the world's single superpower on the cheap is astonishing." A decade earlier he had insisted that the laws of geopolitical gravity would force America down. Yet higher and higher it flew.

In myriad ways, this triple dominance—ideological, military, and economic—was self-reinforcing. America's "end of history" confidence that every country could sustain democracy and respect human rights boosted the arguments for military action in Panama, Haiti, Bosnia, and Kosovo. And when military action worked—sparking jubilation among the Panamanians, Haitians (at least initially), Bosnians, and Kosovars—

that ideological confidence only grew. The post–cold war economic boom, like the post–World War II economic boom, created a new abundance of resources, which quieted those who had warned that America couldn't afford far-flung military commitments. And the boom also subtly altered Americans' understanding of what had triumphed at history's end. In 1989, Fukuyama had merely heralded the ideological triumph of free-market democracy. Now, by 1999, with the European and Japanese versions seemingly discredited, many in Washington believed that *American-style* free-market democracy—with its signature antipathy to high taxes and government regulation—was history's final destination. By Bill Clinton's second term, what Americans called "globalization" was in the rest of the world increasingly being called "Americanization." It meant more than the spread of capitalism and democracy. It meant the spread of a particularly American brand of democracy, symbolized by Clinton's political consultants, who introduced U.S.-style campaigning across the globe; and it meant a particularly American brand of capitalism, spread by American management consultants, economics professors, and investment bankers, with help from the U.S.-dominated IMF and World Bank. By 1999, Fukuyama's democratic triumphalism had morphed into American triumphalism. The "end of history" was not just an idea anymore; it was a place.

Some countries had trouble getting with the program, of course. There were political laggards like Milosevic and the Hutu *genocidaires* who threatened to destabilize whole regions. There were also economic laggards in Latin America and East Asia whose late-1990s meltdowns almost took down the whole global economy. For many non-Americans, these economic crises suggested basic problems with the ultracapitalism America was spreading. But in turn-of-the-millennium Washington, the general assumption was that the Latin Americans and Asians had gotten into trouble not because they had followed the American gospel but because they had not followed it devoutly enough. In any case, the debate soon subsided because American-led economic intervention set things right (at least it looked that way from Washington) just as American-led military intervention had in Bosnia and Kosovo. After the East Asian crisis eased, *Time* pictured Secretary of the Treasury Robert Rubin, his deputy Lawrence Summers, and Federal Reserve chairman Alan Greenspan above the words "The Committee to Save the World."

During the cold war, many Americans had seen the world as defined by two huge, opposing magnets: America and the Soviet Union, democracy and communism, good and evil. Each was a powerful attractive force, pulling nations and even individuals its way. No one could be sure which way the smaller magnets would go.

The cold war's end, however, had eliminated the evil magnet, and with it the grand historical uncertainty. Americans no longer related quite as easily to the movie *Star Wars,* with its story of two opposing armies: the light and dark sides of the force. To read the most influential 1990s commentators—columnist Thomas Friedman, for instance, or the *Economist* magazine—was to see the world not as dueling magnets, but as an escalator, with democratic capitalism at the top. Some nations were further along than others, of course. The journey was occasionally bumpy, and some nations were even trying to climb down. But the futility of the effort would force them to eventually turn around and begin trudging upward, in the same direction as everyone else.

The 1990s had shown that America—and not any rival democracy or international body—stood on the escalator's highest step. And they had shown that America could reach down and corral countries going in the wrong direction before they did too much damage to themselves and others. America did not have to wait for history's escalator to gradually move nations in the right direction; it could speed up the process, if need be by force.

This was not Fukuyama-ism; it was ultra-Fukuyama-ism. The escalator metaphor fit his theory, but he had never said that America could easily crank up the gears. To him, the 1990s hadn't proved that military intervention was easy; they had proved that it was hard. After all, to secure its military victories, the United States had been forced to station troops indefinitely in Bosnia, Kosovo, Haiti, and Iraq, and there was every reason to fear that once they left, things would spin back out of control. When it came to promoting democracy, Fukuyama argued, "there are very sharp limits to American resources and patience." In fact, "given that the United States can build democracy in a poor, agrarian country only through massive intervention over the course of a couple of generations, it is probably better in most cases not to try." The better strategy was simply to trade with authoritarian countries and thus bring them into the world economy, which might gradually produce a middle class that demanded political reform. Countries needed to evolve toward democracy at their own pace, he argued. The world wasn't as malleable as it seemed.

But Fukuyama's anxieties had little resonance in an era overflowing with optimism. "We are fortunate to be alive at this moment in history," declared Bill Clinton in 2000, in his final State of the Union address. "Never before has our nation enjoyed, at once, so much prosperity and social progress with so little internal crisis and so few external threats." A poll taken by the Pew Research Center on the cusp of the millennium found that "Americans are near unanimous in their confidence that life will get better for themselves, their families and the country as a whole." Majorities predicted that the twenty-first century would bring cures for cancer and AIDS, a cleaner environment, and people living in space. A slightly smaller number, 44 percent, predicted Christ's return to earth.

Clinton compared the mood to his youth in the early 1960s, before Vietnam, the last time anything seemed possible. The political writer David Brooks reached back further, comparing young Americans at the twentieth century's end to young Americans at the twentieth century's dawn, before World War I, the last time a generation truly believed in the inevitability of progress and the goodness of man. Neither mentioned the historical corollary: that both times before, it had all ended in tears.

CHAPTER FIFTEEN

FATHERS AND SONS

In 1995, Irving Kristol published his fourth autobiographical essay. (That was nothing: Norman Podhoretz would soon publish his fourth autobiographical *book*.) In general, the essay's tone was jolly, even triumphant. America had won the cold war; American capitalism was booming; and even with a Democrat in the White House, right-wing ideas shaped the Washington debate. "I have much to be cheerful about," declared Kristol, now seventy-five years old. "I deem the neoconservative enterprise to have been a success."

But in the final paragraph, Kristol's tone darkened, and he offered a kind of warning. "I am well aware that the unanticipated consequences of ideas and acts are often very different from what was originally intended," he noted, somewhat cryptically. "That, I would say, is the basic conservative axiom, and it applies to conservatives as well as liberals and radicals."

If Kristol feared any particular unanticipated consequence of neoconservatism's success, he didn't say. Yet even as he wrote, his prophecy was coming true. Neoconservatism's progeny, buoyed by the very success Kristol was celebrating, were fashioning a foreign policy vision sharply at odds with his own. In retrospect, 1995 was a hinge year on the intellectual right, a year in which one generation's voice began to fade and another's began to rise. In the early 1990s it had been mostly the men and women of Kristol's generation who spoke, and what they advocated, for the most part, were limits on American power. In the late '90s their ideological offspring replied, and what they denounced, for the most part, were limits on American power. There was an odd time lag to the discussion. It was as if a group of parents had written their children a letter about the kind of nation America should be after the cold war, and waited for them to grow old enough to respond. And when the children did grow old enough, they read the letter and tore it to shreds.

Of the older generation, the most important foreign policy thinkers were Kristol and Jeane Kirkpatrick, two tough old birds. Intellectually, their defining feature was their hostility to grand efforts to remake the world. At formative stages in their youth, each had bumped up against totalitarianism, and the collisions had left permanent scars. For Kristol, it had occurred at City College in the late 1930s, where he and a few Trotsykist dissenters squared off against hordes of thuggish, brain-dead Stalinoids. For Kirkpatrick, it had come in the early 1950s, when she worked at the State Department interviewing refugees who brought with them harrowing tales from Stalinist Eastern Europe and Maoist China. From these experiences, Kristol and Kirkpatrick drew a similar lesson: Totalitarianism was monstrous because it was unrealistic. Stalin and Mao wanted to create perfect societies; but perfect societies required perfect people, and since people were by nature imperfect, the communists took it upon themselves to perfect them by force, which required crushing the spirit and cracking the skull.

When the New Left came along in the 1960s, Kristol and Kirkpatrick opposed it for the same reason: because they believed it was trying to remake the United States along utopian lines, a project they were sure would end in chaos and blood. They defended America's existing institutions and practices—its universities, cities, families, political system, and cold war foreign policy—not because they believed those institutions and practices were flawless, but because they believed they were pretty darn good considering what human beings were really like. It was no surprise that they both admired Reinhold Niebuhr; their foreign policy realism stemmed from their moral realism. Once, late in life, Kirkpatrick spent an afternoon with a friend and her two small children. Over hot chocolate, the children suggested a game of Sorry! Kirkpatrick agreed, and then trounced them. Asked afterward by their mother why she hadn't let the kids win, Kirkpatrick replied, "I'm merciless when it comes to Sorry! It's the real world."

Like other ultra-realists, Kirkpatrick and Kristol often read Niebuhr with one eye closed, ignoring his warning that if communism was messianic, self-righteous anticommunism could be messianic, too. But in their minds, global containment was not messianic; it was defensive. The communists were the utopians, trying to forcibly remake the world in their image. America was simply holding the line. "Global political instability," Kirkpatrick insisted, "stems from the fact that the Soviet Union is frankly, proudly, a revolutionary power." The United States, by contrast, employs a "minimal use of force for limited, defensive purposes."

Whether or not this defensive, anti-utopian interpretation of the cold war was correct, it proved very important when the cold war ended. When the Soviet empire crashed, Kirkpatrick and Kristol let out a sigh of relief and declared that it was time for America to become, in Kirkpatrick's words, "a normal country in a normal time." The cold war, they argued, had been necessary, and even heroic, but also costly and unnatural. The presence of a revolutionary global superpower had forced the United States to become a defensive global superpower, expending blood and treasure in every continent on earth. Now, with its survival no longer threatened, America could climb down from the barricades. In Kirkpatrick's mind, the United States was like a man forced to defend his village against an invading army. The battle had gone on for a long time, and summoned within him great reservoirs of resourcefulness and courage. But it had also forced him to neglect hearth and home. His crops lay untended; his house was in disrepair; he barely knew his family anymore. Now the marauders had laid down their arms. Waking up the next morning, he should not strap on his armor and go looking for new armies to slay.

Kristol and Kirkpatrick saw no reason whatsoever for America to try to dominate the post–cold war world. First, they didn't think America had the money. They had been willing to pay any price to contain Soviet power, but like Paul Kennedy, they believed the price had been high. When it came to foreign policy, Kirkpatrick argued, there was no longer any need for "expansive, expensive, global purposes," which was a good thing, because the United States lacked "boundless resources" and could not "continue to live beyond our means."

If America lacked the money for world dominance, it also lacked the will. Kirkpatrick and Kristol were convinced that if American elites tried to keep sending the nation's cash—not to mention its sons and daughters—around the world in the absence of a mortal threat, the America public would rise up. They urged disbanding NATO, withdrawing most U.S. troops from Europe (and perhaps Asia as well), cutting the defense budget, and preparing for a multipolar world. "There are theorists who would happily burden us with the mission of monitoring and maintaining a 'balance of power' among other nations, large and small, in Europe, the

Middle East, Asia, etc.," wrote Kristol. "We are just not going to be that kind of imperial power. . . . [T]he American people violently reject any such scenario."

Thirdly, if America lacked the money and will for global dominance, it also lacked the wisdom. Kirkpatrick and Kristol did not want the United States to try to convert the world to its political ideology; that had been the Kremlin's ugly game. The democratic transformations of the 1980s had convinced younger conservatives like Elliott Abrams and Paul Wolfowitz that people around the world wanted democracy, and that America could help them get it. But Kirkpatrick and Kristol were both in their sixties when the cold war ended, and the new optimism about global democracy did not penetrate their hardened intellectual shells. They weren't at all sure that the desire for freedom was universal, and even if it was, they were adamant that trying to turn that desire into reality was beyond America's capacity. In 1990, at a conference of former cold warriors, Kirkpatrick congratulated the Bush administration for standing by while Moscow tried to crush Lithuania's fledgling democracy. "Americans do not know at this stage what is best for the Soviet people," she explained. "Any notion that the United States can manage the changes in that huge, multinational, developing society is grandiose. It is precisely the kind of thinking about foreign policy which Americans need to unlearn." On a later panel, Kristol denounced efforts to promote democracy in Ukraine. "The function of the United States is not to spread democracy all over the world," he insisted, adding that historically such efforts had mostly failed. When someone suggested that the Reagan era offered grounds for optimism, Kristol called the supposed "world-wide sweep of democracy" a mirage. "It is just too good to be true," he insisted. "The world is not like that. It is not going to happen. Something will screw it up."

When Saddam Hussein invaded Kuwait, Kirkpatrick's first instinct was that it was not America's problem. Like George Kennan at the dawn of the cold war, before containment was inflated by success, she defined America's interests narrowly because she had grave reservations about its resources. It was unfair, she argued, for the world to expect America to fight Saddam when "it is believed almost everywhere outside the US that the US is a world leader in serious economic decline."

For a while, she urged giving sanctions time to work. By November, when they still had not, and President Bush had staked his credibility on restoring Kuwait, she reconciled herself to war. But she still opposed sending U.S. ground troops, arguing that America should leave the land fighting to Arab nations and merely bomb from the air. If Saddam fell, she suggested, perhaps the Arabs could manage Iraq's reconstruction. The implication was clear: If America took the lead in fighting Saddam, it would find the price too high, and if America tried to dominate the Middle East, it would be drinking from a poisoned cup.

Kristol was less ambivalent: America needed Gulf oil, and so he supported an American war. But once the United States pushed Saddam out of Kuwait, he heaped scorn on those who advocated democratic regime change in Iraq itself. Unlike Wolfowitz, who believed the Shia and Kurdish rebels were fighting for freedom, Kristol saw them the way George H. W. Bush and Colin Powell did: as secessionists and fanatics fighting for nothing nobler than the chance to visit upon Iraq's Sunni Arabs what the Sunni Arabs had visited upon them. Kristol's view of the Arab world was suffused with old-fashioned conservative pessimism about the capacity of people in exotic lands to build democracy. Fukuyama-style universalists, he declared, find "it close to impossible to understand that the world is populated by other people who are really different from us."

When Bill Clinton took office, Kirkpatrick and Kristol took a generally dim view of his humanitarian wars. "Military intervention in the internal affairs of another state is undesirable," wrote Kirkpatrick in 1995, in denouncing the U.S. occupation of Haiti. "Overturning another government by force or threat of force is still more objectionable—except if there are truly vital American interests and lives at stake." On the former Yugoslavia, Kristol made skeptical noises about military intervention early on, and then ignored the subject. Kirkpatrick supported lifting the arms embargo and even bombing the Serbs, but remained extremely wary of putting U.S. peacekeepers on the ground.

Neither Kirkpatrick nor Kristol was an isolationist. But they were comfortable with a world where other powers—be they European, Asian, or Arab—supervised their corners of the globe, so long as those powers did not pursue a global, revolutionary mission like the Soviet Union's. And they had little interest in America telling other countries how to govern themselves, especially at gunpoint. Their belief that efforts to make society conform to an abstract ideal usually failed, whether in Mao's China or John Lindsay's New York, infused them with the same unsentimental skepticism about America's ability to reshape the world that Colin Powell had learned on his tours in Vietnam. Kirkpatrick and Kristol saw neither

American interests nor American ideals as universal. They believed that the United States would be better off, most of the time, if it let other nations be.

In the early 1990s, Kirkpatrick's and Kristol's foreign policy vision held sway on the intellectual right. It was shared not only by older neoconservatives like Nathan Glazer, Owen Harries, and James Q. Wilson, but by traditional conservatives like *National Review*'s William F. Buckley and the columnist George Will. Patrick Buchanan, then still a respectable figure on the mainstream right, was even more hostile to a global American role. Not every older conservative foreign policy intellectual opposed American dominance, to be sure. Norman Podhoretz, for instance, was more bellicose and moralistic. But in 1996 even Podhoretz admitted that "only a tiny" number of neoconservatives supported "expansive Wilsonian interventionism." A study of right-wing intellectuals published the same year noted their "strengthening commitment to realism," "narrowed view of American vital interests," and "general reluctance to crusade."

"There are no normal times." With those words, written in 1991 and aimed straight at Jeane Kirkpatrick, the younger conservative generation fired its first shot.

The marksman was columnist Charles Krauthammer, an acid-tongued ex-psychiatrist from Montreal, and a man young enough to be Kirkpatrick's son. What Kirkpatrick meant by "normal times" was that since America no longer faced a superpower with messianic ambitions and the muscle to carry them out, it no longer needed to man barricades all over the globe. What Krauthammer meant by "There are no normal times" was that if America climbed down from those barricades the world would go to hell in a handbasket and from out of the chaos new and equally dangerous threats would arise. If Kirkpatrick saw America as a man who could abandon his patrol because invaders no longer threatened his village, Krauthammer warned that once he did, new marauders, seeing that the village was defenseless, would soon begin their attack.

Who were these new marauders? Krauthammer gave two answers. Answer number one was what he called "weapons states": small, tyrannical third-world countries like Iraq and North Korea that through the miracle of high technology could build chemical, biological, or nuclear weapons and thus threaten America's allies or America itself. Answer number two was essentially: "Who knows and it doesn't really matter." By patrolling the world, Krauthammer argued, America preserved global stability. If it stopped, the result would be global chaos, and from that chaos all manner of currently imperceptible dangers would emerge. You didn't have to know who would attack the village, only that if you stopped guarding it sooner or later someone would.

At first glance, Krauthammer was merely proposing that America keep doing what it had done during the cold war. Kirkpatrick wanted to come home; he wanted to stay on patrol. But in truth he was arguing for something far more aggressive than that. During the cold war, after all, America had tolerated "weapons states": not small ones with one or two warheads, but giant ones like the Soviet Union and China with hundreds or thousands of them. We had stood by as hostile powers developed huge nuclear arsenals, and we had responded with a policy of deterrence: warning that if they used their nukes, we would, too. Had Krauthammer simply wanted America to maintain the same level of security it enjoyed during the cold war, he would have proposed that the United States do the same thing now: deter any new "weapons state" with the threat of nuclear retaliation.

But Krauthammer was not proposing that. America, he argued, must "confront" and "if necessary, disarm" the weapons states. In other words, it must use force—or the threat of force—to make sure the arsenals were never built (and for Krauthammer, that meant not only nuclear arsenals, but even chemical and biological ones). This required doing a lot more than simply standing outside the village deterring invading armies. It required going around to neighboring villages looking for armies about to form, and attacking (or threatening to attack) them before they did. What Krauthammer was proposing was a doctrine of preventive (or what George W. Bush would wrongly call "preemptive") war. And since he wasn't merely worried about "weapons states" but also about the more amorphous threat of global instability—which could break out in any strategically significant region not dominated by the United States—Krauthammer was essentially suggesting that America encamp outside any significant village, even if there was no evidence that it was raising an army, just to make sure no one got any big ideas. During the cold war, America had patrolled large swaths of the globe, but it had still acknowledged that there were big, important regions where Moscow and its allies held sway. For the most part, America had been content to live in a bipolar world. Now, Krauthammer was arguing, America must dominate every important region on earth. "The alternative to unipolarity," he declared, "is chaos."

Krauthammer's argument gained force as a result of the Gulf War. As usual in the development of hubris bubbles, it was only once things that formerly looked hard—like liberating Kuwait—had been made to look easy that people set their sights higher. Had America proved militarily unable to keep Saddam from gobbling his neighbor, few would have taken seriously Krauthammer's proposal for a new war, inside Iraq itself, to rid Saddam of his unconventional weapons.

The Gulf War boosted Krauthammer's confidence that America had the money and muscle for preventive wars. But he worried about public will. Kristol and Kirkpatrick insisted that average Americans would never permit the United States to play global policeman in the absence of a grave foreign threat, and Krauthammer half agreed. Yes, he conceded, Americans inclined toward isolationism. They could be roused from that isolationism, however, by the chance to accomplish great and moral deeds. "Americans will venture abroad to do right things," he argued, "but only to do right things. Otherwise they would rather stay home." This was not how the older generation saw it. In Kirkpatrick and Kristol's view, Americans had fought the cold war not because they wanted to remake the world but because they considered themselves in danger from communism. Krauthammer turned that argument on its head. Because Americans were naturally isolationist, he suggested, they would bury their heads in the sand *even when they were in danger.* The only way to get them to fight in their own defense was to tell them they were remaking the world.

Krauthammer was making a subtle point. He wasn't arguing that America should fight wars to make the world a better place. He was arguing that unless Americans *believed* their wars were making the world a better place they would not fight in their own self-protection. And for him, self-protection required American hegemony. Dominance was Krauthammer's goal; idealism was the means of convincing Americans to achieve it.

Herein lay the key difference between the younger, dominance-oriented conservatives and the liberal hawks. For the liberal hawks, spreading American power was worthwhile if it spread democracy and human rights. For the younger conservatives, by contrast, spreading human rights and democracy was worthwhile if they spread American power. Both groups were more confident than their ideological forefathers that democratic values and American might generally went hand in hand. But their criteria for war differed. Liberal hawks supported humanitarian interventions in strategically unimportant places like Haiti and Rwanda, though they might cloak those interventions in the language of national interest to win public support. Dominance conservatives supported cold-blooded, national-interest interventions like the Gulf War, though they might cloak them in moral garb to win public support. Such garb was crucial, Krauthammer argued, because otherwise the American people might not play along. It was an argument reminiscent of the pre-Vietnam years, when foreign policy elites often justified presidential duplicity as a way of protecting Americans from the consequences of their isolationism. Now, as memories of Vietnam faded, and another hubris bubble grew, Krauthammer was making that argument again.

If Krauthammer's attack on Kirkpatrick marked the first big sign that younger conservative intellectuals saw the world differently than their elders, the second came from within the Bush administration itself. Every two years, the Pentagon published something called the Defense Planning Guidance (DPG), a statement of America's foreign policy goals aimed at helping military planners decide where to spend money and deploy forces. In 1991, the document was unusually important, since with the cold war over, no one was quite sure what America's new foreign policy goals actually were.

Secretary of Defense Dick Cheney delegated the job to his third in command, Paul Wolfowitz, who delegated it to an aide named I. Lewis "Scooter" Libby, who delegated it to a guy on his staff named Zalmay Khalilzad, who consulted with Richard Perle and a few others and then produced a draft. The goal of American foreign policy, Khalilzad declared, should be to "prevent any hostile power from dominating a region whose resources would, under consolidated control, be sufficient to generate global power." At first blush—like Krauthammer's essay—this sounded rather defensive. After all, preventing a "hostile power" (the Soviet Union) "from dominating a region whose resources would, under consolidated control, be sufficient to generate global power" (Europe, East Asia, or the Middle East) had been U.S. policy in the cold war. And just six months before Khalilzad wrote his draft, the United States had shown how this principle might work in the post–cold war age: Saddam had invaded Kuwait, thus potentially dominating the Persian Gulf, and the United States had rallied a coalition to repel him. The goal seemed familiar; it even seemed affordable, since other countries would presumably share America's interest in preventing one power from dominating a key region of the world and thus help pick up the tab, as they had in the Gulf.

But Khalilzad's draft also said this: "The U.S. may be faced with the question of whether to take military steps to prevent the deployment or use of weapons of mass destruction" (WMD). Preventive war to stop countries from acquiring WMD was not a continuation of U.S. strategy in the cold war; it was a sharp break. And it offered a window into the very peculiar way in which Khalilzad's paper defined foreign "domination" of a key region. For Khalilzad, hostile nations could "dominate" key regions not just by conquering their neighbors but also by taking an "aggressive posture" internally—in other words, by developing weapons of mass destruction. By that standard, preventing Soviet "domination" of Europe in the late 1940s would have required more than merely defending countries like Italy, Britain, and France; it would have required preventive war to make sure Moscow never acquired a nuclear bomb. Khalilzad's argument was a lot like Krauthammer's: Superficially, he was merely outlining a strategy to protect America against foreign threats. But the definition of what constituted a threat had been so radically lowered that protecting America now required crippling—if not toppling—any hostile regime in any important corner of the globe. Why America's allies would help pay for *that*, Khalilzad didn't say.

When the White House learned of the draft through a leak to the *New York Times*, it flipped out. President Bush, who was in his late sixties, two decades older than Cheney, Wolfowitz, Libby, and Khalilzad, was comfortable talking about resistance to aggression, as he had during the Gulf War. But like Kirkpatrick and Kristol, he was wary of proactive moves to extend the frontiers of American power, as evidenced by his reluctance to dismember the Soviet Union, intervene in the former Yugoslavia, or help depose Saddam. He pleaded with reporters not to "put too much emphasis on leaked reports, particularly ones that I haven't seen."

Wolfowitz and Khalilzad feared that their careers were over. But when Cheney finally read the draft, he loved it. It was the clearest sign yet that he and the rest of the Bush administration didn't see the world the same way. Unlike Bush, he had wanted to pry the Soviet Union apart and opposed running the Gulf War through the UN. Unlike Powell, he had opposed any significant post–cold war cuts in defense spending and supported shooting down Saddam's helicopters. Now he wanted America to dominate every important region on earth. He told Libby to rewrite the document in subtler language but preserve its basic argument. Congratulations, he told Khalilzad. "You've discovered a new rationale for our role in the world."

Krauthammer loved the Defense Planning Guidance. So did two former Reagan administration officials, both even younger than him: Robert Kagan and William Kristol. Kagan was son of a conservative Yale classicist. Kristol was the son of Irving Kristol. With their rise to intellectual stardom, the intraconservative debate became an argument not only between generations, but between a father and son, except that increasingly, only one side spoke. In the late 1990s, Irving Kristol fell mostly silent as his son eviscerated his vision of America's role in the world.

If the first two major statements of dominance conservatism—Krauthammer's 1991 essay and the 1992 Defense Planning Guidance—came in the wake of America's victory in the Gulf, Kristol and Kagan's came in 1996, soon after America's triumph in Bosnia. Their basic argument was similar: that Americans should not be lulled into complacency by the absence of foreign threats, because unless the United States dominated every important region of the world, such threats would arise soon enough. But Kristol and Kagan added two important wrinkles. First, they embraced humanitarian war. The Defense Planning Guidance had largely ignored the topic, and Krauthammer was openly hostile. While he wanted to cloak hardheaded, security-based interventions in moral language to win public support, he opposed going to war on humanitarian grounds alone. "In a country with strong isolationist tendencies," he wrote, "you do not squander blood and treasure on teacup wars." Kristol and Kagan, by contrast, incorporated Bosnia into their argument for American dominance. Humanitarian wars, they argued, combated public isolationism by showing Americans the good their power could do. As a result, they primed the pump for the more strategically important battles to come.

The second innovative feature of Kristol and Kagan's call for American dominance was the person in whose name they issued it: Ronald Reagan. If foreign policy hubris consists of thinking that you are merely applying the lessons of the past while actually expanding them as the result of success, Kristol and Kagan's doctrine of "neo-Reaganism" was the great intellectual example of the post–cold war age.

That Kristol and Kagan invoked Reagan is no surprise. Politically, he was their god. For them, as for Elliott Abrams, Paul Wolfowitz, and Francis Fukuyama, the Reagan presidency was a formative life experience.

Kristol, who had joined the Reagan administration at age thirty-two and worked as chief of staff to Secretary of Education William Bennett (before staying into the Bush administration to do the same job for Vice President Dan Quayle), was particularly influenced by Reagan's *political* success. Kristol had no background in foreign affairs. He was an operator and a partisan, mostly interested in shaping media debates and getting Republicans elected. He was, noted one administration colleague, "a man of action, not introspection," a man who "understood it was all a game." The lesson of the Reagan years, he told a friend during the 1990s, was that when Americans focused on foreign policy, Republicans won.

In the Kristol-Kagan duo, Kristol was Art Garfunkel: He played little role in crafting the lyrics but helped to amplify the sound. Kagan was less partisan and more substantive: He was, in fact, the most gifted conservative foreign policy thinker of the post–cold war age.

Kagan's career and Jeane Kirkpatrick's were in important ways mirror-images. He had entered the Reagan administration just as she was leaving, when she was fifty-eight and he was less than half that. Abrams, who had been her protégé, became his mentor. Nicaragua, one of her passions, became his as well. And it was on Nicaragua that Kagan's intellectual assault on Kirkpatrick began.

She had argued famously in "Dictatorships and Double Standards" that Carter's refusal to support Somoza as the necessary evil—his utopian faith that democracy was possible everywhere, right away—had ushered the Sandinistas to power. Kagan disagreed. After leaving Reagan's State Department, in fact, he wrote a nine-hundred-page book making exactly the opposite claim. Carter's real failure, Kagan insisted, had been to not dump Somoza *earlier*, when there was still a nonviolent, non-Marxist alternative. The implication of Kirkpatrick's argument was that Reagan should stick by pro-American tyrants like Ferdinand Marcos and Chile's Augusto Pinochet, which at first he did. The implication of Kagan's argument, by contrast, was that Reagan's great accomplishment had been switching course and helping to push those dictators aside. For Kagan, Reagan's greatness lay in his rejection of Kirkpatrick's moral pessimism—his faith that democracy could indeed take root everywhere soon. Reagan, wrote Kagan, "completely reversed the grim pronouncements of . . . traditional conservative Republicans. It was a message of optimism, rather than of despair; it pointed to opportunities rather than dangers. Reagan did not have to tell Americans that support for brutal dictatorships was a necessarily evil, nor that the struggle with communism was endless."

In 1997 Kagan made his critique of Kirkpatrick even more explicit. He took to the pages of *Commentary*, where in 1979 she had written "Dictatorships and Double Standards," to publish an essay titled "Democracies and Double Standards." Kirkpatrick, he noted, had denied that democracy could emerge "anytime, anywhere, under any circumstances." But she was wrong; the intervening two decades had proved that it could.

For Kagan and Kristol, being a "neo-Reaganite" meant believing that democracy could triumph anytime, anywhere, and that its triumph would likely extend American power. But it also meant believing that America could intervene aggressively to hasten those triumphs. As one critic later put it, Fukuyama was a Marxist: He believed history was moving inexorably toward America's democratic ideals. Kagan and Kristol were Leninists: They believed in speeding history up.

In Kagan and Kristol's rendition, Reagan was a Leninist, too. He had, they wrote, "refused to accept the limits on American power imposed by the domestic political realities that others assumed were fixed." There was some truth to that: Reagan did spend far more on defense, and do far more to support anticommunist rebels, than any predecessor. But in other ways—the most important ways, in fact—he adhered scrupulously to the limits imposed by Vietnam. He never sent U.S. forces to El Salvador, Nicaragua, or Cuba, as many conservative commentators proposed, or even to Panama, despite Abrams's pleas. In Lebanon, his one nation-building campaign, he responded to a terrorist attack by quickly pulling the plug. When it came to direct military action to promote American ideals and power, Reagan wasn't a Leninist at all. He was, in fact, more cautious than either George H. W. Bush or Bill Clinton, both of whom Kagan and Kristol considered wimps.

Kagan also claimed it was Reagan's "ideological confrontation" that had helped to bring down the Soviet empire. And at times, Reagan certainly had been ideologically confrontational. But in his final years in office, it was Reagan's *lack* of ideological confrontation—his willingness to abandon his evil-empire rhetoric and negotiate the deepest arms reductions of the cold war, long before Gorbachev allowed freedom for Eastern Europe—that led prominent conservatives to compare him to Neville Chamberlain. In fact, as Norman Podhoretz noted bitterly at the time, Reagan began moving away from ideological confrontation in 1984, before Gorbachev even took power. And it was that very lack of ideological confrontation—not Reagan's initial defense buildup and rhetorical belligerence—that helped Gorbachev convince Kremlin hard-liners that the U.S.S.R. could abandon its Eastern European safety vest without fearing U.S. aggression.

Reagan had indeed dreamed of regime change in the Soviet Union, of a world where America was ideologically and geopolitically dominant. But his efforts to bring it about had been highly constrained, both by the trauma of Vietnam, which made large-scale military action politically treacherous, and by his own very personal terror of war. Kristol and Kagan, by contrast, were writing after Panama, the Gulf War, and Bosnia had largely buried the Vietnam syndrome and made war less politically frightening. They also lived in a post–cold war world in which America was squaring off not against another superpower, but against a series of second- and third-rate powers, which seemed like easy prey for America's awesome military machine. War just didn't scare them as much. In truth, Kristol and Kagan were less "neo-Reaganites" than "ultra-Reaganites." If Lyndon Johnson had taken George Kennan's ethic of toughness and stretched it to the point of elephantiasis, now Kristol and Kagan were taking Reagan's ethic of dominance and doing the same. They shared Reagan's basic goals; they just abandoned the limits that had constrained his efforts to achieve them.

Like the Camelot intellectuals of the late 1950s and '60s, Kristol and Kagan suffered from generational envy. Too young to have played major roles in the great anticommunist struggle, they were to the cold war what John F. Kennedy's generation had been to World War II: the junior officers, often relegated to watching events from belowdeck. Irving Kristol and Jeane Kirkpatrick were not nostalgic for the cold war. They were a bit like Dwight Eisenhower in the 1950s: They had seen enough of heroism and its awful price. Normalcy struck them as a pretty attractive thing. But the younger conservative intellectuals, who feared normalcy and its cor-responding lack of potential for heroism, were intensely nostalgic. If Kennedy had tried to hang Churchill's paintings in the White House, Kristol and Kagan were constantly lighting votive candles to Reagan. Four years after their call for a "Neo-Reaganite" foreign policy, they edited a book called *Present Dangers*—the title being a play on Podhoretz's 1980 polemic against détente, *Present Danger*, which had influenced Reagan. "Surely we are entitled to hope for another Reagan," wrote Kristol in 1997. "Who will be this era's Ronald Reagan?" asked his new magazine, the *Weekly Standard*, in 1998. The answer was implicit: whoever listened to them.

Camelot intellectuals like Bundy and Rostow had often compared the late 1950s to the late '30s: a moment of drift and peril requiring a heroic new generation to save freedom in what Kennedy called its "hour of maximum danger." Now Kristol and Kagan announced that "the current situation is reminiscent of the mid-1970's." If Eisenhower had been Neville Chamberlain, Bill Clinton was now Jimmy Carter. But if comparing the 1950s to the '30s was bad, comparing the 1990s to the '70s was even worse. Kagan and Kristol were sure the Clinton years were a period of grave and mounting danger, but they couldn't even quite identify what the mounting danger was. Like Krauthammer, they mentioned "weapons states" with WMD, threw in a resurgent Russia and a potentially expansionist China, and suggested that taken together, these threats were as fearsome as the U.S.S.R. (leading Irving Kristol's close friend Owen Harries to reply that "a bad head cold plus a skin rash plus a slipped disk does not add up to a brain tumor"). But Kristol and Kagan's fallback answer was that the danger was no particular enemy at all; it was America's unwillingness to recognize that it was in danger. "The ubiquitous post–Cold War question—where is the threat?—is thus misconceived," Kristol and Kagan explained. "In a world in which peace and American security depend on American power and the will to use it, the main threat the United States faces now is its own weakness."

It was a remarkable act of intellectual jujitsu. What Kristol and Kagan were proposing was brazenly aggressive: They didn't want merely to contain anti-American regimes; they wanted to overthrow them, and not through the slow and subtle infiltration of global capitalism, but through ideological warfare and quite possibly outright force. As they themselves declared, the United States should pursue "benevolent global hegemony." Yet global hegemony, they insisted—echoing Krauthammer and the Defense Planning Guidance—was merely defensive; it was ultimately no different than repelling the marauders at the gates because any region that America did not dominate would ultimately represent a mortal threat. Essentially, their slogan was "Give me dominance or give me death."

It took a special kind of person to gaze out at America in the Clinton years—a country with a defense budget bigger than all its competitors' combined—and see weakness, but Kristol and Kagan were not comparing post–cold war America to any other country at any other time; they were comparing it to what they considered possible. Like the proponents of NSC 68, they assumed America's economic resources were nearly infinite. (Of the fifteen essays in *Present Dangers*, not one discussed how to pay for the massive increase in military spending that the book proposed.) Because they had come of age during a great wave of global democratization, they tended to believe that democracy could arise quickly anywhere on earth. And because they had seen America win wars so easily, the prospect of military confrontation did not fill them with dread. Kristol and Kagan wanted to replicate the triumphs of their elders: to conquer new Soviet Unions and usher in their very own 1989, and they planned to do it the way they imagined that Reagan had: by confrontation and force. They were the very thing that Irving Kristol and Jeane Kirkpatrick feared most: intellectuals with a messianic vision to remake the world and little sense that national or human fallibility might stand in their way. The whole thing was drenched in irony. Irving Kristol had always associated utopianism with the left; it was the monster he had spent his career trying to slay. Now, in the twilight of his life, it had returned, on the right, championed by his own son. Unanticipated consequences indeed.

By decade's end, the dominance conservatives had congealed into a potent network, which brought policymakers and pundits together in common cause. Abrams, who now also called himself a "neo-Reaganite," penned an essay for *Present Dangers*, as did Wolfowitz. Both men also wrote for the *Weekly Standard* and helped Kristol and Kagan establish a think tank, the Project for the New American Century (PNAC), whose other founding associates included Scooter Libby, Zalmay Khalilzad, Donald Rumsfeld, and Dick Cheney. (Just as significant were the people who did not join: Irving Kristol, Jeane Kirkpatrick, Owen Harries, William F. Buckley, George Will, and Colin Powell.) In the late 1990s, PNAC called for overthrowing Slobodan Milosevic, signaling that for the dominance conservatives, going to war to stop genocide in Bosnia and Kosovo was not enough; without democratic regime change there could be no "full and complete victory" in the Balkans. The bar for going to war kept going down, and the bar for achieving victory kept going up. But in 2000, complete victory was achieved: A democracy movement forced Milosevic from power. That left PNAC free to focus on overthrowing Saddam Hussein.

The intellectual infrastructure for global dominance was now in place. With the U.S. economy booming, money no longer appeared to be a problem, and no one in the PNAC circle worried that America lacked the wisdom. The final obstacle was the one that Irving Kristol and Jeane Kirkpatrick had seen as their ultimate safeguard: the American people. Did they have the will?

The dominance conservatives feared the answer was no. The Republican Congress, after all, was rather isolationist. Clinton had managed to take the country to war in the Balkans nonetheless, but his administration seemed to assume that in an era devoid of grave threats, the public would only support wars that got no Americans killed. For the PNAC crowd, this was evidence of a severe cultural rot. Sounding a lot like Arthur Schlesinger in 1958, they declared that Americans had grown soft. "We have become a nation obsessed with risk avoidance and safety," wrote Kristol's *Weekly Standard* colleague David Brooks, who delved most deeply into this alleged malaise. Americans, Brooks claimed, had "replaced high public aspiration with the narrower concerns of private life." It was up to government to call them to a higher purpose, one that required sacrifice and strain. Four decades earlier, Schlesinger had denounced America's "vacation from public responsibilities"; Krauthammer now spoke of a "holiday from history." Schlesinger had lamented the "decline of greatness"; Brooks decried "post-greatness America." To remedy the problem, he and Kristol offered a new vision: "National Greatness Conservatism." The phrase was cribbed from Theodore Roosevelt, the scourge of softness who had inspired JFK's fifty-mile hikes.

When it came to domestic policy, national-greatness conservatism was paper thin. Like Schlesinger, both Kristol and Brooks wanted government to lead heroic missions at home as well as abroad. But unlike Schlesinger, they were conservatives who believed that, domestically at least, government usually screwed things up. They disliked the extreme hostility to government that had taken hold on the post–cold war right, which they believed offered Americans no goal nobler than being left alone, but they struggled to explain what large domestic initiatives they actually trusted government to undertake. Brooks suggested constructing grand public buildings like the Library of Congress, a project unlikely to rouse Americans from their e-trades, or privatizing Social Security, Medicare, and education, which hardly seemed like a grand public mission at all. Krauthammer added that America should journey to Mars. But as the conservative columnist Jonah Goldberg noted, all this "was very hard to understand except as public relations." Brooks and Kristol had diagnosed what they believed was a cultural malady, and a political one, too, since the Gingrich Congress's strident hostility to government was proving unpopular. But their bid to solve it through inspiring government-led projects kept running into a fundamental problem: As conservatives, they still basically believed that when government messed around in the free market, it usually made things worse.

The only domestic character-building project that gained any traction was the campaign to impeach Bill Clinton. For the writers at the *Weekly Standard*, Clinton embodied the cultural degradation that was preventing America from achieving true greatness. "He is supremely self-seeking, in every conceivable sense," wrote one of Kristol's *Standard* colleagues about the president. "His politics, too, are based on self-interest; the stroking and stoking of many small appetites. He does not lead citizens but rapacious consumers, whose sense of grievance he tries to exaggerate. His mode is to focus on the small glitch and call it a crisis, so that he can step in to cushion still further the already soft edges of boomer life."

The impeachment crusade, however, ended in disaster. The American people, who generally liked Clinton's small-bore, low-strain policies and wanted no part of a jihad against the era in which they lived, punished Republicans at the polls in the midterm elections of 1998. Rather than striking a blow against America's cultural decay, the impeachment struggle became, in Kristol's view, further evidence of it. Rather than serving as the great public cause for which the national-greatness conservatives hungered, it simply reaffirmed how desperately Americans needed such a cause, how badly America needed to become something other than what it was: a pampered, indulgent, dull place.

Fukuyama had seen all this coming. "The end of history will be a very sad time," he had written back in 1989. "The willingness to risk one's life for a purely abstract goal, the worldwide ideological struggle that called forth daring, courage, imagination, and idealism will be replaced by economic calculation, the endless solving of technical problems, environmental concerns, and the satisfaction of sophisticated consumer demands." After returning home with the battle won, the warrior might spend some happy time cultivating his crops, fixing his house, loving his wife, and enjoying the soft comforts of home. If he had grown old fighting, and years of hardship had sapped the animal energies of youth, he might even be content to pass his remaining days in this state of repose. But what about his son? Would he not feel the call to struggle, to danger, to greatness?

Perhaps most Americans did not feel this pull. For them, safety and prosperity were not yet boring. They still had roofs to repair and fields to sow. But Kristol and his colleagues found all this tinkering tedious, even degrading; unlike Clinton, they took no joy in helping Americans with such mundane work. And they searched for signs that average Americans might be growing tired of it as well. How else, they asked, to explain the wave of World War II nostalgia that hit America in the late 1990s, culminating with Tom Brokaw's bestseller, *The Greatest Generation*, and the movie *Saving Private Ryan*—the palpable envy with which younger Americans discussed the savage trials of their parents and grandparents? The dominance conservatives were sure that something was stirring, if only it could be harnessed.

After all, there were still places that history had not yet tamed. Those places could be depicted as a menace to everything Americans held dear, perhaps as great a menace as any that had come before. And even better, because the grand ideological contest was over, and history's escalator was inexorably moving America's way, the outcome of this new battle was preordained. It was the best kind of battle, the kind you could not lose. In the summer of 2000, Jonah Goldberg, the young conservative who had derided Brooks's and Kristol's national-greatness vision as mere public relations, hit upon a mission of his own: Invade Africa. "I mean going in—guns blazing if necessary—for truth and justice," he explained. "I am quite serious about this. . . . We should spend billions upon billions doing it. We should put American troops in harm's way. We should not be surprised that Americans will die doing the right thing." It was a curious pronouncement, a kind of eruption of the dominance conservative id. But there was an honesty to it, a recognition that if the problem was deep boredom born of profound success, the solution did not lie in building big libraries and tinkering with Social Security. It would take something grander, and bloodier, too.

CHAPTER SIXTEEN

SMALL BALL

National-greatness conservatives had a presidential candidate in 2000, and his name wasn't George W. Bush. It was John S. McCain, III, a man who had been surfing America's waves of hubris and tragedy since he was in his teens. They were, in a sense, the story of his life.

McCain had graduated from the U.S. Naval Academy in 1958, the last time America seemed invincible. He was a navy flier, restless and brash, a cocky kid at a cocky time. The Camelot gang, reared on the lessons of Munich, was itching for its own showdown with totalitarianism in the jungles of the developing world. And like them, McCain—the son and grandson of admirals who served in World War II—yearned to fight the dark-skinned Hitlers of his age. It never occurred to him that the Hitlers might win. "I believed," he later explained, "that militarily we could prevail in whatever conflict we were involved in."

Vietnam changed that. Five years in a Hanoi prison turned McCain's hair white. By the time he returned home, he could no longer raise his arms above his head, and he no longer believed America could vanquish every foe. "The American people and Congress now appreciate that we are neither omniscient nor omnipotent," he told the *Los Angeles Times*. "If we do become involved in combat, that involvement must be of relatively short duration and must be readily explained to the man in the street in one or two sentences." Colin Powell couldn't have said it better himself.

In the 1980s, McCain won a seat in Congress, where he proved a Reaganite in the true sense of the word: slow on the trigger when it came to military force. Like Powell he was appalled by the 1982 intervention in Lebanon, which he saw as Vietnam writ small. In 1987 he opposed putting U.S. flags on Kuwaiti oil tankers for fear the United

States would be drawn into war with Iran. When Iraq invaded Kuwait three years later, his first instinct was anything but bellicose. "If you get involved in a major ground war in the Saudi desert, I think [public] support will erode significantly," he warned. "Nor should it be supported. We cannot even contemplate, in my view, trading American blood for Iraqi blood."

On the Gulf War, McCain—like Jeane Kirkpatrick—eventually came around. But he still opposed marching to Baghdad, and every other war of the early 1990s. He railed against Clinton's intervention in Haiti and sponsored legislation to pull U.S. troops from Somalia. And "on Bosnia," wrote a hostile Robert Kagan in 1995, "Senator John McCain led the Republican attack, warning that any use of military power there would result in another failure like Vietnam or Lebanon."

McCain would later say that Srebrenica changed him: All those Bosnians herded to the rape chamber and the slaughterhouse while UN peacekeepers skulked away. But in fact, days after the massacre, he still denied that "our security is so gravely threatened in Bosnia that it requires the sacrifice, in great numbers, of our sons and daughters." What changed McCain wasn't Serb barbarism; it was American success. It was only once NATO warplanes had pounded the Serbs into submission—once it became clear that Bosnia wasn't Vietnam—that he grew talons and became a hawk.

By the 2000 presidential primaries he had come full circle, back to the jaunty confidence of his youth. On Kosovo, while many Republicans wrung their hands, McCain called the Clinton administration's air war timid and urged invading on the ground. He hooked up with his former detractors at the *Weekly Standard* and, at their urging, proposed what he called a "21st century . . . Reagan Doctrine." Just as President Reagan had allegedly overthrown the evil empire by force of arms, President McCain would now overthrow its pipsqueak heirs: Iraq, North Korea, and Iran. Once a Reaganite, McCain was now an ultra-Reaganite, which meant he was reenacting the Gipper's foreign policy—minus the dread of war.

Equally exciting, in Kristol, Brooks, and Kagan's eyes, McCain wrapped his policy agenda in the rhetoric of national greatness. He railed against the tranquilizer of affluence, which had left Americans snoozing lazily on the couch, too fat, happy, and selfish to do anything big, like fight another cold war. The *Standard* compared him to Reagan, Teddy Roosevelt, and, surprisingly for a conservative magazine, JFK. It was 1958 all over again, and after decades of Vietnam-induced caution, John Mc-Cain had rediscovered a lost part of himself.

McCain's chief rival for the Republican nomination, Texas governor George W. Bush, didn't have such strong foreign policy views, in large part because foreign policy hadn't directly influenced his life. While McCain had been rethinking American omnipotence between torture sessions at the Hanoi Hilton, Bush had been sipping banana daiquiris and playing midnight water polo at a swanky Houston singles complex called Chateau Dijon. He too had been a brash, restless kid, eager to match the heroism of his greatest-generation dad. But unlike McCain, he had discovered no grand cause into which to channel his swagger. He never found his own World War II, and even when he tried to walk his father's path in smaller ways, he usually stumbled and fell.

The generational contrast was downright painful. George H. W. Bush had excelled at Andover; George W. almost flunked out. H.W., even as a teenager, had been the family star, dutiful and upright; the teenage W., by contrast, had a mortifying habit of walking up to older women at church and asking whether they still enjoyed sex. (He also relished a good fart joke in mixed company, a predilection that continued into the White House.) H.W. was an accomplished athlete, captaining the baseball team at Yale, as his father had before him; W. was a benchwarmer on the junior varsity, and commissioner of an intramural stickball league. H.W. drank in moderation; W. drank to excess, repeatedly driving drunk with his younger siblings in the car. In his early twenties, H.W. married the daughter of a prominent local family; at the same age, W. tried to do the same thing, but his fiancée broke off the engagement. H.W.'s exploits as a fighter pilot in World War II were the stuff of family legend; during Vietnam, W. hoped to be a pilot as well, but never made it off American soil, serving instead with other trust funders in the "champagne unit" of the Texas National Guard. H.W. started an oil company with a Spanish name (Zapata) and made a million by the time he was thirty; W. started one, too (Arbusto), but it went belly-up.

By the 1970s, Bush's life was a shambles. Well into his thirties, he still didn't have a steady job. He was drinking almost every night and had downsized from Chateau Dijon to a dingy apartment above a garage. After one particularly ignominious professional failure, his father confessed himself "disappointed." At a state dinner, W. introduced himself to Britain's Queen Elizabeth II as the family's black sheep.

Eventually, with middle age fast approaching, Bush found himself—which meant, in large part, no longer trying to be his dad. In 1977 he ran for Congress, a decision that violated his father's code: that no Bush should enter politics before ensuring his family's financial security first. But for W. the experience was empowering. Even though he lost, he discovered a talent for good-ol'-boy glad-handing that his father lacked. In 1985 he quit drinking and embraced evangelical Christianity, rejecting his father's high WASP Episcopalianism in favor of a more emotional, populist faith, which helped him forge ties to social conservatives who distrusted his dad. (Bush's newfound religious fervor also helped him turn the tables on his judgmental parents: After his conversion, W. told his mother that unless she became born-again, too, she wouldn't get into heaven.) Then, in 1989, he became managing partner of the Texas Rangers baseball team, finally finding a job in which—at least in his own mind—he excelled.

By the early 1990s, Bush had developed a signature personal and professional style. Instead of biding his time, playing by the rules, and carefully preparing the ground, as his father had done, he would seize the initiative, aim high, and follow his gut. At a time when complex statistical analysis was transforming the way front offices analyzed baseball, Bush disregarded the advice of his scouts and judged players less on the numbers than on their character. Lacking his father's penchant for dispassionate analysis, he fell back on the emotional intelligence that he considered his strength. He sized up players quickly and made bold, unorthodox decisions, like trading Sammy Sosa to the Chicago White Sox for Harold Baines. In the pickup games he organized for team owners, Bush played the same way. "I remember him striking out a lot," one observer remembered. He took "wild swings with lots of muscle. But he was swinging so hard, trying so hard, he didn't take the chance to watch the ball."

Sosa went on to hit 608 home runs after Bush traded him, while Baines only hit another 195, and in Bush's five years as managing partner, the Rangers hugged the bottom of their division. But Bush considered his baseball years a success. He sold his stake in the Rangers for $18 million, finally making himself a man of means. And the job raised his public profile, setting the stage for another political bid.

If Bush was growing more confident in his aggressive, intuitive style, he was also emboldened by seeing his father's more cautious approach fail when it mattered most. In W.'s mind, his dad lost his 1992 reelection bid because he had grown preoccupied with the minutiae of governance and offered the country no rousing agenda for a second term. "The vision thing matters," W. later told Bob Woodward, in explaining what he had learned from his father's defeat. In politics, the younger Bush decided, victory goes to the bold.

In 1994, Bush told his parents that he wanted to run for governor of Texas. "You can't win," barked his mother, worried that he would siphon off contributions from his more studious brother Jeb, who was seeking Florida's top job. But W. turned out to be the best campaigner in the family, and his grand, if vague, campaign slogan, "What Texans Can Dream, Texans Can Do," appealed to the state's outsize image of itself. By 2000, the family screw-up was the front-runner for the Republican nomination for president. His chief competitor was John McCain.

At first glance, the Bush and McCain campaigns were utterly different. McCain attacked corporate interests; Bush seemed at times to be a virtual front man for them. Bush was most comfortable discussing "compassionate conservatism," which mostly meant standardized tests in schools and government money for religious charities, topics in which McCain showed little interest. McCain spoke in considerable detail, and with great passion, about foreign policy. Bush couldn't publicly name the leaders of India or Pakistan, and in private didn't know that Germany was a member of NATO. His foreign policy statements often seemed guided by no coherent principle at all except the desire to be against whatever Bill Clinton was for.

But rhetorically, the Bush campaign actually shared a theme with McCain, and with JFK as well: the need for Americans to find a cause larger than the pursuit of mere material things. Bush's slogan was "Prosperity with a Purpose." As he told the convention that nominated him, "Prosperity can be a tool in our hands used to build and better our country, or it can be a drug in our system dulling our sense of urgency, of empathy, of duty. Our opportunities are too great, our lives too short, to waste this moment. So tonight, we vow to our nation we will seize this moment of American promise. We will use these good times for great goals." This was a slap at Clinton, of course, who to conservatives embodied purposeless prosperity and degraded, hedonistic ease. But it was also, more subtly, a slap at Bush's father, who Bush believed had wasted his presidency pushing paper. If elected president, Bush vowed, he would not play "small ball." (The Gulf War, which had once turned the knuckles of America's leaders white, was now "small ball." It showed how much national confidence had swelled in the decade since the end of the cold war.) Bush's own term in office would be, in his baseball-heavy vocabulary, a "game changer."

So if McCain was speaking the language of national greatness, Bush was, too. It was just harder to know what he thought greatness meant. For a Republican, he talked a lot about poverty—about kids betrayed by bad schools and lives disfigured by drugs. But like the national-greatness intellectuals, he was handicapped in proposing bold solutions by his distrust of government. On foreign policy, his answers mostly left Kristol and Kagan cold. While they believed Clinton had been timid in his use of military force, Bush seemed to score him for being too aggressive. He denounced the intervention in Haiti, talked about removing U.S. peacekeepers from the Balkans, and warned that American forces were being spread too thin.

But Bush also struck another chord, which echoed, if not Kristol and Kagan, then at least Charles Krauthammer. He opposed the Clinton interventions, he explained, because they were small ball, a distraction from the main event. "We must be selective in the use of our military precisely because America has other great responsibilities that cannot be slighted or compromised," he declared in his first foreign policy address of the campaign. What were those great responsibilities, which might require America to unsheathe its sword? Paul Wolfowitz, one of Bush's foreign policy advisers, spoke—as Krauthammer and McCain did—about toppling dictatorships that were developing WMD. But other advisers said nothing of the sort. It was impossible to know exactly what Bush had in mind, or whether he had anything in mind at all. Unlike McCain, he did not have a bold foreign policy agenda. He had something more inchoate: a taste for boldness, a disdain for merely managing problems, a sense that his political and personal calling lay in epic, world-altering things, if only he could figure out what they were. In 2000, it was only an instinct, and on foreign policy it had little intellectual scaffolding. But it went to the core of who George W. Bush was, who he had decided to be. And in ways that only became clear later, it fit the moment, since a whole cadre of younger conservative intellectuals, emboldened by more than a decade of American military, ideological, and economic triumph, were growing bolder, too. The observers who assumed that W. would be a dumber, looser version of his father, content to "amble into history" with a cigar in one hand and an O'Doul's in the other, didn't understand the man—and the moment—at all.

Among those who didn't understand the new president was his secretary of state, Colin Powell. The years out of government had been good to

Powell. His memoir had sold more than a million copies, and polls rated him among the most admired people on earth. Seeking to bask in his reflected glow, both Bush and McCain had promised during the campaign to appoint him to high office. And when Bush eventually did, observers mused that he might become the most powerful secretary of state ever. Watching the press conference where his selection was unveiled, in which Powell answered question after question with effortless command while his new boss fidgeted in the background, the *Washington Post* noted that "Powell seemed to dominate the President-elect." In the *New York Times*, Thomas Friedman added, "I sure hope Colin Powell is always right in his advice to Mr. Bush because he so towered over the President-elect . . . that it was impossible to imagine Mr. Bush ever challenging or over-ruling Mr. Powell on any issue."

Powell had not been a close adviser during the campaign, and he didn't know Bush well. But he knew his father—knew and admired him—and assumed that the apple hadn't fallen far from the tree. Certainly W.'s vague and vaguely realist campaign pronouncements didn't suggest a radical change. Bush had, to be sure, distanced himself from his dad during the campaign, and invoked Reagan's name instead, but that didn't worry Powell. After all, he knew the real Reagan, the guy who wouldn't even invade Panama. All in all, Powell's return to government looked like the capstone to a glittering career. Facing the cameras as he stepped back onto the public stage, he had every reason to believe that when it came to foreign policy, he would be regent to a dutiful boy king.

The Bush administration's first National Security Council meeting began on January 30, 2001, at 3:35 P.M. exactly. A good omen, Powell thought. The new president, it turned out, was as fanatical about punctuality as he was. He was also an early riser and a stickler for proper dress, never entering the Oval Office except in suit and tie. The NSC meeting was carefully choreographed, each official in his or her assigned seat, each speaking in turn, by order of rank. Powell was impressed. It was good to have Republicans back in charge.

But there were other, darker omens, whose significance Powell did not yet fully grasp. To Bush's right sat old frosty himself, Dick Cheney, now vice president. In the half century since the NSC was created, vice presidents had rarely participated. Cheney, however, would attend almost every session, even taking part by videoconference when he was on the road. He also sat in on the president's morning CIA briefing; that was unprecedented, too. (In fact, the CIA first briefed Cheney alone, which allowed him to instruct the briefers to emphasize certain points with the president and downplay others—in other words, to stack the deck.) Cheney had also headed Bush's transition team—again, highly unusual. For top cabinet posts, he had chosen the candidates Bush interviewed, and while Bush made the final decision, he ratified Cheney's recommendations every time. The result was an administration teeming with vice presidential Mini-Me's. National Security Advisor Condoleezza Rice was Bush's close friend. But she was sandwiched between Cheney's own powerful adviser on national security, Scooter Libby, who in the first Bush administration had outranked her, and her deputy, Stephen Hadley, who had worked for Cheney at the Pentagon and whom Powell's close ally Richard Armitage called the vice president's "mole." The new treasury secretary was Paul O'Neill, a longtime Cheney associate. At Defense was Cheney's mentor, Donald Rumsfeld, a renowned bureaucratic shark. (Powell had wanted Pennsylvania governor Tom Ridge, a less formidable figure, but Cheney pushed hard for Rummy, his old boss from the Nixon and Ford administrations, a man Henry Kissinger—Henry Kissinger!—once called the "most ruthless" government official he had ever met.) Rumsfeld's Pentagon deputy was Wolfowitz, another former Cheney subordinate, and a former boss and mentor to both Hadley and Libby. Cheney even stocked Powell's own State Department with moles, including Undersecretary John Bolton, who would spend Bush's first term spreading anti-Powell gossip, and Cheney's own thirty-six-year-old daughter, Elizabeth.

At the January 30 meeting, Powell spoke first, warning that violence between Israel and the Palestinians was spiraling out of control, and urging that the administration push the parties back to the negotiating table. Nothing doing: Cheney and Rumsfeld declared the peace process a waste of time and shifted the conversation to Iraq. Powell was ready: He suggested refining the sanctions that the UN had imposed after the Gulf War, making them more enforceable and less onerous for the Iraqi people. But Rumsfeld said sanctions would never stop Iraq from acquiring WMD and proposed using U.S.-patrolled no-fly zones to bomb Iraq instead. The conversation then turned to possible covert action to topple Saddam, and eventually adjourned. The Bush administration was only days old but it already had two Iraq policies: Powell was working to contain Saddam; Rumsfeld—with Cheney's patronage—was working to overthrow him.

At a "Principals" meeting (an NSC meeting the president does not attend) ten days later, Rumsfeld went further. It would be convenient, he mused, if the Iraqis shot down a plane enforcing the no-fly zone;

that would justify a major U.S. air strike. (It would be a kind of aerial Tonkin Gulf.) Wolfowitz suggested dramatically expanding support for Iraqi opposition forces, perhaps even helping them establish a provisional government in Iraq's Shia south. "Imagine what the region would look like without Saddam and with a regime that's aligned with U.S. interests," Rumsfeld declared. "It would change everything in the region and beyond it."

Change everything: That's what Cheney, Rumsfeld, and Wolfowitz were back in government to do. Even in the first Bush administration, Cheney and Wolfowitz had been the house gamblers, eager to dismember the U.S.S.R. rather than stabilize it, and to help the Shia topple Saddam, even at the risk of Iranian influence or civil war. On their way out the door, they had sketched a blueprint for global dominance, which contemplated preventive war to ensure American hegemony over every important region of the world. And after eight years on the sidelines, watching even the feckless Clintonites win war after war, they were even more confident of what American power could do. If Bush had an instinct for boldness, Cheney and company had at least the beginnings of a plan.

In imagining a transformed Iraq, flipped from adversary to client, which would further America's dominance of the Middle East, and the world, Cheney, Rumsfeld, and Wolfowitz believed they were walking in Reagan's footsteps. Like John McCain, William Kristol, and Robert Kagan, Wolfowitz had spent the late 1990s advocating a new "Reagan doctrine." In Afghanistan, Nicaragua, and Angola, he argued, Reagan's aid to anticommunist rebels had helped bring the Soviet Union to its knees. In the Balkans, Bill Clinton had gone even further, not only aiding the Croats, the Bosnians, and later the Kosovars but also bombing the Serbs on their behalf. (The dominance conservatives had an odd relationship to Clinton: They bashed him mercilessly, but when it came to foreign policy, they stood on his shoulders.) The lesson seemed clear: By offering money, weapons, and air support to the opponents of Saddam, the United States could turn Iraq from a hostile tyranny to a democratic ally.

Cheney and Rumsfeld were less concerned about the democratic part, but for Wolfowitz, Islamic democracy had become something of a passion. After his success in helping to oust Ferdinand Marcos, he had become Reagan's ambassador to Indonesia, where, in his final days in his post, he had given a controversial speech urging Jakarta to embrace democracy. At the time it seemed quixotic, but in 1999, Indonesia—the world's largest Muslim country—held its first free presidential election.

It underscored what Wolfowitz called "a remarkable phenomenon of our time—the triumph of democracy in country after country, including some with no previous history of democratic rule." And this ideological optimism left him susceptible to the charms of one Ahmed Chalabi, the leader of the opposition group the Iraqi National Congress (INC), who spent the 1990s telling Wolfowitz, Cheney, and anyone else who would listen that pro-American democracy could triumph in Iraq, too. Chalabi, the scion of one of Iraq's wealthiest families, an elegant man with a University of Chicago mathematics Ph.D., knew how to market his movement to Americans. Its very name, the INC, evoked Nelson Mandela's ANC, which had brought democracy to South Africa. And Chalabi's most important intellectual ally, a rumpled Brandeis professor named Kanan Makiya, had authored a democratic charter for Iraq that was explicitly modeled on the work of Eastern European dissidents like Vaclav Havel. Wolfowitz was a smart man: He didn't believe Chalabi and Makiya could turn Iraq into the Czech Republic overnight. But from the Philippines to Panama to Indonesia to Serbia, he had spent the previous decade and a half watching Fukuyama's prophecy come true. So maybe Iraq wouldn't be the Czech Republic. Surely, however, it could be Romania, the Eastern European country with the nastiest dictator and the bumpiest path from communism to freedom. Romania, that was Wolfowitz's preferred analogy—not perfection, but one hell of a game changer nonetheless.

What Wolfowitz and his "neo-Reaganite" allies seemed not to understand was that in imagining that military aid and air support for the INC could usher in a pro-American, democratic Iraq, they were going well beyond what Reagan (or Clinton, for that matter) had actually done. As frequently happens during hubris bubbles, they were invoking the successes of the past to justify something far more ambitious in the present. Of the three countries where the "Reagan Doctrine" had been tried—Afghanistan, Nicaragua, and Angola—America's proxies had won military victory in only the first. And in Afghanistan, that had happened largely because the occupying Soviet army withdrew—a precedent with little relevance to Saddam's Iraq, where there was no foreign occupier. (Nor was Afghanistan exactly a victory for democracy, since America's allies, the mujahideen, were mostly theocrats.) The Clinton analogies were equally misleading, since the Bosnia war was aimed at preserving a regime (and its people), not overthrowing one. It was true that in 2000, fifteen months after the Kosovo war, Slobodan Milosevic did fall, but he was toppled by peaceful revolution, not U.S. bombs or Kosovar troops.

Wolfowitz had every reason to be optimistic about the long-term spread of global democracy. Many new countries had indeed embraced representative government in the 1980s and '90s, but with the exception of Panama, none of those democratic revolutions had come primarily through the barrel of an American gun. (And even Panama had held a democratic election seven months before the United States invaded. After deposing Noriega, the United States simply installed the man Panamanians had already elected.) What Wolfowitz was proposing for Iraq—an American-proxy war to produce a pro-American democracy—was not the "Reagan Doctrine." It was the "Reagan Doctrine" on steroids.

It all made Colin Powell extremely queasy. When he heard people talk about forcibly remaking whole regions of the world, his mind sometimes drifted back to his days as a second lieutenant, stationed in West Germany in the late 1950s. He had been drinking with some other up-and-coming young officers, and they were all feeling pretty good about themselves when their captain decided to take them down a notch. "You guys think you're really hot stuff," he bellowed. "You've got all your little troops in bed and you're sitting here patting yourselves on the back about how you solved all the world's problems today. And you're going to go home and lie down with your little wives and sleep well. But you know what? While you're sleeping, the world is going to get fucked up all over again and you're going to have to get up in the morning and start at the beginning." Powell didn't believe in game changers. For him, foreign policy leadership, like military leadership, mostly consisted of holding things together, keeping all hell from breaking loose, making incremental change. He had no problem with small ball. It was small ball—realistic expectations, meticulous attention to detail—that kept your men alive. If Bush's sports vocabulary was baseball, Powell's was football, a rougher game, and more complex, too. "Success doesn't always come with a deep pass," he liked to say. "Sometimes it's a ground game . . . building up slowly but surely, layer by layer."

Powell didn't have much patience for the Iraqi exiles. If Wolfowitz likened them to the Nicaraguan Contras, the Bosnians, and the Afghan mujahideen, Powell viewed them through the lens of South Vietnam, where America's local allies had promised much, delivered little, and gotten a lot of his buddies killed. He viewed Wolfowitz's proposal for a liberated enclave in the Iraqi south the same way William Crowe had viewed Elliott Abrams's proposal for a liberated enclave in Panama: as the kind of militarily ludicrous suggestion you got from people who had spent their twenties in think tanks, not foxholes. For Powell, Iraq was a problem—not an existential crisis, not a golden opportunity, just a problem to be managed. At his confirmation hearings, he had called Iraq a "broken, weak country." In February he declared that containment was working, and that Saddam posed no threat to his neighbors. It was an extraordinary statement considering that Cheney, Wolfowitz, and their intellectual allies were declaring containment dead, and Iraq as grave a danger as the U.S.S.R. But Powell downplayed the threat from Saddam in part because he feared the costs of trying to eliminate it. Less confident that America had the power, wisdom, and money to midwife a democratic, pro-American Iraq, he was content to muddle through. And muddling through was exactly what Cheney, Rumsfeld, and Wolfowitz were determined not to do.

It was same argument that had pitted Crowe against Abrams more than a decade earlier: whether to tolerate adversaries or topple them. And in the Bush administration's early months, Iraq was only one of the places it played itself out. In March, with the South Korean president about to arrive in town, Powell told the *Washington Post* that the Bush administration would continue Clinton's position of negotiating with the Stalinist North. When the *Post* story came out the next day, he got an irate call from Rice: He had misstated the administration's position. The new policy was not negotiation; it was regime change.

Powell was stunned. The cabinet had never met on North Korea; if coercive regime change was the new line, no one had asked his opinion. To Powell's staff, it smelled like a Cheney power play: a policy hatched in secret and then wielded like a club to show that Powell was not as powerful as everyone thought. It was classic Cheney. Whenever he started a new job, he liked to take down some big game just to prove who was in charge. At the Pentagon in the early 1990s, he had taken down generals. Now that Cheney was vice president, the big game was Colin Powell.

It was ironic. During the Clinton years, Powell had moaned about the length of meetings, about Clinton's inability to make up his mind, about the fact that even junior officials demanded their say. Now information was so tightly held, and decisions so briskly made, that even he was not always in the room. After Rice's call, he sheepishly told the press that he had misstated administration policy. "Sometimes," he explained, "you get a little far forward on your skis."

Even the fights Powell won were painful. A few weeks after the Korea humiliation, a U.S. spy plane collided with a Chinese jet off the coast of

China and made an emergency landing on Chinese soil. Before returning the crew, Beijing demanded an apology and a halt to U.S. spying over what it claimed were its territorial waters. For a week Powell negotiated for the crew's release, finally securing it with a classic diplomatic fudge, a letter that in English offered no apology but did so in the Chinese translation. Powell was ecstatic; this was high-level, cool-headed problem-solving, just what he did best. But not everyone was pleased. The White House refused to let him go on television to trumpet his achievement. Kristol and Kagan, whose anti-Chinese rhetoric had been growing louder and louder, called the non-apology apology a "national humiliation" and implied that Cheney and Rumsfeld disapproved. Then, a few weeks later, Bush told ABC News that America would defend Taiwan in the event of a Chinese attack, junking several decades of carefully crafted diplomatic ambiguity aimed at keeping the United States out of a Pacific war. Wolfowitz and PNAC had called for just such a declaration in 1999, and Powell aides even suspected that top Pentagon officials were pushing Taiwan to declare independence, a move that might well prompt Beijing to strike. Was this Bush's effort at a game changer, a bold, impulsive move aimed at punishing the Chinese for their temerity and showing once and for all that in East Asia, the United States was in charge? The White House walked Bush's statement back, but the *Standard* said Bush had known exactly what he was saying, and Cheney seemed to go out of his way to reaffirm Bush's original remark. That same month, the Pentagon offered Taiwan its largest arms sale in a decade.

Had Powell been kneecapped again? It was hard to be sure, since Cheney—a man known for driving around his home state of Wyoming and only speaking to fellow passengers once or twice per hour—said little, at least when Powell was around. Rumsfeld was opaque, too, often speaking in what Powell called the "third-person passive once removed"— with statements like "one would think" and "one might expect." Powell's deputies were equally baffled. When they went to meetings with Hadley, Libby, Wolfowitz, and Undersecretary of Defense for Policy Douglas Feith, they often suspected that their Pentagon and White House colleagues knew things they didn't, that the real decisions were being made someplace else. (Only later would they learn that Cheney's office sometimes used the National Security Agency to spy on State Department officials when they traveled abroad, and that when Powell's aides received e-mails, blind copies were frequently sent to the vice president's staff.)

What was clear, by the summer of 2001, was that unnamed colleagues were gunning for Powell in the press. One article declared that he was "on a short leash . . . nothing more than a silent symbol or a messenger boy." In anonymous quotes, administration officials suggested that he had lost a step. "Powell's megastar wattage looks curiously dimmed, as if someone has turned his light way down," declared a September 10 cover story in *Time*. For Kristol and Kagan, who were unhappy that the Bush administration was not hewing consistently to a "neo-Reaganite line," Powell had become enemy number one.

In ways Powell hadn't fully appreciated, Washington had changed since the end of the cold war. In Congress, moderate Republicans like himself were now rare. The right had also built a much stronger intellectual and journalistic network, symbolized by magazines like the *Standard*, think tanks like PNAC, nationally syndicated radio talk show hosts like Rush Limbaugh, and talking-head TV channels like Fox News, which had not even existed in the Reagan years. They provided Cheney, Rumsfeld, and Wolfowitz an echo chamber that Powell lacked. That Powell was having problems with the media was surprising; he had long been considered a master of the press. But his appeal was personal, not ideological. As a pragmatist, distrustful of theory, his perspective lacked appeal for the conservative journals, think tanks, and chat shows that specialized in big, aggressive, Manichean visions. In his years out of power, American dominance had become, for a whole generation of conservative commentators, an ideology. And Powell, for all his political savvy, was less formidable in this more ideological age.

As the summer of 2001 faded into fall, Powell was growing weary; Bush II was not going as planned. But he still had one ace in the hole: the American people. It was the very thing that drove the national-greatness conservatives mad. In an era without obvious threats, the public would only support wars in which barely any Americans got killed. Inside the Beltway, the dominance conservatives were winning the bureaucratic battle, but in the country at large, Powell's low-key, low-cost, low-blood foreign policy still enjoyed the upper hand.

If Powell took solace from his public standing, he also retained hopes for Bush himself. In his more optimistic moments, he considered the president a work in progress, with potential to grow in the job. Perhaps as Bush grew more confident, Powell thought, he would bring the Cheney-Rumsfeld-Wolfowitz nexus under control. In a sense, Powell was right: Bush hadn't fully settled into the presidency. He had pushed through new policies on education, taxes, stem cells, and funding for religious charities, but his agenda, in his first eight months, was hardly epic or stirring; it just wasn't an epic or stirring time. Even Bush's "successes," chief speechwriter Michael Gerson later acknowledged, "had a quality of randomness, disconnected from larger purposes." At times Bush seemed to be coasting, spending long stretches at his Texas ranch. Powell hoped that when the president found his feet, and his true calling, he would become a wiser, more diligent, more responsible steward of the nation's fortunes, more his father's son. After eight months, he still didn't really know the man at all.

CHAPTER SEVENTEEN

THE OPPORTUNITY

"We have to think of this as an opportunity." It was an odd thing for President Bush to say at 9:30 P.M. on September 11, 2001, with thousands of New Yorkers and Virginians freshly dead beneath mountains of twisted concrete. But he kept saying it. "This is a new world. . . . This is an opportunity," he told a National Security Council meeting two days later. And then again, in an interview: "I will seize the opportunity."

Opportunity for what? When it came to the Muslim world, Bush did not yet know. South Asia and the Middle East were not regions with which he had deep experience, to put it mildly. Before entering the White House, he had been to the Middle East exactly once, in 1998, on a brief swing designed to burnish his foreign policy credentials, and he had never traveled to the subcontinent at all. Asked about the Taliban by *Glamour* magazine in 2000, he had responded with a blank stare, and then, after being reminded that they ruled Afghanistan, chuckled, "I thought you said some band." As late as January 2003, according to visitors to the Oval Office, he seemed unaware that Iraq was divided between Sunni and Shia.

But it was precisely because Bush knew so little about the greater Middle East that he so easily made it a mirror of his views about the United States. The terrorists, he decided, had attacked America because they considered it weak, and they considered it weak because Americans lacked moral purpose. "[T]here is the image of America out there that we are so materialistic, that we're almost hedonistic, that we don't have values, and that when struck, we wouldn't fight back," he explained. Al Qaeda had created "a mythology of American weakness and decadence," explained Gerson. "President Bush set out radically to change this course of events and smash this myth."

It was a remarkable act of projection. To be sure, Osama bin Laden disdained Western culture, and had once called America the "weak horse" as a way to rally his troops. But he hadn't toppled the Twin Towers because he saw America as weak; he had toppled them because America had grown so ferociously strong, because its victories in the cold war and Gulf War had made it the undisputed master of the Middle East, patron to the Arab and Israeli leaders whom bin Laden loathed, and military defiler of Saudi Arabia's sacred soil. If bin Laden had been truly emboldened by weakness, he would have attacked his old foes in Russia on 9/11. They were the ones who had shown real decadence in the 1990s, watching in a vodka-induced stupor as their entire empire collapsed. Instead, bin Laden had attacked the Soviets after they invaded Afghanistan in the early 1980s, when their empire pushed deep into the Muslim world. Now he was attacking America, the world's new behemoth, because it held Muslim lands in its grasp.

It was not bin Laden who had spent the Clinton years bemoaning American weakness and decadence; it was the national-greatness conservatives. And it was not bin Laden who had railed against Clinton's prosperity without purpose, his blend of 1960s depravity and political small ball; it was George W. Bush. Now, after 9/11, Bush turned that critique into a deeply solipsistic explanation for why three thousand Americans lay dead. Clinton, he declared, had invited attack by responding to past provocations with "pinprick" strikes that merely "pounded sand." America had emboldened bin Laden with a policy of what Rumsfeld called "reflexive pullback." Now that would change: Bush's anti-terror policies would be bold, consequential, decisive. As he told members of Congress soon after 9/11, "I'm not going to fire a two-million-dollar missile at a ten-dollar empty tent and hit a camel in the butt."

As a description of Clinton's actual foreign policy, this was sheer fantasy. Clinton's "reflexive pullback" had extended NATO to the borders of the former U.S.S.R., produced a military budget larger than all of America's competitors combined, and left awed foreigners comparing the United States to ancient Rome. Among Clinton's "pinprick" strikes had been Operation Desert Fox in 1998, a four-day campaign conducted without UN approval, in which the United States launched more cruise missiles against Iraq than it had during the entire Gulf War, killed 1,400 Iraqi troops, and—U.S. weapons inspectors later learned—frightened Saddam into aborting his efforts at producing WMD. In reality, American power had expanded mightily in the Clinton years; it just hadn't expanded as much as the national-greatness conservatives wanted, since, engorged by more than a decade of military, ideological, and economic triumph, they believed America could, and should, vanquish every important adversary on earth.

If Bush's interpretation of 9/11 fit the mounting hubris of the time, it also fit his personal history. The parable of big goals replacing small ball, and moral responsibility replacing self-indulgence, was not merely his narrative for the "war on terror"; it was his narrative for his entire life. As a man, he had meandered aimlessly and self-destructively, then found a higher purpose by overcoming the evil of alcoholism. And now, as president, the story was playing itself out again. In his first eight months in office, he had searched in vain for his presidential calling; now evil had arrived and made that calling clear. He told aides that 9/11 gave his fellow baby boomers a chance to invest their lives with deeper meaning, to show the kind of spirit that their parents had shown in World War II. "In our grief and anger," he told a joint session of Congress nine days after the attacks, "we have found our mission and our moment." At the White House, amid the anguish and fear, there was exhilaration. "No more stilted generational summonses, no more made-up 'callings,'" declared White House speechwriter Matthew Scully after Bush's speech. "Here, finally, was the real thing—a real calling with real heroism." That's what Bush meant, at least initially, by 9/11's "opportunity." It was the opportunity to infuse an aimless presidency, and a pampered age, with heroism, purpose, greatness.

As usual, Powell was the odd man out. When the towers fell, he was in Lima for a meeting of the Organization of American States, listening to Peru's president plead for a reduction in U.S. cotton tariffs. As a result, he was not in Washington that evening to help shape Bush's address to the nation, in which the president declared that in the new "war on terror," the enemy would be not merely the terrorists who had struck Virginia and New York, but the regimes that harbored them.

Powell disliked the formulation. At a cabinet meeting the next morning, he tried to narrow the focus, declaring that America should concentrate "first on the organization that acted yesterday." But Cheney pushed back. "To the extent we define our task broadly," he replied, "including those who support terrorism, then we get at states." On September 13, the dispute spilled into public view. "It's not just simply a matter of capturing people and holding them accountable, but removing the sanctuaries, removing the support systems, ending states who sponsor terrorism," Wolfowitz declared at a Pentagon briefing. Once again Powell tried to dial things back. "We're after ending terrorism," he told reporters, "and if there are states and regimes—nations—that support terrorism, we hope to persuade them that it is in their interest to stop doing that. But I think 'ending terrorism' is where I would leave it and let Mr. Wolfowitz speak for himself."

Behind the newfangled war-on-terror lingo, it was the same debate that had been playing itself out since Panama: negotiation or regime change. Wolfowitz wanted to "end" terror-supporting regimes; Powell wanted to persuade them to change their ways and thus restrict the war to Al Qaeda itself. Wolfowitz's strategy would extend the frontiers of American dominance, creating new clients in countries where the United States did not currently hold sway. Powell's was cheaper in both money and blood.

In the days immediately following 9/11, the debate was about two countries: Afghanistan and Iraq. For Cheney, Wolfowitz, and Rumsfeld, it was Iraq that really mattered. For starters, they genuinely believed that Saddam had had a hand in the attack, and they didn't much care that the administration's terrorism experts thought they were nuts. (On the subject of Iraq and terrorism, Wolfowitz was used to people thinking him nuts. He believed Saddam had masterminded the 1993 World Trade Center bombing, and for a time even suspected him of orchestrating the 1995 bombing in Oklahoma City. Wolfowitz-watchers considered it an intellectual eccentricity, like an otherwise rational man who decides carrot juice can cure cancer.) Cheney, Rumsfeld, and Wolfowitz also considered Afghanistan a bad place to fight a war: the mountainous terrain made ground operations difficult, and the country was so primitive that there was barely anything worthwhile to bomb. Finally, and most important, Afghanistan wasn't a game changer. A geopolitical backwater, it lacked strategic position or critical resources (a polite way of saying oil). "If the war [on terror] does not significantly change the world's political map," Rumsfeld wrote in a memo to Bush, "the U.S. will not achieve its aim"— and attacking Afghanistan didn't do that. It was like invading Haiti: a lot of work for little reward.

Powell thought they were in cloud-cuckoo-land. "What the hell, what are these guys thinking about? Can't you get these guys back in the box?" he moaned to Joint Chiefs of Staff chairman Henry Shelton. Cheney, Rumsfeld, and Wolfowitz wanted to go after Iraq because doing so might reshape the Middle East. Powell wanted to go after Afghanistan because that's where the terrorists were. He wanted to fight Al Qaeda; they believed that to effectively fight Al Qaeda, America must expand its hegemony in the Arab world. In his mind, it was a bait and switch.

Initially, Bush split the difference. He too believed Saddam was involved in 9/11, but he knew he couldn't prove it. He loved the idea of doing something big. As Gerson put it, "All of his instincts tended toward a single ambition: a desire to reshape the security environment we found in the world, rather than endlessly responding to escalating dangers and attacks"—and he feared that merely getting bin Laden and his cronies was not big enough. But still, he could see Powell's point. Americans wanted bin Laden's head on a stick; Afghanistan had to be first. Anything else risked a public revolt. "Start with bin Laden, which Americans expect," he told his advisers. "And then if we succeed, we've struck a huge blow and can move forward." It was a shrewd insight into the politics of hubris. If America won easily in Afghanistan, then the public would grow more malleable on Iraq. After overthrowing the Taliban, overthrowing Saddam would no longer seem so hard.

But in the days after 9/11, even overthrowing the Taliban seemed hard. In 2002, once American confidence had swelled further, toppling the regime would be remembered as easy and obvious, a mere appetizer before the Mesopotamian main course. But it didn't look that way at the time. Initially the military wanted no part of Afghan regime change, which it assumed would require tens of thousands of troops. In fact, when Shelton addressed Bush's war cabinet on September 15, two of the three military options he proposed involved no U.S. ground forces at all. Afghanistan was a "graveyard for the interests of great powers," cautioned the *New York Times*, "a general's nightmare and guerrilla commander's fantasy." U.S. allies were even more squeamish. French foreign minister Hubert Védrine warned that bin Laden had set a "diabolical trap" for the United States, luring it into a war it could not win. His German counterpart worried that invading Afghanistan might "create more instability." When administration representatives traveled to Russia, a country that knew something about Afghan wars, a Kremlin official told them, "With regret, I have to say you're really going to get the hell kicked out of you."

Powell feared the Russians were right. Like his old colleagues in the military, and like the CIA, he wanted to blow up some Al Qaeda bases and lure the Taliban from their alliance with bin Laden, not overthrow them. "It is not the goal at the outset to change the regime," he told a Principals meeting on September 23, "but to get the regime to do the right thing." It was a lot like his debate with Wolfowitz at the close of the Gulf War. Once again Powell preferred sticking with the barbaric regime he knew to deposing it and potentially getting mired in civil war.

For the dominance conservatives, Powell was on the wrong side of history yet again. "It is deeply troubling to see the secretary of state be-gin to wobble," wrote Charles Krauthammer in late September. "Eleven years ago, then-President Bush overrode Powell's resistance to fighting Saddam," wrote William Kristol. "Bush was vindicated in doing so. Will the current President Bush follow Powell's lead? Or will Bush lead and de-mand that Powell follow?" Like Wolfowitz, both Krauthammer and Kris-tol would have been happy for America to go after Saddam right away, but they understood Bush's logic. If America didn't successfully overthrow the Taliban first, there would be no war in Iraq.

On Afghanistan, Powell lost. To Bush, merely lobbing missiles meant pounding sand, killing camels, Clinton stuff. "We'll attack with missiles, bombers *and* boots on the ground," he told the NSC on September 17. "Let's hit them hard. We want to signal this is a change from the past."

Ironically, while Bush believed he was breaking with Clinton, he was actually building on Clinton's example. America's Afghan war strategy, devised largely by the CIA, involved massive U.S. airpower, and a small number of U.S. trainers, arming and advising the Northern Alliance so they could defeat the Taliban on the ground. It was a lot like what the Clinton administration had done in the Balkans, where the United States had as-sisted first Croat, and then Kosovar, ground forces while NATO pounded Serb forces from the air. And it was because this "Clinton Doctrine" (or "Reagan Doctrine," in Wolfowitz's terminology, since the Republican bloodlines had to be kept pure) had worked in Bosnia and Kosovo that Bush pushed it further in the Hindu Kush. It was another example of how success begets ambition. Commentators would later say that the 9/11 attacks made overthrowing the Taliban inevitable, that doing less was in-conceivable after the murder of three thousand on American soil. But had Kosovo turned into Vietnam, or even Somalia, doing less would have been highly conceivable. September 11 was crucial to America's decision to in-vade Afghanistan, of course; without the attacks, invading would never have been seriously discussed. But without the self-confidence engendered by a decade of military, economic, and ideological triumph, policymakers might well have chosen the safer and more limited military response that Powell and many in the military and intelligence community preferred. When it came to the Afghan War, September 11 was the match. American self-confidence was the dry tinder that burst into flame.

At first, everything went wrong. Starting in early October, America bombed Afghanistan for days, and then weeks, but the Taliban didn't buckle. Efforts to foment a rebellion among Pashtuns in the Afghan south failed miserably, as a key anti-Taliban exile was captured just hours after crossing with his fighters onto Afghan soil. "Like an unwelcome specter from an unhappy past, the ominous word 'quagmire' has begun to haunt conversations among government officials and students of foreign policy, both here and abroad. Could Afghanistan become another Vietnam?" mused a late October news analysis in the *New York Times*. With the Northern Alliance looking amateurish and ineffectual, and American air strikes having little apparent effect, the military began making plans for the dreaded land invasion—fifty thousand troops, for starters. To Powell, it just made the Vietnam analogy all the more apt: When bombing and local allies failed, the next step was always U.S. troops on the ground. He began to protest, telling a Principals meeting that "I'd rule out the United States going after the Afghans [on the ground], who have been there for 5,000 years." It was one of Powell's verbal tics. When he wanted America to stay out of someone else's war—whether in Lebanon, Bosnia, or now Afghanistan—he claimed the people there had been fighting for hundreds or thousands of years. It was a window into his view of the world, his sense that problems were often incorrigible and that at least in some places, not only was history not ending, but it didn't progress much at all.

It was a crucial moment, those days just before and after Halloween 2001, and it set the stage for much that was to come. The Europeans were skittish; the *New York Times* was talking quagmire; Powell seemed to want to cut America's losses. Bush hated quitters; his natural instinct, when challenged, was to double down. He began privately cursing the media, suggesting that they wanted him to fail. The *Weekly Standard* attacked Powell as defeatist and demanded that Bush send in the army. At one point Bush took Cheney aside and asked him if he still believed they would win. Cheney said he had absolute faith.

Then, suddenly, the tide turned. On November 2, an American air strike destroyed a famed transmission tower that the Soviets had tried for years to hit, but never could. The smart bombs that had worked such wonders in the 1990s were finally working their magic again. Outside the town of Mazar-i-Sharif, a U.S. Special Forces soldier on horseback spotted a clump of Taliban fighters, tapped their coordinates into his laptop, and summoned an armed pilotless drone—the newest gadget in America's high-tech arsenal—along with a B-52 bomber. Nineteen minutes later, the Taliban fighters were dead.

Quickly, the Taliban began to crumble. Between November 9 and

November 12, the Northern Alliance increased its share of the country from 15 percent to 50 percent. On November 13, Afghanistan's capital, Kabul, fell. Kites, an Afghan passion banned by the Taliban as un-Islamic, started to dot the sky. Indian pop music, also formerly banned, soon blared from taxi radios and boom boxes. In Kabul's main stadium, which the Taliban had used to publicly execute those who defied religious law, soccer games broke out. Women began removing their burkas; men shaved their beards. Girls, doomed under the Taliban to a life of illiterate, beastlike submission, began—often for the first time in their lives—to attend school. By Christmas, Afghanistan had sworn in a new leader, Hamid Karzai, not a Northern Alliance warlord, but an apparently decent and modern man with brothers who ran restaurants in Chicago, San Francisco, Baltimore, and Boston, a democrat. His new government did not merely replace tribe with tribe; it represented Afghan's ethnic diversity. Karzai himself was a Pashtun. In January, at Bush's State of the Union address, he sat next to first lady Laura Bush, along with a female member of the new Afghan cabinet.

It was, in a sense, the culmination of all the success America had achieved since 1989. Militarily, the accomplishment was breathtaking, exceeding even what the United States had done in Panama, the Gulf, Bosnia, and Kosovo. From a standing start, America had gone to war in a forbidding, landlocked country half a world away, a legendary graveyard of empire that had brought the mighty U.S.S.R to its knees. It had done so with a ground force of 316 Special Forces troops and 110 CIA agents. And in just over a month, it had brought down the enemy regime at a total cost of $1 billion to $2 billion, less than a single B-2 bomber. In the wake of Afghanistan, suggested an article in the conservative magazine *National Review,* "we face the prospect of being able to dominate the world at the touch of a button—and at virtually no cost in casualties." It was a dream come true.

Ideologically, the victory was just as heady. Even more than Panama or the Balkans, Afghanistan was a place that by history, culture, location, and economics seemed condemned to despotism, yet its people were embracing democracy with tears of joy. Afghan's new leaders rushed to tell Americans the same thing that the Panamanian priest had told them more than a decade before: that their real country was a free, decent, modern place, concealed from the world by a small, monstrous faction. The Taliban were not Afghanistan's true face; they were its mask. And America was not Afghanistan's occupier, bringing alien, unwelcome ideas, but its liberator, allowing it to become, once again, its truest self. A country that only weeks before had been marching backward toward the seventh century now seemed like just another young democracy, hopping onto Fukuyama's escalator, wobbling a bit, but peering up toward the light.

Geopolitically, the war extended American power deep into Central Asia, just as the Gulf War had in the Middle East and Bosnia had in Eastern Europe. By 2002, not only was the U.S. Army deployed across Afghanistan, but American troops and planes were nestled at bases in Kazakhstan, Uzbekistan, Tajikistan, and Kyrgyzstan, in the former U.S.S.R.

For the Bush administration and its intellectual allies, who only days before had been accused of leading the nation into a second Vietnam, vindication was sweet and reinforced one of the central tenets of dominance conservatism, not to mention George W. Bush's life: that glory goes to those with ambition and faith, not the faint of heart. "Critics of the U.S. war in Afghanistan have been wrong about virtually everything," crowed an article in the *Weekly Standard*. At a press conference in late November, Rumsfeld taunted the press, declaring that from the beginning the Afghan War had gone as planned, but that thanks to negative reporting, "It looked like we were in a—all together now—QUAGMIRE!" In the wake of Bush's success, the press grew more quiescent; a *Saturday Night Live* skit showed reporters so intimidated by Rumsfeld that they would not ask any questions at all. In December, *People* named the sixty-nine-year-old defense secretary one of the sexiest men of the year, after which Bush dubbed him "Rumstud."

Politically, Bush looked like quite a stud himself. September 11 had sent his approval rating into the stratosphere, as Americans rallied around the flag, but it was the Afghan victory that kept it there. By the fall of Kabul, according to Gallup, he enjoyed the support of 87 percent of Americans. On September 10, he had been an accidental president, widely considered out of his depth. Now he dominated Washington as few presidents ever had.

As Bush grew stronger, so did the presidency itself. As Gerald Ford's thirty-four-year-old chief of staff in the dying days of Vietnam, Dick Cheney had run a White House under siege, as the hubris of toughness collapsed amid an orgy of congressional and journalistic recrimination. Deeply scarred by the ordeal, Cheney had devoted his career to the restoration of the imperial presidency of the early cold war. Now 9/11 and Afghanistan gave him the power to make that dream come true. In January 2002, with no input from Congress, he steamrolled Powell and convinced

Bush that the combatants the United States had picked up on the Afghan battlefield should not enjoy the protections of the Geneva Convention. In February, at Cheney's insistence, the White House went to federal court to deny the General Accounting Office (GAO) access to records of an energy task force he had led. The Bush administration, wrote the GAO's head, "seeks to work a revolution in separation of powers principles, one that would drastically interfere with Congress's essential power to oversee the activities of the executive branch." Bush and Cheney won the case.

By every measure, it looked like an awesome performance. The war in Afghanistan was not over, to be sure. In a rugged mountainous region called Tora Bora, roughly fifteen hundred Al Qaeda fighters had escaped across the border into Pakistan. They would need to be mopped up. But few worried about that in the thrilling winter of 2001. "The initial reaction to 9/11," Gerson later remembered, "had been successful beyond expectation, and perhaps beyond precedent." And Bush was determined not to rest on his laurels, as his father had done after the Gulf War. There was even greater glory to come. "They won in Afghanistan when everybody said it wouldn't work, and it's got them in a euphoric mood of cockiness," trembled one Powell ally, "and anyone who now preaches any approach of solving problems with diplomacy is scoffed at. They're on a roll."

CHAPTER EIGHTEEN

THE ROMANTIC BULLY

Why did America invade Iraq? The Bush administration's answer went something like this: Before September 11, Saddam Hussein was a problem. He had attacked his neighbors; he was building weapons of mass destruction (or so it seemed); he was defying UN resolutions; he cavorted with terrorists. Then came 9/11, and America's leaders peered into the abyss. If terrorists could kill three thousand Americans with boarding passes and box cutters, imagine what they could do with weapons of mass destruction. The anthrax attacks a month later underscored the danger. So the Bushies asked the obvious question: Who might supply terrorists with WMD? Who had the motive and the means? Saddam!

America invaded, in other words, out of fear. September 11 showed that the terrorist threat was graver than previously understood, and so America took graver steps to confront it. It's not hard to see why Bush officials favored this argument: It made the Iraq War sound defensive. But it ignored something crucial: Fears don't exist in isolation. They tend to rise and fall depending on what people think they can do about them. In Robert Kagan's words, "A man armed only with a knife may decide that a bear prowling the forest is a tolerable danger. . . . The same man armed with a rifle, however, will likely make a different calculation."

Had September 11 happened in 1977, or 1983, or 1989, America would have been more like the man with the knife. Back then, Vietnam still haunted discussions of military force; America's post–cold war triumphs had not yet exorcised the ghost. In the 1970s and '80s, nothing in America's recent experience suggested that America could successfully invade and occupy a large, distant country like Iraq (especially after it had just invaded and occupied another large, distant country: Afghanistan). After all, the United States hadn't even managed to rescue the hostages in Iran in 1980 or protect its Marines in Beirut in 1983. And if America's leaders doubted their capacity to rid the world of Saddam, they would have likely hesitated before describing him as a mortal threat. Why stoke public fear unless you can put it to rest?

Bush, by contrast, stoked public fear relentlessly. In the run-up to war with Iraq, he claimed that while deterrence had worked against the Soviet Union and China, it couldn't work against dictators like Saddam. As an argument about threats, that made little sense: There was little in Saddam's record to suggest that he was less deterrable than Stalin or Mao. But Bush encouraged Americans to *believe* that the threat was greater because he thought he could eliminate it. During the cold war, America had chosen deterrence because preventive war against the Soviet Union or China was too frightening to seriously contemplate. Preventive war against Iraq, however, was not so frightening, especially given America's recent run of military success. Top Bush officials thought America could invade and occupy Iraq with relative ease. And so they depicted Saddam, and his presumed weapons, as an intolerable, undeterrable threat.

The timing of Bush's decision to go to war illustrates the point. In October 2001, when Afghanistan looked like a quagmire, there had been little talk inside the White House about a second war. When Rumsfeld raised the issue at an NSC meeting on October 9, he was rebuffed. But when the Taliban fell in mid-November, the mood quickly changed. Eight days after the capture of Kabul, Bush told Rumsfeld to develop a strategy for toppling Saddam. By February 2002, the U.S. military was shifting forces from Afghanistan to the Persian Gulf. The Iraqi threat had not grown as a result of America's apparent victory over the Taliban, but the Bush administration's self-confidence had.

That confidence came in three parts, the first—and least appreciated—of which was economic. In 2002, as the White House prepared for a second war, the budget surplus of the 1990s turned to deficit. But the dominance conservatives, like the authors of NSC 68, did not fear deficits. After all, Reagan had racked up huge ones fighting the cold war, and although Paul Kennedy and others had predicted catastrophe, those deficits had disappeared by century's end. For the Bush administration's "neo-Reaganites," the moral of the story was that deficits were a small and temporary price to pay for vanquishing America's foes. At a meeting in November 2002, when Secretary of the Treasury Paul O'Neill warned that America was "moving toward a fiscal crisis," Cheney interrupted. "Reagan," he declared, "proved deficits don't matter."

Rather than raising taxes to reduce the deficit and pay for America's two wars, the Bush administration cut them in 2002, and then again in 2003. The tax cuts, it insisted, would largely pay for themselves as the economy boomed, and so would Iraq, whose vast oil wealth would fund its own postwar reconstruction. When White House economic adviser Lawrence Lindsey suggested that toppling Saddam might cost $200 billion, he was reprimanded, then fired. Top administration officials derided his suggestion as ludicrous. After all, toppling the Taliban had cost less than 1 percent of that.

If the Bushies were brimming with economic confidence, their military confidence was equally high. It wasn't just that America had spent the last decade winning wars—since Bosnia, it had been winning wars with barely any U.S. ground troops. From this, Donald Rumsfeld drew a striking conclusion: The era of massive land armies was over. The future of war belonged to the lean, fast, and ultrahigh-tech. In his mind, Afghanistan proved the point. Before the war, General Tommy Franks, head of Central Command, had told him that overthrowing the Taliban would require tens of thousands or even hundreds of thousands of GIs, a hulking force that would take nine months to fully deploy. Instead the CIA had parachuted in one hundred or so agents carrying laptops and suitcases of cash, and, with help from the U.S. Air Force, the Northern Alliance, and a few hundred Special Forces troops on horseback, they had taken down the Taliban in little more than a month.

It was more than the hypercompetitive Rumsfeld could bear. His doctrine had apparently been vindicated, but his CIA rivals were getting the glory. A haughty man in the best of times, he developed an epic disdain for the army's top brass, whom he considered timid, conformist, and flat-out dumb. He felt special contempt for Franks, no one's idea of an intellectual, a man whose favorite movie was *The Nutty Professor*. As a result, in early December, when Franks said invading Iraq would require close to four hundred thousand troops, Rumsfeld shot him down, remarking that "I'm not sure that that much force is needed given what we've learned coming out of Afghanistan." Over the next fifteen months, he forced the troop number down by almost a third.

When William Crowe or Colin Powell ran the Joint Chiefs of Staff, America's generals had not been so easily bullied. But by 2002, after more than a decade of successful wars, the civilian-military balance of power had shifted. The military's warnings no longer made civilians' teeth chatter. "America's senior generals have opposed nearly every intervention that the United States has undertaken in the post–cold war era," wrote William Kristol and the *New Republic*'s Lawrence Kaplan. (The implication was too obvious to require spelling out: Every time, the generals had been wrong.) In the summer of 2002, Bush was seen toting a book called *Supreme Command*, which argued that the greatest wartime leaders regularly overruled their generals. And with his backing, Rumsfeld developed a relationship with the uniformed military that approximated the relationship between a batterer and his spouse. "Shut up. I don't want any excuses. You are through and you'll not have time to clean out your desk if this is not taken care of," he barked at one admiral. Asked once by the president for his opinion, Franks replied jokingly, "I think exactly what my secretary thinks, what he's ever thought, what he will ever think, or whatever he thought he might think." In early 2002, when Bob Woodward asked Chairman of the Joint Chiefs Richard Myers to describe Rumsfeld's personality, the highest-ranking general in the U.S. military simply buried his head in his arms.

For Rumsfeld and Cheney, economic and military confidence was enough. They did not much care whether Iraq became a democracy after Saddam fell, as long as it served America's will. Their foreign policy strategy, like their management style, was directed less at hearts and minds than at the lower regions of the anatomy, in the belief that if intense pressure was applied there, heart and mind would surely follow. Where "there was no room for idealism or sentimentality," remarked one White House lawyer, "you'd find the vice president there."

But for Bush, idealism did play a role. For him, ideological confidence—his belief that Iraq could become a democracy and a model for the region, a shining Arab city on a hill—was as important as military and economic confidence. In this regard, the president was a more contradictory figure than his vice president and secretary of defense. On the one hand, he believed as they did in establishing dominance and inspiring fear. Bush may have become a committed Christian in middle age, but he had been a bully all his life. You could see it in his habit of assigning nicknames, which established a subtle supremacy over those around him. (Bush's subordinates were never encouraged to give him nicknames in return.) It was evident during his dad's presidential campaigns, in which he played the role of enforcer. And even family members quietly testified to his thuggery. Once during an intramural basketball game at Harvard Business School, W. had brazenly elbowed the opposing captain in the mouth, nearly sparking a brawl. Many years later, the recipient of that blow met Jeb and recounted the incident. George, his younger brother explained, "truly enjoys getting people to knuckle under."

So it was little surprise that Bush attributed 9/11 to the fact that America wasn't sufficiently frightening. But that was only half the story. If the president was part bully, he was part romantic, too. Cheney and Rumsfeld were not huggers, and they did not openly weep. But Bush did both, often. As public policy, compassionate conservatism may have been thin gruel, but for Bush emotionally, it was real. Running for reelection as Texas governor, he had encountered a skinny fifteen-year-old black kid named Johnny Baulkmon in a juvenile prison. "What do you think about us?" Baulkmon asked, catching the governor off guard. "I think you can succeed," Bush responded. "The state of Texas still loves you all. We haven't given up on you," and as he spoke, he almost began to cry. For Bush, the author Robert Draper notes, the incident produced "a strange euphoria." For weeks he spoke about it to anyone who would listen, not only in public—where the encounter became part of his stump speech—but also in private. His imagined communion with the kid became, in his own mind, the raison d'être of his campaign. (For the kid, the experience was less revelatory. Asked about the experience years later, Baulkmon—now an adult criminal—said Bush "doesn't care about anything but himself. He's complete trash, a horrible evil person.")

After 9/11, Bush began to feel a bond with people suffering oppression in the Muslim world similar to the bond he had felt with Johnny Baulkmon. It started with the heartrending scenes from liberated Kabul. After that, Bush started seeking out Iraqis who had endured Saddam's horrors, and upon hearing their stories, he often broke down. Bush's critics would later call his freedom rhetoric cynical, a pretext hatched once it became clear that Saddam had no WMD. But many of those close to the president saw it the other way around: For Bush personally, they believed, WMD was the pretext for war and democratic transformation the real cause. "For Bush," wrote Press Secretary Scott McClellan, "removing the 'grave and gathering danger' that Iraq supposedly posed was primarily a means for achieving the far more grandiose objective of reshaping the Middle East as a region of peaceful democracies." Cheney told confidants in March 2003, "Democracy in the Middle East is just a big deal for him [Bush]. It's what's driving him." And a White House aide told the *Standard* in February that Iraqi democracy is "what animates him. It's on his heart, his mind, his agenda. This is what he wants to talk about." It is certainly true that Bush didn't stress democracy as a public justification for war until after Saddam was gone, but that may be more because he believed it wouldn't convince the American people than because it didn't convince him.

That Bush believed Iraq capable of dramatic transformation was not surprising; dramatic transformation was the story of his life. His victory over alcohol had not been slow, messy, and complex; he had freed himself in one epic act of will. That had been the template for compassionate conservatism. When it came to poverty, Bush showed scant interest in structural conditions and incremental improvements. He believed people could radically improve their lives if they changed their hearts. He believed the vehicles for that change would be religious charities, the same "armies of compassion" that had changed his life. And now, after 9/11, he envisioned the U.S. military as the greatest army of compassion of all. In 2002, Bush exhibited no more curiosity about the cultural and historical roots of Iraqi tyranny than he had exhibited about the cultural and historical roots of American poverty. Rather, he identified with ordinary Iraqis suffering under Saddam and believed them capable of transforming their lives, just like Johnny Baulkmon—and like Bush himself—if only someone gave them a chance.

For a conservative, Michael Gerson noted, Bush had a view of human nature that was uncommonly bright, and his brand of Christianity striking untroubled by original sin. Like the New Left activists who rebelled against the ethic of toughness in the 1960s, and like his hero, Ronald Reagan, Bush believed that people were generally better than their governments. And he believed this was true of people of all nations and creeds, from the inner-city criminal to the Iraqi peasant. Bush's critics sometimes called him a Christian crusader, leading a "clash of civilizations" against Islam. But in reality, Bush's faith made him a universalist, a fervent believer that all God's children were basically the same, and basically good. "The human heart desires the same good things, everywhere on Earth," he declared in February 2003. "In our desire to be safe from brutal and bullying oppression, human beings are the same. In our desire to care for our children and give them a better life, we are the same. For these fundamental reasons, freedom and democracy will always and everywhere have greater appeal than the slogans of hatred and the tactics of terror."

"Always and everywhere"? In the age of Hitler and Stalin, when democracy looked like a fading force and men like Reinhold Niebuhr and Arthur Schlesinger understood that totalitarianism also had a claim on the human heart, declaring that freedom was humanity's inevitable choice would have sounded bizarre. But Bush's personal optimism mingled with the ideological optimism of an era in which country after country had overthrown tyranny. As he prepared for war with Iraq, Bush peppered his speeches with the language of democratic peace theory and of Fukuyama's end of history. Just as democracies did not war with each other, he argued, democracies would not incubate terrorism. And democracy was possible in the Middle East, because democracy was the universal yearning of all humanity. "The 20th century ended with a single surviving model of human progress," Bush told cadets at West Point. "For most of the twentieth century, the world was divided by a great struggle over ideas: destructive totalitarian visions versus freedom and equality," declared his administration's National Security Strategy. "That great struggle is over."

The war with Iraq had not yet begun, but ideologically the great global struggle was already over. The implication was clear. Ultimately America could not fail in Iraq, because American ideals were the ultimate destination of all humankind. History itself was moving America's way; George W. Bush was simply giving it a push.

Had Bush been a more introspective man, he might have pondered the tension between the two sides of his nature: the bully who wanted to frighten Arabs and Muslims into submission and the romantic who wanted to liberate them from bondage. Had he known more about the Middle East, he might have recognized that to people with a history of imperial subjugation, whose parents and grandparents had been subjugated by idealists carrying Maxim guns, this twin mission would look both familiar and ominous.

But there is no evidence that he did. In Bush's own life, after all, virtue and power had gone hand in hand; when he quit drinking he became a better person and a more successful one, too. He saw recent American history the same way. Under Reagan, he believed, America had done well by doing good. From Latin America to East Asia to Eastern Europe, the United States had pursued dominance and democracy at the same time; now America just needed to apply that template to the Middle East. The older, darker conservative view, born in the totalitarian age and embodied by Jeane Kirkpatrick and Irving Kristol—which stressed the limits of good intentions and the incorrigibility and inscrutability of far-off lands—had less resonance in an era that had witnessed so much moral and geopolitical triumph. And it was particularly alien to the sunny, unreflective mind of George W. Bush.

That's why Chalabi and his Iraqi National Congress proved so useful. They told the Bushies that post-Saddam Iraq would be pro-American, even pro-Israeli, and yet democratic as well. They squared the circle that Bush needed squared: between American dominance and Iraqi freedom. By depicting himself as the heir of Lech Walesa and Vaclav Havel, Chalabi allowed the "neo-Reaganites" to imagine Iraq as the reincarnation of Poland, Czechoslovakia, or at least Romania after the cold war: countries where democratic transformation and pro-American transformation had occurred in tandem. He allowed them to disregard warnings that in the Arab world, a region with fewer democratic traditions than Eastern Europe and much fiercer hostility to the United States, making Iraq an American client would be hard, making it a stable democracy would be harder, and making it both would be hardest of all.

The Iraqi exiles—who had been in Washington long enough to know what the Bushies wanted to hear—insisted not only that all this was possible, but that it would be cheap. In the run-up to war, the administration and its supporters sometimes cited the examples of Germany and Japan, where America had toppled hostile dictatorships and built pro-American democracies in their place. But those analogies, if taken seriously, would have implied a long and costly postwar occupation—exactly what the Bush administration was insisting Iraq would not require. The more influential, and more comforting analogy, was to the Reagan Doctrine, which Wolfowitz had written about so much in the 1990s. In this model, America didn't need to occupy and govern Iraq. It just needed to install its allies, who would do the occupying and governing themselves. In the 1980s, those allies had been the Contras and the Afghan mujahideen. In the 1990s, they had been the Bosnians and the Kosovars. And after 9/11, they had been Hamid Karzai and the Northern Alliance. As Bush himself told a reporter, America would do in Iraq "the same [thing] we did in Afghanistan—it's a blueprint, a model."

The Afghan example was particularly reassuring because America seemed to have birthed a fledgling democracy with very few troops, thus reconciling Bush's dream of democratic transformation with Rumsfeld's hatred of nation building. And with Chalabi's help, top Bush officials hoped to do the same in Iraq. In February 2003, when General Eric Shinseki famously warned that occupying Iraq would require several hundred thousand U.S. troops, Wolfowitz said he was wrong because "we are training free Iraqi forces to perform functions of that kind," by which he meant the Iraqi National Congress. Wolfowitz and other top Pentagon civilians hoped to establish the INC as a provisional government-in-exile even before the war began, with its own U.S.-trained army, and then hand it the reins soon after Saddam fell. "[W]e had in mind our recent experience in Afghanistan," explained Douglas Feith, "where the United States immediately recognized an interim government of Afghans and never became an occupying power."

For Rumsfeld and Cheney, the INC's democratic bona fides were secondary. It mattered less that they were democrats than that they were ours. But for Wolfowitz, who had become captivated by the idea of Islamic democracy during his time as ambassador to Indonesia and after watching Iraq's Shia rise up against Saddam in 1991, those bona fides mattered a lot. More than Rumsfeld or Cheney, he shared Bush's passion for spreading freedom, and because he did, he gained influence beyond his station. Wolfowitz, noted the *Weekly Standard*'s well-connected White House reporter Fred Barnes, did more to shape Bush's post-9/11 thinking than any other adviser. They were an odd pair: the bookish, Jewish defense intellectual and the swaggering, faith-based president. But they had both come to see dictatorship as unnatural, alien to the basic yearnings of humankind. In his discussion of Iraq, Wolfowitz resisted the terms *occupation* and *nation building*, preferring *liberation* instead. It was a revealing choice. *Nation building* implies constructing something new; *liberation* implies freeing something that already exists. "'Export of democracy' isn't really a good phrase," Wolfowitz told an interviewer. "We're trying to remove the shackles on democracy." Wolfowitz's faith in liberation—his belief that if repressive structures were torn down, humanity's innate goodness would break free—put him philosophically closer to the New Left of the 1960s, which also believed that human beings were fundamentally superior to the societies in which they lived, than to older conservatives like Kirkpatrick and Irving Kristol. For Wolfowitz, it was as if democracy already existed in the hearts and minds of ordinary Iraqis. This nascent democracy sat captive within the walls of Saddam's prison-state. All America had to do was turn the key.

For Colin Powell, by contrast, Saddam's Iraq was not a prison; it was a vase. It would not unlock, releasing the democracy inside; it would shatter, bloodying America's hands. When discussing Iraq, Powell liked to cite the "Pottery Barn rule": You break it, you own it. (In fact, Pottery Barn had no such rule and accused Powell of slandering the company and driving away business.) To Wolfowitz and Bush, Saddam was Iraq's jailer; to Powell, he was its glue.

By the summer of 2002, Powell was no longer the commanding figure he had been when the administration began. He was still popular among his foreign counterparts, but admiration was now tinged with pity. To get something done with the Americans, diplomats whispered, you had to deal with the barbarians at the White House. That's where the power lay.

Powell doubted that Rice, whom he considered far too eager to please, was telling Bush about the dangers of war, and he was sure that Cheney wasn't. In the conservative press, the mantra repeated again and again was that it was "hard to imagine" how toppling Saddam could leave America and Iraq worse off. (It was another revealing turn of phrase. People who had grown accustomed to seeing dictatorship give way to democracy, not chaos, and had little memory of an American war gone horrifically wrong, found such things "hard to imagine.") But Powell could imagine it; Vietnam and Lebanon were like shrapnel lodged in his brain. In early August 2002, on a flight back from Asia, he resolved to tell Bush everything that might go wrong.

On the evening of August 5, he dined with Bush and Rice at the White House. "When you hit this thing," he warned, "it's like crystal glass. . . . [I]t's going to shatter. There will be no government. There will be civil disorder." He cited Iraq's lack of democratic traditions, the war's potentially massive economic costs, and its destabilizing impact on America's Arab allies. It was a grim litany, a clear call for Bush to reconsider his path. And then Powell reached the punch line, the advice he had been waiting months to give: Ask for UN support. Huh? It didn't make sense. Many of the horrors Powell was outlining were intrinsic to any invasion of Iraq; they were likely even if the UN did give its blessing. Perhaps Powell believed the Security Council would resist the push for war and make Bush back down—in other words, that the French, Russians, and Chinese would do his work for him. But by making an argument about *how* to go to war, not *whether* to go, Powell implicitly conceded the central point.

It was eerily similar to his encounter with Bush's father before the Gulf War. Once again he had summoned his courage, built up a head of steam, and then swerved at the last minute to avoid a collision. Once again he had refused to go baldly into opposition, to put his reputation, and his job, on the line. Those close to him believed it just wasn't in his nature. His great strength—which set him apart from Cheney, Wolfowitz, and Bush—was that he lacked certainty. He was not an ideologue; he distrusted abstractions; he saw shades of gray. But precisely because he did lack certainty, he could not say with total conviction that the war was a mistake. It must have crossed his mind that he had warned of disaster three times before—before the Gulf War, Bosnia, and Afghanistan—and each time disaster had not struck. Three times presidents had ignored his advice, and three times everything had worked out fine, both for America and for him. If a decade of military triumph had made the hawks too willing to disregard his warnings, perhaps it had made him too comfortable seeing them disregarded.

If Powell did not go baldly into opposition, neither did the leaders of the Democratic Party. Before the Gulf War, the Democratic leadership had lined up in loud and virtually unanimous opposition. But by 2002, after a decade of military triumph, opposing an American war seemed like an excellent way to kill your political career. Everyone knew the parable of Georgia Senator Sam Nunn, who by forcefully opposing the Gulf War had sunk his chances of reaching the White House. Nor did it escape notice that of the ten Senate Democrats who had backed the Gulf War, two of them—Joe Lieberman and Al Gore—had ended up on the 2000 Democratic presidential ticket. When the resolution authorizing the Iraq War came before Congress in October 2002, Democrats with safe seats and without national ambitions generally voted no. But most of the party hierarchy—and almost every Democrat thinking seriously of running for president, from John Kerry to John Edwards to Joseph Biden to Hillary Clinton—voted yes.

During the debate over the Gulf War, congressional Democrats had endlessly invoked Vietnam. But this time, even though America was contemplating a far more ambitious mission, Vietnam analogies were scarce. In his 1991 speech opposing the Gulf War, Massachusetts Senator Kerry, who had launched his political career protesting Vietnam, mentioned it ten times. In his 2002 speech supporting war against Iraq, he mentioned Vietnam only once.

For many Washington Democrats in 2002, there was something passé about citing Vietnam. It was the foreign policy equivalent of wearing tie-dye. It pegged you as a relic of a bygone age, too traumatized by the past to see that military force now worked. This was particularly true among younger Democrats. Fifty-eight percent of House Democrats ages forty-five or younger voted to authorize the Iraq War, compared to only 35 percent over forty-five. (In the Senate, there were only two Democrats younger than forty-five, and both voted yes, compared to 56 percent of older Senate Democrats.) Age forty-five was a useful dividing line because someone who was under forty-five in 2002 would have been under fifteen in 1972, meaning they would likely have been too young to be directly shaped by Vietnam. Even among older Democrats, who did remember that war, its impact had been dulled by America's post–cold war military success. But among younger Democrats, who had seen nothing but military success, Vietnam's impact was fainter still.

As a result of their experience, younger Democrats were also particularly inclined to believe that military force and liberal values could go hand in hand. That faith had begun during Panama and the Gulf War, when liberals learned that the American military could shoot straight, that it could do more than napalm villages; it could win wars. It had grown during Bosnia and Kosovo, when liberals saw that not only could America win, but it could win in liberalism's cause, that the fight for human rights truly could be waged through the barrel of a gun.

For many liberal intellectuals in the 1990s, Sarajevo had been the perfect city: cosmopolitan, multi-ethnic, secular, and fighting for its life. And Milosevic's Serbs had been the perfect enemy: racist, sexist, bloodthirsty. *Fascist!* The word gained currency on the Clinton-era left; it evoked a spirit of World War II–era leftist hawkishness long thought dead. Then came 9/11. Now another multi-ethnic, cosmopolitan, culturally left-wing city was under attack, and it was the very city where many of the liberal hawks lived: New York. Like Sarajevo, it had been gouged by people in love with purity, willing to kill to prevent the mongrelization of races and religions, the very things liberals loved. If liberal hawks had seen Milosevic as the face of a resurrected European fascism, climbing out of history's grave, they coined a parallel term to describe Al Qaeda and the Taliban: *Islamofascism*. Although later associated with the neocons, the term was actually coined by Paul Berman and Christopher Hitchens, two of the most eloquent left hawks. Its function was to explain that America's new terrorist enemy—like its Balkan enemy a few years earlier—resided on the ideological right, and that fighting it was thus more than kosher; it was the great liberal duty of the age.

When liberal hawks said that America was fighting fascism, they meant it in part as a rebuke to Fukuyama. Berman attacked "the deluded, triumphalist atmosphere of 1989," when Fukuyama declared that democracy had won. Totalitarianism, Berman insisted, had never really died; Americans had just failed to recognize it in its new, Islamic garb. But

Berman and the other liberal hawks owed more to Fukuyama than they cared to admit. Al Qaeda and the Taliban may indeed have been totalitarian, but unlike the totalitarianisms of the mid-twentieth century, they posed no serious ideological challenge to democracy. The Taliban was a mud-hut pariah regime, and while bin Laden enjoyed some prestige in the Muslim world for having bloodied the American Goliath, his vision of society stirred few Middle Eastern hearts and minds. Barely anyone believed that Al Qaeda and the Taliban offered an economic model that could outperform democratic capitalism, as many had believed about fascism and communism during the Depression. The jihadists were not only militarily weak, they were ideologically weak. Like Milosevic, they were totalitarians in a post-totalitarian age

Some liberal hawks tacitly acknowledged this. As Berman himself wrote, in "the revolutions of 1989 . . . the notion that one or another race or culture or religion is hopelessly allergic to liberal ideas—this notion did pretty much explode." In other words, America's new anti-totalitarian struggle differed fundamentally from the struggles against Nazism and Stalinism because 1989 had shown that liberal democracy could penetrate every society on earth. Mid-century intellectuals like Niebuhr and Morgenthau had not been nearly so confident that democracy was the world's universal creed. They had hoped merely to contain Soviet totalitarianism and perhaps soften it, but held out little hope that it could be wiped from the earth. Berman and the liberal hawks, by contrast—like Bush and Wolfowitz—believed in America's ideological dominance; they saw their anti-totalitarian struggle as a kind of ideological mopping-up operation, with the ultimate outcome preordained by the trajectory of history itself. The war on terror was less a sequel to the anti-totalitarian struggles of the twentieth century than an epilogue. There were a few more pages to write, some new characters and plot twists, but everyone already knew how the story would end.

For the liberal hawks, like the dominance conservatives, Afghanistan added to the ideological swagger. Before the war, some left-wing doves had warned that intervening would bring military disaster and moral horror. Noam Chomsky, the anarchist-linguist who decades earlier had invoked Randolph Bourne to oppose Vietnam, warned that if America attacked the Taliban millions might die. But America attacked, and instead of mass murder there was celebration, and an embryonic democracy. As suspected, the new Islamic totalitarianism was not very strong at all.

Thus emboldened, many liberal hawks urged war against Saddam.

As in Bosnia, the UN—whose weapons inspectors had left Iraq in 1998—seemed feckless. So once again liberal hawks turned to American power. "Multilateral solutions to the world's problems are all very well, but they have no teeth unless America bares its fangs," wrote Michael Ignatieff, another influential liberal hawk. Or as one younger liberal foreign policy hand put it, "I can't say with a straight face that it's fine to go around the UN for Kosovo and not do it in Iraq."

But Iraq was not Kosovo, or even Afghanistan. In the Balkans, the casus belli had been genocide, an unfolding moral emergency. In Afghanistan, whatever the human rights side benefits, the casus belli had been self-defense. In Iraq, by contrast, not only was there no imminent threat; there was no genocide, either. Some liberal hawks argued that Saddam's rule constituted a moral emergency in and of itself. But by that standard so did Burma, North Korea, Zimbabwe, and a host of other nasty dictatorships. Intellectually, the liberal hawks who went from backing war in the Balkans and Afghanistan to backing war in Iraq were not walking a treadmill; they were climbing a ladder.

So it was that a significant portion of the left, whether because of political ambition or moral ambition or both, waved the Bush administration on as it sped toward war. And so did many on the older, anti-dominance right. Jeane Kirkpatrick, now seventy-six years old, privately worried that trying to violently remake Iraq was utopian, and would bring not freedom, but chaos. But she never said so publicly, where it might have made a difference. She was personally close to the Cheneys and to many other leading hawks. Perhaps she thought public dissent would give aid and comfort to the enemies of her friends, or perhaps, after a decade of victorious wars, she, like Powell, had lost the confidence of her convictions. For whatever reason, she not only failed to publicly object; she flew—at the Bush administration's request—to Geneva to defend the war before the UN Human Rights Commission. Later she insisted that she had not defended Bush's decision to invade Iraq, only his right to do so under international law. It was a subtle distinction, and utterly lost in the clamor for war.

Irving Kristol, now eighty-two, said even less. With his son leading an assault on the foreign policy axioms that he had espoused for much of his life, the man sometimes called the godfather of neoconservatism wrote barely a single word about what many were calling the neocon war. One writer heard him mutter that his son was trying to turn George W. Bush into Napoleon. Others heard rumors that he was privately critical of what the *Standard* wrote. But publicly, he said virtually nothing at all.

Another father also held his tongue. In August 2002, George H. W. Bush's former national security advisor and close confidant, Brent Scowcroft, publicly denounced the impending war. Scowcroft would later claim that the elder Bush shared his views, but the elder Bush himself stayed silent. In January 2003, the former first lady, Barbara Bush, approached ex-Democratic senator David Boren at a dinner and asked, "Are we right to be worried about this Iraq thing?" Boren answered yes. "Well, his father is certainly worried and is losing sleep over it," she replied. "He's up at night worried." Boren suggested that the former president talk to his son, but she answered, "He doesn't think he should unless he's asked." For his part, W. told Fox News that he didn't frequently consult his father on foreign policy, because his dad didn't have the latest intelligence.

One conservative who did publicly object was Francis Fukuyama, the Daedalus of the dominance age, inventor of a doctrine that was soaring beyond his control. In 1998, he had signed a letter—along with William Kristol, Robert Kagan, Elliott Abrams, and Paul Wolfowitz—urging that America pursue regime change in Iraq. But in 2002, when they went a step further—urging that America pursue regime change with U.S. troops— he grew skittish. The Iraq War, Fukuyama warned in December 2002, is "an immensely ambitious exercise in the political re-engineering of a hostile part of the world," exactly the kind of exercise that once made conservatives tremble. "The United States is not good at either implementing or sticking to such projects over the long run," he added, picking up one of Irving Kristol's favorite themes. And "it is not at all clear that the American public understands it is getting into an imperial project." Eventually, Fukuyama argued, liberal democracy could sprout in the Middle East, as it had in other formerly authoritarian parts of the world. And when that happened, America would be safer. But that growth would have to be organic. Spilling Iraqi blood in a preventive war, he insisted, would not fertilize the soil.

Fukuyama did not just publish these arguments; he said them to Paul Wolfowitz's face. The two men had been friends for thirty-five years, since Fukuyama's days as an undergraduate at Cornell. Wolfowitz had given Fukuyama his first government job in the 1970s, then hired him again during the Reagan years, and later brought him to teach at Johns Hopkins's School of Advanced International Studies, where Wolfowitz was dean. After 9/11, Wolfowitz convened three study groups to make suggestions for the broad direction of the war on terror, and asked Fukuyama to head one. One day in January 2003, Fukuyama went to an office building in Arlington, Virginia, to present his findings. Jihadist terrorism, he argued, was not another great totalitarian foe like communism or fascism, and America should not treat it as such. The important thing was not to overreact, not to take military action that alienated Muslims, thus strengthening Al Qaeda's inherently weak hand. Instead America should rely on diplomacy, intelligence gathering, law enforcement, and patience. History was on its side, but history could not be rushed.

Fukuyama was arguing against making the war on terror a grand, heroic project, a successor to the cold war; he was urging small ball. It was an assault on the basic premises underlying Bush and Wolfowitz's war on terror, phrased in the language of a more cautious conservatism, forged in a more humble age. More than many conservatives, who only mumbled their fears, Fukuyama was showing intellectual courage. His was the only presentation Wolfowitz attended. The deputy defense secretary listened and then exited the room without comment. The two never discussed Fukuyama's presentation again.

From liberals to conservatives to economists to generals, by 2003 many of the people who in the past might have resisted war had either fallen silent or been shoved aside. Without them, Colin Powell was a very vulnerable man. "Someday when you're retired and I'm retired," he told Democratic senator Joseph Biden, "I'll tell you about all the pressure I've been put under over here."

On January 13, Bush called Powell into the Oval Office, sat down in front of the fireplace, and announced, "I really think I'm going to have to do this." Powell asked if he was sure. It was a silly question: Bush was always sure. "Are you with me on this?" Bush asked. "I'm with you, Mr. President," came the reply. The entire conversation lasted twelve minutes. Bush had still never asked Powell whether he thought war was wise. "I didn't need his permission," he later explained.

Powell was on the hook, which in the Bush administration was a dangerous place to be. On January 25, Bush asked him to publicly make the case that Saddam was hiding WMD. "You have the credibility to do it," Bush explained. He was right: According to one poll, Americans trusted Powell over the president on Iraq by a margin of almost three to one. Cheney put it more aggressively. "You've got high poll ratings," he said, jabbing his finger into Powell's chest; "you can afford to lose a few points."

This, in Bush and Cheney's mind, was what Powell was there for: not to make Iraq policy but to sell it. He was—there was no gentler way to say it—being used.

Three days later, the White House sent Powell a forty-eight-page, single-spaced memo drafted, Powell suspected, in Cheney's office. It was filled with falsehoods, like the claim that 9/11 ringleader Mohammed Atta had met Iraqi intelligence officials in Prague. Powell tossed the document aside and on Saturday, February 1, parked himself—along with key aides—at the CIA, where he began looking for the truly convincing evidence on Iraqi WMD, the stuff that you didn't have to be a fanatic to believe. Tensions ran high, as Cheney's staff—whom Armitage privately called "the Gestapo"—kept pushing to restore allegations that officials from State and the CIA considered garbage. As Powell's chief of staff, a no-nonsense retired colonel named Lawrence Wilkerson, later admitted, "We were beginning to get leery of our own presentation." But Powell could not afford to get too leery. By telling Bush he would support him on the war, he had bolted the door behind him. So he did what he always did: He managed the situation, making the best of the circumstances he was in. After a while, he grew more comfortable with the presentation. He had never bailed out before, and things had always turned out okay.

Just before 10:30 A.M. on a frigid Wednesday, February 5, Powell strode into the UN Security Council chamber, dressed in a dark suit and red tie, and made the case for preventive war. "Every statement I make today," he declared, "is backed up by sources, solid sources. These are not assertions. What we are giving you are facts and conclusions based on solid intelligence." Even members of Powell's own staff doubted that was true.

In the United States, commentators pronounced the speech a triumph. The *Washington Post* called it "irrefutable" (though, in fact, UN weapons inspectors began refuting it almost instantly). The *New York Times* called it "the most powerful case to date." A *Newsweek* poll found that the percentage of Americans supporting war jumped almost instantly. The White House was overjoyed: Good old Colin Powell had finally shown he was on the team. But Powell's wife, who had traveled with him to New York, was overcome with apprehension, a premonition of bad things ahead. His daughter, who heard the presentation on the radio, thought her father did not sound like himself. When Powell's talk was over, Wilkerson left the UN for his Manhattan hotel room, where he fell into a restless sleep and awoke in despair. It was, he would later say, "the lowest moment of my life." When he returned to Washington, he ordered plaques for everyone who had worked on the speech. But when Powell handed them out, he noticed that his chief of staff hadn't ordered one for himself. I don't want one, Wilkerson explained.

At 10:15 P.M., on Wednesday, March 19, 2003, President Bush addressed the nation from the Oval Office. The invasion of Iraq had begun. In the days before the Gulf War, his father had found it difficult to sleep; he had lost weight and occasionally struggled to breathe. The decision to send young Americans to their death, he wrote to his children, "lingers and plagues the heart." Before the Iraq War, by contrast, his son let those around him know that he was sleeping well. "There is no doubt in my mind we're doing the right thing," he insisted. "Not one doubt." Just before the cameras rolled, as he prepared to tell the nation it was at war, Bush pumped his fist and said, "Feels good."

The beginning of the Iraq War resembled the beginning of the Afghan War: Progress was alarmingly slow. While Saddam's regular troops largely melted away, the United States met fierce resistance from Iraqi paramilitaries in civilian clothes who hid among the local population and launched raids on America's supply lines, which U.S. forces lacked sufficient numbers to defend. On television, retired generals began saying that the Bush administration had attacked with too few troops, and that as a result, things were going dangerously wrong.

Then Rumsfeld's light but ultrafast force reached Baghdad, and everything changed. On April 9, only days after the TV generals had warned of a quagmire, Saddam's regime collapsed. In the city's main square, Marine Corporal Edward Chin attached a hook from his tank to a statue of the tyrant and hauled it down, as locals cheered. The next day, a group of distraught Iraqis led Lieutenant Colonel Frank Padilla and his battalion to a forbidding, walled compound. When the Marines broke through its outside gate, 150 disheveled, beaten, and malnourished children—some as young as seven—rushed into the arms of the Iraqis waiting outside. It was a children's prison where Saddam kept the sons and daughters of families he considered disloyal. The Iraqi parents showered the Marines with kisses. If this wasn't liberation, nothing was.

In an age of intoxicating military victories, it was the most intoxicating of all. Toppling Saddam had taken three weeks and left 108 U.S. soldiers dead from hostile fire, fewer than in the Gulf War. None of the parade of horribles that Powell, Fukuyama, Scowcroft, and others predicted had come true. The Arab street had not risen up; Saddam had not torched his oil fields or attacked Israel; Iraq's neighbors had not intervened. There was looting, to be sure, but top Bush officials and their intellectual allies generally shrugged it off. "It is hard to be overly troubled by the sight of Iraqis looting the homes and offices of leading Baathists," wrote Max Boot in the *Weekly Standard*. "Why shouldn't the people take back a few of the regime's ill-gotten gains?"

Commentators struggled to comprehend the magnitude of America's military achievement. "The stunning success of the 'combat portion' of Operation Iraqi Freedom challenges any understanding based upon previous military history," declared one article in the *Standard*. "One gets the impression that U.S. military dominance is now so overwhelming that the rules of conflict are being rewritten," added David Brooks. "We could be entering the age of decapitating wars, in which the United States can change evil regimes without widespread loss of life." General Tommy Franks declared that America's ultrasophisticated technology meant that it could now see the battlefield with the "kind of Olympian perspective that Homer had given his gods."

Ideologically, American dominance seemed equally profound. "In the images of celebrating Iraqis," Bush declared on May 1, from the deck of the aircraft carrier *Abraham Lincoln*, "we have also seen the ageless appeal of human freedom. . . . Men and women in every culture need liberty like they need food and water and air." For Brooks, this was what national greatness was all about: a heroic leader leading a heroic people in heroic deeds. The post–cold war generation had found its calling. The Iraq War, Brooks wrote, is "what the United States is on earth to achieve."

The dominance conservatives began imagining a new global order, redesigned to reflect America's epic might. Kristol said it was time to consider withdrawing from the UN; America would have more power, and more legitimacy, without it. Another article in the *Standard* suggested pulling out of the G-8, the world's club of major industrial democracies, because it gave antiwar countries like France and Germany too much say. Krauthammer went furthest of all, arguing that even NATO, by refusing to endorse the war, had made itself worthless. If the post–cold war era had begun with visions of a new, Wilsonian world of international cooperation and international law, the opposite had now come to pass. America bestrode the world unfettered, Gulliver freed from the Lilliputians' chains.

Bush's approval rating, which had drooped into the mid-50s in March, spiked back up to 77 percent. Donald Rumsfeld basked in the adulation as well. On April 14, Michael Jordan's final home game in the NBA,

Rumsfeld presented Jordan with an American flag at Washington's MCI Center and got a standing ovation even more thunderous than the greatest basketball player of all time. Tommy Franks and CIA director George Tenet did well from the war, too. After retiring from the army, Franks signed a multimillion-dollar book contract, and pocketed a million more in speaking fees. In 2004, Bush awarded both him and Tenet the Presidential Medal of Freedom.

The only top administration official who didn't prosper from Saddam's fall was Colin Powell. Before the war, he had been merely vulnerable. Now he was expendable, and the jackals circled. On April 22, former House Speaker Newt Gingrich, a longtime ally of Cheney's, and a member of Rumsfeld's Defense Policy Board, charged Powell's State Department with a "deliberate and systematic effort to undermine the President's policies." Asked by reporters whether he agreed with Gingrich's charge, Rumsfeld dodged the question. Bush also declined to defend his secretary of state.

In mid-April, Cheney invited Wolfowitz, Scooter Libby, and an old Ford administration colleague and war booster named Kenneth Adelman to a celebratory dinner at his house with their wives. They talked about what a mistake it had been to halt the Gulf War in 1991 and marveled at how easy taking Baghdad had ultimately proved. They traded affectionate stories about their friend and ally Rumsfeld, and they toasted the president. Then someone mentioned Powell, and they all laughed.

CHAPTER NINETEEN

I'M DELIGHTED TO SEE
MR. BOURNE

"The scenes we've witnessed in Baghdad and other free Iraqi cities belie the widespread early commentary suggesting that Iraqis were ambivalent or even opposed to the coalition's arrival in their country." Donald Rumsfeld was lecturing the press, two days after Saddam's fall. Behind him, photos flashed across a Pentagon screen. "Iraqis share laugh with a U.S. Army soldier," read the caption beneath one. "Jubilant Iraqis cheer U.S. Army soldiers" read another. "Happy Iraqis pose with a U.S. Army soldier," read a third.

A reporter piped up. "Mr. Secretary, you spoke of the television pictures that went around the world earlier of Iraqis welcoming US forces with open arms. But now television pictures are showing looting and other signs of lawlessness. . . ."

Rumsfeld responded philosophically. "I think the way to think about that is that if you go from a repressive regime . . . and then you go to something other than that—a liberated Iraq—that you go through a transition period. And in every country, in my adult lifetime, that's had the wonderful opportunity to do that, to move from a repressed dictatorial regime to something that's freer, we've seen in that transition period there is untidiness."

The reporter tried again: "Do you think that the words 'anarchy' and 'lawlessness' are ill-chosen. . . ."

Rumsfeld cut him off. "Absolutely. I picked up the newspaper today and I couldn't believe it. I read eight headlines that talked about chaos, violence, unrest. And it was just Henny Penny—'the sky is falling.' I've never seen anything like it! And here is a country that is being liberated, here are people who are going from being repressed and held under the thumb of a vicious dictator, and they're free!"

Sixty-two hundred miles away, in Baghdad, newly liberated Iraqis were liberating their government of refrigerators, desks, chairs, mattresses, even a plastic Santa Claus. From the house of Saddam's son Uday, vandals snatched liquor, guns, lewd paintings, and white Arabian horses. From the house of one of Saddam's first cousins, they seized a battery-powered model Ferrari, a parachute, and a motorized water scooter. One enterprising burglar managed to haul away a boat. All in all, looters gutted seventeen of Iraq's twenty-three government ministries, burning many to the ground. The total cost of the rampage: $12 billion.

It seemed like a paradox. In Afghanistan and now Iraq, the American military had ventured halfway across the globe and toppled hostile governments at low cost and lightning speed, awing the world. Yet this global Goliath, which used Google-age technology to ensure that even as Iraq's government fell its public infrastructure remained largely unscathed, could not stop unarmed Iraqi civilians from plundering that infrastructure like the armies of Genghis Khan.

Part of the answer was that America had not really tried to stop them. To Powell's chagrin, Rumsfeld had kept the invading force ultra-lean, fast enough to sprint to Baghdad in record time but too small to squat there and police a city of more than six million people, where American troops were the only law. What's more, those troops were given little guidance about what to do once Saddam fell, largely because Rumsfeld and General Tommy Franks didn't want them to do much of anything except turn around and go home. When an anguished colonel pleaded with Lieutenant General David McKiernan, commander of U.S. ground forces, that "You have got to stop this. . . . [E]verything's being destroyed," McKiernan replied in cold fury: "I don't ever want to hear that from your lips again. This is not my job."

It was the job of a breezy retired general named Jay Garner, head of the Pentagon's newly created Office for Reconstruction and Humanitarian Assistance, which held its first major planning meeting less than a month before the war began, and arrived in Baghdad only after looters had already picked Iraq's government clean. Garner's office, confided Britain's top diplomat in postwar Iraq, "is an unbelievable mess." Lawrence Diamond, a Stanford sociologist dispatched to Iraq to help build democracy, found the Pentagon's postwar efforts so appalling that he rifled through the California penal code searching for an appropriate indictment. He settled on "criminal negligence."

So America's failure to control postwar Baghdad was partly a matter of choice. But it was a matter of ability, too. The American military was built for fast wars and bloodless occupations, occupations of stable, peaceful countries like Germany, South Korea, and Japan, and small ones like Bosnia and Kosovo, where America's allies contributed most of the troops. The GIs who found themselves running (or not running) Baghdad in the weeks after Saddam's fall were mostly trained to kill enemy soldiers, not stop looters, restock ransacked hospitals, and direct traffic. The 1.3 million-person active-duty U.S. Army contained only 16,000 military police and civil affairs officers (with another 47,000 in the National Guard and Reserves). Garner's office, which was supposed to run the occupation's civilian side, boasted barely any Arabic-speakers, partly because Rumsfeld and his top aides distrusted regional experts, whom they suspected of ideological disloyalty, but mostly because there were only fifty or sixty fluent Arabic-speakers in the entire U.S. diplomatic corps. Garner himself was so ignorant of the country he was supposed to rebuild that when someone mentioned Ayatollah Ali al-Sistani, Iraq's most powerful man, he responded with a blank stare. Successfully occupying Iraq, as opposed to merely invading it, required lots of deployable bodies, lots of local knowledge, and lots of patience, traits that had not been required during America's flurry of post–cold war wins, and which the American military—and indeed, American society—did not possess.

In truth, the scene was not as paradoxical as it appeared. The American military was extraordinarily good at some things and not very good at others—and those others proved crucial when Saddam fell. It was like watching Michael Jordan play baseball. In the summer of 2003, as American soldiers in full body armor cursed and sweltered under Baghdad's 110-degree sun, America's military confidence, which had been growing like a wave for a decade and a half, began to break. America's ideological confidence would soon follow, and then its economic confidence after that. The hubris of dominance was beginning to come apart.

If postwar Iraq confounded the Bush administration's military expectations, it confounded its ideological expectations as well, and the two were intertwined. For his part, Rumsfeld didn't much care how Iraqis governed themselves with Saddam gone. The invasion had toppled one enemy, scared others, and vindicated his vision of a light, agile, high-tech military—or, at least, vindicated it if you considered the postwar irrelevant, which Rumsfeld basically did. For Bush and Wolfowitz, however, what happened in post-Saddam Iraq mattered a great deal. They opposed nation building not out of indifference but out of faith, not because they didn't care whether democracy arose in Iraq but because they believed it would arise spontaneously, without lots of U.S. money or troops. For Wolfowitz, in particular, Chalabi's exiles were crucial. In late 2002, he proposed training them to fight alongside U.S. troops, thus perpetuating the idea that, as in Nicaragua, the Balkans, and Afghanistan, America's local allies were actually liberating themselves, with the United States playing only a supporting role. Chalabi's fans in the media often talked about "Free Iraqi" forces, an allusion to the "Free French" who had fought alongside America and Britain in World War II, and a memo by Assistant Secretary of Defense Peter Rodman, which Rumsfeld forwarded to Bush's entire war cabinet, made the analogy explicit. "Had FDR and Churchill actually imposed an occupation government" in France, Rodman argued, "the Gaullists would have been neutered." The implication was clear: Chalabi was Iraq's de Gaulle, and with him and his army in charge, the United States could quickly draw down its forces and watch an indigenous, pro-American Iraqi democracy blossom.

Chalabi certainly had de Gaulle's ego, but in most other respects he and his men proved dismally unable to play the role the Bushies had assigned them. As often happens in a hubris bubble, policymakers who thought they were merely learning the lessons of the past actually stretched and distorted them. In Afghanistan in the 1980s, Bosnia and Kosovo in the '90s, and Afghanistan again after 9/11, a nationalist fighting force had already existed. In Iraq, by contrast, Wolfowitz and his allies simply assumed that one *should* exist because it fit a historical pattern that they believed had brought great success. (Iraq did have one indigenous rebel force, the Kurdish Peshmerga, but they fought for Kurdistan, not Iraq, which was exactly the problem.) At Wolfowitz's insistence, in the months before the war, the Pentagon began hastily raising a prefabricated "Free Iraqi" army. Chalabi promised 10,000 men; the U.S. military planned for 6,000. Ultimately, after crippling problems with recruitment, screening, and instruction, the force dispatched to Iraq totaled 73.

In mid-April, the United States flew Chalabi and some supporters to the outskirts of the southern Iraqi city of Nasariyah, where they took part in a political assembly designed to form the nucleus of a post-Saddam regime. But to the surprise and dismay of his American backers, Chalabi turned out to have virtually no local following. He hadn't lived in Iraq since 1958, during which time Islamists had largely supplanted the secular elite from which he hailed. Moreover, he didn't exactly endear himself to his countrymen once he hit the ground. When Baghdad fell, Chalabi's men quickly claimed a series of mansions favored by one of Saddam's sons. Then they appropriated some of Saddam's SUVs, which they allegedly sold overseas for a fat profit. Then fellow Iraqis began accusing them of stealing reconstruction funds. In 2004, Iraqi and American forces raided Chalabi's offices on charges of embezzlement, theft, kidnapping, and passing classified U.S. intelligence to Iran. Polls showed him with the lowest approval ratings of any Iraqi politician, including Saddam.

In Iraq, it turned out, Chalabi was not the man he had appeared to be in Washington. To his boosters at the American Enterprise Institute, the Pentagon, and the *Weekly Standard*, he had seemed like the archetypal end-of-history figure: a man dedicated to putting his nation on the escalator to democracy, an Arabic version of the leaders who had toppled dictators from Poland to Panama and South Africa to South Korea. But back in Iraq, he became ideologically inscrutable, a man of shadowy dealings, hidden loyalties, and dark misdeeds. And in this sense, he typified the Bush administration's experience with Iraq more generally, a place that—like Vietnam and like Europe during World War I—grew more alien the deeper America waded in.

In May, the White House replaced Garner with Paul Bremer, a man even more culturally ignorant and even more ideologically self-assured. Bremer, who had never before served in the Arab world, saw Saddam's regime the way Bush and Wolfowitz did: as an ugly skin concealing the pro-American democracy trapped inside. So being a can-do guy (critics preferred the phrase "control freak"), Bremer boldly tore off the remaining scab. Four days after arriving in Iraq, in his first public act as head of the newly formed Coalition Provisional Authority (CPA), he banned thirty thousand members of Saddam's Baath Party from holding jobs in Iraq's new government. Assuming—wrongly—that Baathists dominated Saddam's armed forces, he followed up a week later by dissolving the Iraqi military, interior ministry, and presidential guards. Then he abolished all restrictions on foreign imports and began rapidly privatizing the two hundred state-owned companies that had formed the backbone of the Iraqi economy. Soon, more than a million government employees were out of work. "We must make it clear to everyone that we mean business: that Saddam and the Baathists are finished," Bremer wrote to President Bush. "The dissolution of his chosen instrument of political domination, the Baath Party, has been very well received."

Meanwhile, throngs of angry former soldiers and government workers

began to congregate at the CPA's gates. "We will start ambushes, bombings and even suicide bombings. We will not let the Americans rule us
in such a humiliating way," declared one fired Iraqi officer. "The only
thing left for me is to blow myself up in the face of tyrants," said another.
Bremer vowed that he would not be blackmailed by terrorists. Around the
same time, the first roadside bomb hit an American Humvee driving from
the Baghdad airport. Later, those bombs would gain a name: improvised
explosive devices (IEDs).

   Once again Iraq was deviating from the ideological script. Ever since
Panama, a central premise of dominance foreign policy had been that
anti-American dictators were alien to the societies they ruled and that
pro-American rebels represented the true popular will. But in Iraq it was
Chalabi who lacked deep roots in the political soil, and Saddam's army
and political party that proved far more entrenched than Bremer and his
fellow ideologues had imagined. If the Iraqi military and Baath Party were
Iraq's ugly outer skin, they were also—as Powell had warned—its glue.
Many Sunnis, in particular, had become Baathists as a requirement for
employment. And even more had joined the army, which in Sunni com-
munities was a key source of status and jobs. With the army dissolved,
large numbers of Sunnis suddenly found themselves unemployed, humiliated, and heavily armed. When Bremer ripped off Iraq's governing skin,
he gashed a large segment of Iraqi society in the process, and opened a sore
that soon began gushing blood.

Soon Americans and Iraqis were engaged in a darkly comic, savagely
violent, dialogue of the deaf. In the center of Baghdad, behind seventeen-
foot-high concrete walls topped with razor wire, the CPA created the
Green Zone, an Epcot America where the televisions played Fox, the
radios blared classic rock, the recreation officers taught yoga and salsa
dancing, and the stores sold Cheetos, Dr. Pepper, booze, protein pow-
der, and T-shirts reading "Who's Your Baghdaddy?" At the cafeteria in
Saddam's former Republican Palace, well-mannered South Asian workers
served cheeseburgers, hot dogs, grits, fried chicken, and freedom fries.
The menu, which had a distinctly southern flavor, included large quantities of pork, which Iraqi Muslims might have found offensive to prepare.
But that wasn't an issue, since Iraqis weren't permitted to work in the dining hall for fear they would poison the food.

   When Americans ventured into "the red zone"—otherwise known as
Iraq—they passed through the looking glass, into a parched, dust-brown

riddle of a country where American logic often seemed turned upside down. When the Americans set about building an army to replace the one they had disbanded, they dubbed it the New Iraqi Corps, or NIC, only to later learn that in Iraqi Arabic *nic* resembles the word for "fuck." When a visiting administration official toured Iraq's streets, he was pleased to see kids flash him the thumbs-up sign, only to be told that in Iraq a raised thumb was the equivalent of a raised middle finger. U.S. officials talked incessantly about freedom. (At one CPA briefing, an Iraqi journalist asked an American general why U.S. helicopters flew so low to the ground, scaring local children. "What we would tell the children of Iraq," the general replied, "is that the noise they hear is the sound of freedom.") But Arabic-speakers noticed that when Iraqis spoke back, they talked less about "freedom" (*hurriya*)—which according to George W. Bush all people desired like food and water—than "justice" (*adil*), which had a more confrontational ring. "Marines are from Mars, Iraqis are from Venus," wrote one young major in an e-mail to friends. "I started to realize," noted the *Washington Post*'s Pulitzer Prize–winning correspondent, Anthony Shadid, "how little any of us—journalists, policy makers, citizens—really understood about Iraq."

The assumptions of ideological dominance were proving wrong, with consequences that undermined America's military dominance as well. Between the spring and fall of 2003, the number of insurgent attacks against occupying forces and Iraqi sympathizers tripled, to roughly one thousand per month. Mortar attacks on the Green Zone became a near-daily occurrence, and Bremer's staff began carrying guns to their offices. In September, 94 percent of Iraqis said Baghdad was more dangerous than it had been under Saddam.

As Iraq grew more savage, the U.S. military grew more isolated. Troops hunkered down in heavily fortified mini Green Zones across the country, periodically venturing out in full body armor, accompanied by Abrams tanks, Bradley fighting vehicles, and Apache attack helicopters for raids against suspected insurgents. But the more raids they conducted, the bloodier the insurgency got. Given America's ignorance of Iraq, its military strategy was the equivalent of putting a blindfolded man in a room filled with fragile objects, handing him a flyswatter, and telling him to smash the buzzing sound. For every insurgent that U.S. forces captured or killed when they stormed through Iraqi neighborhoods, they enraged dozens of ordinary Iraqis whose homes were violated, damaged, or destroyed, and whose relatives were arrested, injured, or killed. Realizing that they lacked the information to hunt insurgents effectively, America's military leaders began a frantic search for better intelligence. U.S. troops rounded up tens of thousands of detainees and dispatched them for interrogation. But here again, America's vaunted military was not up to the job. The endless shipments of detainees overwhelmed the small number of military police assigned to handle them, turning the military prison system into a black hole into which innocent Iraqis disappeared for weeks, months, or even years. Worse, the stress of handling far too many detainees under dangerous and chaotic conditions, while being pressured to gather more and more intelligence, led to a grotesque moral breakdown, as ill-trained and traumatized prison guards brutalized their wards in ways that turned the stomach of the world.

By spring 2004, when this breakdown made Abu Ghraib prison a household name, the battle for Iraqi hearts and minds had turned into a rout. A Gallup poll found that only 5 percent of Iraqis believed the United States was in Iraq "to assist the Iraqi people" and only 1 percent thought it was there to establish democracy. By contrast, 50 percent believed America's primary mission was "to rob Iraq's oil." In March, four U.S. contractors were kidnapped and murdered in the Sunni stronghold of Fallujah; their bodies were mutilated and hanged from the city's main bridge, as locals cheered. Days later, a rebel Shia militia—led by the virulently anti-American and exceedingly popular cleric Moqtada al-Sadr— rose up as well, seizing control of the vast Baghdad slum known as Sadr City. Suddenly a large chunk of Sunni Iraq and a large chunk of Shia Baghdad were in insurgent hands. Between the fall of 2003 and the fall of 2004, insurgent attacks tripled again, to roughly three thousand a month. U.S. troops began writing their blood type on the inside of their helmets before going out on patrol.

Underlying the ethic of dominance had been an assumption of mutual innocence, the belief that beneath every tyrant lay a decent, democratic people, and that America could be the agent of their liberation because its own government was so committed to decency and democracy around the world. But by 2004, after the mutilations in Fallujah and the molestations at Abu Ghraib, Americans and Iraqis looked at each other with cold, un-innocent eyes. Average Americans increasingly saw Iraqis as ungrateful, bloody-minded savages, and average Iraqis saw Americans as haughty, rapacious occupiers. What's more, some Americans began to suspect that the Iraqis were right: that the experience of occupation was debasing America's military, and its entire government. The Iraqis believed America was defiling Iraq, and growing numbers of Americans believed Iraq was corrupting America's soul.

During Vietnam, Lyndon Johnson's brave public face had concealed private agony. He had raged against his predicament: a war he didn't want to fight but thought he couldn't afford to lose; an ethic of foreign policy toughness that was supposed to protect his beloved domestic agenda but was swallowing it instead. George W. Bush was different: He didn't agonize; he believed. In late 2003 and early 2004, as Iraq descended into hell, he gave Americans platitude-laden pep talks, which boiled down to "Buck up, we're going to win." And in private, to the amazement of White House visitors, he said essentially the same thing. "It takes an optimistic person to lead. Who's going to follow a leader who says, 'Follow me, things are going to get worse?'" he told aides. "I'm the calcium in the backbone." Calcium in the backbone—Bush loved the phrase and repeated it often. The implication was clear: Doubt, even private doubt, perhaps especially private doubt, makes you spineless. "If you're weak internally," he insisted, "this job will run you all over town."

Bush's harshest critics often assumed he didn't believe his own words, that he was simply lying, which was understandable given his tendency to say things like "We found the weapons of mass destruction. We found biological laboratories. . . . [F]or those who say we haven't found the banned manufacturing devices or banned weapons, they're wrong. We found them"—as he did in late May 2003. But the accusations of duplicity partly missed the point. According to Chief of Staff Andrew Card, Bush genuinely believed that Saddam had possessed WMD, not just in 2003 and 2004, but as late as 2006. If Bush was lying, he was first and foremost lying to himself.

Far from admitting error, in fact, Bush was doubling down. Soon after Saddam's fall, he created the Greater Middle East Initiative, a White House office meant to promote freedom across the Muslim world. It was led by Elliott Abrams, the man who a decade and a half earlier had helped to inaugurate dominance foreign policy by pushing to invade Panama. Then, in January 2005, Bush delivered an inaugural address that employed variations of "freedom" and "liberty" forty-four times, including nine in the last two paragraphs alone. "History," he declared, "has a visible direction." And America would help push it forward, "with the ultimate goal of ending tyranny in our world."

For a moment, events actually seemed to be moving that way. In

January, an impressive 59 percent of Iraqis went to the polls to elect a National Assembly. In February, the Cedar Revolution brought an end to Syria's three-decade occupation of Lebanon. In March, the Tulip Revolution brought democracy to Kyrgyzstan. Palestinians held relatively free elections in the Gaza Strip and West Bank. Like the toppling of Saddam Hussein in spring 2003, the democratic breakthroughs of spring 2005 offered a tantalizing glimpse of a transformed world. But once again it proved a mirage. In the Gaza Strip, Palestinians elected the theocratic, virulently anti-Israeli militia Hamas—which led the Bush administration, in flagrant violation of its democratic sermonizing, to privately urge its moderate Palestinian allies to try to overturn the election results by force. In Lebanon, the coalition that had risen up against Syrian rule fractured along largely religious lines, with the theocratic, virulently anti-Israeli militia Hezbollah emerging as the nation's most powerful force. In Iraq, Sunnis largely boycotted the vote and multi-ethnic, secular parties were crushed. The insurgency raged on.

The democracy escalator, it turned out, was not progressing ever upward. Not in the Arab world, where people kept electing medievalists, sectarians, and thugs. Not in theocratic Iran, which was exploiting Iraq's weakness to challenge American dominance of the Persian Gulf. Not in Latin America, where Hugo Chávez's illiberal, anti-American populism was metastasizing across the continent. Not in Russia, which under Vladimir Putin was marching brazenly down the escalator toward autocracy while ordinary Russians cheered. And, most important, not in China, whose dictatorship Americans had once assumed was living on borrowed time. Now Chinese tyranny seemed stable, prosperous, and confident—an alternative model for the world. Beijing "no longer . . . tries to cultivate influence in distant African countries as it did in the 1960s," Fukuyama had written in 1989, in explaining that America no longer had global ideological competitors. But by Bush's second term, Beijing was cultivating widespread influence in Africa, and across the globe. One academic declared it "the end of the End of History."

In 2006, Jeane Kirkpatrick died. Her final years had been rough. She had struggled to finish a book on post–cold war foreign policy, a task that grew harder as her faculties diminished and her strength ebbed. Among top officials at the American Enterprise Institute (AEI), where she had once been a towering force, she became a figure of ridicule. When discussing which scholar to recommend for a prestigious commission in 2005, one

AEI administrator joked, "Maybe we'll send Jeane," prompting snickers among colleagues. Eventually, after almost three decades, AEI booted her out the door without so much as a farewell party.

But the book did finally appear, posthumously, with the help of a friend. From the grave, Kirkpatrick lashed out against the Iraq War and planted a final flag for the older, grimmer conservatism that institutions such as AEI had abandoned. "Democracy requires security as a prerequisite," she declared. "That is why, throughout history, if the single force of political stability in a region is removed without critical institutions in place to fill the resulting vacuum of power, the security of societies and their budding institutions will be precarious at best." It was "Dictatorships and Double Standards" all over again, an attack on the assumption that beneath every tyrant was a democracy waiting to flower. Kirkpatrick, noted her co-author, had learned her politics in the shadow of the Holocaust. Unlike George W. Bush, she believed "that we are born not as angels, but more like animals," pulled as powerfully toward darkness as toward light.

The book itself garnered only modest attention. But by 2006, the year of Kirkpatrick's death, her ideas were gaining new life, even among some of the younger conservatives who had once scorned them. "I used to see the world as a landscape of rolling hills," wrote David Brooks. "People everywhere seemed to want the same things: to live in normal societies, to be free, to give their children better lives. Now it seems that was an oversimplified view of human nature." Conservatism, declared George Will, had become infected by a missionary, crusading spirit, and it was time for an exorcism. "Three years ago," he wrote, "the [Bush] administration had a theory: Democratic institutions do not just spring from a hospitable culture, they can also create such a culture. That theory has been a casualty of the war that began three years ago."

In March, Fukuyama joined the counterattack, publishing *America at the Crossroads*, a kind of eulogy for Kirkpatrick and Irving Kristol's older, more cautious neoconservatism. The dominance conservatives, he argued, had betrayed their ideological parents and left a once-proud intellectual tradition in ruins. Kagan responded with a book called *The Return of History and the End of Dreams*, which claimed that it was actually Fukuyama who had been too quick to count dictatorship out. "The autocratic tradition," Kagan insisted in a companion article, "has a long and distinguished past, and it is not as obvious as it once seemed that is has no future." All of a sudden, the most important intellectuals on the foreign policy right were loudly denying that democracy was on the march, and attacking each other for ever having been so naïve as to believe such a thing.

"I'm delighted to see Mr. Bourne with us today." Casey Blake, a Columbia University history professor, was welcoming Randolph Bourne back to his alma mater, which was a little odd, given that Bourne had been dead for more than eighty years. But surveying the crowd of antiwar students in the fall of 2004, Blake detected Bourne's presence. The tiny, bent figure with the acid pen had been rescued from oblivion once before, during Vietnam, when anti-toughness intellectuals like Marcus Raskin and Noam Chomsky needed a muse. Now, Blake was suggesting, he had returned again. In 2001, activists opposed to the war on terror had founded the Randolph Bourne Institute. Earlier in 2004, the *Atlantic*'s James Fallows had invoked Bourne to protest the Bush administration's wartime propaganda. Now Columbia was sponsoring a symposium on Bourne's new relevance. And Bourne wasn't the only ghost of hubris past to haunt American foreign policy during the Iraq years. At around the same time, a political scientist and former army officer named Andrew Bacevich—who would become one of the war's most influential critics—set out to revive the memory of Bourne's old professor, Charles Beard.

"If Bourne's ghost hovers around us," declared the historian Robert Westbrook, one of the speakers at the Columbia symposium, "it is less William Kristol, Richard Perle and Paul Wolfowitz who would interest him than the likes of Paul Berman, Christopher Hitchens, and Michael Ignatieff, left-wing intellectuals who echoed the arguments of [John] Dewey and other pro-war progressives as they rushed to support the war in Iraq as above all a war for human rights." Bourne would have known what to do with such people, Westbrook suggested. He would have gutted the liberal hawks as mercilessly as he gutted their intellectual forefathers during World War I.

By 2004, the gutting was well under way on the intellectual left. Before Iraq, what had united liberal hawks and dominance conservatives was the belief that the spread of human rights and the spread of American military power generally went hand in hand. That had been the lesson of the Balkans, where American bombers helped stop genocide, and of Afghanistan, where GIs escorted little girls to school. But in Iraq, many liberals again learned the lesson of Bourne and the New Left: that war was not an instrument of democracy, but a menace to it. The "war-technique," wrote Westbrook, quoting Bourne, "determines its own end—victory, and government crushes out automatically all forces that deflect, or threaten to deflect, energy from the path of organization to that end." In the opinion of many liberals, that was exactly what America's government had done in Iraq. To justify the war it had hyped the threat of WMD, and to prosecute the war it had sodomized innocents at Abu Ghraib. And since the "war on terror" threatened to go on forever, the "war-technique" threatened to eat away at American democracy until it was an empty shell. If influential conservatives were turning against dominance foreign policy because they doubted war could bring democracy to Iraq, many liberals were turning against it because they believed the war in Iraq was corroding democracy in the United States.

The hubris of dominance, like the hubris of reason and the hubris of toughness before it, had relied on faith in political authority. Public trust and public malleability had permitted elites to march the nation to war. But in the late Bush years, Iraq generated a distrust of political authority unseen since Vietnam. Hollywood began pumping out TV shows and movies that pictured Islamic terrorists as mere pawns of the world's true villains, who resided at the White House, Pentagon, and CIA. In *Syriana* (2005), the CIA assassinates a reformist Arab prince trying to free his nation from America's imperial death grip. In the Bourne trilogy (no relation to Randolph), a latter-day James Bond uses quick wits, martial arts, and high-tech gadgets to defeat the bad guys and get the girl—except that in this case, the bad guys are his own former bosses at the CIA. In *Mission: Impossible III* (2006), another supertough, supercool intelligence agent battles a dastardly international arms smuggler, only to learn that the smuggler is a front man for neoconservatives eager to provoke war in the Middle East. Even the television series *24*, which initially depicted the war on terror as a brutal but necessary struggle, in 2006 swerved into antigovernment paranoia. The show's first president, the wise and honest David Palmer, is succeeded by Charles Logan, who tries to provoke a terrorist attack against America so he can seize Central Asia's oil.

If, in Iraq's wake, filmmakers saw Washington as a treacherous place, so did a new generation of liberal activists dedicated to exposing the fiction that America's governing elite was either wise or benign. Just as Vietnam had helped create the adversary journalism of the 1970s—unwilling to take official statements at face value, always on the prowl for corruption, disdainful of the deference previously accorded people in power—Iraq also created a brazen, in-your-face journalistic culture, this time on the Internet. On the "netroots" blogs of the late Bush years, profanity was common, insults were rampant, and the guiding assumption was that the "Washington establishment" (a term that covered pundits as well as politicians) was stupid, pompous, predatory, and corrupt. Iraq, of course, was Exhibit A. The netroots, explained Matt Stoller, one of its stars, "is a group of people who came to the web because we felt betrayed by a system we formerly trusted." The most successful liberal bloggers, he noted, win their audiences by exposing the fact "that power and authority was built on silly illusions."

As in the 1920s and the 1970s, this revolt against political authority was, above all, a revolt against presidential authority. In 2007, in a sign that the Internet's adversarial culture was permeating the journalistic establishment, the *New York Times* announced that it would no longer attend the White House Correspondents' Association Dinner, which had become a symbol of the media's incestuous relationship with the executive branch. In 2005 and 2006, Congress placed the first real restrictions on Bush's ability to detain and torture "war on terror" suspects. It also squashed his domestic agenda, rebuffing White House efforts to alter immigration law, partially privatize Social Security, and extend tax cuts. "What you have seen," said Republican senator Lindsey Graham, "is a Congress, which has been AWOL through intimidation or lack of unity, get off the sidelines and jump in with both feet."

The judiciary also awoke from its slumber. In 2004, after three years of docility, the Supreme Court ruled that judges could decide whether non-Americans were wrongfully imprisoned at the U.S. naval base at Guantánamo Bay, Cuba. Two years later, it deemed the military tribunals that the Bush administration had established there unconstitutional. In 2007, the U.S. Court of Appeals for the District of Columbia Circuit ordered the Bush administration to release information about Gitmo detainees. And in 2008, in a fourth straight legal slap, the Supreme Court ruled that detainees could challenge their detention in federal court. "The irony," noted former Reagan administration lawyer Bruce Fein, "is that the president has now ended up with lesser powers than he would have had if they had made less extravagant, monarchial claims."

Finally, the public itself delivered the most crushing blow to Bush's once-imperial presidency, handing the Democrats control of both houses of Congress in the landslide midterm elections of 2006. Rumsfeld was fired. On his last night as defense secretary, the man who in 2003 had received a louder ovation than Michael Jordan took his family to an upscale

Washington restaurant. It was not exactly a triumphant scene. "I'm not serving a war criminal," shouted the chef, who only relented after another cook agreed to prepare Rumsfeld's meal.

For his part, Cheney now registered lower public approval ratings than Michael Jackson after he was accused of pedophilia and O. J. Simpson after he was tried for murder. In Bush's second term, his network of allies and moles—Rumsfeld, Wolfowitz, John Bolton, and, most important, Chief of Staff Scooter Libby, who was sentenced to thirty months in prison for lying under oath—largely came apart, substantially reducing the vice president's influence. Thirty pounds overweight, suffering from gout, and the victim of no less than eight cardiac episodes during his time in the White House, the most powerful vice president in history was now less fearsome than pitiable.

Bush, who after 9/11 had boasted the highest presidential approval rating in American history, was now the most disliked president on record. Pollsters found that attaching his name to even popular proposals turned the public against them. Even within the military, a Republican bastion, he could not escape public fury over the war. "See? It's not worth it," yelled a family member at Brooke Army Medical Center in San Antonio during Bush's visit, pointing to their disfigured relative. At the National Naval Medical Center in Bethesda, Maryland, the mother of a dying soldier burst into tears, then screamed at the president, "This is not your daughter!" In the summer of 2008, Bush became the first sitting president since Lyndon Johnson not to attend his party's presidential convention.

At first, Bush's descent and Iraq's went hand in hand. In February 2006, a Sunni attack on a Shia shrine in the town of Samarra triggered an orgy of sectarian slaughter horrifying even by Iraqi standards. Cities, neighborhoods, even streets were "ethnically cleansed." People on the wrong side of the sectarian divide ended up with limbs hacked off or drill holes in their head.

In May, with domestic opinion turning hard against the war, Congress tried to seize control. It created an outside panel, the Iraq Study Group, led by George H. W. Bush's former secretary of state, James Baker, and stocked with former Bush I officials. It was the foreign policy equivalent, one panelist quipped, of a family intervention. The Study Group proposed withdrawing tens of thousands of U.S. combat troops from Iraq and initiating negotiations with the anti-American dictatorships in neighboring Syria and Iran. Militarily, the group was suggesting, America could not impose its will on Iraq, and ideologically it could not impose its will on the Middle East. For the first time in the dominance era, a prominent, bipartisan group of foreign policy elites was demanding the retrenchment of American power.

In 1968, when the titans of the foreign policy establishment told Lyndon Johnson that the jig was up in Vietnam, he cursed and moaned, but ultimately conceded defeat. But in 2006, when they delivered the same news to George W. Bush, he told his father's consiglieres to go to hell. He made no bold diplomatic overtures to Damascus and Tehran, and instead of withdrawing troops from Iraq, he added them, in what became known as "the surge." He would keep the troops in Iraq, Bush vowed, even if his only supporters were his wife, Laura, and Barney, the dog.

Most foreign policy experts, most politicians, and most ordinary Americans reacted with a mixture of disbelief and rage. Soon, however, something extraordinary happened: The situation in Iraq began to improve. By the time additional U.S. troops arrived in 2007, Iraq's Sunnis had come to a strange and painful realization: America was their last, best hope. Ever since the Iraq War began, Sunni leaders had deluded themselves into believing that if they drove out the U.S. occupiers they could reestablish their historic dominance over Iraq's Shia majority. In the civil war of 2006, however, that dream died as Shia militias expelled Sunnis from much of Baghdad. Suddenly, the 160,000 U.S. troops in Iraq no longer looked like the barrier standing between Sunnis and supremacy; they looked like the barrier standing between Sunnis and extermination. As a weapon against the Americans, many Sunni leaders had found Al Qaeda useful. But now those leaders wanted U.S. protection, which required casting the jihadists aside. This they did with little remorse, since the jihadists had alienated even conservative Sunnis with their nasty habit of maiming and killing Iraqis for alleged violations of Islamic law.

The timing was most fortuitous. At the very moment Sunnis realized they needed American protection, American troops started flooding their cities and towns. The United States not only had more forces in Iraq as a result of the surge but also began deploying them differently, sending troops out of their isolated bases to live among ordinary Iraqis, as dictated by the counterinsurgency doctrine embraced by General David Petraeus. By late 2007, Sunni tribesmen were turning en masse against Al Qaeda and toward the United States.

This, in turn, changed Shia politics. When Sunni violence was rampant, many Shia had turned to Moqtada al-Sadr's Mahdi Army for protection. But over time, the Mahdi Army had degenerated into a criminal gang, offering protection of a Soprano-esque sort while muscling in on local businesses. When Sunni violence declined, the Mahdi Army's services became less appreciated, and when a resurgent Iraqi army—backed by more U.S. troops—began challenging Sadr's men for control of the Shia streets, the Mahdi Army lost public support. Realizing that he could summon neither the military firepower nor the popular enthusiasm necessary to defeat the Americans, Sadr called a cease-fire and then dropped out of sight.

It was a stunning turnaround: Attacks on U.S. soldiers, which had totaled roughly 1,500 a week when the surge began, dropped to around 150 a week by spring 2009. But the larger truth remained: Dominance foreign policy was dead. The ethic of dominance had been premised on the belief that the United States could win wars fast, and with minimal outlays of American cash or blood. But by 2009, an occupation that was supposed to last six months was entering its sixth year, and "the quiet consensus emerging among many people who have served in Iraq," noted military expert Thomas Ricks, was that the United States would still be fighting there in another six. In 2002, White House economic adviser Lawrence Lindsey had been fired for suggesting that the war might cost $200 billion; by 2009, according to one estimate, it had cost America $3 trillion. On September 11, 2001, Al Qaeda had taken three thousand American lives; by 2009, the war in Iraq had taken more than four thousand.

Petraeus's counterinsurgency strategy had proved brilliant, but it was also slow, costly, painful, and morally ambiguous—exactly what the Bush administration had promised the Iraq War would not be. The United States had not vanquished the Sunni insurgents; it had cut deals with them. Iraqis who killed Americans had not received retribution; they had received paychecks. The American military had shown that it could fight an insurgency, but the Iraqi insurgents had offered the world an instruction manual for grounding the American juggernaut: Bog the Americans down in a protracted, gruesome guerrilla war and bleed them until their ADD-afflicted public demands that their leaders turn off the TV. Even if you don't force the Americans out, you will force them to the bargaining table, where they will shower you with money, which you can use for the power struggle that begins in earnest when they leave.

The Taliban got the message loud and clear. In 2008 and 2009, even as Americans grew more optimistic about Iraq, they grew increasingly despondent about Afghanistan. Exploiting the Bush administration's refusal to launch a serious nation-building effort in 2002 and 2003, the Taliban and Al Qaeda had come roaring back, seizing large chunks of the Afghan south and sizable portions of neighboring Pakistan. In the United States, views of Afghanistan were coming full circle. Right after 9/11, many foreign policy commentators had warned that building a functioning, liberal, pro-American regime there was impossible: the country was too backward, too savage, too xenophobic. By 2002, that pessimism had been largely forgotten as Americans celebrated their rapid overthrow of the Taliban. By 2009, however, all the old clichés—"quagmire," "graveyard of empires"—were back. Vietnam analogies, which had been so out of style when America went to war in Iraq, were suddenly everywhere. Commentators who had once noted fondly that Hamid Karzai had brothers who were restaurateurs in Boston and Baltimore now discovered that another brother was one of Afghanistan's biggest drug kingpins. Americans were again speaking about Afghanistan in the language of despair.

By 2009, even policymakers who believed victory in Afghanistan was possible were defining it far more modestly than they had in the hubris-swollen days of 2002 and 2003. For America's top military leaders, the lesson of Iraq was that the United States would have to pay a far higher price for far humbler goals. On Iraq, Petraeus called himself a "minimalist." The United States, explained his strategic adviser, Major General David Fastabend, will have "to settle for far less than the vision that drove it to Baghdad." And when Petraeus was named head of Central Command, which gave him authority over the Afghan War, he encouraged this minimalism there, too. Even as the new administration of President Barack Obama sent more troops to Afghanistan, it signaled a willingness to cut deals that allowed the brutal, misogynistic Taliban a share in power. The Taliban, Obama officials bluntly conceded, would never be vanquished.

In both Mesopotamia and the Hindu Kush, the dream of ideological dominance was fading. After a massively rigged Afghan election in 2009, few Americans still saw Karzai as a committed democrat. And in Iraq, noted the Brookings Institution's Kenneth Pollack, "Many Iraqis (and Americans) believe [Prime Minister Nuri al-] Maliki intends to make himself a new dictator." American policymakers still hoped that democracy would survive in Afghanistan and Iraq, of course, but the bald truth was that ensuring its survival was no longer part of America's mission. In both countries, America now defined "victory" not as the creation of governments that served as ideological models for the Muslim world, or even governments that served as bases for U.S. power, but merely governments that could prevent their territory from incubating attacks against the United States. In defining victory down, U.S. officials backhandedly accepted Kirkpatrick's old logic: that in some parts of the world, the lesser evil was the best America had a right to expect.

If America's military and ideological dominance was ebbing, its economic dominance soon followed. In the late 1990s, when Governor George W. Bush was pondering a presidential run, Lawrence Lindsey had warned him that a bubble was developing on Wall Street as the result of inflated high-tech stocks. "There's a good chance it'll burst while you're president," Lindsey cautioned. Bush took little heed. In fact, as president he occasionally needled Lindsey on the subject. "How's that bubble doing, Lindsey?" he joked, as mortgage-backed securities replaced Internet stocks as the darling of Wall Street and the stock market soared ever higher. "Gonna burst any day now?"

In the fall of 2008, it did. Even before the meltdown, commentators had warned that U.S. economic power was waning. From 2001 to 2008, America's share of world GDP declined every year. America remained the world's largest consumer, the primary place where Chinese factories disgorged their mountain of manufactured goods, and the largest recipient of foreign investment, the primary place where Chinese and Middle Eastern investment funds parked their vast savings. All this investment subsidized the profligate habits of America's government and people, both of whom lived beyond their means. Even before the crash, however, experts warned that all this borrowing was sapping America's geopolitical muscle. As the Council on Foreign Relations' Bradley Setser noted in a 2008 paper, "a debtor's capacity to project military power hinges on the support of its creditors." One of dominance foreign policy's critical, if generally unspoken, assumptions was that the United States could economically sustain its aggressive military posture. But that aggressive posture, according to Setser, now required the forbearance of China, which served as America's bank.

Then came the financial deluge. In the short term, investment in the United States actually rose, since in a time of panic America seemed a safer bet than most other places. But to forestall an economic depression, Washington began printing money wildly—and plunged massively into debt. In financial circles, observers grew increasingly pessimistic about the long-term value of the dollar, sparking fear that Asian and Middle Eastern investors might stop buying as many greenbacks, which might force

Americans to begin living within their means—a frightening prospect for a populace and a government that have not done so in a very long time.

And if the financial crisis threatened to drain America of cash, it also drained it of prestige. In the late 1990s, America's ideological dominance had rested not only on the belief that history's escalator was carrying the world toward democracy, but that it was carrying the world toward American-style capitalism as well. But that trajectory looked far less inevitable after the financial crash, when America—which had spent the past two decades telling the world to deregulate its financial markets—began struggling to reregulate its own. "The teachers now have some problems," noted Chinese vice premier Wang Qishan drolly. French president Nicolas Sarkozy was blunter: "Le laisser-faire, c'est fini."

It was a good moment to be Paul Kennedy. In the 1990s he had been a figure of intellectual sport, the poor sap who bet against the United States just before it left the world in its dust. But the critics weren't laughing anymore. In 2006, China published a new edition of *Rise and Fall of the Great Powers*, accompanied by an eight-part television series, largely cribbed from its pages. In 2008, the American Political Science Association hosted a panel to mark the twentieth anniversary of Kennedy's book. Its title: "Is the United States in Decline Again?"

With a roar, American intellectuals shouted yes. In January 2008, in a *New York Times Magazine* cover story, the New America Foundation's Parag Khanna heralded the birth of the "Non-American world." Four months later, *Newsweek*'s Fareed Zakaria published *The Post-American World*. And in 2009, *National Journal*'s Paul Starobin came out with *After America*, which lionized Paul Kennedy for being ahead of his time. "Kennedy was savaged in certain quarters," noted Starobin in a companion article. "And then he was ridiculed. . . . [But] these days, Kennedy is looking less like a heretic and more like a prophet."

"American civilization," wrote Starobin, "has reached the end of its long ascendancy in the world." And many ordinary Americans seemed to agree. A 2009 Pew Research Center poll found that only 36 percent of Americans believed their children would have a better life than them, down 19 points from 1999. Among Indians, by contrast, the figure was 78 percent, and among Chinese, 89 percent.

Among the American pessimists was George Kennan. His great realist contemporaries—Morgenthau, Lippmann, and Niebuhr—had all died during the Vietnam War or its aftermath, and spent their final years in political despair. Kennan, however, had lived to see the miracle of the cold war's end, and having seen it, he told family and friends that he was now content to die. But he didn't die. He lived on, past the fat, carefree '90s, past 9/11, and all the way to Iraq, a war that filled him with dread. "I take an extremely dark view of all this—see it, in fact, [as] the beginning of the end of anything like a normal life for all the rest of us," he wrote to his nephew on the eve of the war. "What is being done to our country today is surely something from which we will never be able to restore the sort of country you and I have known." He asked that the letter be destroyed, but it never was. Having lived to see his greatest hopes realized, he had now lived to see them dashed again. A decade and a half after tipping his hat to Wilson, his pessimism had been vindicated after all. In 2005 he died, like the others, in political despair.

THE BEAUTIFUL LIE

"We need some great failures . . . we ever-successful Americans—conscious, intelligent, illuminating failures."
—LINCOLN STEFFENS

Easy for him to say. It's hard to be illuminated by failure in a country where success is a national religion. It's hard to say "no, we can't" in a country that spent its first century conquering a continent and its second conquering much of the world. The American character, wrote Arthur Schlesinger, Sr., "is bottomed upon the profound conviction that nothing in the world is beyond its power to accomplish."

But that conviction is a lie. Much is beyond our power to accomplish, especially when it comes to the world beyond our shores. When America's leaders fall in love with the lie, when success convinces them it might actually be true, when they forget we fly on wings of wax, the gods take their revenge.

Yet if they denounce the lie, expose its absurdity, declare themselves deaf to its infantile charms, they exile themselves from our political tradition. Intellectuals may warn that we are too ignorant and overstretched to remake the world; ordinary Americans may quietly acknowledge that we will never fully vanquish our foes—in the same way they quietly acknowledge that politicians will always have affairs. But woe to the leader who speaks those heresies out loud. "My fellow Americans, we must lower our sights, reconcile ourselves to evils that have no remedy, admit that the world will not bend to our will"—the words cannot be spoken. You cannot stand up in church and denounce the virgin birth. Privately, we speak in different tongues, but officially, can-do-ism is the only vocabulary we have.

So the lie is politically essential. But more than that: It's morally essential. In bad times, it helps us to look beyond the dismal present, to find hope in the unseen. Arthur Schlesinger, Sr.'s, analysis of the American character, after all, was delivered during the winter of 1942, when Nazi flags flew from the Atlantic to the Volga and the Arctic to the Sahara. His optimism was less an invitation to overreach than a counsel against despair. For the French and British in the 1940s, or the Bosnians and Kosovars in the 1990s, American can-do-ism was a very welcome thing.

What America needs today is a jubilant undertaker, someone like Franklin Roosevelt and Ronald Reagan who can bury the hubris of the past while convincing Americans that we are witnessing a wedding, not a funeral. The hubris of dominance, like the hubris of reason and the hubris of toughness before it, has crashed against reality's shoals. Woodrow Wilson could not make politics between nations resemble politics between Americans. Lyndon Johnson could not halt every communist advance. And we cannot make ourselves master of every important region on earth. We have learned that there are prices we cannot pay and burdens we cannot bear, and our adversaries have learned it, too. We must ruthlessly accommodate ourselves to a world that has shown, once again, that it is not putty in our hands.

For starters, that means remembering that we did not always believe we needed to dominate the world in order to live safely, profitably, and ethically in it. In the decade and a half after the Soviet empire fell, dominance came so easily that we began to see it as the normal order of things. We expanded NATO into East Germany, then into Eastern Europe, then onto former Soviet soil, while at the same time encircling Russia with military bases in a host of Central Asian countries that once flew the hammer and sickle. We established a virtual Monroe Doctrine in the Middle East, shutting out all outside military powers, and the Bush administration set about enforcing a Roosevelt Corollary, too, granting itself the right to take down unfriendly local regimes. In East Asia we waited expectantly for China to democratize or implode and thus follow Russia down the path to ideological and strategic submission. And we stopped thinking about Latin America much at all, since we took it as a virtual fact of nature that no foreign power would ever again challenge us in our backyard.

We were like the warrior guarding his village who suddenly finds that the enemy has abandoned the battlefield, leaving vast tracts of territory undefended, and so takes them for his own, since the acquisition apparently involves little risk and cost. And once those lands have been incorporated, he sees that even more is available: The inhabitants offer little resistance and even appear pleased to join the realm. And as his domain extends outward, the warrior begins to see its new size less as a choice than a necessity: the bare minimum necessary to keep his family safe. The old borders, which he once deemed sufficient, now strike him as frighteningly exposed. In fact, he comes to suspect that even his current territory is inadequate; he has grown so used to expansion that mere stasis strikes him as a form of retreat. And meanwhile, the lands just beyond his domain are no longer so welcoming or unguarded, and mutinies have broken out in some of his recent acquisitions. Fulfilling his obligations is no longer so effortless and the resources at his disposal are no longer so plentiful. His challenge is to step back from the border skirmishes that now consume his time and try to recover the more disciplined habits of mind that guided him in the days before the recent windfall, because the days of windfall are now clearly gone.

If the men and women who shape American foreign policy conduct this intellectual audit they will discover a sharp discontinuity between some of today's widely held assumptions and the assumptions of successful American policymakers in eras past. After 9/11, in the name of fighting terror, the Bush administration declared war or cold war on Iraq, Iran, Syria, the Taliban, Hezbollah, and Hamas, virtually every significant regime and militia in the greater Middle East that did not kiss our ring. And in its pursuit of regional dominance, it claimed that it was merely doing in the Muslim world what past generations had done in Europe and Asia. But that's not right. Franklin Roosevelt did not wage World War II so America could be the world's sole superpower, or even Europe's. He wanted Four Policemen; unipolarity was Hitler's goal. And FDR did not wage war against all the enemies of freedom. He allied with Stalin to defeat Hitler and Tojo. Similarly, during the cold war America did not take on the entire communist world, except for a period of hubristic intoxication that began with McCarthyism and culminated in Vietnam. In the late 1940s we made common cause with the communists in Belgrade, and in the 1970s and '80s we made common cause with the communists in Beijing, all to contain the communists we feared most, who resided in Moscow. George Kennan saw the purpose of containment as ensuring that no single power controlled the world's centers of economic and military might, not ensuring that that single power was the United States.

How could our forefathers have been so cowardly and immoral? Stalin was a monster; so was Mao. And they both had nuclear weapons aimed at us. Why did we live with that sword of Damocles? Why did we accept their dominion over billions of souls? Once upon a time, the answer was obvious: because we lacked the power not to. Franklin Roosevelt knew the American people would not sacrifice their sons by the thousands to keep Eastern Europe from Soviet hands. During Korea, Harry Truman blundered into war with Beijing and realized that in Asia too the price of denying America's communist foes a sphere of influence was far too high. Even Ronald Reagan proved so reluctant to challenge Soviet control over Poland in the early 1980s that conservative commentators cried betrayal. In different ways, all these presidents understood that in foreign policy, as in life, there are things you may fervently desire but cannot afford. And in foreign policy, that recognition is even more important, since you are not merely spending other people's money; you are spilling other people's blood.

In our time, these tragic choices have been largely airbrushed from public memory. World War II has become the story of America single-handedly saving Europe from Nazi totalitarianism, even though U.S. troops didn't hit the beaches of Normandy until our Soviet totalitarian allies—who lost ninety men for every one of ours—had already turned the tide. The cold war has become the story of America's triumph over communism, even though we played off one communist giant against the other. As a result, many policymakers and pundits have come to see prioritizing among adversaries as immoral and un-American, a view exemplified by the Bush administration, which spurned Iran's offers of help against Al Qaeda, did little to turn Damascus against Tehran, waited until 2007 to exploit divisions between nationalist and jihadist insurgents in Iraq, and made no meaningful distinction between Al Qaeda and the Taliban in Afghanistan. The assumption was that a nation as powerful as ours did not need to choose.

Now the days of reckoning have arrived. Our commitments have grown massively since the cold war's end, but our resources—economic, military, and ideological—are not what they once appeared. In Walter Lippmann's phrase, American foreign policy is "insolvent." Our obligations exceed our power. Yet like a homeowner unwilling to sell his house for $500,000 because he once believed it worth $1 million, we cling to illusions born in easier times. We lack even the vocabulary for hard choices. In public debates over foreign policy, commentators talk endlessly about combating foreign threats and promoting American values, but much less about safeguarding American interests. Without a prior conception of interests, however—a conception of what the world must look like for America to be safe, prosperous, and free—intelligently deciding what constitutes a threat is impossible. And without a prior conception of interests, promoting American values is an infinite project. Talking about threats makes us feel tough and talking about values make us feel virtuous, but only talking about interests forces us to acknowledge the limits of our ability to be either tough or virtuous. This discomfort with the language of interest is a symptom of America's post–cold war inability to prioritize. We remain in thrall to a series of assumptions about American omnipotence that, if not challenged, threatens to drive our foreign policy deeper into the red.

The first assumption concerns terrorism. After September 11, America's leaders warned that Al Qaeda would likely strike again on American soil and with even greater force. Partly this was a natural response to the agony of the moment. But the decision to describe terrorism in apocalyptic terms—to see 9/11 not as the limit of Al Qaeda's capacities but as a mere warm-up for the real horror to come—also stemmed from America's extraordinary post–cold war confidence. We defined the terrorist threat as virtually unlimited because we believed we had virtually unlimited resources to fight it. Today, by contrast, our resources are gravely strained, and yet in Washington, official rhetoric about terrorism remains almost as apocalyptic as it was eight years ago, even though Al Qaeda has not struck the United States—or any other country—on nearly the scale of 9/11. Obviously, jihadist terrorism remains a threat. Yet our refusal to reevaluate the severity of that threat in the face of new information carries a cost. Since, officially, the terrorist threat has not been downgraded, homeland security and defense spending remain politically sacrosanct, despite a budget deficit that has swelled dramatically. And since, officially, the terrorist threat has not been downgraded, it remains politically treacherous to eschew the most aggressive counterterrorist actions—drone attacks in Pakistan, for instance, which kill large numbers of civilians and spark hatred of the United States—even if those actions harm American security more than they help. In their unwillingness to revise official rhetoric about terrorism in the face of eight years' worth of evidence that Al Qaeda is not—thank God—the danger Americans once imagined, American politicians resemble Lyndon Johnson and Dean Rusk, who in the mid-1960s spoke about the communist threat as if Moscow and Beijing were still allies and Stalin was still alive.

The second assumption concerns nuclear deterrence. In 2002, George

W. Bush declared the concept dead. In the post-9/11 world, he insisted, threats of nuclear retaliation were not only useless against Al Qaeda; they were also useless against anti-American dictatorships with WMD. Analytically the argument made little sense. Why, exactly, would deterrence not work against Iraq, Iran, and North Korea? Because they were led by ruthless dictators who espoused fanatical ideologies, aided terrorists, and butchered their own people—unlike, say, Joseph Stalin or Mao Zedong, whom America successfully deterred during the cold war? The logic behind deterrence does not rely on a regime's decency; it relies on a regime's desire to stay in power, something a nuclear exchange would put at risk. In fact, even as the Bush administration repudiated deterrence, it began quietly deterring North Korea, which it had permitted to produce several nuclear bombs.

There is no guarantee that deterrence will work always and everywhere, of course. Although Tehran's theocratic rulers have no history of suicidal behavior (they have never given their terrorist allies biological or chemical weapons, for instance), America and the world would certainly be better off if they didn't get the bomb. But America and the world would have been better off had Stalin, Mao, and Kim Jong Il never gotten the bomb, too; we relied on nuclear deterrence not because it was ideal but because the military alternatives were ghastly. Now the military alternatives are ghastly again. An American or Israeli strike would have grave consequences for the two wars America is already fighting on Iran's border: in Afghanistan and Iraq. If commentators who urge military action against Tehran are willing to imperil success in those conflicts to prevent an Iranian bomb, so be it. But pro-war commentators rarely make that argument, because it would require painful choices among competing priorities. And in Washington today, such choices are rarely publicly acknowledged. Seven years after the invasion of Iraq, the illusions of omnipotence remain: We are so powerful that we do not need to choose.

The third area in which these illusions warp our foreign policy debate concerns Russia. Officially, the United States remains committed to admitting Georgia and Ukraine into NATO, thus continuing the eastward expansion that began when the Soviet Union fell. The logic behind this expansion was that Washington and its allies were so strong, and Moscow was so weak, that we could push closer and closer to Russia's border without ever paying a price. But in 2008, when it invaded Georgia, Russia showed that it could indeed exact a price. Since NATO is a mutual defense pact, admitting Georgia would mean that if Russia attacked again, the United States would be obligated to send young men and women from Cleveland to fight and die for Gori, something the American people would never tolerate.

The United States has been in this kind of situation before. During the cold war, Washington and Moscow struck a de facto bargain over Finland, the U.S.S.R.'s neighbor to the northwest. The United States tacitly accepted Soviet influence over Finnish foreign policy while the Soviets allowed Finland to remain democratic and self-governing in its domestic affairs. The arrangement was ugly and unfair. It required acknowledging that Finland, because of its size and location, would not enjoy the full privileges of sovereignty. But America's leaders knew that they would be doing the Finns no favors by encouraging them to poke the Kremlin in the eye, since America would not protect them from the Soviet response. Today a similar arrangement is likely the best America can expect when it comes to Georgia and Ukraine. But acknowledging that is unpleasant and so most politicians do not. Instead they perpetuate the fiction that we can keep pushing the frontiers of American hegemony outward, even though it requires writing the foreign policy equivalent of bounced checks.

There is a reason politicians do this: In our political culture, publicly acknowledging that something is beyond America's power is perilous. The president who presides over a nuclear Iran or concedes a Russian sphere of influence in parts of the former U.S.S.R. or admits that we will never fully vanquish jihadist terrorism will be, politically, a marked man. He will be accused of betraying Americanism itself, of having desecrated the church of optimism. The aftershocks may not be as great as the ones that upended American politics after we conceded China in 1949 or fled Vietnam in 1975, but they will be tumultuous nonetheless. Ideologues and opportunists alike will insist that the talk of limits was an illusion, that the only limits that mattered were the limits of presidential will. They will insist that everything would have been possible had only real Americans roamed the corridors of power. If only the men and women in the White House had truly believed.

To survive this onslaught—to truly lay the hubris of dominance to rest—Barack Obama will need to redefine our national faith, to decouple American optimism from the project of American global mastery. He will need to find new, more manageable missions that draw upon that beautiful American lie: that nothing is beyond our reach.

First, that means redirecting American can-do-ism inward. During the Bush years, a strange inversion occurred: America's government grew more ambitious overseas even as it grew more complacent at home. When it came to protecting the environment, improving health care, or overseeing Wall Street, President Bush oozed fatalism even as he invaded two countries and vowed to end tyranny on earth. This gets things backward. In politics, as in life, we should be most ambitious in those spheres where we have the most power. There are limits to the federal government's capacities at home as well, of course: limits of both money and knowledge. But we know better how to rebuild New Orleans than how to rebuild Afghanistan, more about how to regulate the U.S. financial system than how to establish one in Iraq. The main purpose of foreign policy is to serve, in Lippmann's words, as the "shield of the republic": to protect our democracy so it can thrive. If American power swells overseas but the quality of life for Americans deteriorates at home, then American foreign policy has failed.

The Bush administration's combination of domestic complacency and overseas grandeur stemmed partly from ideology. But it also stemmed from a particular allocation of American resources. Defense now accounts for well over half of the federal government's discretionary budget, close to five times as much as discretionary spending on health care and education combined. All that money is both a cause and an effect of the fact that Americans think more highly of the military than of other institutions of government. The U.S. military is, to be sure, a remarkable institution, more than worthy of pride. But practically and psychologically, we rely on it too much. As our diplomatic capacity has withered, the military—whose budget is ten times that of the State Department—has become our primary instrument for interacting with the world. As our welfare state has withered, the military has become one of our few effective instruments for economic security and upward mobility at home. It has also become an island of discipline, self-sacrifice, and public-spiritedness in a culture that is materialistic, hedonistic, and guilty about it. We are like a high school that cheers lustily every Friday night for its champion football team, thus distracting itself from its overcrowded classrooms, mediocre test scores, and dismal chess club. And when we do acknowledge that the school is in disrepair, our solution is to ask the football coaches to start purchasing computers and establishing after-school math programs because they have all the money and are the only ones we trust to get anything done.

Now the days of easy gridiron triumph are over, and we must find new sources of school spirit. We must focus more attention on—and devote more money to—the other ways in which we compete with the world:

from educational standards to environmental quality to scientific achievement to public health. We must remember that the military and geopolitical triumphs we commemorate were, at their core, economic triumphs. We won World War II as much on the factory floor—where American workers churned out more and better planes and tanks than their German and Japanese counterparts—as on the battlefield. And we won the cold war because America and its democratic allies made capitalism stable and humane while the Soviets never made communism creative and dynamic. We must retell our narrative of national success as a narrative of economic and technological success. We must remember the things we admired about our country when militarism and Americanism were not so deeply intertwined.

But redirecting American ambition inward will not, by itself, reconcile Americans to the new limits on our power. Americans want their government to project optimism and energy at home, but they also want to be strong overseas. Obama's challenge is to show that strength and dominance are not the same thing. In this regard, he should learn from Ronald Reagan, who scrupulously avoided Vietnam-type military interventions yet found symbolic ways—like awarding the Medal of Honor to overlooked Vietnam hero Roy Benavidez—to make Americans feel proud and strong. As Reagan understood, foreign policy debates are often cultural debates in disguise. Whether it was the Swift Boat attack on John Kerry in 2004 or the attacks on Obama himself for not wearing a flag lapel pin, liberals are often vulnerable during such debates because conservatives cast their criticisms of U.S. policy as evidence of lack of patriotism. Politically, the answer is not for liberals to abandon such criticism, which itself represents a form of patriotism, but rather to leaven patriotic criticism with patriotic affirmation. That means more than ritual incantations about flag and country; it means aggressively challenging those who unfairly deride the United States and its institutions. From a purely foreign policy perspective, publicly confronting Mahmoud Ahmadinejad or Hugo Chávez when they malign the United States may seem useless, or even counterproductive. Likewise, there may be no purely foreign policy rationale for publicly confronting universities that ban military recruiters from campus. But there is no such thing as pure foreign policy; foreign policy requires domestic support. If Barack Obama does not want to be Jimmy Carter, if he does not want Americans to equate his restraint with their humiliation, he must be as aggressive as Reagan in finding symbolic balm for America's wounded pride.

Even this, however, is not enough. Great presidents reconcile a respect for limits not merely with an affirmation of American pride, but with an affirmation of American ideals. Despite everything, we remain a missionary nation with an enduring desire to repair and redeem the world. That missionary zeal, which has at times produced delusion and catastrophe, can be a force for extraordinary progress if it is tempered by two kinds of humility.

The first is the humility of time. What distinguished George W. Bush in 2003—and Woodrow Wilson in 1917—from more successful presidents was not their belief that liberty could spread around the world; it was their millenarianism, their belief that through the wars they launched it could spread in one big bang. We need more patience. In general, America cannot bludgeon dictatorships into democratizing. In fact, we cannot even deny large authoritarian regimes like those in Russia and China influence over their smaller neighbors. In the short term, our strategy should resemble the vision of great-power cooperation across ideological lines that animated Franklin Roosevelt during World War II. In a world threatened by global warming, financial panic, and swine flu, in fact, the need for such cooperation is even greater now than it was in FDR's day. And because none of the current great powers espouses a revolutionary ideology aimed at overturning the current international order, as Nazi Germany and the Soviet Union did, the opportunity for cooperation is greater, too.

But if cooperation with authoritarian powers requires realism, we should be short-term realists with long-term, nonrealist dreams. Like Roosevelt and Harry Truman, we should seek to enmesh undemocratic regimes in an open, rule-based, global trading system, because capitalism can slowly and subtly create pressures for the rule of law. It is capitalism, after all, that has turned China from a totalitarian state into an authoritarian one, thus allowing over one billion people a degree of personal freedom unimaginable in Mao's day. In addition, like Roosevelt and Truman, we should try to convince governments to sign declarations—about freedom of speech, freedom of religion, due process, and free elections—that rebuke their own behavior. Critics will dismiss these declarations as worthless because the words will fall far short of current reality, and the organizations created to monitor them will be weak, and America will continue to do business with some of the worst offenders. But forcing regimes into hypocrisy will be precisely the point. In a world where technology gives citizens the power to organize as never before, ordinary people may be able to wield these charters as weapons against an unjust status quo, as dissidents in

the Soviet bloc did after the Helsinki Accords, and as Americans themselves did when they protested the Bush administration's violations of
the international ban on torture. Activists may even band together across
borders in ways scarcely imaginable in the pre-Internet age. "Ideals," wrote
Reinhold Niebuhr, "have a way of taking vengeance upon the facts which
momentarily imprison them." When it comes to ugliness in far-off places,
America usually lacks the power to vindicate those ideals in the here and
now. But we can help plant seeds that others may one day reap.

For that to happen, however, the declarations must bind us as well.
Precisely because freedom and dignity are universal values, they are not
ours alone to define. George W. Bush not only lacked the humility of
time; he lacked the humility of place. His universalism was oddly parochial. The freedom agenda, as the Bush administration defined it, meant
pressure for democratic elections, but only if they produced the outcomes
we wanted; pressure for human rights, but not in war-on-terror allies
like Israel, Saudi Arabia, and Pakistan; and a refusal to even contemplate the possibility that America's own anti-terror policies might violate
the principles we preach. What non-Americans know—and Americans
should admit—is that our national interests and prejudices will always
taint our pursuit of universal ideals. We are not some deracinated global
umpire, sitting in judgment over humankind; we are one of the players
on the field.

It is this recognition that our idealism is tainted by self-interest that
should make us pause and pause and pause again before unilaterally invading tyrannical nations on the assumption that their people will thank
us for it. Even if we genuinely believe that we are acting from altruistic
motives, the people whose country we invade will generally be more suspicious, especially if they have been on the receiving end of armed Western altruism before, and especially once an American soldier shoots their
cousin or breaks down their door. And even if we are initially welcomed as
liberators, countries with awful governments usually have awful political
cultures, which do not change easily or quickly. Unless we are willing to
stay long enough to change those political cultures—and Americans are
not known for our patience, especially when it comes to costly endeavors
in far-off lands—we will be building castles in the sand.

George W. Bush was right that people want freedom and dignity for
themselves, but they also want freedom and dignity for their nation, and
they don't always draw a bright line between the two. This is particularly
true for people whose nations have suffered the indignity of imperial rule,

nations that wield far more global influence today than they did a generation or two ago. The shift in power from countries that were once colo-
nizers to countries that were once colonized constitutes one of the most
profound trends in world politics, and any project that looks like idealism
to us but imperialism to them will collide disastrously with the nationalism of peoples whose historical memory has left them with a chip on their
shoulders about Western governments telling them what to do.

Often, our ignorance leaves us unable to even understand why the
chip is there. Listening to the foreign policy debate in Washington, one
would not know that the United States was once a quasi-imperial power
in China, or that we helped overthrow an elected prime minister of Iran.
As a result, Americans are often genuinely confused when Iranians and
Chinese—including Iranians and Chinese deeply committed to human
rights—bristle at our lectures. For such a lustily nationalistic people, we
remain curiously blind to the nationalism of others.

The challenge is to convince postcolonial regimes that their quest
for greater dignity and power ultimately rests on providing their people
greater freedom, and that America can be their partner in that effort.
In this regard, we should keep pushing to reshape global bodies like the
UN Security Council, World Bank, and International Monetary Fund
so their composition no longer reflects the imperial age. We should stop
demanding that third-world nations halt their pursuit of nuclear weapons
without simultaneously acknowledging our own responsibility to curb our
arsenals. We should accept that as a rich nation, which industrialized with
little heed to its effect on the environment, we bear more of the initial
burden of curbing global warming. And we should acknowledge that the
struggle against global poverty—a struggle that looms far larger in countries where millions go hungry—is also a struggle for human rights, and
that in this struggle, America can do far more.

All these efforts differ from the hubris of dominance in one fundamental way: Rather than merely requiring that other nations change, they
require that we change as well. They combine greater modesty in our demands of others with more strenuous demands of ourselves. They require
that instead of imagining that we stand, self-satisfied, on history's highest
rung, we see ourselves as fellow travelers on a path whose course we alone
do not set, and along which we have far to climb.

Lurking behind many of America's current anxieties is the fear that
what will follow hubris is decay; that we are in the autumn of our power;
that other nations will soon occupy history's center stage. But we have

been through this before. In 1969, with America neck-deep in Vietnam, the historian C. Vann Woodward wrote that "history has begun to catch up with Americans, the fabled immunity from frustration and defeat has faltered. . . . America has at last encountered problems that are difficult to reconcile with traditional myths of indomitable optimism." Four decades later, we can see that although communism won the war for Vietnam, capitalism won the war for East Asia, and democracy has won stirring battles there, too. The trauma of our defeat gradually faded, yet American optimism endured, not because we clung to the impossible dream of global containment but because we abandoned it and created space for other, better dreams to take its place. Now another generation of Americans must jettison our visions of invincibility. We should do so joyfully, for the recognition that no collection of mortals can impose its will on an unruly globe is not a sign of decay, but of wisdom. And tempered by wisdom, American optimism is—and always will be—one of the great wonders of the world.

ACKNOWLEDGMENTS

While writing this book, I often felt that I was living my subject—that I had embarked upon my own personal act of hubris. Again and again, I wandered into an intellectual thicket, lost my way, and called out for help. This book only exists because brilliant and generous people—many of them initially strangers—answered my pleas.

I met the first Good Samaritan, ironically, at a bar mitzvah. The University of Virginia's Melvyn Leffler is among the most eminent historians of twentieth-century American foreign policy alive. I met him over hors d'oeuvres at a South Carolina synagogue, and grasping this thin reed, asked if he would chair the Council on Foreign Relations study group for my book in Washington, D.C. Not only did he agree, but he undertook the task with extraordinary diligence—repeatedly driving up from Charlottesville, reading successive drafts, suggesting numerous sources, painstakingly correcting my mistakes, and allowing me to see, up close, how a great historian thinks. In return, he got a couple of meals at Asian restaurants, and this paragraph of thanks, which barely begins to repay my debt of gratitude.

What Mel Leffler is to the historical study of American foreign policy, Fareed Zakaria is to public commentary on American foreign policy: the best. When I asked him to chair the study group for my book in New York, he had every reason to say no, given the crushing demands on his time. He said yes because he found the subject interesting, and because he is a mensch. And every time we discussed the book, he asked the big, hard questions that forced me to dig myself out from underneath the avalanche of historical minutiae and figure out what I was trying to say.

Lloyd Ambrosius of the University of Nebraska and David Steigerwald of Ohio State, two men I e-mailed cold and still have never met in person, devoted hours to helping me understand Woodrow Wilson.

American University's Robert Beisner corrected my misperceptions of the Truman years, while his wife, the classicist Valerie French, offered insights into the Greeks. I spent several long afternoons with George Washington University's James Hershberg, shamelessly cribbing from his encyclopedic knowledge of Kennedy, Johnson, and Vietnam. David Greenberg of Rutgers tutored me on Nixon, Reagan, and everything in between. Princeton's Gary Bass taught me most of what I know about humanitarian intervention in the 1990s. Elliot Leffler was my guide to Aeschylus. Robert Brustein warned me against lazy writing and buoyed my spirits with his praise. Christopher Chen and Adam Stempel helped me confirm a hunch about the generational skew among politicians and pundits over the Iraq War. Alexander Star, a brilliant editor, and Amanda Fazzone, a brilliant copyeditor, each read the manuscript with great care and skill. Columbia's Robert Jervis suggested dozens of useful modifications and sources. Stephen Winch generously supported my work. So did Roger Hertog, who despite considering me hopelessly left-wing, has for many years been a cherished professional and personal guide.

In addition, I worked on this project with three remarkable recent college graduates, who served, in succession, as my research assistants. The first was William Evans, a political junkie with the soul of a scholar who has forgotten more about Theodore Roosevelt than I will ever know. The second was Conor Savoy, who became an invaluable partner in trying to understand the Vietnam years, and was a delight to have around. The third was Jamie Holmes, who worked long and sometimes frantic days in the project's final months, and exhibited an astounding level of organizational ability, analytical skill, determination, and good humor. Between them, Will, Conor, and Jamie shaped every page of this book and made my work life a pleasure for three years. I will watch with pride the great things they do in the years ahead.

In molding their research into a book, I was extremely lucky to work, once again, with my agent, Tina Bennett, and with Tim Duggan at HarperCollins. At critical moments, I relied on Tina's signature combination of intellectual sophistication and professional savvy. And Tim shepherded me through the writing and editing process with insight, diligence, and an infectious enthusiasm. Allison Lorentzen at HarperCollins patiently and expertly answered dozens of e-mails.

Many other people also gave generously of their time and talent. They include Peter Ackerman, Spencer Ackerman, Madeleine Albright, Eric Alterman, William Antholis, Jeremy Bash, Robyn Bash, Warren Bass,

Michelle Baute, Amanda Beck, Rand Beers, Daniel Benjamin, Paul Berman, Casey Blake, John Morton Blum, Laura Blumenfeld, Nina Blustein, Stacy Bosshardt, Spencer Boyer, Nell Breyer, Howard Brick, Clive Brock, Christian Brose, Tina Brown, Frank Calzon, Derek Chollet, Conor Clarke, Steve Clemons, Warren Cohen, Steven Cook, Matthew Continetti, Kristin Cullinson, Greg Craig, Robert Dallek, Jacquelyn Davis, Joy Demenil, James Denton, Alan Dershowitz, Carolyn Dershowitz, John Patrick Diggins, E. J. Dionne, Trish Dorff, Ross Douhat, Colin Dueck, Edward Felsenthal, Michael Flamm, Nic Fokas, Richard Fox, Annabelle Friedman, Caroline Friedman, Peter Friedman, David Frum, Francis Fukuyama, John Lewis Gaddis, Adam Garfinkle, Carl Gershman, Allen Gerson, Mark Gerson, Todd Gitlin, James Goldgeier, Mahen Gunaratna, Jonah Goldberg, Andrew Gundlach, Leigh Gusts, Jacquelyn Hardy, Breton Harned, Owen Harries, Marie Hastreiter, Melvin Heineman, Michael Hirsh, Richard Holbrooke, Peter Holquist, Deneen Howell, Isabell Hull, Lena Hull, Heather Hurlburt, Walter Isaacson, Maurice Isserman, Alan Jones, Katherine Jordan, John Judis, Erez Kalir, Lawrence Kaplan, Paul Kennedy, Thomas Kleine-Brockhoff, David Knoll, William Kristol, Charles Kupchan, Mark Lagon, Julie Lascar, Nicholas Lemann, Meagan Liaboe, Robert Lieber, Matt Lieppe, Kelli Long, Oliver Lough, Brian Lowe, Marc Lynch, Joe Eugene McCarraher, Joseph McCartin, Cara McCarty, Doyle McManus, Joe McReynolds, Walter Russell Mead, Dmitri Mehlhorn, Shelton Metcalf, Susan Moeller, Sylvia Moss, John Mueller, Siddhartha Mukherjee, Donald Mullen, Joshua Muravchik, Gita Murthy, Henry Nau, Richard Niemi, Lia Norton, Brendan O'Connor, Dan O'Hara, Priscilla Painton, Davide Panagia, Richard Pells, Richard Plepler, Ramesh Ponnuru, Sasha Pulakow-Suransky, Romesh Ratnesar, Jonathan Rauch, Kathy Reich, John Renehan, Berenice Ronthal, Lamiya Rahman, Marcus Raskin, Jonathan Rauch, Patrick Rigby, Gideon Rose, Mark Schmitt, Howard Schuman, Peter Scoblic, Stephen Sestanovich, Brad Setser, Aaron Silverman, Anne-Marie Slaughter, David Sloan, Michael Sofer, Michael Smith, Tom Smith, Jose Sorzano, Marcia Sprules, Connie Stagnaro, Allan Stam, Paul Starobin, Alexandra Starr, Sally Steinberg, Craig Stern, Strobe Talbott, Ray Takeyh, Paul Testa, Francoise Thiese, Nicholas Thompson, Stephen Trachtenberg, Nicholas Valentino, Reed Van Beveren, Don Vieira, Peter Wallsten, Baruch Weiss, Maureen White, Steven Whitfield, Leon Wieseltier, George Will, Marshall Wittmann, Adam Wolfson, Matthew Yglesias, and Barron YoungSmith. Needless to say, the book's flaws are not theirs, but mine.

Beyond relying on exceptional individuals, I relied on exceptional institutions. The Council on Foreign Relations, where I was a Senior Fellow from 2007 to 2009, is a phenomenal place to write a book. The only problem with being surrounded by excellent facilities, skilled librarians, stimulating colleagues, and free food is that if your book is lousy, you have no one to blame but yourself. For creating this atmosphere, and allowing me to partake in it, I am grateful to Richard Haass, Gary Samore, James Lindsay, Nancy Roman, and Kay King. I am also grateful to the Smith-Richardson Foundation, and particularly Mark Steinmeyer, for supporting my work at CFR.

I completed the book at the New America Foundation, where I am privileged to be a Schwartz Senior Fellow. New America is a vibrant, iconoclastic, entrepreneurial place, and the atmosphere is infectious. I am grateful to Steve Coll and Andrés Martinez for giving me the chance to come there and to Faith Smith for making me feel so welcome. As this book was winding down, I also began an association with the City University of New York, where I am Associate Professor of Journalism and Political Science, and my bosses there, Stephen Shepard, Judith Watson, Matthew Goldstein, William Kelly, Joe Rollins, and Ruth O'Brien, have been extremely supportive as well.

Now the hard part: my family. My parents, Doreen Beinart and Julian Beinart, read the manuscript and offered comments and encouragement. But they also contributed to the book in a more fundamental way. As a child, I saw America through their immigrant eyes, and glimpsed the combination of profound gratitude and ironic detachment they felt toward their adopted land. From them, I learned that irony is not the enemy of devotion. To the contrary, a willingness to chuckle occasionally at national pieties keeps patriotism from becoming unthinking and insincere. The American right has long believed that seeing America ironically makes America weaker. But long before I first encountered the writings of Reinhold Niebuhr, my parents taught me what he taught the nation: that the root of the word *irony* is iron.

My sister, Jean Stern, provided advice on graphic elements of the book. But, more important, she provided what she has always provided: wisdom and love. One of my greatest hopes for my own son is that he gains as much joy from his sister as I have gained from my mine.

My wife's parents, Marlene and Arthur Hartstein, have—since I first married into their family—treated me like a son. And I've relied particularly heavily on their generosity and affection during the writing of this book. My sister-in-law, Lieutenant Colonel Bonnie Hartstein, and her husband, Gal Shweiki, contributed to the book as well, mostly through the power of their example. Since September 11, 2001, they have lived the history that I only write about. I only wish all of America's wartime leaders had handled their burdens with as much integrity, dedication, and grace.

There is simply no way I can adequately thank my wife, Diana Hartstein Beinart, for everything she has endured over the past three years. Writing a book is like inviting an old college buddy to live in your house. You see him as a pleasant, stimulating companion; your spouse ends up picking up his sweat socks. Day after day, in the most mundane ways and in the most profound, Diana has shouldered the burdens that come with young children, an old house, and an addled husband, all the while excelling in her own, very demanding, career. And she's done so with the same warmth, determination, and charisma that made me fall in love with her almost a decade ago.

Finally, this book is dedicated to the smallest and biggest people in Diana's and my lives: Ezra Beinart, age four, and Naomi Beinart, age two. I used to think that writing books gave me joy. Now I have a better understanding of the word. A while back, Ezra informed me that he was writing a book of his own. When I asked what it was about, he replied, "Orange hubris." How lucky I am, I thought. My days of thinking about hubris are almost over. But my days of thinking about orange hubris have only just begun.

NOTES

Introduction

1   *"tyrants to which men are ever subject":* William Graham Sumner, *War and Other Essays*, ed. Albert Galloway Keller (New Haven, CT: Yale University Press, 1911), 36.

1   Republicans *over the age of sixty-five:* According to Pew, 94 percent of Republicans under the age of thirty backed the Iraq War, compared to only 65 percent of Republicans over sixty-five. By contrast, 61 percent of Democrats under thirty backed the war compared to only 33 percent over sixty-five. "Generations Divide over Military Action in Iraq," Pew Research Center for the People & the Press, October 17, 2002, http://people-press.org/commentary/?analysisid=57. A later Pew poll, in February 2003, once again found that Americans under thirty were more pro-war than Americans over sixty-five, although this time the gap was only eight points. "Public Favors Force in Iraq, but U.S. Needs More International Backing," Pew Research Center for the People & the Press, February 20, 2003, http://people-press.org/reports/pdf/173.pdf.

2   *revealed the same generational skew:* At my request, two enterprising Yale undergraduates, Chris Chen and Adam Stempel, surveyed the views of influential bloggers, television and radio pundits, and print columnists on the Iraq War between September 1, 2002, and March 15, 2003. To gather their cohort of influential bloggers they used a list published in June 2003 by the *Online Journalism Review* (http://www.ojr.org/ojr/glaser/1056050270.php), disregarding those bloggers whose views on Iraq could not be discerned. For the print columnists, they surveyed all the signed op-eds written about the war in the *New York Times, Washington Post,* and *Wall Street Journal* between September 1, 2002, and March 15, 2003. (If a columnist's opinion was ambiguous, or the columnist was a government official or member of Congress, he or she was not included.) For television and radio pundits, they surveyed the views of the hosts of the twenty-five most-watched television talk shows and most-listened-to radio talk shows, according to the Nielsen television ratings and Arbitron radio ratings closest to October 2002 (http://www.mediabistro.com/tvnewser/original/0409_q3ratings.pdf; http://www.talkers.com/talkhosts.htm). (If a television or radio host did not offer explicit views on the war, he or she was discarded.) They also surveyed the views of a list of influential Iraq-specific pundits compiled by *Slate* in 2003 (http://slate.msn.com/id/2080099/). Where a search of Wikipedia pages, personal profiles, Library of Congress listings, and dates of college graduation did not reveal a commentator's age, he or she was discarded. Ultimately, Chen and Stempel found that 72 percent of the bloggers, op-ed writers, and television and radio pundits under the age of forty-five supported the war, compared to only 60 percent of those over forty-five. Forty-five was chosen as the cutoff date because someone who was forty-five in 2003 would have been only fifteen when the final U.S. ground troops left Vietnam in 1973, and thus, likely too young to have been heavily

influenced by that war. The raw data that Chen and Stempel used is available at http://sites
.google.com/site/icarussyndromebook.

2 *"vital interests of the republic":* Arthur Schlesinger, Jr., "White Slaves in the Persian Gulf," *Wall Street Journal*, January 7, 1991, A14.

2 *"30 years ago for intervention in Vietnam":* Arthur Schlesinger, Jr., "How to Think About Bosnia," *Wall Street Journal*, May 3, 1993, A16.

2 *"they should have reflected on Vietnam":* Arthur Schlesinger, Jr., "Are We Trapped in Another Vietnam?" *The Independent*, November 2, 2001, 5.

3 *until Nazi troops reached French soil:* Arthur Schlesinger, Jr., *A Life in the Twentieth Century: Innocent Beginnings, 1917–1950* (New York: Houghton Mifflin, 2000), 231–32.

3 *leading, perhaps, to wisdom:* Schlesinger's famous book on cycles of reform and reaction in American politics is *The Cycles of American History* (New York: Mariner, 1999).

3 *defined it as insolence toward the gods:* An alternative theory, associated with the classicist N.R.E. Fisher, argues that while the Greeks were certainly interested in godly retribution against humans who displayed excessive pride, the word *hubris* or *hybris* actually meant something different: an act of aggression by one person against another designed to inflict shame or assert superiority. See N.R.E. Fisher, *Hybris: A Study in the Values of Honour and Shame in Ancient Greece* (Warminster, England: Aris & Phillips, 1992).

3 *hacks him to death in the bath:* Aeschylus, *1: The Oresteia*, trans. David R. Slavitt (Philadelphia: University of Pennsylvania Press, 1998).

3 *"but not on good advice / he'd overrule all gods":* Aeschylus, *Persians*, trans. Janet Lembke and C. J. Herington (New York: Oxford University Press, 1981), 76–77.

4 *plunges to his death, into the sea:* Ovid, *Metamorphoses*, trans. David R. Slavitt (Baltimore: Johns Hopkins University Press, 1994), 154–55.

5 *would bring forth into the world:* My own book *The Good Fight: Why Liberals—and Only Liberals—Can Win the War on Terror and Make America Great Again* (New York: Harper-Collins, 2006) is somewhat guilty of this.

7 *refused to join us in Vietnam:* Sylvia Ellis, *Britain, America and the Vietnam War* (Westport, CT: Praeger, 2004), 116; Fredrik Logevall, "De Gaulle, Neutralization, and American Involvement in Vietnam, 1963–1964," *Pacific Historical Review* 61, no. 1 (1992): 91–93; Fredrik Logevall, *Choosing War: The Lost Chance for Peace and the Escalation of War in Vietnam* (Berkeley: University of California Press, 1998), 156–60; David Kaiser, *American Tragedy: Kennedy, Johnson, and the Origins of the Vietnam War* (Cambridge, MA: Belknap, 2000), 315.

8 *had never done in its entire history:* Jennifer K. Elsea and Richard F. Grimmett, "Declarations of War and Authorizations for the Use of Military Force: Historical Background and Legal Implications," *Congressional Research Service,* March 8, 2007, http://fas.org/sgp/crs/natsec/RL31133.pdf.

9 *offered by America's recent past:* For a detailed discussion of how policymakers can effectively use historical analogies, see Richard E. Neustadt and Ernest R. May, *Thinking in Time: The Uses of History for Decision Makers* (New York: Free Press, 1986).

10 *hated and feared universal doctrines:* "There are very few general observations which can be made about the conduct of states which have any absolute validity at all times and in all cases," Kennan wrote. "The few that might have such validity are almost invariably to be found in the realm of platitude. If this absolute validity is lacking, the chances are that the utterance in question will some day rise to haunt us in a context where it is no longer fully applicable. If, on the other hand, the utterance remains in the realm of platitude, then there is all the more reason why we should not associate ourselves with it." John Lewis Gaddis, *Strategies of Containment*, rev. ed. (New York: Oxford University Press, 2005), 51–52.

10 *on Russia's northwest border:* For a list of Kennan's diplomatic postings, see David Mayers, *George Kennan and the Dilemmas of US Foreign Policy* (New York: Oxford University Press, 1988), xiii–xiv.

10 *American foreign policy during the Russian Revolution:* Nicholas Thompson, *The Hawk and The Dove* (New York: Henry Holt, 2009), 9, 158, 166; Mayers, *George Kennan and the Dilemmas of US Foreign Policy*, 31.

10 *be clones of their Soviet counterparts:* For a discussion of the destruction of the "China hands" and its effect on Vietnam policy, see David Halberstam, *The Best and the Brightest* (New York: Ballantine, 1992), 103–4, 188–91; and E. J. Kahn, Jr., *The China Hands: America's Foreign Service Officers and What Befell Them* (New York: Penguin, 1972).

11 *never before been posted to the Arab world:* Michael R. Gordon and Bernard E. Trainor, *Cobra II: The Inside Story of the Invasion and Occupation of Iraq* (New York: Vintage, 2007), 545.

11 *go home and prepare for the future:* For an argument about the shifting relationship between Xerxes and the chorus, see Harry C. Avery, "Dramatic Devices in Aeschylus' Persians," *American Journal of Philology* 85, no. 2 (1964): 180–83.

11 *"not a guide by which to live":* Evan Thomas, *Robert Kennedy* (New York: Simon & Schuster, 2000), 286–87; "The Politics of Restoration," *Time*, May 24, 1968, http://www.time.com/time/magazine/article/0,9171,844434,00.html.

PART I: THE HUBRIS OF REASON

Chapter 1: A Scientific Peace

15 *no one ever called* him *cold and removed:* John Milton Cooper, Jr., *The Warrior and the Priest: Woodrow Wilson and Theodore Roosevelt* (Cambridge, MA: Harvard University Press, 1983), 246–47; Alexander L. George and Juliette L. George, *Woodrow Wilson and Colonel House: A Personality Study* (New York: John Day, 1956), 16.

15 *"a ten-cent mackerel in brown paper":* Walter LaFeber, *The American Age,* 2nd ed. (New York: Norton, 1994), 270.

15 *"I didn't know whether God or him was talking":* Eric F. Goldman, *Rendezvous with Destiny: A History of Modern American Reform* (New York: Ivan R. Dee, 2001), 213.

15 *they often talked back:* John M. Mulder, *Woodrow Wilson: The Years of Preparation* (Princeton, NJ: Princeton University Press, 1978), 273.

15 *"His thoughts and mine are one":* Godfrey Hodgson, *Woodrow Wilson's Right Hand: The Life of Colonel Edward M. House* (New Haven, CT: Yale University Press, 2006), 10.

16 *"even when he was cutting your throat":* Hodgson, *Woodrow Wilson's Right Hand,* 2; Margaret MacMillan, *Paris 1919: Six Months That Changed the World* (New York: Random House, 2003), 17.

16 *Colonel House wanted to meet him outside:* Ronald Steel, *Walter Lippmann and the American Century* (New York: Transaction, 1999), 17, 20, 107, 109, 117, 129; Charles Forcey, *The Crossroads of Liberalism: Croly, Weyl, Lippmann and the Progressive Era, 1900–1925* (New York: Oxford University Press, 1961), 88–89, 109.

16 *a project of the most fearsome secrecy:* Steel, *Walter Lippmann,* 127; Lawrence E. Gelfand, *The Inquiry: American Preparations for Peace, 1917–1919* (New Haven, CT: Yale University Press, 1961), 26.

16 *five million Germans were either wounded or dead:* James McRandle and James Quirk, "The Blood Test Revisited: A New Look at German Casualty Counts in World War I," *Journal of Military History* 70 (2006): 688; Thomas J. Knock, *To End All Wars: Woodrow Wilson and the Quest for a New World Order* (New York: Oxford University Press, 1992), 32.

16 *four thousand villages had simply ceased to exist:* LaFeber, *The American Age,* 316.

17 *he returned to the White House and wept:* Arthur Link, *Woodrow Wilson and the Progressive Era 1910–1917* (New York: Harper, 1954), 282.

17 *they would call it simply "the Inquiry":* Gelfand, *the Inquiry,* 41.

17 *what House called a "scientific peace":* Lloyd Ambrosius, *Woodrow Wilson and the American Diplomatic Tradition: The Treaty Fight in Perspective* (New York: Cambridge University Press, 1987), 32.

17 *historians insist, was not one thing:* See, for instance, Daniel T. Rodgers, "In Search of Progressivism," *Reviews in American History* 10 (1982): 113–32.

18 *a fourth branch of government:* Forcey, *The Crossroads of Liberalism,* 6–8; Goldman, *Rendezvous with Destiny,* 189; Sidney Kaplan, "Social Engineers as Saviors: Effects of World War I on Some American Liberals," *Journal of the History of Ideas* 17 (1956): 355.

18 *handing them control over the economy:* John Patrick Diggins, *Thorstein Veblen: Theorist of the Leisure Class* (Princeton, NJ: Princeton University Press, 1999), 20–24, 194–96.

18 *to the world beyond America's shores:* Edward M. House, *Philip Dru: Administrator, A Story of Tomorrow 1920–1935* (New York: B. W. Huebsch, 1912), http://www.gutenberg.org/dirs/etext04/8phlp10h.htm.

18 *"like the individual, must go to school":* Forcey, *The Crossroads of Liberalism*, 43–44.

18 *that made them act like beasts:* For Veblen, for example, prehistoric society was "characterized by considerable group solidarity. . . . The prime requisite for survival under the conditions would be a propensity unselfishly and impersonally to make the most of all the material means at hand and a penchant for turning all resources of knowledge and materials to account to sustain the life of the group." Quoted in David W. Noble, *The Progressive Mind, 1890–1917* (Minneapolis: Burgess, 1981), 50. On Beard's belief that absent the distorting effect of unregulated capitalism people would incline toward cooperation, see Richard Hofstadter, *The Progressive Historians: Turner, Beard, and Parrington* (New York: Knopf, 1968), 175–77.

18 *evil resided in the world, not in man:* Noble, *The Progressive Mind*, 76.

18 *"lost innocence of the race":* Christopher Lasch, *The New Radicalism in America, 1890–1963: The Intellectual as a Social Type* (New York: Knopf, 1965), 143–4.

19 *keeping himself awake by reading Tolstoy:* Cooper, *The Warrior and the Priest*, 27–35; William Henry Harbaugh, *Power and Responsibility: The Life and Times of Theodore Roosevelt* (New York: Farrar, Straus & Cudahy, 1961), 47–56; Edmund Morris, *The Rise of Theodore Roosevelt* (New York: Ballantine, 1980), 321–25.

20 *recognition of their union:* For conditions in the anthracite mines, see Harbaugh, *Power and Responsibility*, 169; James Abrams, "Anthracite Mining Unionism and the UMW: An Oral History," *Pennsylvania History* 58 (1991): 330–37; Perry K. Blatz, "Local Leadership and Local Militancy: The Nanticoke Strike of 1899 and the Roots of Unionization in the Northern Anthracite Field," *Pennsylvania History* 58 (1991): 278–97.

20 *lackeys bloodied workers into submission:* The historian John Blum calls Roosevelt's actions during the coal strike "an extraordinary departure from the protection that presidents had long afforded to management." John M. Blum, *The Progressive Presidents: Roosevelt, Wilson, Roosevelt, Johnson* (New York: Norton, 1989), 40.

20 *without anyone's getting shot:* Robert H. Wiebe, "The Anthracite Strike of 1902: A Record of Confusion," *Mississippi Valley Historical Review* 48 (1961): 229–51; Jonathan Grossman, "The Coal Strike of 1902—Turning Point in U.S. Policy," *Monthly Labor Review*, October 1975, http://www.dol.gov/oasam/programs/history/coalstrike.htm; Forcey, *The Crossroads of Liberalism*, 67.

20 *and to shut down slaughterhouses if they did:* Blum, *The Progressive Presidents*, 39–44; Michael McGerr, *A Fierce Discontent: The Rise and Fall of the Progressive Movement in America, 1870–1920* (New York: Free Press, 2003), 161–63.

21 *order might break down once more:* "Behind the apparent optimism of the belief in the inevitability of progress," writes one analyst of Wilson's thought, "there was the perception of a demand that society be driven forward lest it should again split apart into warring camps." Niels Aage Thorsen, *The Political Thought of Woodrow Wilson* (Princeton, NJ: Princeton University Press, 1988), 15. See also Mulder, *The Years of Preparation*, 12, 29; Knock, *To End All Wars*, 3; Anthony Gaughan, "Woodrow Wilson and the Legacy of the Civil War," *Civil War History* 43 (1997): 225–43.

21 *the production of unbiased expertise:* Gaughan, "Woodrow Wilson and the Legacy of the Civil War," 225–43; Thorsen, *The Political Thought of Woodrow Wilson*, 12, 126–29, 192.

22 *"at our leisure as they become necessary":* Mulder, *The Years of Preparation*, 34, 38–39, 41–42, 51–52, 56, 66, 82–83, 270; Arthur S. Link, "His Presbyterian Inheritance," in *The Higher Realism of Woodrow Wilson*, ed. Arthur S. Link (Nashville, TN: Vanderbilt University Press, 1971), 16; Knock, *To End All Wars*, 4.

22 *in the history of American politics:* Thorsen, *The Political Thought of Woodrow Wilson*, 28.

22 *"if he could speak to enough of them":* Ray Stannard Baker, *Woodrow Wilson: Life and Letters* (New York: Greenwood, 1968), iii, 173.

22 *the most dazzling first terms in American history:* "During his first term," writes John Milton Cooper, "Wilson compiled a spectacular, possibly unmatched, record of legislative and party leadership." Cooper, *The Warrior and the Priest,* 229.

23 *doubled since the Civil War:* Link, *Woodrow Wilson and the Progressive Era,* 35; Karen E. Schnietz, "The 1916 Tariff Commission: Democrats' Use of Expert Information to Constrain Republican Tariff Protection," *Business and Economic History* 23 (1994): 179.

23 *political rather than scientific criteria to make loans:* Allan H. Meltzer, *A History of the Federal Reserve* (Chicago: University of Chicago Press, 2003), 69–71; Blum, *The Progressive Presidents,* 69–70; Link, *Woodrow Wilson and the Progressive Era,* 45, 48–49; Bruce Champ, "The National Banking System: Empirical Observations," Working Paper 07-19, Federal Reserve Bank of Cleveland, December 2007, 26–29, http://www.clevelandfed.org/research/work paper/2007/wp0719.pdf.

23 *the first banking reorganization in almost fifty:* Schnietz, "The 1916 Tariff Commission," 178–79; Bruce Champ, "The National Banking System: A Brief History," Working Paper 07-23, Federal Reserve Bank of Cleveland, December 2007, http://www.clevelandfed.org/ research/workpaper/2007/wp0723.pdf.

23 *averted a potentially bloody nationwide strike:* Link, *Woodrow Wilson and the Progressive Era,* 59, 61–63, 67–74; Blum, *The Progressive Presidents,* 72–74, 78–79; Cooper, *The Warrior and the Priest,* 251.

24 *an unshakable faith in the number 13:* Hodgson, *Woodrow Wilson's Right Hand,* 211; Link, *Woodrow Wilson and the Progressive Era,* 59–61. For Wilson's appalling views of blacks, see Lloyd E. Ambrosius, "Woodrow Wilson and The Birth of a Nation: American Democracy and International Relations," *Diplomacy and Statecraft* 18 (2007), 689–718.

24 *a serious economic downturn in twenty-five:* Some might call the Spanish-American War traumatic, but it cost the United States a mere 385 combat deaths, with another 1,662 Americans wounded. U.S. troops kept fighting off and on in the Philippines until the 1930s, but after the election of 1900 those skirmishes were never a major domestic political issue. Allan R. Millett and Peter Maslowski, *For the Common Defense: A Military History of the United States of America* (New York: Free Press, 1994), 284–301; E. Berkeley Tompkins, "Scylla and Charybdis: The Anti-Imperialist Dilemma in the Election of 1900," *Pacific Historical Review* 36 (1967): 143–61; Fred H. Harrington, "The Anti-Imperialist Movement in the United States, 1898–1900," *Mississippi Valley Historical Review* 22 (1935): 211–30.

24 *soon be a thing of the past:* Robert H. Wiebe, *The Search for Order 1877–1920* (New York: Hill & Wang, 1967), 141; McGerr, *A Fierce Discontent,* 116.

24 *"the inevitability of progress, in the perfectibility of man":* Steel, *Walter Lippmann,* 45.

25 *"even here in this life upon earth":* William T. R. Fox, "Interwar International Relations Research: The American Experience," *World Politics* 2 (1949): 68.

25 *"illogical passion in us all?":* Forcey, *The Crossroads of Liberalism,* 224–25.

25 *"possibility of amelioration was retained":* J. A. Thompson, "American Progressive Publicists and the First World War, 1914–1917," *Journal of American History* 58 (1971): 369.

25 *"and human rights, shall prevail":* Noble, *The Progressive Mind,* 184.

25 *"we used to call the 'balance of power'":* Inis L. Claude, *Power and International Relations* (New York: Random House, 1988), 82.

26 *"because we are a tough bunch":* MacMillan, *Paris 1919,* 23.

26 *"the balance of power in its own favor":* David Steigerwald, *Wilsonian Idealism in America* (Ithaca, NY: Cornell University Press, 1994), 55.

26 *"political fortunes for our own benefit":* Lloyd E. Ambrosius, *Wilsonianism: Woodrow Wilson and His Legacy in American Foreign Relations* (New York: Palgrave Macmillan, 2002), 37.

26 *external aggression and internal disorder:* Mark T. Gilderhus, *Pan American Visions: Woodrow Wilson in the Western Hemisphere 1913–1921* (Tucson: University of Arizona Press, 1986), 49.

27 *preventing meddling by European powers:* Walter H. Posner, "American Marines in Haiti, 1915–1922," *Americas* 20 (1964): 231–66; Bruce J. Calder, "Caudillos and Gavilleros versus the United States Marines: Guerrilla Insurgency during the Dominican Intervention, 1916–1924," *Hispanic American Historical Review* 58 (1978): 649–75.

27 *"South American republics to elect good men!":* Wiebe, *The Search for Order,* 247.

27 *South American banks and businesses from Europe:* Gilderhus, *Pan American Visions,* 37.

28 *world governed by reason, not force:* Hodgson, *Woodrow Wilson's Right Hand,* 231.

28 *"a part of impartial mediation":* Blum, *The Progressive Presidents,* 86.

28 *"friendship and disinterested service":* Cooper, *The Warrior and the Priest,* 273.

29 *less than two to one:* Paul Kennedy, *The Rise and Fall of the Great Powers: Economic Change and Military Conflict from 1500 to 2000* (New York: Vintage, 1989), 199, 200, 203.

29 *"and not as it ought to be":* Ambrosius, *Woodrow Wilson and the American Diplomatic Tradition,* 85.

29 *Huns were growing too strong:* Cooper, *The Warrior and the Priest,* 154.

30 *"Central America?" he asked in 1915:* Henry A. Kissinger, *Diplomacy* (New York: Simon & Schuster, 1994), 42.

30 *if Germany went after that next:* Cooper, *The Warrior and the Priest,* 277–78.

30 *by sinking whatever tried to pass:* David Stevenson, *Cataclysm: The First World War as Political Tragedy* (New York: Basic Books, 2004), 199–214.

30 *when they heard the news:* Cooper, *The Warrior and the Priest,* 288.

30 *Lodge and TR were already there:* William C. Widenor, *Henry Cabot Lodge and the Search for an American Foreign Policy* (Los Angeles: University of California Press, 1980), 185–198; Harbaugh, *Power and Responsibility,* 474–476.

31 *"and the like," Wilson instructed House:* Lloyd E. Ambrosius, *Wilsonian Statecraft: Theory and Practice of Liberal Internationalism During World War I* (Wilmington, DE: SR Books, 1991), 40, 49.

32 *"utterances since the Monroe Doctrine":* Steel, *Walter Lippmann,* 96.

32 *"a town without a brothel":* LaFeber, *The American Age,* 294–95.

32 *"prevail is the desire of enlightened men everywhere":* Ambrosius, *Wilsonian Statecraft,* 113–14.

32 *running low on subs anyway:* Paul Birdsall, "Why We Went to War," in *The Shaping of Twentieth Century America,* ed. Richard Abrams and Lawrence Levine (Boston: Little, Brown, 1965), 317; Blum, *The Progressive Presidents,* 91–92; Stevenson, *Cataclysm,* 210–12, 254–56.

32 *his government was doing just that:* David Kennedy, *Over Here: The First World War and American Society* (Oxford: Oxford University Press, 2004), 5.

32 *"American has set foot on the continent":* LaFeber, *The American Age,* 294; Stevenson, *Cataclysm,* 264–66.

33 *broke off diplomatic relations with Berlin:* Link, *Woodrow Wilson and the Progressive Era,* 268.

33 *killing fifteen Americans:* "Thrown from a Small Boat," *New York Times,* March 20, 1917, 1.

33 *a wave of public support for war:* Link, *Woodrow Wilson and the Progressive Era,* 271–73.

33 *the moderate, pro-Western Alexander Kerensky:* Stevenson, *Cataclysm,* 247–54.

33 *"our terms for our democracy and civilization":* John Dewey, "In a Time of National Hesitation," in *John Dewey: The Middle Works,* vol. 10, 1899–1924, ed. Jo Ann Boydston (Carbondale: Southern Illinois University Press, 1980), 259.

33 *"democratic revolution the world over":* Steel, *Walter Lippmann and the American Century,* 113.

34 *"fruits of this experiment":* John Dewey, "America in the World," in *John Dewey: The Middle Works,* vol. 11, 72.

34 *America was at war:* Woodrow Wilson, "War Message," speech to Joint Session of Congress, Washington, D.C., April 2, 1917, http://wwi.lib.byu.edu/index.php/Wilson%27s_War_Message_to_Congress; "Must Exert All Our Power," *New York Times,* April 3, 1917, 1; Robert H. Ferrell, *Woodrow Wilson and World War I, 1917–1921* (New York: HarperCollins, 1985), 1; Link, *Woodrow Wilson and the Progressive Era,* 282.

35 *designing the postwar world:* At the Paris Peace Conference, Wilson claimed that he had never seen the secret treaties, a claim that his press secretary, Ray Stannard Baker, reasserted in his early biography of Wilson. But this is probably untrue. After America entered the war, British foreign secretary Arthur Balfour traveled to Washington to confer. During a series of meetings with Colonel House, Balfour explained the content of the secret agreements concluded by the French, British, Italians, and Russians. Balfour also promised to provide

copies of these agreements to Wilson and House. Godfrey Hodgson notes that House and Wilson corresponded about the secret agreements well before Lenin's government published them in late 1917. Hodgson, *Woodrow Wilson's Right Hand*, 144–47.

35 *the Inquiry's headquarters night and day:* Gelfand, *The Inquiry*, 37, 39, 41–42, 45, 103, 121; Steel, *Walter Lippmann*, 129.

35 *were thus largely ignored:* Gelfand, *The Inquiry*, 33, 187–90.

36 *"the policy of the United States":* Ambrosius, *Wilsonianism*, 101.

36 *than did most members of Congress:* Gelfand, *The Inquiry*, 120–21, 243, 317, 330–31.

36 *secret treaties dividing up the spoils of war:* Arno J. Mayer, *Wilson vs. Lenin: Political Origins of the New Diplomacy 1917–1918* (New York: Meridian, 1969), 279.

36 *"and nothing else will do":* Steel, *Walter Lippmann*, 129, 133.

37 *"at half past twelve":* Knock, *To End All Wars*, 142.

37 *determined by reason, not force:* Mayer, *Wilson vs. Lenin*, 267–68; Knock, *To End All Wars*, 143–44; LaFeber, *The American Age*, 309–10.

37 *"great hope of mankind which we are trying to realize":* Steel, *Walter Lippmann*, 141–43.

37 *sneaked across the border into Sweden:* MacMillan, *Paris 1919*, 157–58; E. Eyck, "The Generals and the Downfall of the German Monarchy 1917–1918," *Transactions of the Royal Historical Society* 2 (1952): 47–67; D. K. Buse, "Ebert and the German Crisis, 1917–1920," *Central European History* 5 (1972): 234–55; Klaus Epstein, "Wrong Man in a Maelstrom: The Government of Max of Baden," *Review of Politics* 26 (1964): 215–43.

37 *on the basis of the Fourteen Points:* Steel, *Walter Lippmann*, 149–50; David Fromkin, *In the Time of the Americans: The Generation That Changed America's Role in the World* (New York: Vintage, 1996), 260.

38 *no subordinate would steal the glory:* Fromkin, *In the Time of the Americans*, 267–69; MacMillan, *Paris 1919*, 3.

38 *cut through the gray sky:* Ferrell, *Woodrow Wilson and World War I*, 138; Steel, *Walter Lippmann*, 151; Knock, *To End All Wars*, 194; MacMillan, *Paris 1919*, 3.

38 *the largest crowd in French history:* Fromkin, *In the Time of the Americans*, 284–85; Knock, *To End All Wars*, 194–95; MacMillan, *Paris 1919*, 15–16.

38 *surrounded by sacred candles:* MacMillan, *Paris 1919*, 15; Ferrell, *Woodrow Wilson and World War I*, 139–40.

38 *"he became a Messiah":* Kennedy, *Over Here*, 356; Goldman, *Rendezvous with Destiny*, 262; John B. Judis, *The Folly of Empire: What George W. Bush Could Learn from Theodore Roosevelt and Woodrow Wilson* (New York: Scribner, 2004), 95–96; Knock, *To End All Wars*, epigraph.

38 *"and yet again it might be a tragedy":* MacMillan, *Paris 1919*, 27.

Chapter 2: The Frightening Dwarf

39 *and the Bible by kindergarten:* Carl Resek, "Introduction," in Randolph Bourne, *War and the Intellectuals: Collected Essays, 1915–1919*, ed. Carl Resek (Indianapolis: Harper & Row, 1999), viii–ix; Eric J. Sandeen, "Bourne Again," *American Literary History* 1 (1989): 491–92; Bruce Clayton, *Forgotten Prophet: The Life of Randolph Bourne* (Baton Rouge: Louisiana State University Press), 1, 8.

39 *the men who changed his life:* Resek, "Introduction," viii, ix–x.; Roderick Nash, *The Nervous Generation: American Thought, 1917–1930* (Chicago: Elephant Paperbacks, 1970), 36–38; Lasch, *The New Radicalism in America*, 76–79.

39 *progressivism's greatest historian:* Ellen Nore, *Charles A. Beard: An Intellectual Biography* (Carbondale: Southern Illinois University Press, 1983), 4–12.

39 *its greatest philosopher:* Charles F. Howlett, *Troubled Philosopher* (Port Washington, NY: National University Publications, 1977), 59–60; Robert Westbrook, *John Dewey and American Democracy* (Ithaca, NY: Cornell University Press, 1991), 1; Goldman, *Rendezvous with Destiny*, 156–58.

39 *who transcended their selfish desires:* Randolph Bourne, "Twilight of Idols," in Bourne, *War and the Intellectuals*, 60; Westbrook, *John Dewey and American Democracy*, 195.

39   *democracy and anarchy to organization:* Noble, *The Progressive Mind*, 28.

40   *"the incarnation of reason":* Kaplan, "Social Engineers as Saviors," 350–51.

40   *job at the fledgling* New Republic: Resek, "Introduction," xi.

40   *might transform the country and the world:* Steel, *Walter Lippmann*, 109.

40   *publishing his political writing at all:* Forcey, *The Crossroads of Liberalism*, 249, 283.

40   *"a concert among the nations":* Frederic A. Ogg and Charles A. Beard, *National Governments and the World War* (New York: Macmillan, 1919), 570.

40   *"systematic utilization of the scientific expert":* John Dewey, "What Are We Fighting For?" in *John Dewey: The Middle Works*, vol. 11, 99, 104.

40   *laboratories for tinkering with human behavior:* Bourne, "Twilight of Idols," 56.

40   *" from the path of organization to that end":* Westbrook, *John Dewey and American Democracy*, 208; Randolph Bourne, "A War Diary," in Bourne, *War and the Intellectuals*, ed. Resek, 41.

41   *"permitted to do their own thinking":* Kennedy, *Over Here*, 59–60; McGerr, *A Fierce Discontent*, 290.

41   *the causes and purposes of the war:* Thomas C. Kennedy, *Charles Beard and American Foreign Policy* (Gainesville: University of Florida Press, 1975), 32–33; Nore, *Charles A. Beard*, 74.

41   *"the more violent appeals to passion":* John Dewey, "What America Will Fight For," in *John Dewey: The Middle Works*, vol. 10, 273; Clayton, *Forgotten Prophet*, 223.

41   *it took a jury twenty-five minutes to acquit:* Walter I. Trattner, "Progressivism and World War I: A Reappraisal," *Mid-America* 44 (1962): 137; Kennedy, *Over Here*, 68, 80; Knock, *To End All Wars*, 133.

41   *"including Dewey mad, drove Bourne sane":* Clayton, *Forgotten Prophet*, 4.

42   *"they had been waiting for each other":* Randolph Bourne, "The War and the Intellectuals," in Bourne, *War and the Intellectuals*, ed. Resek, 11; Bourne, "Twilight of Idols," 60.

42   *"because they knew this was an illusion":* Bourne, "A War Diary," 41, 45.

42   *that his nemesis be fired:* Louis Filler, *Randolph Bourne* (Washington, DC: American Council on Public Affairs, 1943), 113; Resek, "Introduction," xiii; Clayton, *Forgotten Prophet*, 229, 232–35; Westbrook, *John Dewey and American Democracy*, 212, 233–35.

42   *influenza just weeks after the war's end:* Sandeen, "Bourne Again," 504; Resek, "Introduction," xiii; Clayton, *Forgotten Prophet*, 255–56.

42   *no journal would print his work:* Lasch, *The New Radicalism in America*, 74.

43   *a newly unified Germany:* MacMillan, *Paris 1919*, 63.

43   *keep vigil even from the grave:* Gregor Dallas, *At the Heart of a Tiger: Clemenceau and His World 1841–1929* (New York: Carroll & Graf, 1993), 111–12; MacMillan, *Paris 1919*, 27–28, 33.

44   *were made in German factories:* MacMillan, *Paris 1919*, 26, 28, 31–32; Gerhard Weinberg, "The Defeat of Germany in 1918 and the European Balance of Power," *Central European History* 2 (1969): 256–57.

44   *"it is desolation, it is death":* MacMillan, *Paris 1919*, 22, 28, 185.

44   *did not want to be swayed by emotion:* Eventually, in March, Wilson did make a brief battlefield tour. But it did not mollify the insulted French. MacMillan, *Paris 1919*, 22.

44   *fewer "hysterical" French:* MacMillan, *Paris 1919*, 27, 168.

44   *"only disinterested people at the Peace Conference":* MacMillan, *Paris 1919*, 9.

44   *"we are here to see you get nothing":* Ferrell, *Woodrow Wilson and World War I*, 135.

44   *"And a very good thing too!":* Steigerwald, *Wilsonian Idealism in America*, 77; Ambrosius, *Woodrow Wilson and the American Diplomatic Tradition*, 54; MacMillan, *Paris 1919*, 23–24, 33.

45   *the economic balance of power with Berlin:* MacMillan, *Paris 1919*, 170–73, 191; Sally Marks, "The Myths of Reparations," *Central European History* 11 (1978): 254–55.

45   *But Wilson rejected that, too:* Ambrosius, *Wilsonianism*, 68–69; Ambrosius, *Woodrow Wilson and the American Diplomatic Tradition*, 75–77.

45   *"of the public opinion of the world":* Ambrosius, *Woodrow Wilson and the American Diplomatic Tradition*, 78.

45   *the Allies could occupy it for fifteen years:* Ambrosius, *Wilsonianism*, 72.

46   *already pledged to do under the League:* Ambrosius, *Wilsonianism*, 71.

46 *"over the heads of their rulers":* Ray Stannard Baker and William E. Dodd, eds., *The Public Papers of Woodrow Wilson: War and Peace,* vol. 5 (New York: Harper & Brothers, 1927), 259; E. H. Carr, *The Twenty Years' Crisis, 1919–1939: An Introduction to the Study of International Relations* (New York: Palgrave, 2001), 32–34; Knock, *To End All Wars,* 162.

46 *rather than the Royal Navy:* LaFeber, *The American Age,* 315.

47 *covered them over in disgust:* Kennedy, *Over Here,* 357. Wilson also threatened to appeal to the people of France against their government's hard line on Germany, predicting that if he did, Clemenceau's government would fall. Fromkin, *In the Time of the Americans,* 322.

47 *"a world governed by intrigue and force":* Ambrosius, *Wilsonian Statecraft,* 113–114.

47 *not as punitive as legend suggests:* As Margaret McMillan writes, "When historians look, as they have increasingly been doing, at the other details, the picture of a Germany crushed by a vindictive peace cannot be sustained. Germany did lose territory; that was an inevitable consequence of losing the war. If it had won, we should remember, it would have certainly taken Belgium, Luxembourg, parts of the north of France and much of the Netherlands. The Treaty of Brest-Litovsk showed the intentions of the German supreme command for the eastern frontiers. Despite its losses Germany remained the largest country in Europe west of the Soviet Union between the wars. Its strategic position was significantly better than it had been before 1914. With the reemergence of Poland, there was now a barrier between it and the old Russian menace. In place of Austria-Hungary, Germany had only a series of weaker and quarreling states on its eastern frontier. As the 1930s showed, Germany was well placed to extend its economic and political sway among them." MacMillan, *Paris 1919,* 481–82.

47 *when they exited the war in May 1918:* McMillan, *Paris 1919,* 161, 481–82; Weinberg, "The Defeat of Germany in 1918 and the European Balance of Power," 260; Stevenson, *Cataclysm,* 323–24.

47 *"our own security in this hemisphere":* "The Political Scene," *New Republic,* March 22, 1919, 14.

48 *nearly all backed U.S. entrance into the League:* Ambrosius, *Woodrow Wilson and the American Diplomatic Tradition,* 91, 146.

48 *were certain to vote no:* The Irreconcilables were themselves divided into three categories: a group of isolationists who opposed all binding U.S. commitments overseas, a group of idealists or pacifists who thought the League didn't go far enough toward disarmament and decolonization, and a group of realists who considered the League utopian. Ralph A. Stone, "The Irreconcilables' Alternatives to the League of Nations," *Mid-America* 49 (1967): 165–71; Robert J. Maddox, *William E. Borah and American Foreign Policy* (Baton Rouge: Louisiana State University Press, 1969), 62, 64.

48 *with the fury of lovers scorned:* Steel, *Walter Lippmann,* 163.

48 *could take America to war:* Historians have long debated whether the Senate's "strong Reservationists," led by Henry Cabot Lodge, were sincere in their reservations, or were simply using them as cover to defeat the League. David Mervin portrays Lodge as deeply opposed to U.S. membership in the League but unwilling to say so publicly for political reasons. But Lloyd Ambrosius and William Widenor make a convincing case that Lodge would have swallowed U.S. membership, with reservations, in return for America's ratification of the French security treaty, about which he cared passionately. They cite numerous examples of Lodge declaring, in private, that he could support U.S. entry into the League if, as he told the novelist Edith Wharton in early 1919, it is "so constituted that it will . . . not endanger or imperil the United States." David Mervin, "Henry Cabot Lodge and the League of Nations," *Journal of American Studies* 4 (1971): 210; Widenor, *Henry Cabot Lodge and the Search for an American Foreign Policy,* 308–11; Ambrosius, *Woodrow Wilson and the American Diplomatic Tradition,* 137.

48 *dropped dead of a heart attack:* Harbaugh, *Power and Responsibility,* 518–19.

48 *"brothers under the skin":* Hans Morgenthau, *In Defense of the National Interest* (New York: Knopf, 1951), 29.

49 *not Lincoln the impartial reconciler:* Widenor, *Henry Cabot Lodge and the Search for an American Foreign Policy,* 19–21, 216, 217, 284.

49 *"but as one of the allies":* Ambrosius, *Woodrow Wilson and the American Diplomatic Tradition,* 86.

49 *"impossible for her to break out again":* Widenor, *Henry Cabot Lodge and the Search for an American Foreign Policy,* 298.

49 *"by a statute or a written constitution":* Ambrosius, *Woodrow Wilson and the American Diplomatic Tradition,* 86; Widenor, *Henry Cabot Lodge and the Search for an American Foreign Policy,* 351.

50 *"count his personal fortunes in the reckoning":* Knock, *To End All Wars,* 251, 259; MacMillan, *Paris 1919,* 490; Kenneth S. Davis, *FDR: The Beckoning of Destiny, 1882–1928* (New York: G. P. Putnam's Sons, 1972), 565, 583.

50 *emerged to watch his successor sworn in:* Fromkin, *In the Time of the Americans,* 348, 350; Kennedy, *Over Here,* 361; Ferrell, *Woodrow Wilson and World War I,* 169; Knock, *To End All Wars,* 262–63.

50 *Wilson's European crusade as one big waste:* Ambrosius, *Woodrow Wilson and the American Diplomatic Tradition,* 187, 252.

50 *"the lonely citadel of his soul":* Hodgson, *Woodrow Wilson's Right Hand,* 215–56; George and George, *Woodrow Wilson and Colonel House,* xv.

51 *the two-thirds needed for ratification:* Knock, *To End All Wars,* 263–64.

51 *with France, and so it died as well:* Ambrosius, *Woodrow Wilson and the American Diplomatic Tradition,* 212–14; Ambrosius, *Wilsonianism,* 141–42.

51 *"discount all the more violent appeals to passion":* John Dewey, "What America Will Fight For," in *John Dewey: The Middle Works,* vol. 10, 273.

51 *for "pro-German" leanings himself:* Carol Signer Gruber, "Academic Freedom at Columbia University, 1917–1918: The Case of James McKeen Cattell," *AAUP Bulletin* 58 (1972): 297, 301–3; Westbrook, *John Dewey and American Democracy,* 117, 222.

52 *"inculcate disrespect for American institutions":* Carol S. Gruber, *Mars and Minerva: World War I and the Uses of Higher Learning in America* (Baton Rouge: Louisiana State University Press, 1975), 189; Nore, *Charles A. Beard,* 78–79.

52 *"unconventional views in political matters":* Goldman, *Rendezvous with Destiny,* 259–60.

52 *while their teacher wept:* Nore, *Charles A. Beard,* 81; Goldman, *Rendezvous with Destiny,* 259–60; William Summerscales, *Affirmation and Dissent: Columbia's Response to the Crisis of World War I* (New York: Teachers College Press, 1970), 94–96.

52 *banned some of his books from its training camps:* Kennedy, *Charles Beard and American Foreign Policy,* 38; Nore, *Charles A. Beard,* 84.

52 *"amid howling gales of passion":* Westbrook, *John Dewey and American Democracy,* 210.

52 *who had died just weeks before:* Kennedy, *Over Here,* 90; John Dewey, "The Cult of Irrationality," in *Characters and Events: Essays in Social and Political History,* ed. Joseph Ratner (New York: Henry Holt, 1929), 587–91; Westbrook, *John Dewey and American Democracy,* 210, 239.

52 *where he spent the next two years:* Howlett, *Troubled Philosopher,* 39; Westbrook, *John Dewey and American Democracy,* 239–40, 256.

52 *to build a world free of force:* Westbrook, *John Dewey and American Democracy,* 260.

53 *"its own necessities and experiences":* Nore, *Charles A. Beard,* 99–100; Kennedy, *Charles Beard and American Foreign Policy,* 45, 49–50.

53 *Paris was now on its own:* Ambrosius, *Wilsonianism,* 93, 104.

53 *required the Germans to substantially disarm:* Hans W. Gatzke, "Russo-German Military Collaboration During the Weimar Republic," *American Historical Review* 63 (1958): 565–97.

53 *paying more than that themselves:* Sally Marks, "Reparations Reconsidered: A Reminder," *Central European History* 2 (1969): 359–61, 365; Gustav Stolper, Karl Hauser, and Knut Borchardt, *The German Economy: 1870 to the Present* (New York: Harcourt, 1967), 83; Marks, "The Myths of Reparations," 237.

54 *which Germans swallowed with barely a whimper:* Although the Treaty of Versailles has gone down in historical memory as savagely harsh, and the Allies' treatment of Germany after World War II is, by contrast, remembered as generous and enlightened, the truth is in

some ways the reverse. After World War II, Germany, along with Berlin, was divided and occupied by the Allied powers, with full civil authority only being restored in 1949 (and reunification not coming until 1990). While the reparations imposed on Germany after World War I have become synonymous with sadistic excess, they were at least restricted to a particular figure (with the real figure, as noted above, far below that). After World War II, by contrast, the Potsdam Agreement set no reparations limit; it simply allowed each occupying power to expropriate from its occupation zone as it saw fit. The Soviets and French responded by treating their occupation zones as cash cows, stripping them of vast amounts of industrial machinery, supplies, and other materials, which they shipped back home. Finally, it is true that the victors dispossessed Germany of slightly less land in 1945 than in 1919 (10 percent of Germany's prewar total compared to 13 percent), but much of the post–World War I territory was populated by non-German speakers, while after World War II German speakers comprised the vast bulk of people living on land parceled out to Germany's neighbors. MacMillan, *Paris 1919*, 465; Joseph B. Schechtman, "Postwar Population Transfers in Europe: A Survey," *Review of Politics* 15 (1953): 151.

54   *their dreams of European hegemony:* Historians once saw World War I as a mere accident, stumbled into by powers that did not truly want war. But that view has been significantly undermined in recent decades by the German historian Fritz Fischer, who argued in his 1961 book, *Germany's Aims in the First World War,* that a wide swath of the German political elite had since the early 1900s been eager for war with Russia, which they believed would give Germany hegemony over the European continent. Germany, Fischer argues, was not dragged into World War I by its weak ally, Austria-Hungary, but actually pushed a hesitant Hapsburg empire to make extreme demands of the Serbs following the assassination of Archduke Ferdinand, demands that Berlin knew would likely lead to war. In recent years, historians have gone even further, suggesting that German leaders desired war even though they knew Britain would probably enter the fray, and even though they knew the conflict might be long and painful. See, for instance, Keir A. Lieber, "The New History of World War I and What It Means for International Relations Theory," *International Security* 32, no. 2 (2007): 155–91; Frederick A. Hale, "Fritz Fischer and the Historiography of World War One," *History Teacher* 9 (1976): 258–79.

54   *to seize German mines and steel plants as collateral:* Marks, "The Myths of Reparations," 241.

54   *whose popularity was beginning to build:* Stolper, Hauser, and Borchardt, *The German Economy,* 74–89.

55   *"utterly brutal and insane":* Warren Cohen, *Empire Without Tears* (Philadelphia: Temple University Press, 1987), 92.

55   *a litany of senators denounced French militarism:* Royal J. Schmidt, *Versailles and the Ruhr: Seedbed of World War II* (The Hague: Martinus Nijhoff, 1968), 210–12.

55   *American enforcement of Versailles:* Betty Glad, *Charles Evans Hughes and the Illusions of Innocence: A Study in American Diplomacy* (Urbana: University of Illinois Press, 1966), 223; Melvyn P. Leffler, *The Elusive Quest: America's Pursuit of European Stability and French Security, 1919–1933* (Chapel Hill: University of North Carolina Press, 1979), 83; Jolyon P. Girad, "Congress and Presidential Military Policy: The Occupation of Germany, 1919–1923," *Mid-America* 56 (1974): 211–19.

55   *could now default with virtual impunity:* Frank Costigliola, "The United States and the Reconstruction of Germany in the 1920s," *Business History Review* 50 (1976): 493; Marks, "The Myths of Reparations," 249.

55   *"a disinterested position in relation to international affairs":* Robert H. Ferrell, *Peace in Their Time: The Origins of the Kellogg-Briand Pact* (New Haven, CT: Yale University Press, 1952), 131.

56   *all won the Nobel Prize:* Kissinger, *Diplomacy,* 272–76.

56   *the two-front strategy on which French security relied:* Ferrell, *Peace in Their Time,* 65; Ambrosius, *Wilsonianism,* 106–7.

56   *the Maginot Line:* Ferrell, *Peace in Their Time,* 49–50; Ambrosius, *Wilsonianism,* 107; MacMillan, *Paris 1919,* 176, 481; Kissinger, *Diplomacy,* 280.

56 *"are still absolutely in control"*: John C. Farrell, "John Dewey and World War I: Armageddon Tests a Liberal's Faith," *Perspectives in American History* 9 (1975): 329.

57 *that there was nobility in war:* John Dos Passos, "The Scene of Battle," in *The Culture of the Twenties,* ed. Loren Baritz (Indianapolis: Bobbs-Merrill, 1970), 5–11; Henry May, *The Discontent of the Intellectuals: A Problem of the Twenties* (Chicago: Rand McNally, 1963), 21–22; Cohen, *Empire Without Tears,* 45; Robert Osgood, *Ideals and Self-Interest in America's Foreign Relations* (Chicago: University of Chicago Press, 1953), 330.

57 *"the Department of Peace":* J. Chalmers Vinson, "Military Force and American Policy, 1919–1939," in *Isolation and Security: Ideas and Interests in Twentieth Century American Foreign Policy,* ed. Alexander DeConde (Durham, NC: Duke University Press, 1957), 57.

57 *250,000 Americans inquired about taking part:* Ferrell, *Peace in Their Time,* 25; Charles De Benedetti, "The $100,000 American Peace Award of 1924," *Pennsylvania Magazine of History and Biography* 98 (1974): 224–49.

57 *how barbaric human beings had once been:* Osgood, *Ideals and Self-Interest in America's Foreign Relations,* 324–25.

57 *It passed the Senate, 74–0:* Cohen, *Empire Without Tears,* 46–47; Maddox, *William E. Borah and American Foreign Policy,* 86–87.

57 *naval spending dropped by one-third:* Glad, *Charles Evans Hughes,* 270; Phillips Payson O'Brien, *British and American Naval Power: Politics and Policy, 1900–1936* (Westport, CT: Praeger, 1998), 149–76, 180–81, 230–36.

58 *a de facto belligerent on France's side:* Ferrell, *Peace in Their Time,* 62, 64, 73, 80.

58 *that war became almost unthinkable:* Westbrook, *John Dewey and American Democracy,* 260–71.

59 *at the rate of six hundred per day:* Howlett, *Troubled Philosopher,* 87–88; Ferrell, *Peace in Their Time,* 33–35, 232–33, 239.

59 *in the history of international affairs:* Ferrell, *Peace in Their Time,* 208, 218; Osgood, *Ideals and Self-Interest in America's Foreign Relations,* 348; Farrell, "John Dewey and World War I," 331.

Chapter 3: Twice-Born

60 *won only 4 percent of the vote:* Richard Wightman Fox, *Reinhold Niebuhr: A Biography* (Ithaca, NY: Cornell University Press, 1996), 135–36; Reinhold Niebuhr, *Moral Man and Immoral Society: A Study in Ethics and Politics* (Westminster, MO: John Knox, 2002), xxviii.

60 *scores to settle from World War I:* Daniel F. Rice, *Reinhold Niebuhr and John Dewey: An American Odyssey* (Albany: State University of New York Press, 1993), 3–4; Martin Halliwell, *The Constant Dialogue: Reinhold Niebuhr and American Culture* (New York: Rowman & Littlefield, 2005), 60.

61 *"out-and-out loyalty":* Fox, *Reinhold Niebuhr,* 3–21, 28–29, 42, 45, 52–56; William G. Crystal, "'A Man of the Hour and of the Time': The Legacy of Gustav Niebuhr," *Church History* 49 (1980): 417.

61 *He called Wilson a dupe:* Fox, *Reinhold Niebuhr,* 53, 58–59.

61 *"as potential criminals" by their adopted nation:* Reinhold Niebuhr, "What the War Did to My Mind," *Christian Century,* September 27, 1928, 1161–62.

61 *"a child of the age of disillusionment":* Fox, *Reinhold Niebuhr,* 72, 74, 79, 83, 99; Niebuhr, "What the War Did to My Mind," 1161.

62 *joining the Socialist Party:* Fox, *Reinhold Niebuhr,* 94–100, 109–10.

62 *less than one-third its level in 1929:* Lynn Dumenil, *The Modern Temper: American Culture and Society in the 1920s* (New York: Hill & Wang, 1995), 305.

62 *garbage dumps for something to eat:* Goldman, *Rendezvous with Destiny,* 321.

62 *the time for pedagogy is done:* Dewey's foremost biographer, Robert Westbrook, argues that Niebuhr's attack was unfair. "Although at an earlier stage in Dewey's career Niebuhr's contention that he hoped to persuade the powerful to admit the injustice of their rule and to relinquish their power might have had some force, by the early thirties it was misplaced.

Dewey's call for an 'intelligent' politics was not a plea to the oppressed to abandon the effort to match power with power in favor of reasoning with the 'dominant economic interests' but rather an appeal to them to wage their struggles intelligently." Westbrook, *John Dewey and American Democracy*, 526–27. Niebuhr's best biographer, Richard Wightman Fox, concurs, writing that "Niebuhr was not making a close study of his opponent's thought; he was constructing an ideal-type opponent who was easy to take down." Fox, *Reinhold Niebuhr*, 136–37.

62 *educate those in power to remedy them:* Lincoln Steffens, *The Autobiography of Lincoln Steffens* (Berkeley, CA: Heyday Books, 2005), 374–75.

62 *revolution was the only option left:* John Chamberlain, *Farewell to Reform: The Rise, Life and Decay of the Progressive Mind in America* (Chicago: Quadrangle, 1932), 306–10.

62 *they were voting communist for president:* David M. Kennedy, *Freedom from Fear: The American People in Depression and War, 1929–1945* (New York: Oxford University Press, 1999), 222; Arthur M. Schlesinger, Jr., *The Crisis of the Old Order, 1919–1933* (New York: Houghton Mifflin, 2003), 436–37.

63 *had made him a multimillionaire:* Joan Hoff Wilson, *Herbert Hoover: Forgotten Progressive* (Boston: Little, Brown, 1975), 4–17.

63 *mention that he became president:* Australian Prospectors and Miners Hall of Fame, "Herbert Hoover," http://mininghall.com/MiningHallOfFame/HallOfFameDatabase/Inductee. php?InducteeID=1164.

63 *"want[ing] nothing . . . from Congress [except] efficiency":* Hoff Wilson, *Herbert Hoover*, 37, 43.

64 *"began his career in California and will end it in heaven":* Hoff Wilson, *Herbert Hoover*, 44; Richard Norton Smith, *An Uncommon Man: The Triumph of Herbert Hoover* (New York: Simon & Schuster, 1984), 80.

64 *hoover, meaning "to help":* Hoff Wilson, *Herbert Hoover*, 46–47.

64 *"Hooverize when it comes to loving you!":* Hoff Wilson, *Herbert Hoover*, 59–60.

64 *would ever shed their murderous ways:* Selig Adler, *The Uncertain Giant: 1921–1941; American Foreign Policy Between the Wars* (New York: Macmillan, 1965), 126.

65 *they could not process all the mail:* Charles Chatfield, *For Peace and Justice* (Knoxville: University of Tennessee Press, 1971), 160.

65 *destroyers, cruisers, and submarines:* O'Brien, *British and American Naval Power*, 212–14.

65 *"until she has to be thrashed again":* Adler, *Uncertain Giant*, 128–29; Cohen, *Empire Without Tears*, 104–5; Zara Steiner, *The Lights That Failed: European International History, 1919–1933* (Oxford: Oxford University Press, 2005), 589–91; Henry Blumenthal, *Illusion and Reality in Franco-American Diplomacy 1914–1945* (Baton Rouge: Louisiana State University Press, 1986), 158–59.

65 *the Reichstag's second-largest party:* Steiner, *Lights That Failed*, 642; E.J. Feuchtwanger, *From Weimar to Hitler: Germany, 1918–1933* (New York: St. Martin's Press, 1993), 337.

65 *which Versailles explicitly banned:* Frank Costigliola, *Awkward Dominion: American Political, Economic, and Cultural Relations with Europe, 1919–1933* (Ithaca, NY: Cornell University Press, 1984), 235.

65 *"I never did":* Ambrosius, *Wilsonianism*, 108.

66 *by a young hypernationalist:* Steel, *Walter Lippmann*, 288; Adler, *Uncertain Giant*, 140–41; Wilfrid Fleisher, "Premier Is Shot by Tokyo Fanatic; Feared to Be Dying," *New York Times*, November 14, 1930, 1; James B. Crowley, *Japan's Quest for Autonomy: National Security and Foreign Policy, 1930–1938* (Princeton, NJ: Princeton University Press, 1966), 78.

66 *army and navy were firmly in control:* Crowley, *Japan's Quest for Autonomy*, 168–75.

66 *Japan walked out:* Myung Soo Cha, "Did Takahashi Korekiyo Rescue Japan from the Great Depression?" *Journal of Economic History* 63 (2003): 127–44; Adler, *Uncertain Giant*, 146.

66 *"peace among other nations by force":* Hoff Wilson, *Herbert Hoover*, 205; Elting E. Morrison, *Turmoil and Tradition: A Study of the Life and Times of Henry L. Stimson* (Boston: Houghton Mifflin, 1960), 382; Henry L. Stimson and McGeorge Bundy, *On Active Service in Peace*

and War (New York: Harper & Row, 1948), 244; J. Chalmers Vinson, "Military Force and American Policy, 1919–1939," 57.

66 *did not commission a single new warship:* Cohen, *Empire Without Tears*, 106.

67 *armaments be cut by at least a quarter:* Steiner, *Lights That Failed*, 781.

67 *taken to the White House by police escort:* Associated Press, "Says Hoover Plan Saved Arms Parley," *New York Times*, November 17, 1932, 8; Chatfield, *For Peace and Justice*, 161.

67 *was downright terrifying:* Steiner, *Lights That Failed*, 767.

67 *"which it was not prepared to make?":* Steiner, *Lights That Failed*, 777; Blumenthal, *Illusion and Reality*, 165, 181.

67 *a plurality of the vote:* Cohen, *Empire Without Tears*, 121–22; Leffler, *Elusive Quest*, 274–301; H. Arthur Steiner, "The Geneva Disarmament Conference of 1932," *Annals of the American Academy of Political and Social Science* 168 (1933): 214–17; Marlies Ter Borg, "Reducing Offensive Capabilities: The Attempt of 1932," *Journal of Peace Research* 29 (1992): 147–53; Blumenthal, *Illusion and Reality*, 180–87.

67 *and raised it to the wind:* Robert Dallek, *Franklin D. Roosevelt and American Foreign Policy, 1932–1945* (New York: Oxford University Press, 1995), 336; Fromkin, *In the Time of the Americans*, 41–42; Davis, *The Beckoning of Destiny*, 152; Steven Casey, *Cautious Crusade: Franklin D. Roosevelt, American Public Opinion and the War Against Nazi Germany* (New York: Oxford University Press, 2001), xxiii.

68 *then becoming governor of New York:* Blum, *Progressive Presidents*, 108–9; Fromkin, *In the Time of the Americans*, 40, 42; Dallek, *Franklin D. Roosevelt and American Foreign Policy*, 8.

68 *a potential German threat:* Robert A. Divine, *Roosevelt and World War II* (Baltimore: Johns Hopkins University Press, 1969), 50; Willard Range, *Franklin D. Roosevelt's World Order* (Athens: University of Georgia Press, 1959), 20; Davis, *The Beckoning of Destiny*, 383–94; Dallek, *Franklin D. Roosevelt and American Foreign Policy*, 9–10.

68 *"handing out to a gullible public":* Dallek, *Franklin D. Roosevelt and American Foreign Policy*, 9–10.

68 *"merely a beautiful dream, a Utopia":* Range, *Franklin D. Roosevelt's World Order*, 29, 31, 168.

68 *essential to a nation's power:* Davis, *The Beckoning of Destiny*, 71–72, 123; Dallek, *Franklin D. Roosevelt and American Foreign Policy*, 7–8; Robert Osgood, *Ideals and Self-Interest in America's Foreign Relations*, 411; Alfred T. Mahan, *The Influence of Seapower upon History* (Boston: Little, Brown, 1890).

69 *every summer from ages seven to fifteen:* Davis, *The Beckoning of Destiny*, 102; Dallek, *Franklin D. Roosevelt and American Foreign Policy*, 3.

69 *the Western Hemisphere would offer no refuge:* Davis, *The Beckoning of Destiny*, 389.

69 *urging louder cheering at football games:* Davis, *The Beckoning of Destiny*, 164–65.

69 *her former assistant, Lucy Mercer:* Fromkin, *In the Time of the Americans*, 291–93; James MacGregor Burns and Susan Dunn, *The Three Roosevelts: Patrician Leaders Who Transformed America* (New York: Atlantic Monthly Press, 2001), 84, 155; Jean Edward Smith, *FDR* (New York: Random House, 2007), 153, 159; Conrad Black, *Franklin Delano Roosevelt: Champion of Freedom* (New York: PublicAffairs, 2003), 96; "Livingston Davis Commits Suicide," *New York Times*, January 12, 1932, 18.

70 *classmates called him behind his back:* Davis, *The Beckoning of Destiny*, 152.

70 *paralyzed from the waist down by polio:* Davis, *The Beckoning of Destiny*, 648–50. In 2003, in the *Journal of Medical Biography*, several medical experts speculated that, in fact, FDR's disease may have been not polio, but Guillain-Barré syndrome. Armond S. Goldman, Elisabeth J. Schmalstieg, Daniel H. Freeman, Jr., Daniel A Goldman, and Frank C Schmalstieg, Jr., "What Was the Cause of Franklin Delano Roosevelt's Paralytic Illness?" *Journal of Medical Biography* 11 (2003): 232–40.

70 *"a medieval torture chamber":* Davis, *The Beckoning of Destiny*, 668–69.

70 *then, somehow, found it again:* Davis, *The Beckoning of Destiny*, 655.

70 *"twice-born man":* Davis, *The Beckoning of Destiny*, 669, 677–78, 680.

70 *Livingston Davis, shot himself:* "Livingston Davis Commits Suicide," 18.

71 *defend America's home waters:* Range, *Franklin D. Roosevelt's World Order*, 95; Divine, *Roosevelt and World War II*, 54–55; Robert A. Divine, *Second Chance: The Triumph of Internationalism in America During World War II* (New York: Atheneum, 1967), 19; Adler, *Uncertain Giant*, 152–53, 157; Dallek, *Franklin D. Roosevelt and American Foreign Policy*, 15–16, 54, 75.

71 *"intellectual parents of the New Deal":* Warren I. Cohen, *The American Revisionists* (Chicago: University of Chicago Press, 1967), 133; Kennedy, *Charles A. Beard and American Foreign Policy*, 73.

71 *retreat at the first sign of foreign resistance:* Blumenthal, *Illusion and Reality*, 183, 223; Kissinger, *Diplomacy*, 292–93, 302–5, 378.

71 *a naval blockade if he refused:* Fromkin, *In The Time of the Americans*, 335–36; Dallek, *Franklin D. Roosevelt and American Foreign Policy*, 102–3.

72 *"preserve our own oasis of liberty":* Kennedy, *Freedom from Fear*, 386.

72 *hurled essentially the same charge:* Kennedy, *Freedom from Fear*, 388; Adler, *Uncertain Giant*, 160–66; Richard W. Leopold, "The Problem with American Intervention, 1917: An Historical Retrospect," *World Politics* 2 (1950): 412–13, 418–20; Dallek, *Franklin D. Roosevelt and American Foreign Policy*, 102; Goldman, *Rendezvous with Destiny*, 377.

72 *US intervention in World War I a mistake:* Kennedy, *Freedom from Fear*, 394; Hadley Cantril, *Public Opinion, 1935–1946* (Princeton, NJ: Princeton University Press, 1951), 201.

72 *they did so at their own risk:* Adler, *The Uncertain Giant*, 173–75, 182.

73 *"I also learned much of what not to do":* Dallek, *Franklin D. Roosevelt and American Foreign Policy*, 152.

73 *ensure a reelection landslide:* Jane Karoline Vieth, "The Diplomacy of the Depression," in *Modern American Diplomacy*, ed. John M. Carroll and George C. Herring (Wilmington, DE: Scholarly Resources, 1996), 88.

73 *"Western Hemisphere will not be attacked":* Divine, *Roosevelt and World War II*, 16.

73 *about an international "quarantine":* Divine, *Roosevelt and World War II*, 17–18; Dallek, *Franklin D. Roosevelt and American Foreign Policy*, 148.

73 *Congress threatened impeachment:* Adler, *Uncertain Giant*, 190; Vieth, "The Diplomacy of the Depression," 91.

73 *screamed the* Wall Street Journal: Travis Beal Jacobs has argued that, contrary to early historical interpretations, popular reaction to the quarantine speech was not particularly negative. But there is no doubt that the speech stirred isolationist anger, and rightly or wrongly, FDR interpreted that isolationism as the dominant public mood. Travis Beal Jacobs, "Roosevelt's 'Quarantine Speech,'" *Historian* 24 (1962): 488. See also Dorothy Borg, "Notes on Roosevelt's 'Quarantine' Speech," *Political Science Quarterly* 72 (1957): 405–33.

73 *"even hostile and resentful ears":* Kissinger, *Diplomacy*, 379.

73 *except in cases of foreign attack:* Adler, *Uncertain Giant*, 195–96; Chatfield, *For Peace and Justice*, 283–85; Vinson, "Military Force and American Policy, 1919–1939," 80.

73 *"trying to lead and to find no one there":* Vieth, "The Diplomacy of the Depression," 91–92; Burns and Dunn, *The Three Roosevelts*, 357.

74 *offering only the mildest of protests:* Vieth, "The Diplomacy of the Depression," 92.

74 *"of whom we know nothing":* Telford Taylor, *Munich: The Price of Peace* (New York: Doubleday, 1979), 884.

74 *Moscow would not act alone:* Blumenthal, *Illusion and Reality*, 242–43; Taylor, *Munich*, 99–102.

74 *Daladier replied, "With what?":* Kissinger, *Diplomacy*, 293, 302–3, 308–9; Blumenthal, *Illusion and Reality*, 243–44.

74 *lacked the planes to defend London:* Taylor, *Munich*, 801–2, 807, 835–38; Kissinger, *Diplomacy*, 312–14.

75 *a two-word cable: "Good man":* Dallek, *Franklin D. Roosevelt and American Foreign Policy*, 161–66; Kissinger, *Diplomacy*, 314; Divine, *Roosevelt and World War II*, 21; Vieth, "The Diplomacy of the Depression," 94.

75 *their fate and Europe's were intertwined:* Divine, *Roosevelt and World War II*, 25; Kissinger, *Diplomacy*, 383.

75 *Events would have to speak for themselves:* For FDR, writes the historian Robert Dallek, America "would have to enter the fighting with a minimum of doubt and dissent, and the way to achieve this was not through educational talks to the public or strong Executive action, but through developments abroad which aroused the country to fight." Dallek, *Franklin D. Roosevelt and American Foreign Policy,* 267. John Milton Cooper adds that "the chief lesson Roosevelt drew from Wilson was not to pitch his appeals to people too high or to expect too much in the way of educating the public. The deviousness and self-deception in his policies toward the warring powers from 1939 through 1941 reflected that lesson." Cooper, *The Warrior and the Priest,* 360–61.

75 *Britain and France decided to fight:* Divine, *Second Chance,* 29.

75 *"picking up again an interrupted routine":* Dallek, *Franklin D. Roosevelt and American Foreign Policy,* 197–98; Fromkin, *In the Time of the Americans,* 484–88.

76 *"ruled by force in the hands of a few":* Dallek, *Franklin D. Roosevelt and American Foreign Policy,* 199, 215; Fromkin, *In the Time of the Americans,* 489.

76 *carried them away in their own ships:* Divine, *Roosevelt and World War II,* 29.

76 *"unpitying masters of other continents":* Divine, *Roosevelt and World War II,* 31.

76 *"she was expecting salvation":* Blumenthal, *Illusion and Reality,* 260–61.

76 *had surrendered twenty-two years before:* Cantril, *Public Opinion, 1935–1946,* 971–72; Adler, *Uncertain Giant,* 229.

76 *military bases in the Western Hemisphere:* Divine, *Roosevelt and World War II,* 33–35.

76 *which essentially provided them for free:* Dallek, *Franklin D. Roosevelt and American Foreign Policy,* 248–49, 255–60.

77 *was now a rabid isolationist:* Osgood, *Ideals and Self-Interest in America's Foreign Relations,* 379.

77 *prevent them from falling into Nazi hands:* Robert L. Beisner, *Dean Acheson: A Life in the Cold War* (New York: Oxford University Press, 2006), 14; Fromkin, *In the Time of the Americans,* 527; Dallek, *Franklin D. Roosevelt and American Foreign Policy,* 245; Blum, *The Progressive Presidents,* 149; Divine, *Roosevelt and World War II,* 43; Kissinger, *Diplomacy,* 389–90.

77 *he had already decided to enter the war:* Dallek, *Franklin D. Roosevelt and American Foreign Policy,* 530; Casey, *Cautious Crusade,* 11, 14–15.

78 *would return to haunt the nation:* Adler, *Uncertain Giant,* 259; Divine, *Roosevelt and World War II,* 48–49; Cantril, *Public Opinion, 1935–1946,* 977; Dallek, *Franklin D. Roosevelt and American Foreign Policy,* 285–86.

78 *"control and domination of the seas":* Divine, *Roosevelt and World War II,* 44.

78 *"from the effects of a Nazi victory":* Osgood, *Ideals and Self-Interest in America's Foreign Relations,* 425.

78 *cutting off Tokyo's supply of oil:* Walter LaFeber, *The Clash: A History of U.S.-Japan Relations* (New York: Norton, 1997), 192–200.

78 *would unify the country behind war:* Dallek, *Franklin D. Roosevelt and American Foreign Policy,* 303–10.

79 *"than he had appeared in a long time":* Dallek, *Franklin D. Roosevelt and American Foreign Policy,* 311.

79 *few epithets were worse:* Fox, *Reinhold Niebuhr,* 193–94.

79 *"as effectively as that can be done":* Fox, *Reinhold Niebuhr,* 149.

80 *end up in the same place politically:* Howard Brick, *Daniel Bell and the Decline of Intellectual Radicalism* (Madison: University of Wisconsin Press, 1986), 35–36, 55–56.

80 *the same phrase FDR's uncle had applied to him:* Daniel Bell, *The End of Ideology: On the Exhaustion of Political Ideas in the Fifties* (New York: Free Press, 1962), 300.

81 *"it is still a civilization":* Justus D. Doenecke, "Reinhold Niebuhr and His Critics: The Interventionist Controversy in World War II," *Anglican and Episcopal History* 64 (1995): 461.

81 *whether democracy could ever really work:* See, for instance, Walter Lippmann, *Public Opinion* (New York: Harcourt Brace, 1922); Walter Lippmann, *The Phantom Public* (New Brunswick, NJ: Transaction, 1993); and Walter Lippmann, *Liberty and the News* (Princeton, NJ: Princeton University Press, 2008).

81 *"have a very good mind":* Steel, *Walter Lippmann,* 291, 382.

81 *Dewey felt he was reliving a nightmare:* Westbrook, *John Dewey and American Democracy*, 512; Howlett, *Troubled Philosopher*, 145.

82 *Beard wrote to a friend:* Kennedy, *Charles Beard and American Foreign Policy*, 93, 105, 120–21, 126–27, 152.

82 *"what is beginning is too much for me":* Westbrook, *John Dewey and American Democracy*, 512.

82 *"because it is not pure enough":* Richard W. Fox, "Reinhold Niebuhr and the Emergence of the Liberal Realist Faith, 1930–1945," *Review of Politics* 38 (1976): 257–58.

82 *"tyranny, sadism, and human defilement":* Nore, *Charles A. Beard*, 183, 200.

82 *FDR's foreign policy a Jewish plot:* Clayton, *Forgotten Prophet*, 265; W. A. Swanberg, *Dreiser* (New York: Scribner, 1965), 461–65.

Chapter 4: I Didn't Say It Was Good

83 *the most expensive film yet made:* Thomas J. Knock, "'History with Lightning': The Forgotten Film *Wilson*," *American Quarterly* 29 (1976): 529.

83 *tribute to the former president:* Divine, *Second Chance*, 57, 168, 213.

83 *even made him seem friendly:* Knock, "'History with Lightning,'" 535, 538.

83 *his blood pressure spiked to 240 over 130:* FDR saw the movie while meeting with Churchill in Quebec. Divine, *Second Chance*, 169–70; Knock, *To End All Wars*, 272; John Morton Blum, *V Was for Victory: Politics and American Culture During World War II* (New York: Harcourt Brace, 1976), 8; United States Department of State, *Foreign Relations of the United States: Conference at Quebec, 1944* (1944), http://digicoll.library.wisc.edu/cgi-bin/FRUS/FRUS-idx?type=goto&id=FRUS.FRUS1944&isize=M&submit=Go+to+page&page=292; Howard G. Bruenn, "Clinical Notes on the Illness and Death of President Franklin D. Roosevelt," *Annals of Internal Medicine* 72 (1970): 587.

83 *"somewhere within the rim of his consciousness":* Robert E. Sherwood, *Roosevelt and Hopkins: An Intimate History* (New York: Harper, 1948), 227.

84 *it was the twenty-sixth:* Forrest Davis, a journalist FDR repeatedly took into his confidence, says this outright: "In his management of foreign affairs, Franklin Roosevelt follows more closely the path of his distant cousin, Theodore, than of Woodrow Wilson." Forrest Davis, "Roosevelt's World Blueprint," *Saturday Evening Post*, April 10, 1943, 110.

84 *between America, Britain, and France:* William Clinton Olson, "Theodore Roosevelt's Conception of an International League," *World Affairs Quarterly* 29 (1959): 346; Cooper, *The Warrior and the Priest*, 333.

84 *"to be realistic":* Divine, *Second Chance*, 43–44; Sherwood, *Roosevelt and Hopkins*, 359–60.

85 *would control the Asian mainland:* Warren F. Kimball, *The Juggler: Franklin Roosevelt as Wartime Statesmen* (Princeton, NJ: Princeton University Press, 1991), 86; Range, *Franklin D. Roosevelt's World Order*, 178; Robert L. Messer, "World War II and the Coming of the Cold War," in *Modern American Diplomacy*, ed. Carroll and Herring, 125–126.

85 *"too many nations to satisfy":* George Schild, "The Roosevelt Administration and the United Nations: Re-Creation or Rejection of the League Experience?" *World Affairs* 158 (1995): 29. The descriptions of Molotov are Churchill's. Raymond H. Anderson, "A Lifetime of History: Vyacheslav M. Molotov, Close Associate of Stalin, Is Dead in Moscow at 96," *New York Times*, November 11, 1986, B1, B7.

85 *"the military force to uphold it":* Davis, "Roosevelt's World Blueprint," 20, 110.

85 *which nations should exist, and where:* The State Department, at the urging of Undersecretary Sumner Welles (one of the strongest Wilsonians in the Roosevelt administration), did launch a postwar planning effort involving regional experts inside and outside government, but it had far less impact given Roosevelt's realist proclivities.

86 *as much at Newfoundland:* Schild, "The Roosevelt Administration and the United Nations," 30; Divine, *Second Chance*, 43–44; Sherwood, *Roosevelt and Hopkins*, 359–60.

86 *by 1944, 72 percent did:* Divine, *Second Chance*, 68; Knock, "'History with Lightning,'" 540.

86 *"blow off steam":* Divine, *Second Chance*, 158; Kimball, *The Juggler*, 86.

86 *respiratory problems and in chronic pain:* Kimball, *The Juggler*, 88, 238.

86 *demanding that they call it a sheep:* Fromkin, *In the Time of the Americans*, 603.

87 *most powerful player in Eastern Europe:* Messer, "World War II and the Coming of the Cold War," 118, 124–25.

87 *revolt against the additional loss of life:* Gaddis, *Strategies of Containment*, 6–7; Messer, "World War II and the Coming of the Cold War," 111; Dallek, *Franklin D. Roosevelt and American Foreign Policy*, 433–34.

87 *to defeat Hitler, Mussolini, and Tojo:* Goldman, *Rendezvous with Destiny*, 390.

88 *in his upcoming reelection bid:* Dallek, *Franklin D. Roosevelt and American Foreign Policy*, 436; Divine, *Roosevelt and World War II*, 92.

88 *"and we reject it":* Dallek, *Franklin D. Roosevelt and American Foreign Policy*, 439.

89 *"in favour of the great":* Schild, "The Roosevelt Administration and the United Nations," 31; Leland M. Goodrich, "From League of Nations to United Nations," *International Organization* 1 (1947), 8–10; Divine, *Roosevelt and World War II*, 66; Evan Luard, *A History of The United Nations, Volume 1: The Years of Western Domination, 1945–1955* (New York: St. Martin's, 1982), 44.

89 *the body afforded large nations:* Divine, *Second Chance*, 196–99.

89 *"Hull knows the whole story":* Forrest Davis, "What Really Happened at Teheran," 46; Divine, *Second Chance*, 201.

89 *"Americans might be shocked":* Daniel Yergin, *Shattered Peace* (Boston: Houghton Mifflin, 1977), 59; Dallek, *Franklin D. Roosevelt and American Foreign Policy*, 479; Joseph M. Siracusa, "The Night Stalin and Churchill Divided Europe: The View from Washington," *Review of Politics* 43 (1981): 381–409; Kimball, *The Juggler*, 162.

90 *"as they affect sentiment in America":* Fromkin, *In the Time of the Americans*, 599; Dallek, *Franklin D. Roosevelt and American Foreign Policy*, 503; Wilson D. Miscamble, *From Roosevelt to Truman: Potsdam, Hiroshima, and the Cold War* (New York: Cambridge University Press, 2007), 54–57.

90 *rarely challenged his postwar efforts:* Dallek, *Franklin D. Roosevelt and American Foreign Policy*, 483–84; James MacGregor Burns, *Roosevelt: The Soldier of Freedom* (New York: Harcourt Brace, 1971), 516, 528–29; Divine, *Second Chance*, 219; John Foster Dulles, *War or Peace* (New York: Macmillan, 1950), 124–25.

90 *"the almost ravaged appearance of his face":* Dallek, *Franklin D. Roosevelt and American Foreign Policy*, 442, 481.

90 *wrapped himself in a large cape to stay warm:* Dallek, *Franklin D. Roosevelt and American Foreign Policy*, 508; Burns, *Roosevelt*, 564.

90 *Manchuria and northern China:* Messer, "World War II and the Coming of the Cold War," 133.

91 *make decisions by majority vote:* Dallek, *Franklin D. Roosevelt and American Foreign Policy*, 466–67; Kimball, *The Juggler*, 98; Divine, *Second Chance*, 276.

91 *exiled prewar Polish leadership in London:* Melvyn P. Leffler, "Adherence to Agreements: Yalta and the Experiences of the Early Cold War," *International Security* 11 (1986): 94–95; Jaime Reynolds, "Communists, Socialists and Workers: Poland 1944–1948," *Soviet Studies* 30 (1978): 517.

91 *"but in fact she had her sins":* LaFeber, *The American Age*, 439–40; Dallek, *Franklin D. Roosevelt and American Foreign Policy*, 516; Vladislav Zubok and Constantine Pleshakov, *Inside the Kremlin's Cold War: From Stalin to Khrushchev* (Cambridge, MA: Harvard University Press, 1996), 10.

91 *"and have always failed":* Fromkin, *In the Time of the Americans*, 618; Franklin D. Roosevelt, "Address to Congress on Yalta," March 1, 1945, http://millercenter.org/scripps/archive/speeches/detail/3338.

92 *Roosevelt did not want a public row:* Richard F. Staar, "Elections in Communist Poland," *Midwest Journal of Political Science* 2 (1958): 202–3; Reynolds, "Communists, Socialists and Workers," 519–22; Dallek, *Franklin D. Roosevelt and American Foreign Policy*, 524–25; Kimball, *The Juggler*, 176–78.

92 *"in an admittedly imperfect world":* Franklin D. Roosevelt, "State of the Union Address," January 6, 1945, http://www.presidency.ucsb.edu/ws/index.php?pid=16595.

92 *more than twenty-six years earlier:* Divine, *Second Chance*, 277; Dallek, *Franklin D. Roosevelt and American Foreign Policy*, 528; Burns, *Roosevelt*, 600.

92 *opening ceremony of the United Nations:* Divine, *Second Chance*, 287; John Lewis Gaddis, *The Cold War: A New History* (New York: Penguin, 2005), 5.

93 *kept their colonial possessions:* Divine, *Second Chance*, 297; UN Universal Declaration of Human Rights, December 10, 1948, http://www.un.org/Overview/rights.html.

93 *deferred seating Russia's puppet government:* Divine, *Second Chance*, 290.

93 *"other is Franklin D. Roosevelt":* Divine, *Second Chance*, 311.

93 *sparked no cheering or applause:* Divine, *Second Chance*, 290, 311–14.

93 *killing eighty thousand in a single night:* Arnold Offner, *Another Such Victory: President Truman and the Cold War, 1945–1953* (Stanford, CA: Stanford University Press, 2002), 92–93.

94 *imprisoned for most of the war:* LaFeber, *The American Age*, 406; Greg Robinson, *By Order of the President: FDR and the Internment of Japanese-Americans* (Cambridge, MA: Harvard University Press, 2001), 250.

94 *"I said it was the best I could do":* Divine, *Second Chance*, 285, 297; Dallek, *Franklin D. Roosevelt and American Foreign Policy*, 521.

94 *not reading highbrow journals:* Richard Hofstadter, *The American Political Tradition and the Men Who Made It* (New York: Knopf, 1948), 319–20.

94 *"to atomic war":* Arthur M. Schlesinger, Jr., *The Vital Center: The Politics of Freedom* (Boston: Houghton Mifflin, 1949), 1–2.

95 *"the Christian doctrine of original sin":* Reinhold Niebuhr, *The Children of Light and the Children of Darkness: A Vindication of Democracy and a Critique of Its Traditional Defense* (New York: C. Scribner's Sons, 1944), 16.

95 *to see the past in prelapsarian terms:* It was during the 1920s, after all, that Sigmund Freud made his American splash. Among the many U.S. intellectuals he influenced during that decade was Walter Lippmann, whose extremely influential books about public irrationality hardly brimmed with optimism about the human condition. See Steel, *Walter Lippmann*, 173, 262; and Dumenil, *The Modern Temper*, 146–47, on Freud's influence on 1920s intellectual life.

95 *tragedy about the human condition:* Richard H. Pells, *Radical Visions and American Dreams: Culture and Social Thought in the Depression Years* (New York: Harper & Row, 1977), 358; Kevin Mattson, *When America Was Great: The Fighting Faith of Postwar Liberalism* (New York: Routledge, 2004), 34–35.

95 *"unwarranted optimism about man":* Schlesinger, *The Vital Center*, 40.

95 *resist the Nazis nonetheless:* I am grateful to Professor Casey Blake for the analogy. Norman Graebner makes a similar point in *The Age of Doubt: American Thought and Culture in the 1940s* (Boston: Twayne, 1991), 4.

95 *might not have won the war?:* Kennedy, *Freedom from Fear*, 653–58, 666; Graebner, *The Age of Doubt*, 51.

95 *"necessary for human conduct":* Mortimer J. Adler, "This Pre-War Generation," *Harper's*, October 1940, http://radicalacademy.com/adlerprewargeneration.htm.

95 *"good and bad, right and wrong":* Mortimer J. Adler, "God and the Professors," Conference on Science, Philosophy and Religion (1941), http://radicalacademy.com/adlergodprofessors.htm; Westbrook, *John Dewey and American Democracy*, 518.

96 *its confrontations with the old world:* Joel H. Rosenthal, *Righteous Realists: Political Realism, Responsible Power, and American Culture in the Nuclear Age* (Baton Rouge: Louisiana State University Press, 1991), 2–7; Greg Russell, *Hans J. Morgenthau and the Ethics of American Statecraft* (Baton Rouge: Louisiana State University Press, 1990), 62, 66; Lewis A. Coser, *Refugee Scholars in America: Their Impact and Their Experiences* (New Haven, CT: Yale University Press, 1984), 219–20; William E. Scheuerman, "Carl Schmitt and Hans Morgenthau: Realism and Beyond," in *Realism Reconsidered: The Legacy of Hans Morgenthau in International Relations*, ed. Michael C. Williams (New York: Oxford University Press, 2007), 62–77; Richard Little, "The Balance of Power in Politics Among Nations," in *Realism Reconsidered*, ed. Williams, 153–54.

96 *"cured by means of reason":* Hans J. Morgenthau, *Scientific Man vs. Power Politics* (Chicago: University of Chicago Press, 1946), 71.

96 *incurable lust for power:* Ghazi A. R. Algosaibi, "The Theory of International Relations: Hans Morgenthau and His Critics," in *Realism Reconsidered*, ed. Williams, 229, 240.

96 *courts that ruled aggression illegal:* Morgenthau, *Scientific Man*, 50–51.

97 *moral norms it did not uphold itself:* Divine, *Second Chance*, 180–81; Steel, *Walter Lippmann*, 405, 426.

98 *"not at all what I always knew":* Walter Lippmann, *US Foreign Policy: Shield of the Republic* (New York: Pocket, 1943), viii.

98 *ate most of his meals alone:* Walter Isaacson and Evan Thomas, *The Wise Men: Six Friends and the World They Made* (New York: Touchstone, 1986), 76–78, 228; Thompson, *The Hawk and the Dove*, 24–26.

98 *they conjured the mother he had never had:* Isaacson and Thomas, *The Wise Men*, 140, 145, 147, 153, 165, 227; Mayers, *George Kennan and the Dilemmas of US Foreign Policy*, 101.

98 *"not a member of its household":* Isaacson and Thomas, *The Wise Men*, 74.

99 *unelected wise men like, well, himself:* On Kennan's prejudices, see Joshua Botts, " 'Nothing to Seek and . . . Nothing to Defend': George F. Kennan's Core Values and American Foreign Policy, 1938–1993," *Diplomatic History* 30 (2006): 839–66.

99 *its own circumstances and needs:* Isaacson and Thomas, *The Wise Men*, 147–51; Rosenthal, *Righteous Realists*, 25–26.

99 *an intimate understanding of the participants:* Michael Joseph Smith, *Realist Thought from Weber to Kissinger* (Baton Rouge: Louisiana State University Press, 1986), 170.

100 *when it meets with some unanswerable force":* X, "The Sources of Soviet Conduct," *Foreign Affairs* 25 (1947): 574.

100 *was losing patience with Moscow:* See Dallek, *Franklin D. Roosevelt and American Foreign Policy*, 550–551; Gaddis, *Strategies of Containment*, 14–15; Arthur M. Schlesinger, Jr., "Origins of the Cold War," *Foreign Affairs* 46 (1967): 24. For an opposing view, see Kimball, *The Juggler*, 180; Miscamble, *From Roosevelt to Truman*, 73–79.

100 *large and brutal neighbor:* John Lewis Gaddis, "Was the Truman Doctrine a Real Turning Point?" *Foreign Affairs* 52 (1974): 388.

101 *the challenges he would face:* Offner, *Another Such Victory*, 22–25.

101 *"are the one in trouble now":* Miscamble, *From Roosevelt to Truman*, ix.

101 *in the war as had the United States:* Gaddis, *The Cold War*, 9.

101 *began to get tough:* Messer, "World War II and the Coming of the Cold War," 128–29.

101 *"the straight one-two to the jaw":* Miscamble, *From Roosevelt to Truman*, 137. In recent years, some historians have suggested that in his memoirs Truman exaggerated how tough he actually was on Molotov during their April 1945 meeting. See Geoffrey Roberts, "Sexing Up the Cold War: New Evidence on the Molotov-Truman Talks of April 1945," *Cold War History* 4 (2004): 105–25. More generally, Wilson Miscamble has argued that Truman's hard line evolved more slowly and fitfully than traditional accounts suggest. *From Roosevelt to Truman*, 87–332. But regardless of the speed at which the change occurred, there is no question that U.S. policy toward the U.S.S.R. grew tougher over the course of 1945.

102 *bread and a pistol under his pillow:* "Gun No. 242332," *Time*, September 12, 1955, http://www.time.com/time/magazine/article/0,9171,893108,00.html.

102 *the longest ever sent from the embassy in Moscow:* Mayers, *George Kennan and the Dilemmas of US Foreign Policy*, 99; Walter L. Hixson, *George Kennan: Cold War Iconoclast* (New York: Columbia University Press, 1989), 32; Miscamble, *From Roosevelt to Truman*, 279.

102 *we were now it:* Gaddis, *Strategies of Containment*, 18–20, 380.

103 *not one postwar world, but two:* Mayers, *George Kennan and the Dilemmas of US Foreign Policy*, 100; Townsend Hoopes and Douglas Brinkley, *Driven Patriot: The Life and Times of James Forrestal* (Annapolis, MD: Naval Institute Press, 1992), 272; Gaddis, *Strategies of Containment*, 56.

103 *he was speaking entirely metaphorically:* Dallek, *Franklin D. Roosevelt and American Foreign Policy*, 472; Miscamble, *From Roosevelt to Truman*, 70–71.

103 *inflaming the Germans:* Mary N. Hampton, "NATO at the Creation: U.S. Foreign Policy, West Germany and the Wilsonian Impulse," *Security Studies* 4 (1995): 619–20.

103 *Europe's suffering to its advantage:* Beisner, *Dean Acheson*, 69–71, 76–89.

105 *"facts which momentarily imprison them":* Mary S. McAuliffe, *Crisis on the Left: Cold War Politics and American Liberals* (Amherst: University of Massachusetts Press, 1978), 65, 160; Smith, *Realist Thought*, 113–14, 130; Fox, *Reinhold Niebuhr*, 59.
105 *FDR created a sleeping beauty:* Fromkin, *In the Time of the Americans*, 647.

PART II: THE HUBRIS OF TOUGHNESS

Chapter 5: The Murder of Sheep

109 *Secretary of State George Marshall:* Thompson, *The Hawk and the Dove*, 61; Hoopes and Brinkley, *Driven Patriot*, 272; Wilson D. Miscamble, *George F. Kennan and the Making of American Foreign Policy, 1947–1950* (Princeton, NJ: Princeton University Press, 1992), 218.
109 *a man they didn't really know at all:* Hoopes and Brinkley, *Driven Patriot*, 15, 36–47.
110 *"we also laughed at Hitler":* Hoopes and Brinkley, *Driven Patriot*, 266–69.
111 *that Stalin, like Hitler, wanted a world war:* Hoopes and Brinkley, *Driven Patriot*, 275; Mayers, *George Kennan and the Dilemmas of US Foreign Policy*, 109–10.
111 *"Lenin's religion-philosophy":* Mayers, *George Kennan and the Dilemmas of US Foreign Policy*, 110–11; Hoopes and Brinkley, *Driven Patriot*, 280.
111 *State Department's Policy Planning staff:* Hoopes and Brinkley, *Driven Patriot*, 278; Miscamble, *George F. Kennan and the Making of American Foreign Policy*, 10.
111 *the byline read simply "By X":* Mayers, *George Kennan and the Dilemmas of US Foreign Policy*, 113; X, "The Sources of Soviet Conduct," 566–82; George Kennan, *Memoirs 1925–50* (Boston: Atlantic Monthly Press, 1967), 354–58.
111 *friends began calling her "Miss X":* Hoopes and Brinkley, *Driven Patriot*, 276; Miscamble, *George F. Kennan and the Making of American Foreign Policy*, 64, 66; Thompson, *The Hawk and the Dove*, 75–76.
111 *to undermine Soviet control in Eastern Europe:* Gregory Mitrovitch, *Undermining the Kremlin: America's Strategy to Subvert the Soviet Bloc, 1947–1956* (Ithaca, NY: Cornell University Press, 2000), 5–46.
112 *unnecessary and probably counterproductive:* Mayers, *George Kennan and the Dilemmas of US Foreign Policy*, 153–55.
113 la guerre froide, *the cold war:* Steel, *Walter Lippmann*, 444–45.
113 *"both in money and military power":* Walter Lippmann, *The Cold War* (New York: Harper & Brothers, 1947), 22–23.
113 *"many views with which I profoundly agreed":* Steel, *Walter Lippmann*, 445; Thompson, *The Hawk and the Dove*, 78.
113 *too much publicity already, refused:* Miscamble, *George F. Kennan and the Making of American Foreign Policy*, 68.
114 *the columnist had gotten his views all wrong:* Thompson, *The Hawk and the Dove*, 79; Miscamble, *George F. Kennan and the Making of American Foreign Policy*, 67; Kennan, *Memoirs 1925–1950*, 359–62.
114 *similar pledges in other corners of the globe:* Robert Jervis, "The Impact of the Korean War on the Cold War," *Journal of Conflict Resolution* 24 (1980): 574.
114 *Kennan's insistence on containment's geographic limits:* Melvyn P. Leffler, *Preponderance of Power: National Security, the Truman Administration, and the Cold War* (Stanford, CA: Stanford University Press, 1992), 304–11.
115 *slashed military spending by 15 percent:* Gaddis, *Strategies of Containment*, 91; Jervis, "The Impact of the Korean War on the Cold War," 568.
115 *And this time, America passed:* Offner, *Another Such Victory*, 138–43; Bruce R. Kuniholm, *The Origins of the Cold War in the Near East: Great Power Conflict and Diplomacy in Iran, Turkey, and Greece* (Princeton, NJ: Princeton University Press, 1980), 321–34.
115 *to defend Greece or Turkey, either:* Beisner, *Dean Acheson*, 54.
115 *the Eastern Mediterranean and the Middle East:* Leffler, *Preponderance of Power*, 123–124, 143; Beisner, *Dean Acheson*, 39.

115 *come to Greece and Turkey's aid:* Thompson, *The Hawk and the Dove*, 71.

116 *Kennan pleaded in a memo to his superiors:* Mayers, *George Kennan and the Dilemmas of US Foreign Policy*, 349.

116 *regained their foreign policy edge:* Offner, *Another Such Victory*, 201–9; Zubok and Pleshakov, *Inside the Kremlin's Cold War*, 46, 57, 127.

116 *an upset reelection victory that fall:* Offner, *Another Such Victory*, 254–66.

117 *"a whole new approach to the affairs of the world":* Miscamble, *George F. Kennan and the Making of American Foreign Policy*, 217.

117 *Davies, a man rather similar to himself:* Kennan openly admitted this, saying that Davies was responsible for "whatever insight I was able to muster in those years into the nature of Soviet policies toward the Far East." Miscamble, *George F. Kennan and the Making of American Foreign Policy*, 215–16.

117 *Davies was the first person he hired:* Miscamble, *George F. Kennan and the Making of American Foreign Policy*, 213, 218; Halberstam, *The Best and the Brightest*, 384; Kahn, *The China Hands*, 54.

117 *"control of the communist movement":* Mayers, *George Kennan and the Dilemmas of US Foreign Policy*, 173. In Moscow, as Soviet archives would later reveal, Stalin was already worrying about exactly that. Zubok and Pleshakov, *Inside the Kremlin's Cold War*, 56.

117 *"pouring sand in a rat hole":* Hoopes and Brinkley, *Driven Patriot*, 307–8; Offner, *Another Such Victory*, 200.

118 *demanded that the latter be brought in line:* See, for instance, Richard Freeland's account of a congressional hearing on March 20, 1947, in which Representative James Fulton of Pennsylvania repeatedly tried to get Acting Secretary of State Dean Acheson to promise that American policy in China would abide by the principles laid out in Truman's Greece and Turkey speech. Richard M. Freeland, *The Truman Doctrine and the Origins of McCarthyism: Foreign Policy, Domestic Politics and Internal Security, 1946–1948* (New York: Knopf, 1972), 112–13.

118 *billions in economic and military aid to Chiang:* Mayers, *George Kennan and the Dilemmas of US Foreign Policy*, 179.

118 *Mao did not pose a grave threat:* Miscamble, *George F. Kennan and the Making of American Foreign Policy*, 227.

118 *then it hit him: treason!:* Halberstam, *The Best and the Brightest*, 117.

119 *had been cut in half:* Ernest R. May, *"Lessons" of the Past: The Use and Misuse of History in American Foreign Policy* (New York: Oxford University Press, 1973), 78–79.

119 *"with a Harvard accent":* Beisner, *Dean Acheson*, 307.

119 *"the Communists took over in China":* Geoffrey S. Smith, "National Security and Personal Isolation: Sex, Gender and Disease in the Cold War United States," *International History Review* 14 (1992): 319; Robert D. Dean, *Imperial Brotherhood: Gender and the Making of Cold War Foreign Policy* (Amherst: University of Massachusetts Press, 2001), 66; Gaddis, *Strategies of Containment*, 240.

119 *that they would not defend Seoul:* Beisner, *Dean Acheson*, 326–28; Bruce Cumings, *Origins of the Korean War, Volume II: The Roaring of the Cataract, 1947–1950* (Princeton, NJ: Princeton University Press, 1990), 420–27; William Stueck, *The Korean War: An International History* (Princeton, NJ: Princeton University Press, 1995), 30.

120 *"Iran, Berlin and Greece":* May, *"Lessons" of the Past*, 77, 81–82.

120 *"continue to take responsibility for it":* Miscamble, *George F. Kennan and the Making of American Foreign Policy*, 324; Stueck, *The Korean War*, 61, 63; Thompson, *The Hawk and the Dove*, 119.

120 *"suspicious foreign government":* Miscamble, *George F. Kennan and the Making of American Foreign Policy*, 326; Thompson, *The Hawk and the Dove*, 123.

121 *to do just that:* Chen Jia, "The Sino-Soviet Alliance and China's Entry into the Korean War," working paper of the Cold War History Project, Woodrow Wilson International Center, Washington, D.C., 1991; Michael Hunt, "Beijing and the Korea Crisis," *Political Science Quarterly* (1992): 453–78.

121 *his men would be home by Christmas:* Stueck, *The Korean War*, 93–94, 97–103; Thompson, *The Hawk and the Dove*, 123; Walter LaFeber, *America, Russia, and the Cold War, 1945–1975* (New York: Wiley, 1976), 121–22; Miscamble, *George F. Kennan and the Making of American Foreign Policy*, 327.

121 *Truman fired him:* Stueck, *The Korean War*, 167–68; David Halberstam, *The Coldest Winter: America and the Korean War* (New York: Hyperion, 2007), 597–606.

121 *"the Munich conference in 1938":* Goran Rystad, *Prisoners of the Past? The Munich Syndrome and Makers of American Foreign Policy in the Cold War Era* (Lund, Sweden: CWK Gleerup, 1982), 37; David McLellan, "Dean Acheson and the Korean War," *Political Science Quarterly* 83 (1969): 32.

121 *flooded with pro-MacArthur mail:* LaFeber, *America, Russia and the Cold War*, 128; Robert Dallek, *Lone Star Rising: Lyndon Johnson and His Times, 1908–1960* (New York: Oxford University Press, 1991), 398–399.

122 *hadn't wrecked America's economy:* Alan Brinkley, *Unfinished Nation: A Concise History of the American People* (New York: Knopf, 1993), 817.

122 *"not do if it wanted to do it":* Gaddis, *Strategies of Containment*, 92.

123 *would not rise significantly as a percentage of GNP:* Leffler, *Preponderance of Power*, 355–60; Gaddis, *Strategies of Containment*, 88, 91.

123 *within the United States itself:* Gaddis, *Strategies of Containment*, 95, 111.

123 *both to others and to ourselves:* On the idea of credibility, and its importance to American foreign policymakers, see Robert J. McMahon, "Credibility and World Power: Exploring the Psychological Dimension in Postwar American Diplomacy," Stuart L. Bernath Memorial Lecture, *Diplomatic History* 15 (1991), 455–71; Daryl G. Press, *Calculating Credibility: How Leaders Assess Military Threats* (Ithaca, NY: Cornell University Press, 2005); and Ted Hopf, *Peripheral Visions: Deterrence Theory and American Foreign Policy in the Third World, 1965–1990* (Ann Arbor: University of Michigan Press, 1994).

123 *"in each and every one of us":* Gaddis, *The Cold War*, 46.

124 *applied them only to the other side:* National Security Council, "NSC–68: United States Objectives and Programs for National Security," April 14, 1950, http://www.fas.org/irp/off docs/nsc-hst/nsc–68.htm.

124 *sedatives, tranquilizers, and shots of insulin:* Hoopes and Brinkley, *Driven Patriot*, 440, 450–51, 460.

124 *he turns the sword on himself:* Sophocles, *Ajax*, trans. John Tipton (Chicago: Flood, 2008).

124 *with a thud, thirteen floors below:* Hoopes and Brinkley, *Driven Patriot*, 465.

125 *when he entered a room:* Halberstam, *The Best and the Brightest*, 381.

125 *into the U.S. government:* Hixson, *George Kennan*, 164–65; Mayers, *George Kennan and the Dilemmas of US Foreign Policy*, 202, 205; Kahn, *The China Hands*, 244–46; Michael T. Kaufman, "John Paton Davies, Diplomat Who Ran Afoul of McCarthy over China, Dies at 91," *New York Times*, December 24, 1999, http://www.nytimes.com/1999/12/24/world/ john-paton-davies-diplomat-who-ran-afoul-of-mccarthy-over-china-dies-at-91.html.

125 *sustaining his family by building furniture:* Kahn, *The China Hands*, 258, 261–62; Hixson, *George Kennan*, 167; Halberstam, *The Best and the Brightest*, 381.

125 *"to be of much practical use to you":* Miscamble, *George F. Kennan and the Making of American Foreign Policy*, 277–78.

Chapter 6: The Problem with Men

126 *it does nothing for eight years:* Herbert S. Parmet, *Eisenhower and the American Crusades* (New Brunswick, NJ: Transaction, 1999), xiii, xvii.

127 *"the sacrifice to capture a worthless objective":* Stephen Ambrose, *Eisenhower: Soldier and President* (New York: Simon & Schuster, 1990), 87–88, 96–97, 181, 189–92, 270.

127 *the least lamentable casualty of war:* Halberstam, *The Coldest Winter*, 629–30; John E. Mueller, "Trends in Popular Support for the Wars in Korea and Vietnam," *American Political Science Review* 65, no. 2 (1971): 360–61; William Stueck, *Rethinking the Korean War: A New Diplomatic and Strategic History* (Princeton, NJ: Princeton University Press, 2002), 157.

127 *should cede much of the combat to South Koreans:* Robert A. Divine, *Foreign Policy and U.S. Presidential Elections, 1952–1960* (New York: Franklin Watts, 1974), 10, 45–46, 53–54, 71, 74, 80.

127 *"the contest ended that night":* Elie Abel, "General in Pledge, First Task Would Be Early and Honorable End of the War," *New York Times*, October 25, 1952, A1; Parmet, *Eisenhower and the American Crusades*, 143; Robert A. Divine, *Eisenhower and the Cold War* (New York: Oxford University Press, 1981), 18–19; Halberstam, *The Coldest Winter*, 626.

128 *where it had been when the war began:* Roger Dingman, "Atomic Diplomacy During the Korean War," *International Security* 13, no. 3 (1988/1989): 85–88; Ambrose, *Eisenhower*, 294–295, 327–331; Stueck, *Rethinking the Korean War*, 171–181; Zubok and Pleshakov, *Inside the Kremlin's Cold War*, 71, 155; Stueck, *The Korean War*, 224, 302; Divine, *Eisenhower and the Cold War*, 27–31; Philip West, "Review: Interpreting the Korean War," *American Historical Review* 94, no. 1 (1989): 80.

128 *"bills at the local department store":* Gaddis, *Strategies of Containment*, 132.

128 *Eisenhower's policies on national defense:* Gaddis, *Strategies of Containment*, 162, 169; Stephen E. Ambrose and Douglas G. Brinkley, *Rise to Globalism: American Foreign Policy Since 1938* (New York: Penguin, 1997), 131; Halberstam, *The Best and the Brightest*, 40.

129 *supposedly kept Latin leaders in power:* John Prados, *Safe for Democracy: The Secret Wars of the CIA* (Chicago: Irvin R. Dee, 2006), 99–123, 166–83, 215, 275–80, 432; Tim Weiner, *Legacy of Ashes: The History of the CIA* (New York: Doubleday, 2007), 94, 171–72; Stephen Kinzer, *Overthrow: America's Century of Regime Change from Hawaii to Iraq* (New York: Times Books, 2006), 136–38.

129 *"blow the hell out of them":* Chen Jian, *Mao's China and the Cold War* (Chapel Hill: University of North Carolina Press, 2001), 167–70, 199–202; Campbell Craig, *Destroying the Village: Eisenhower and Thermonuclear War* (New York: Columbia University Press, 1998), 52, 78–107; Gaddis, *Strategies of Containment*, 147, 167, 169.

130 *prevent both sides from taking the first step:* Craig, *Destroying the Village*, 52–118.

130 *or worse later, refused:* George C. Herring and Richard H. Immerman, "Eisenhower, Dulles, and Dienbienphu: 'The Day We Didn't Go to War' Revisited," *Journal of American History* 71 (1984): 343–63.

130 *U.S. forces were gone:* Gordon H. Chang, "To the Nuclear Brink: Eisenhower, Dulles, and the Quemoy-Matsu Crisis," *International Security* 12, no. 4 (1988): 96–123; H. W. Brands, Jr., "Testing Massive Retaliation: Credibility and Crisis Management in the Taiwan Strait," *International Security* 12, no. 4 (1988): 124–51; Leonard H. D. Gordon, "United States Opposition to the Use of Force in the Taiwan Strait, 1954–1962," *Journal of American History* 72, no. 3 (1985): 637–60; Divine, *Eisenhower and the Cold War*, 55–66, 97–104; Ambrose, *Eisenhower*, 469.

130 *the war a mistake fell by 15 points:* Stephen G. Rabe, "Eisenhower Revisionism," *Diplomatic History* 17, no. 1 (1993): 100; Yuen Foong Khong, *Analogies at War: Korea, Munich, Dien Bien Phu, and the Vietnam Decisions of 1965* (Princeton, NJ: Princeton University Press, 1992), 114.

131 *horrors that made them old before their time:* Bell, *The End of Ideology.*

131 *"the unromantic realm of more-or-less":* Richard H. Pells, *The Liberal Mind in a Conservative Age: American Intellectuals in the 1940s and 1950s* (Middletown, CT: Wesleyan University Press, 1989), 138, 140.

132 *they had experienced almost ceaseless boom:* John Patrick Diggins, *The Proud Decades: America in War and in Peace, 1941–1960* (New York: Norton, 1988), 345; Diane Kunz, *Butter and Guns: America's Cold War Economic Diplomacy* (New York: Free Press, 1997), 66; The Misery Index, "The U.S. Inflation Rate, 1950–1960," http://www.miseryindex.us/iRbyyear.asp?StartYear=1950&EndYear=1960.

132 *"malaise of the privileged":* Pells, *The Liberal Mind in a Conservative Age*, 240.

132 *"and a search for passion":* Bell, *The End of Ideology*, 404, 300.

133 *junior senator from Massachusetts, John Fitzgerald Kennedy:* Arthur Schlesinger, Jr., "The Crisis of American Masculinity," in *The Politics of Hope and The Bitter Heritage* (Princeton, NJ: Princeton University Press, 2008), 292–93, 295, 303.

133 *"were days of intense physical pain":* Edward J. Renehan, Jr., *The Kennedys at War, 1937–1945* (New York: Doubleday, 2002), 7–8; John Hellmann, *The Kennedy Obsession* (New York: Columbia University Press, 1997), 9; Doris Kearns Goodwin, *The Fitzgeralds and the Kennedys* (New York: Simon & Schuster, 1987), 309–12. On Kennedy's poor health, see Robert Dallek, *An Unfinished Life: John F. Kennedy, 1917–1963* (Boston: Little, Brown, 2003), 85–88, 100–5, 397–99, 704–6.

133 *"when he really wasn't":* Kearns Goodwin, *The Fitzgeralds and the Kennedys*, 354; Hellmann, *The Kennedy Obsession*, 9; Thomas, *Robert Kennedy*, 39; Renehan, *The Kennedys at War*, 7, 220.

133 *And he went to war:* Renehan, *The Kennedys at War*, 162.

134 *the coconut shell in a case on his desk:* Dallek, *An Unfinished Life*, 82–88, 95–98; Renehan, *The Kennedys at War*, 223, 256–71.

134 *the real purpose of world affairs: making money:* Dallek, *An Unfinished Life*, 21; Renehan, *The Kennedys at War*, 34, 63, 142.

134 *of succumbing to pressure from Jews:* Dallek, *An Unfinished Life*, 54; Renehan, *The Kennedys at War*, 16, 24, 45, 71, 105, 142.

135 *stay in London with the nation's men:* Renehan, *The Kennedys at War*, 80, 117–18, 160, 175.

135 *Committee Against Military Intervention in Europe:* Renehan, *The Kennedys at War*, 73, 74, 91–92, 182.

135 *the form of government most likely to survive:* Renehan, *The Kennedys at War*, 92, 138.

135 *"don't want to take sides too much":* Renehan, *The Kennedys at War*, 139.

136 *calling his book* Why England Slept: Renehan, *The Kennedys at War*, 139, 153–54; Dallek, *An Unfinished Life*, 65; John D. Fair, "The Intellectual JFK: Lessons in Statesmanship from British History," *Diplomatic History* 30, no. 1 (2006), 129–30.

136 *clear his and his father's names:* Hellmann, *The Kennedy Obsession*, 21. Arthur Krock claimed that Joe Jr. "was seeking to prove by its [his mission's] very danger that the Kennedys were not yellow. That's what killed that boy. . . . And his father realized it." Ronald Steel, *In Love with Night: The American Romance with Robert Kennedy* (New York: Simon & Schuster, 2000), 33.

136 *cowardice, if not treason:* Renehan, *The Kennedys at War*, 254.

136 *not the United States:* John Noble Wilford, "With Fear and Wonder in Its Wake, Sputnik Lifted Us into the Future," *New York Times*, September 25, 2007, http://www.nytimes .com/2007/09/25/science/space/25sput.html.

137 *science labs to math teachers:* Brinkley, *Unfinished Nation*, 885–86; LaFeber, *America, Russia and the Cold War*, 203.

137 *nowhere near as many ICBMs:* Gaddis, *Strategies of Containment*, 182–86; David L. Snead, *The Gaither Committee, Eisenhower and the Cold War* (Columbus: Ohio University Press, 1999), 154–56, 180–81; Peter J. Roman, *Eisenhower and the Missile Gap* (Ithaca, NY: Cornell University Press, 1995), 118–40.

137 *"the politics of fatigue":* Schlesinger, "The New Mood in Politics," in *The Politics of Hope and The Bitter Heritage*, 109.

137 *prime minister who had preceded Chamberlain:* William V. Shannon, "Eisenhower as President," *Commentary* 26, no. 5 (1958): 398.

137 *"like a man, can grow soft and complacent":* Paul Dickson, *Sputnik: The Shock of the Century* (New York: Walker, 2001), 223.

137 *Jack relished the line:* Elie Abel, "Stevenson Called Appeaser by Nixon," *New York Times*, October 17, 1952, 19; Thomas G. Paterson, "Bearing the Burden: A Critical Look at JFK's Foreign Policy," *Virginia Quarterly Review* 54 (1978): 196; Dallek, *An Unfinished Life*, 160, 259.

138 *"have the largest rockets in the world":* Robert Dean, "Masculinity as Ideology: John F. Kennedy and the Domestic Politics of Foreign Policy," *Diplomatic History* 22, no. 1 (1998): 45–46.

138 *"lack of physical fitness is a menace to our security":* John F. Kennedy, "The Soft American," *Sports Illustrated*, December 26, 1960, www.ihpra.org/soft_american.htm.

138 *"the best of our energies and abilities":* John F. Kennedy, *The Strategy of Peace* (New York: Harper & Brothers, 1960), 201, quoted in Dean, "Masculinity as Ideology," 29; Gaddis, *Strategies of Containment,* 232.

138 *Ike's soft, grandfatherly game:* Cooper, *The Warrior and the Priest,* 70; Gaddis, *Strategies of Containment,* 198; Dean, "Masculinity as Ideology," 47; Dean, *Imperial Brotherhood,* 183; Richard Reeves, *President Kennedy: Profile of Power* (New York: Simon & Schuster, 1993), 471–72; Dallek, *An Unfinished Life,* 472.

138 *he wore thermal underwear:* Halberstam, *The Best and the Brightest,* 241.

139 *"by a hard and bitter peace":* John F. Kennedy, "Inaugural Address," January 20, 1961, http://www.jfklibrary.org/Historical+Resources/Archives/Reference+Desk/Speeches/JFK/003POF03Inaugural01201961.htm.

139 *"From Innocence to Engagement":* McGeorge Bundy, "Foreign Policy: From Innocence to Engagement," in *Paths of American Thought,* ed. Arthur M. Schlesinger, Jr., and Morton White (Boston: Houghton Mifflin, 1963), 293–308.

139 *"the world was plastic and the future unlimited":* Paterson, "Bearing the Burden," 203.

139 *David Halberstam called "ultra-realists":* Ambrosius, *Wilsonianism,* 146; C. Wright Mills, *The Power Elite* (New York: Oxford University Press, 1956), 313.

139 *innocence, inaction, and mounting danger:* For an example of Kennedy comparing the 1950s to the '30s, see his November 13, 1959, speech to the annual convention of the Democratic Party of Wisconsin, printed in Kennedy, *The Strategy of Peace,* 193, and quoted in Dean, "Masculinity as Ideology," 45.

140 *one of his paintings in the White House:* Fair, "The Intellectual JFK," 133, 141.

140 *far behind:* On America's military advantage over the Soviet Union when Kennedy took office, see Gareth Porter, *The Perils of Dominance* (Los Angeles: University of California Press, 2006), 8–9, 15; David Rosenberg, "The Origins of Overkill: Nuclear Weapons and American Strategy, 1945–1960," *International Security* 7, no. 4 (1983): 27–71; James G. Hershberg, "The Cuban Missile Crisis," in *The Cambridge History of the Cold War, Volume 2: Crises and Détente,* ed. Melvyn P. Leffler and Odd Arne Westad (New York: Cambridge University Press, 2009), 66.

Chapter 7: Saving Sarkhan

141 *are obsessed with astrology and palm-reading:* William J. Lederer and Eugene Burdick, *The Ugly American* (New York: Norton, 1958), 14–15, 69, 84–85, 91, 109–12, 116, 127, 145, 174–81, 187.

141 *an adviser on Cuba instead:* Jonathan Nashel, *Edward Lansdale's Cold War* (Amherst: University of Massachusetts Press, 2005), 71, 191; Dean, "Masculinity as Ideology," 36–37, 43, 49; Dean, *Imperial Brotherhood,* 172–79; John Hellmann, *American Myth and the Legacy of Vietnam* (New York: Columbia University Press, 1986), 17.

142 *purposely serving him an awful lunch:* Dallek, *An Unfinished Life,* 165, 167; Halberstam, *The Best and the Brightest,* 95.

143 *"the lands of the rising people":* Ambrose and Brinkley, *Rise to Globalism,* 173.

143 *"the 'future' is all that counts":* Bell, *The End of Ideology,* 405.

143 *"on which so much of our destiny hinges":* Nils Gilman, *Mandarins of the Future: Modernization Theory in Cold War America* (Baltimore: Johns Hopkins University Press, 2003), 45.

143 *"He tells us what he's going to do":* Melvyn Leffler, *For the Soul of Mankind* (New York: Hill & Wang, 2007), 176; Dallek, *An Unfinished Life,* 350; David Milne, *America's Rasputin: Walt Rostow and the Vietnam War* (New York: Hill & Wang, 2008), 75; Jack Anderson with Daryl Gibson, *Peace, War and Politics: An Eyewitness Account* (New York: Forge, 1999), 103.

143 *to learn how not to drown:* Gaddis, *Strategies of Containment,* 216; George J. W. Goodman, "The Unconventional Warriors," *Esquire,* November 1961, 130; Dean, "Masculinity as Ideology," 51, 56; Dean, *Imperial Brotherhood,* 187–97.

144 *landed at the proud president's feet:* Goodman, "The Unconventional Warriors," 129; Dean, "Masculinity as Ideology," 50; Joseph Kraft, "Hot Weapon in the Cold War," *Saturday*

Evening Post, April 28, 1961, 88; Halberstam, *The Best and the Brightest*, 123–24; Reeves, *President Kennedy*, 284.

144 *American-backed dictatorship of Fulgencio Batista:* J. David Truby, "Castro's Curveball," *Harper's*, May 1989, 32–34.

145 *foment Marxist revolution across the Americas:* Evan Thomas, *The Very Best Men: The Daring Early Years of the CIA* (New York: Simon & Schuster, 1995), 241, 249; Prados, *Safe for Democracy*, 236–42, 265; Lawrence Freedman, *Kennedy's Wars: Berlin, Cuba, Laos, and Vietnam* (Oxford: Oxford University Press, 2000), 124–27; Reeves, *President Kennedy*, 69–70.

145 *invading Cuba would sabotage that effort:* Freedman, *Kennedy's Wars*, 124.

145 *"Operation Castration":* Thomas, *The Very Best Men*, 237; Dallek, *An Unfinished Life*, 362; Skip Willman, "The Kennedys, Fleming, and Cuba," in *Ian Fleming and James Bond: The Cultural Politics of 007*, ed. Edward P. Comentale, Stephen Watt, and Skip Willman (Indianapolis: Indiana University Press, 2005), 178, 193; Dean, "Masculinity as Ideology," 48.

145 *"a shift from defensive reaction to initiative":* Thomas, *The Very Best Men*, 251; Reeves, *President Kennedy*, 72; Paterson, "Bearing the Burden," 195–96.

146 *he would face a firestorm on his right:* Reeves, *President Kennedy*, 71. Historians are divided over whether Eisenhower, had he still been in office, would have approved the Bay of Pigs operation. Several of his comments to Kennedy before and after the invasion were quite hawkish. Also, having sent U.S. Marines into Lebanon in 1958, Eisenhower was not opposed to military interventions in the third world. Yet he also demonstrated caution during the Dien Bien Phu and Quemoy-Matsu crises. Further, after the Bay of Pigs invasion, he reportedly cautioned Kennedy against future military intervention. Ambrose, *Eisenhower*, 385, 553–54; Divine, *Eisenhower and the Cold War*, 65; Prados, *Safe for Democracy*, 264.

146 *"appeaser of Castro":* Dallek, *An Unfinished Life*, 358.

146 *to try to dissuade Kennedy:* Freedman, *Kennedy's Wars*, 148; Halberstam, *The Best and the Brightest*, 67.

146 *first American military aid to Vietnam:* Thomas J. Schoenbaum, *Waging Peace and War: Dean Rusk in the Truman, Kennedy, and Johnson Years* (New York: Simon & Schuster, 1988), 49–50, 233–34; Robert Mann, *A Grand Delusion: America's Descent into Vietnam* (New York: Basic Books, 2001), 73, 78–79.

146 *so did almost everyone else:* Schoenbaum, *Waging Peace and War*, 291–301, 303–4. Senator William Fulbright, invited to a meeting at the White House to discuss the proposed invasion, did protest on moral grounds, but was dismissed as naïve. Freedman, *Kennedy's Wars*, 132–33; Thomas, *The Very Best Men*, 251–52.

146 *"bold as everybody else":* Paterson, "Bearing the Burden," 202.

147 *a place he often went to fish:* Reeves, *President Kennedy*, 83; Dallek, *An Unfinished Life*, 363; Thomas, *The Very Best Men*, 243, 260–61; Aleksandr Fursenko and Timothy Naftali, *"One Hell of a Gamble": The Secret History of the Cuban Missile Crisis* (New York: Norton, 1997), 99–100.

147 *U.S. attack would bring—refused:* Freedman, *Kennedy's Wars,* 141–44.

147 *would have no choice but to invade:* As Dulles acknowledged in private notes, "We [the leadership of the CIA] felt that when the chips were down, when the crisis arose in reality, any action required for success would be authorized rather than permit the enterprise to fail." Thomas, *The Very Best Men*, 247.

147 *alone in his bedroom, in tears:* Reeves, *President Kennedy*, 93, 95; Thomas, *The Very Best Men*, 263; Prados, *Safe for Democracy*, 263.

147 *the limits of American power:* Only four American airmen died in the Bay of Pigs invasion. Prados, *Safe for Democracy*, 263.

148 *that he was soft:* Reeves, *President Kennedy*, 95; John F. Kennedy, "Address Before the American Society of Newspaper Editors," April 20, 1961, http://www.presidency.ucsb.edu/ws/index.php?pid=8076&st=&st1; Freedman, *Kennedy's Wars*, 337.

148 *might take the whole region down with it:* Reeves, *President Kennedy*, 74; George Herring, *America's Longest War: The United States and Vietnam, 1950–1975* (New York: McGraw-Hill, 2002), 92; Freedman, *Kennedy's Wars*, 294–95.

149 *"conscientious objectors from World War I":* Reeves, *President Kennedy*, 74; Freedman, *Kennedy's Wars*, 295–96.

149 *"You start using atomic weapons!":* Logevall, *Choosing War*, 24–25; Reeves, *President Kennedy*, 111; Dallek, *An Unfinished Life*, 353; Michael Hunt, *Lyndon Johnson's War: America's Cold War Crusade in Vietnam, 1945–1968* (New York: Hill & Wang, 1996), 55.

149 *military alliances with neither east nor west:* Reeves, *President Kennedy*, 75, 115–16; Thomas G. Paterson, J. Garry Clifford, and Kenneth J. Hagan, "JFK: A Can-Do President," in *To Reason Why: The Debate About the Causes of U.S. Involvement in the Vietnam War*, ed. Jeffrey P. Kimball (Philadelphia: Temple University Press, 1994), 183.

149 *"the obscurity that it richly deserves":* Galbraith's comment actually referred to all of Southeast Asia. "The Importance of Obscurity," *Time*, May 6, 1966, http://www.time.com/time/magazine/article/0,9171,901831,00.html. In truth, Laos did not return to complete obscurity after the U.S.-Soviet deal. North Vietnamese forces remained there and steadily expanded sections of the Ho Chi Minh trail, thus helping supply communist forces in South Vietnam. In 1964, fighting resumed between the Pathet Lao and the royal Laotian military. As U.S. involvement in Vietnam deepened, America began massive air strikes against North Vietnamese forces in Laos. The royal government slowly lost control of the country to the better-equipped and better-trained Pathet Lao, and in 1975 the Pathet Lao seized complete control of the country. Herring, *America's Longest War*, 340–41; Stanley Karnow, *Vietnam: A History* (New York: Penguin, 1997), 346–47.

149 *"the United States is in a yielding mood":* Milne, *America's Rasputin*, 123; Reeves, *President Kennedy*, 157.

150 *there would be war:* Reeves, *President Kennedy*, 160–72; Dallek, *An Unfinished Life*, 407; Zubok and Pleshakov, *Inside the Kremlin's Cold War*, 180, 236, 243–53.

150 *backed by the United States:* Reeves, *President Kennedy*, 185, 187, 203; Zubok and Pleshakov, *Inside the Kremlin's Cold War*, 195; Hope Harrison, *Driving the Soviets up the Wall: Soviet–East German Relations, 1953–1961* (Princeton, NJ: Princeton University Press, 2003), 227.

151 *The place to start was Vietnam:* Reeves, *President Kennedy*, 173; Herring, *America's Longest War*, 96.

151 *was increasingly paying the bills:* Herring, *America's Longest War*, 6–29; Khong, *Analogies at War*, 237.

151 *Indochina was lost:* Halberstam, *The Best and the Brightest*, 94; Dallek, *An Unfinished Life*, 167, 185–87.

151 *would unite the country once again:* Herring, *America's Longest War*, 49.

152 *American Friends of Vietnam:* Diane Kunz, "Camelot Continued: What If John F. Kennedy Had Lived?" in *Virtual History: Alternatives and Counterfactuals*, ed. Niall Ferguson (London: Papermac, 1998), 378.

152 *"the finger in the dike":* Mann, *A Grand Delusion*, 200.

152 *"real in powerful men's minds":* Halberstam, *The Best and the Brightest*, 148.

152 *travel to South Vietnam to investigate:* Herring, *America's Longest War*, 80–83; Freedman, *Kennedy's Wars*, 322–23.

153 *America was utopia:* Milne, *America's Rasputin*, 21, 23, 26, 35, 60–63; Gilman, *Mandarins of the Future*, 202.

153 *"becoming a bore":* Milne, *America's Rasputin*, 63.

153 *"something you know nothing about?":* Halberstam, *The Best and the Brightest*, 159.

154 *Lincoln Center for the Performing Arts:* Freedman, *Kennedy's Wars*, 18–19; Halberstam, *The Best and the Brightest*, 162–63, 171–72, 468–70.

154 *it was to answer that question yes:* Maxwell D. Taylor, *The Uncertain Trumpet* (New York: Harper & Brothers, 1959); Halberstam, *The Best and the Brightest*, 162–63.

154 *scare the North into stopping the war:* Halberstam, *The Best and the Brightest*, 169–70; Herring, *America's Longest War*, 97; Freedman, *Kennedy's Wars*, 323–27.

154 *without U.S. soldiers:* Herring, *America's Longest War*, 100; Kai Bird, *The Color of Truth: McGeorge Bundy and William Bundy, Brothers in Arms* (New York: Touchstone, 1998), 222; Freedman, *Kennedy's Wars*, 330–34; Khong, *Analogies at War*, 96.

155 *they excelled at killing dangerous snakes:* Reeves, *President Kennedy*, 335–36, 343; Nashel, *Edward Lansdale's Cold War*, 71, 239; Freedman, *Kennedy's Wars*, 153–59; Prados, *Safe for Democracy*, 298–306.

155 *"the next President of the United States":* Hershberg, "The Cuban Missile Crisis," 67–70; Fursenko and Naftali, *"One Hell of a Gamble,"* 92, 97–98, 139–40; Reeves, *President Kennedy*, 339, 343–47, 350, 368–70, 372.

155 *"and appearances contribute to reality":* Hershberg, "The Cuban Missile Crisis," 72; Leffler, *For the Soul of Mankind*, 175.

155 *would later cast doubt on the claim:* See, for instance, Press, *Calculating Credibility*; Hopf, *Peripheral Visions*; Stephen M. Walt, *The Origins of Alliances* (Ithaca, NY: Cornell University Press, 1987); Jonathan Mercer, *Reputation and International Politics* (Ithaca, NY: Cornell University Press, 1996).

155 *would get his brother impeached:* Reeves, *President Kennedy*, 401.

156 *"ultimately leads to war":* Hershberg, "The Cuban Missile Crisis," 71–73; Reeves, *President Kennedy*, 395–96; John F. Kennedy, "Report to the American People on the Soviet Arms Buildup in Cuba," October 22, 1962, http://www.jfklibrary.org/Historical+Resources/ Archives/Reference+Desk/Speeches/JFK/003POF03CubaCrisis10221962.htm.

156 *to withdraw the missiles in Turkey:* Michael Dobbs, *One Minute to Midnight: Kennedy, Khrushchev, and Castro on the Brink of Nuclear War* (New York: Knopf, 2008), 310–11; Reeves, *President Kennedy*, 400–24; Hershberg, "The Cuban Missile Crisis," 80–83; Fursenko and Naftali, *"One Hell of a Gamble,"* 285–87, 321.

156 *"back away from rash initiatives":* Reeves, *President Kennedy*, 428.

156 *gained seats in the Senate:* Reeves, *President Kennedy*, 426, 429.

157 *"take the wrong lessons away from the crisis":* Reeves, *President Kennedy*, 427.

157 *taken his private heresies even further:* Fursenko and Naftali, *"One Hell of a Gamble,"* 291–93; Dobbs, *One Minute to Midnight*, 45–46; Hershberg, "The Cuban Missile Crisis," 81.

157 *half the South Vietnamese cabinet:* Freedman, *Kennedy's Wars*, 325–26; Herring, *America's Longest War*, 59–60, 77–78.

158 *including weddings and funerals:* Halberstam, *The Best and the Brightest*, 181–82; Herring, *America's Longest War*, 101, 108–9.

158 *killing two adults and six children:* Mann, *A Grand Delusion*, 284.

158 *it was the Vietnamese word for "monk":* Mann, *A Grand Delusion*, 284–86; Reeves, *President Kennedy*, 491.

158 *helped convince U.S. officials that he needed to go:* Freedman, *Kennedy's Wars*, 368; Logevall, *Choosing War*, 45–50.

159 *turned pale and fled silently from the room:* Freedman, *Kennedy's Wars*, 390–95; Reeves, *President Kennedy*, 649.

159 *McNamara's Pentagon until the last minute:* Reeves, *President Kennedy*, 507; Dallek, *An Unfinished Life*, 619.

159 *had taken his life fourteen years before:* John F. Kennedy, "Commencement Address at American University," June 10, 1963, http://www.jfklibrary.org/Historical+Resources/Archives/ Reference+Desk/Speeches/JFK/003POF03AmericanUniversity06101963.htm; Reeves, *President Kennedy*, 447; Bird, *The Color of Truth*, 251–52.

160 *Billings, Montana, and Salt Lake City, Utah:* Reeves, *President Kennedy*, 606–7; Halberstam, *The Best and the Brightest*, 295–96.

160 *"get the American people to reelect me":* Reeves, *President Kennedy*, 444; Logevall, *Choosing War*, 38–49.

160 *maybe he could square the circle:* Logevall, *Choosing War*, 73; Reeves, *President Kennedy*, 600–1; Nashal, *Edward Lansdale's Cold War*, 57.

Chapter 8: Things Are in the Saddle

161 *Lyndon Johnson was its face:* Dallek, *Lone Star Rising*, 7.

161 *"when the parts were Lyndon Johnson's":* Robert Dallek, "Presidential Address: Lyndon John-

son and Vietnam: The Making of a Tragedy" *Diplomatic History* 20, no. 2 (1996): 155; Robert Caro, *Master of the Senate* (New York: Knopf, 2002), 121–22.

162 *"I want his pecker in my pocket"*: Halberstam, *The Best and the Brightest*, 434, 436–37.

162 *how democracy survived:* Halberstam, *The Best and the Brightest*, 444; Rystad, *Prisoners of the Past*, 55; D. B. Hardeman and Donald C. Bacon, *Rayburn: A Biography* (Austin: Texas Monthly Press, 1987), 265–70; Dallek, *Lone Star Rising*, 226–27.

163 *became the industry's champion, and its beneficiary:* Dallek, *Lone Star Rising*, 292, 294, 383.

163 *"terribly shaken":* Dallek, *Lone Star Rising*, 323–24, 399; Halberstam, *The Best and the Brightest*, 590–91; Lyndon B. Johnson, "Nightmares of Crucifixion," in *To Reason Why*, ed. Kimball, 45–46.

163 *time trying to recover them:* Halberstam, *The Best and the Brightest*, 298.

164 *grew and grew:* Logevall, *Choosing War*, 108, 321; Kaiser, *American Tragedy*, 304–5; Herring, *America's Longest War*, 133–39.

164 *decided to take the fight to Hanoi:* Herring, *America's Longest War*, 139; Kaiser, *American Tragedy*, 331.

165 *"repel any armed attacks against the forces of the United States":* Herring, *America's Longest War*, 142–44; Logevall, *Choosing War*, 196–205; Hunt, *Lyndon Johnson's War*, 79.

165 *"would rather not have":* Logevall, *Choosing War*, 205; Kaiser, *American Tragedy*, 336–38; Ambrose and Brinkley, *Rise to Globalism*, 142; Paterson, "Bearing the Burden," 211; David Frum, *How We Got Here: The 70s* (New York: Basic Books, 2000), 4, 39.

166 *its vote was unanimous:* Dallek, *Lone Star Rising*, 84–85; Dallek, "Presidential Address: Lyndon Johnson and Vietnam," 151; Herring, *America's Longest War*, 144; Logevall, *Choosing War*, 198; Kaiser, *American Tragedy*, 336–38.

166 *"there were better ones, golden ones, ahead":* Rystad, *Prisoners of the Past*, 56; Logevall, *Choosing War*, 205; Herring, *America's Longest War*, 145; Halberstam, *The Best and the Brightest*, 423.

166 *"conflict between parents and children is letting up":* Kaiser, *American Tragedy*, 483; Leffler, *For the Soul of Mankind*, 221; Halberstam, *The Best and the Brightest*, 423; "Students: On the Fringe of a Golden Era," *Time*, January 29, 1965, http://www.time.com/time/printout/0,8816,839175,00.html; "Lyndon B. Attitudes," *Time*, September 10, 1965, http://www.time.com/time/printout/0,8816,834259,00.html.

166 *"We can do it all":* Rick Perlstein, *Nixonland: The Rise of a President and the Fracturing of America* (New York: Scribner, 2008), 5.

166 *end poverty for good:* Robert Dallek, *Flawed Giant: Lyndon Johnson and His Times, 1961–1973* (New York: Oxford University Press, 1998), 184; Lyndon B. Johnson, "Inaugural Address," January 20, 1965, http://www.bartelby.org/124/pres57.html.

166 *"I was born in a manger":* Halberstam, *The Best and the Brightest*, 431; "Lyndon B. Attitudes."

167 *"raggedy-assed little fourth-rate country":* Halberstam, *The Best and the Brightest*, 512.

167 *buoying morale in the South:* For a discussion of the Johnson administration's planning for military action in November 1964, see Kaiser, *American Tragedy*, 355–74.

167 *"They'd impeach a President who'd run out, wouldn't they?":* Logevall, *Choosing War*, 145; Kaiser, *American Tragedy*, 319–20.

167 Zero Hour: Summons to the Free: McGeorge Bundy, "They Say in the Colleges . . . ," in *Zero Hour: A Summons of the Free*, ed. Stephen Vincent Benét (New York: Farrar & Rinehart, 1940), 79–116; Bird, *The Color of Truth*, 66–67; Leonard Lyons, "The New Yorker," *Washington Post*, July 20, 1940, 4; Thomas C. Lyons, "Books of the Times," *New York Times*, August 3, 1940, 17.

167 *"the isolationist-interventionist debate":* Bird, *The Color of Truth*, 107; Khong, *Analogies at War*, 197.

168 *"shall not perish from this earth":* Logevall, *Choosing War*, 324–27; Bird, *The Color of Truth*, 306–7; Halberstam, *The Best and the Brightest*, 523.

168 *Johnson himself had counseled caution:* Herring and Immerman, "Eisenhower, Dulles, and Dienbienphu," 353–54; Hunt, *Lyndon Johnson's War*, 76, 101.

168 *close to twice the number in 1952:* Mueller, "Trends in Popular Support for the Wars in Korea and Vietnam," 360–61; Khong, *Analogies at War*, 102.

168   *It had worked so far:* Logevall, *Choosing War,* 360–61.

169   *he was already over the edge:* Logevall, *Choosing War,* 330, 370; Herring, *America's Longest War,* 156–58, 162.

169   *150,000 troops and more intense bombing:* Logevall, *Choosing War,* 367; Herring, *America's Longest War,* 162–63.

169   *"but that he did pay [Dean] Rusk":* Daniel DiLeo, *George Ball, Vietnam and the Rethinking of Containment* (Chapel Hill: University of North Carolina Press, 1991), 25–29; George W. Ball, *The Past Has Another Pattern* (New York: Norton, 1982), 152–53, 365; Kaiser, *American Tragedy,* 350–51, 362, 395–96, 402, 405–6; Logevall, *Choosing War,* 361–62.

170   *"an haughty spirit before a fall":* Steel, *Walter Lippmann and the American Century,* 558–80; Fredrik Logevall, "First Among Critics: Walter Lippmann and the Vietnam War," *Journal of American-East Asian Relations* 4, no. 4 (1995), 351–54, 364–72.

170   *"a fatally unfortunate conclusion":* Mayers, *George Kennan and the Dilemmas of US Foreign Policy,* 277–79; Hixson, *George Kennan,* 227; Khong, *Analogies at War,* 174.

170   *"[as justification] for the present war":* Fox, *Reinhold Niebuhr,* 284–85.

171   *"general discouragement, sometimes bordering on despair":* Jennifer W. See, "A Prophet Without Honor: Hans Morgenthau and the War in Vietnam, 1955–1965," *Pacific Historical Review* 70, no. 3 (2001), 437, 439–40; Bird, *The Color of Truth,* 321; CBS News, "Dean Bundy Meets Professor Morgenthau," in Louis Menashe and Ronald Radosh, eds., *Teach-Ins: USA Reports, Opinions, Documents* (Washington, DC: Praeger, 1967), 198–206; Christoph Frei, *Hans J. Morgenthau: An Intellectual Biography* (Baton Rouge: Louisiana State University Press, 2001), 221.

171   *another hundred thousand troops to South Vietnam:* Khong, *Analogies at War,* 151; Herring, *America's Longest War,* 164–65.

Chapter 9: Liberation

172   *until the fateful day arrived:* Freedman, *Kennedy's Wars,* 142, 145; Prados, *Safe for Democracy,* 262–63; Halberstam, *The Best and the Brightest,* 68.

172   *would no longer be required:* Bird, *The Color of Truth,* 200; Marcus Raskin and Robert Spero, "Ahead of History: Marcus Raskin and the Institute for Policy Studies," in *The Four Freedoms Under Siege: The Clear and Present Danger of Our National Security State,* ed. Marcus Raskin and Robert Spero (Westport, CT: Praeger, 2007), 276; Marcus Raskin, interview with author, February 22, 2008.

173   *where he began causing trouble:* Raskin and Spero, "Ahead of History," 273–74.

173   *retire once the decisions were made:* Bird, *The Color of Truth,* 187, 219; Marcus Raskin, interview with author, February 22, 2008.

174   *"that he won't pee on the floor":* Bird, *The Color of Truth,* 200, 206–7; Marcus Raskin, interview with author, February 22, 2008.

174   *that America try to end the cold war:* James Roosevelt, ed., *The Liberal Papers* (New York: Doubleday, 1962), 17, 27, 58–59, 146, 175, 251–52, 267, 279, 321, 326.

174   *By year's end he was gone:* Bird, *The Color of Truth,* 218–19; Raskin and Spero, "Ahead of History," 277.

174   *"we are intellectually and emotionally spastic":* Halberstam, *The Best and the Brightest,* 426–27; Gilbert Seldes, "'Patient' Diagnosis of 'Dr. Strangelove,'" *New York Times,* April 5, 1964, X7.

174   *House Un-American Activities Committee:* Paul Buhle and William Rice-Maximin, *William Appleman Williams: The Tragedy of Empire* (New York: Routledge, 1995), 115–16; Jonathan M. Wiener, "Radical Historians and the Crisis in American History, 1959–1980," *Journal of American History* 76 (1989): 407.

175   *a generational manifesto that would sell sixty thousand copies:* Allen J. Matusow, *The Unraveling of America: A History of Liberalism in the 1960s* (New York: Harper Torchbooks, 1986), 313; James Miller, *Democracy Is in the Streets: From Port Huron to the Siege of Chicago* (New York: Simon & Schuster, 1987), 13; Todd Gitlin, *The Sixties: Years of Hope, Days of Rage* (New York: Bantam, 1993), 111.

175 *of nuclear bombs:* Howard Brick, *Age of Contradiction: American Thought and Culture in the 1960s* (New York: Twayne, 1998), 149–50; Charles DeBenedetti, *An American Ordeal: The Antiwar Movement of the Vietnam Era* (Syracuse, NY: Syracuse University Press, 1990), 389; Maurice Isserman, *If I Had a Hammer: The Death of the Old Left and the Birth of the New Left* (Urbana: University of Illinois Press, 1993), 149.

175 *College enrollment nearly quadrupled between 1946 and 1970:* Matusow, *The Unraveling of America*, 308.

175 *"know a Communist if they tripped over one":* Herring, *America's Longest War*, 219.

175 *declared the Port Huron Statement:* Students for a Democratic Society, "The Port Huron Statement," June 15, 1962, http://history.hanover.edu/courses/excerpts/111hur.html.

176 *lay down his body for justice and peace:* Isserman, *If I Had a Hammer*, 151–55.

176 *integrate interstate transport across the South:* Raymond Arsenault, *Freedom Riders: 1961 and the Struggle for Racial Justice* (New York: Oxford University Press, 2006), 99–101.

176 *the 1963 March on Washington:* John D'Emilio, *Lost Prophet: The Life and Times of Bayard Rustin* (New York: Free Press, 2003), 37–50, 327–60.

176 *founded on pacifist principles as well:* Akinyele O. Omoja, "The Ballot and the Bullet: A Comparative Analysis of Armed Resistance in the Civil Rights Movement," *Journal of Black Studies* 29 (1999): 560–61.

176 *the following year:* Eric Stoper, "The Student Nonviolent Coordinating Committee: Rise and Fall of a Redemptive Organization," *Journal of Black Studies* 8 (1977): 31; Arsenault, *Freedom Riders*, 451; Tom Hayden, *Reunion: A Memoir* (New York: Random House, 1988), 53–55.

176 *"all men are really good at heart":* Isserman, *If I Had a Hammer*, 127.

177 *"made sonsofbitches of us all":* Marcus Raskin, "A National Security Manager Tries to Explain," in *Essays of a Citizen: From National Security State to Democracy*, ed. Marcus Raskin (Armonk, NY: M. E. Sharpe, 1991), 104.

177 *found his view of human nature too bleak:* Halliwell, *The Constant Dialogue*, 230–36; David J. Garrow, *Bearing the Cross: Martin Luther King, Jr., and the Southern Christian Leadership Conference* (New York: William Morrow, 1986), 45.

178 *"releases the prisoners; abolishes repression":* Douglas Martin, "Norman O. Brown Dies, Playful Philosopher was 89," *New York Times*, October 4, 2002; Matusow, *The Unraveling of America*, 277–79; Brick, *Age of Contradiction*, 134–35.

178 *every one dollar of damage it did to North Vietnam:* Halberstam, *The Best and the Brightest*, 241; Herring, *America's Longest War*, 168, 172–77, 179; Logevall, *Choosing War*, 409–10.

179 *the membership of the SDS quadrupled:* DeBenedetti, *An American Ordeal*, 107–8, 111–12, 132, 150; Matusow, *The Unraveling of America*, 320.

179 *urged Johnson to halt the bombing:* Mueller, "Trends in Popular Support for the Wars in Korea and Vietnam," 363; Thomas, *Robert Kennedy*, 334–35; Herring, *America's Longest War*, 267.

179 *"the greatest purveyor of violence in the world today":* DeBenedetti, *An American Ordeal*, 172.

179 *threw bottles and bags of blood:* Gitlin, *The Sixties*, 254.

179 *a picture of his father mounted on the wall:* Bird, *The Color of Truth*, 323, 372, 397–98, 401.

179 *began to break under the strain:* Deborah Shapley, *Power and Promise: The Life and Times of Robert McNamara* (Boston: Little, Brown, 1993), 408, 483.

179 *her father did not:* DeBenedetti, *An American Ordeal*, 129; Shapley, *Power and Promise*, 354–55.

179 *bursting into tears:* Bird, *The Color of Truth*, 345.

180 *Bundy and McNamara were both gone:* Halberstam, *The Best and the Brightest*, 629; Bird, *The Color of Truth*, 342, 347.

180 *"a fanatic in sheep's clothing," in one detractor's words:* Halberstam, *The Best and the Brightest*, 628; Milne, *America's Rasputin*, 7.

180 *"I think we can hold out longer than that":* Halberstam, *The Best and the Brightest*, 637–38.

180 *antiwar senators were taking orders from the Kremlin:* Dallek, "Presidential Address: Lyndon Johnson and Vietnam," 467.

180 *pulled out his penis and screamed, "This is why!"*: Dallek, *Flawed Giant*, 448, 452, 491, 526.
180 *"I am alive and it has been more torturous"*: Halberstam, *The Best and the Brightest*, 640.
180 *"bitch of a war"*: Gaddis, *Strategies of Containment*, 268.
180 *"A man without a spine"*: Doris Kearns, *Lyndon Johnson and the American Dream* (New York: Harper & Row, 1976), 253.
181 *"the end begins to come into view"*: Herring, *America's Longest War*, 221.
181 *"choked with rubble and rotting bodies"*: Herring, *America's Longest War*, 225–31; Karnow, *Vietnam*, 538–40.
181 *"I thought we were winning the war!"*: Herring, *America's Longest War*, 232; Dallek, *Flawed Giant*, 506.
181 *"it will only be a matter of time before they give in"*: Herring, *America's Longest War*, 229–31; Halberstam, *The Best and the Brightest*, 648.
182 *subordinate it to his own*: Gaddis, *Strategies of Containment*, 256; Herring, *America's Longest War*, 163; Halberstam, *The Best and the Brightest*, 597.
182 *and billed him another five thousand dollars*: Halberstam, *The Best and the Brightest*, 650–51.
182 *who was beholden to whom*: Dallek, *Lone Star Rising*, 39–40.
182 *a plurality of Americans opposed the war*: Beisner, *Dean Acheson*, 633; Matusow, *The Unraveling of America*, 391; Gitlin, *The Sixties*, 293.
182 *"has dampened expansionist ideas"*: Herring, *America's Longest War*, 245–57; Kunz, *Butter and Guns*, 114–19; Walter LaFeber, *Deadly Bet: LBJ, Vietnam, and the 1968 Elections* (Lanham, MD: Rowman & Littlefield, 2005), 55–57; Robert M. Collins, "The Economic Crisis of 1968 and the Waning of the 'American Century,'" *American Historical Review* 101 (1996): 396–422.
183 *"have bailed out"*: Herring, *America's Longest War*, 250. Although a majority of the "wise men" argued that it was time to disengage from Vietnam, a few hawks—such as Maxwell Taylor and Supreme Court justice (and crony of Johnson) Abe Fortas—argued for seeing the war through. Isaacson and Thomas, *The Wise Men*, 702–3.
183 *"in disarray around him"*: Theodore White, *The Making of the President, 1968* (New York: Atheneum, 1969), 102; Gitlin, *The Sixties*, 304; Dallek, *Flawed Giant*, 528.
183 *and perhaps his life itself*: Dallek, *Flawed Giant*, 522–23.
183 *"it was like committing actual suicide"*: Dallek, *Flawed Giant*, 601–5, 613–14.
184 *in the party's gallery of honor*: Halberstam, *The Best and the Brightest*, 658; Dallek, *Flawed Giant*, 573, 617.
184 *"dissolved in the morass of war in Vietnam"*: "Lyndon Johnson, 36th President, Is Dead," *New York Times*, January 23, 1973, 81.
184 *in an essay on intellectuals and war*: Bourne, *War and the Intellectuals*, ed. Resek; Lillian Schlissel, ed., *The World of Randolph Bourne: An Anthology* (New York: Dutton, 1966); Lasch, *The New Radicalism in America*, 69–103, 300–1; Noam Chomsky, *American Power and the New Mandarins* (New York: Pantheon, 1969), 4–8.
184 *a model for post-Vietnam foreign policy*: Marcus Raskin, "The Erosion of Congressional Power," in *Washington Plans an Aggressive War*, ed. Stavins, Barnett, and Raskin, 255; Marcus Raskin, interview with author, February 22, 2008.
184 *Dewey was an inspiration*: Miller, *Democracy Is in the Streets*, 16, 149; Westbrook, *John Dewey and American Democracy*, 547–49.
184 *had rarely been uttered without a sneer*: For instance, "If there is to be a politics of the New Left . . . our work is necessarily structural—and so, *for us*, just now—utopian." C. Wright Mills, "The New Left," in *Power, Politics and People: The Collected Essays of C. Wright Mills*, ed. Irving Louis Horowitz (New York: Oxford University Press, 1963), 254. And "the masses of hungry, aspiring, utopian peoples [are] intervening in history for the first time." Tom Hayden, quoted in Miller, *Democracy Is in the Streets*, 101.
184 *the economic motives for war*: Kennedy, *Charles Beard and American Foreign Policy*, 160, 165–67; H. W. Brands, *What America Owes the World* (New York: Cambridge University Press, 1998), 242–43.
184 *a young, militant historian named Ronald Radosh*: Ronald Radosh, *Prophets on the Right: Profiles of Conservative Critics of American Globalism* (New York: Simon & Schuster, 1975), 51.

184 *a second, more positive look:* See, for instance, Wilson, *Herbert Hoover.*

185 *endemic to American foreign policy and American life:* Herring, *America's Longest War,* xiv, 267.

185 *hoping that LBJ might appoint him ambassador to the UN:* Dominic Sandbrook, *Eugene McCarthy: The Rise and Fall of Postwar American Liberalism* (New York: Knopf, 2004), 120–21, 127, 133.

185 *"I know it didn't, sir":* Sandbrook, *Eugene McCarthy,* 17–18, 132–33, 139, 145, 155.

186 *An old monk told him to seize the day:* Sandbrook, *Eugene McCarthy,* 171.

186 *even though the crowd had not yet arrived:* Sandbrook, *Eugene McCarthy,* 190–91.

186 *stared at the beast until it trotted away:* Steel, *In Love with Night,* 39, 125.

187 *wimps when they pointed out the risks:* Halberstam, *The Best and the Brightest,* 70; Steel, *In Love with Night,* 25–26, 49, 76–78.

187 *"We do not have . . . to be afraid":* Sandbrook, *Eugene McCarthy,* 150, 195; Matusow, *The Unraveling of America,* 421.

Chapter 10: The Scold

188 *the arc of his own career:* Schlesinger, "Vietnam: Lessons of the Tragedy," in *The Politics of Hope and The Bitter Heritage,* 514.

188 *a moratorium on the production of nuclear weapons:* Robert David Johnson, *Congress and the Cold War* (New York: Cambridge University Press, 2006), 146; Benjamin O. Fordham, "The Evolution of Republican and Democratic Positions on Cold War Military Spending," *Social Science History* 31 (2007): 609, 627; Theodore White, *The Making of the President 1972* (New York: Atheneum, 1973), 46.

188 *cut the number of American troops in Europe in half:* Gaddis, *Strategies of Containment,* 318.

188 *"an exercise in futility":* Schlesinger, "Vietnam: Lessons of the Tragedy," in *The Politics of Hope and The Bitter Heritage,* 522.

189 *Potomac Electric Power Company:* Bird, *Color of Truth,* 356; Karnow, *Vietnam,* 471.

189 *"a singularly inept instrument of foreign policy":* Paul C. Warnke and Leslie H. Gelb, "Security or Confrontation: The Case for a Defense Policy," *Foreign Policy* 1 (1970/1971): 15.

189 *"it's going to mean our foreign policy has been idiotic":* "The Real Paul Warnke," *New Republic,* March 26, 1977, 22–25.

189 *accused the new intellectuals of isolationism:* Norman Podhoretz, for instance, wrote that "the new isolationism, then, was what happened to the [New Left] Movement's ideas about the American role in world affairs when they were cleaned up to 'work within the system.'" Norman Podhoretz, *Breaking Ranks: A Political Memoir* (New York: Harper & Row, 1979), 291–92.

189 *born too late to remember Munich:* Anthony Lake, "Introduction," in *The Vietnam Legacy: The War, American Society and the Future of American Foreign Policy,* ed. Anthony Lake (New York: New York University Press, 1976), xxviii.

190 *two of whose three congressmen had opposed World War I:* "Detailed Vote in the House of Representatives on the Passage of the Resolution Declaring War," *New York Times,* April 7, 1917, 4.

190 *which denounced the UN as too weak:* Bruce Miroff, *The Liberals' Moment: The McGovern Insurgency and the Identity Crisis of the Democratic Party* (Lawrence: University Press of Kansas, 2007), 30; Bernard Hennessy, "A Case Study of Intra-Pressure Group Conflicts: The United World Federalists," *Journal of Politics* 16 (1954): 76–95; "Group Asks Drive for World Rule," *New York Times,* November 2, 1947, 50.

191 *"the get tough policy of the Truman administration":* George McGovern, *Grassroots: The Autobiography of George McGovern* (New York: Random House, 1977), 45; Miroff, *The Liberals' Moment,* 123.

191 *"Somehow we have to settle down and live with them":* McGovern, *Grassroots,* 49; Miroff, *The Liberals' Moment,* 36, 132; White, *The Making of the President 1972,* 116–17.

191 *fated to live in a hostile world:* Michael T. Kaufman, "Paul Warnke, 81, a Leading Dove in Vietnam Era, Dies," *New York Times,* November 1, 2001.

191 *"outdated stereotypes of military confrontation and power politics"*: George McGovern, "Acceptance Speech Before Democratic National Convention," July 14, 1972, http://www.4president.org/speeches/mcgovern1972acceptance.htm; John Ehrman, *The Rise of Neoconservatism: Intellectuals and Foreign Affairs, 1945–1994* (New Haven, CT: Yale University Press, 1995), 59.

191 *he quoted them in his autobiography*: McGovern, *Grassroots*, 35.

192 *like "pee[ing] on the floor"*: Bird, *The Color of Truth*, 200.

192 *"those young men will some day curse us"*: Miroff, *The Liberals' Moment*, 38.

192 *an absolute dread of flying a plane*: McGovern, *Grassroots*, 20–23.

193 *"'I killed more people than any of you guys'"*: McGovern, *Grassroots*, 25–28; Miroff, *The Liberals' Moment*, 29–30.

193 *to retrieve American prisoners of war*: "Senator McGovern Presenting His Views on Foreign Policy," *New York Times*, October 6, 1972, 26; Frum, *How We Got Here*, 308; Jonathan Rieder, "Rise of the 'Silent Majority,'" in *Rise and Fall of the New Deal Order*, ed. Steve Fraser and Gary Gerstle (Princeton, NJ: Princeton University Press, 1989), 262.

193 *boasted about eating sheep's testicles*: Bruce J. Schulman, *The Seventies: The Great Shift in American Culture, Society, and Politics* (New York: Free Press, 2001), 23; DeBenedetti, *An American Ordeal*, 239; Gaddis, *Strategies of Containment*, 332; Walter Isaacson, *Kissinger: A Biography* (New York: Simon & Schuster, 1992), 259–67; Herring, *The Longest War*, 291.

194 *rejected by the fancy Manhattan firms*: Tom Wicker, *One of Us: Richard Nixon and the American Dream* (New York: Random House, 1991), 8, 10; Perlstein, *Nixonland*, 20–24.

194 *insinuating that he espoused the "communist line"*: Irwin F. Gellman, *The Contender: Richard Nixon, The Congress Years 1946–1952* (New York: Free Press, 1999), 65, 84–85; Stephen E. Ambrose, *Nixon, Vol. 1: The Education of a Politician, 1913–1962* (New York: Simon & Schuster, 1987), 117–40; Wicker, *One of Us*, 39–50.

194 *"sitting on their fat butts"*: Ambrose, *Eisenhower*, 272; Wicker, *One of Us*, 9, 10.

194 *when gentile boys approached*: Jeremy Suri, *Henry Kissinger and the American Century* (Cambridge, MA: Belknap, 2007), 39–41, 45, 221; Isaacson, *Kissinger*, 26, 30, 33.

195 *"The weak, the old had no chance"*: Jussi Hanhimäki, *The Flawed Architect: Henry Kissinger and American Foreign Policy* (New York: Oxford University Press, 2004), 3–4; Robert Dallek, *Nixon and Kissinger: Partners in Power* (New York: HarperCollins, 2007), 34–35; Isaacson, *Kissinger*, 28, 30, 52.

195 *limit the length of all future submissions*: Suri, *Henry Kissinger and the American Century*, 29; Isaacson, *Kissinger*, 65, 561, 697.

195 *within the hearts of women and men*: This is a major theme of Suri's *Henry Kissinger and the American Century*.

195 *"the kind of guy who would send you to the showers"*: Dallek, *Nixon and Kissinger*, 170–71; Suri, *Henry Kissinger and the American Century*, 207–10; Isaacson, *Kissinger*, 279–80, 393.

195 *to be dangerously soft*: Shulman, *The Seventies*, 24–25; Isaacson, *Kissinger*, 120.

196 *helped spur dissident movements in Eastern Europe*: Hanhimäki, *The Flawed Architect*, 433–37; Raymond L. Garthoff, *Détente and Confrontation: American-Soviet Relations from Nixon to Reagan* (Washington, DC: Brookings Institution Press, 1994), 527–37.

196 *"step on them, crush them, show them no mercy"*: Isaacson, *Kissinger*, 328.

197 *Nixon called it "the madman theory"*: Dallek, *Nixon and Kissinger*, 106–7.

197 *mused that Kissinger required psychiatric care*: Isaacson, *Kissinger*, 262–63, 391.

197 *kept trying to topple the government of South Vietnam*: Hanhimäki, *The Flawed Architect*, 43–46.

197 *ordering that U.S. troops invade Cambodia outright*: Dallek, *Nixon and Kissinger*, 194–200; Perlstein, *Nixonland*, 472.

197 *"free nations and free institutions throughout the world"*: McMahon, "Credibility and World Power," 467.

197 *"like they're going to be bombed this time"*: Dallek, *Nixon and Kissinger*, 385; Isaacson, *Kissinger*, 419.

198 "should the settlement be violated by North Vietnam": Hanhimäki, The Flawed Architect, 256–
     57, 267–68.

198 Congress rejected the aid and South Vietnam fell: Dallek, Nixon and Kissinger, 472–73; Isaacson,
     Kissinger, 642–43; Hanhimäki, The Flawed Architect, 393; Gerald Ford, "Reporting on the
     United States Foreign Policy," April 10, 1975, http://www.presidency.ucsb.edu/ws/?pid=4826.

198 "and you're likely to find an old peace activist": DeBenedetti, An American Ordeal, 304.

198 less than 5 percent by 1977: Gaddis, Strategies of Containment, 319.

199 denounce his boss to reporters: Prados, Safe for Democracy, 425–26, 429–30; Garthoff, Détente and
     Confrontation, 564–65; Gaddis, Strategies of Containment, 329; Isaacson, Kissinger, 683–84; Odd
     Arne Westad, The Global Cold War (New York: Cambridge University Press, 2009), 218–41.

199 Operation Midnight Climax: Prados, Safe For Democracy, 437; Martin A. Lee and Bruce
     Shlain, Acid Dreams: The Complete Social History of LSD (New York: Grove, 1985), 32.

199 America's first ambassador to the People's Republic: C. L. Sulzberger, "Our Next Man in Pe-
     king," New York Times, June 27, 1971, E15.

199 "what is called a civilized country": Kahn, The China Hands, 283–84, 290; U.S. Congress,
     Senate, Hearing before the Committee on Foreign Relations, The Evolution of U.S. Policy
     Toward Mainland China, 92nd Cong., 1st Sess., Report No. 67-891 (Washington, DC: U.S.
     Government Printing Office, 1971), 7.

200 immoral and a tad pathetic: Cecil Currey, Edward Lansdale: The Unquiet American (Bos-
     ton: Houghton Mifflin, 1988), 339; U.S. Congress, Senate, Select Committee to Study
     Governmental Operations with Respect to Intelligence Activities, Alleged Assassination Plots
     Involving Foreign Leaders: An Interim Report, 94th Cong., 1st Sess., Report No. 94-465
     (Washington, DC: U.S. Government Printing Office, 1975), 142.

200 none of his aides owned passports: Gaddis Smith, Morality, Reason and Power: American Diplomacy
     in the Carter Years (New York: Hill & Wang, 1986), 27; Westad, The Global Cold War, 195.

200 he still corrected aides' grammatical mistakes: Kenneth E. Morris, Jimmy Carter: American Mor-
     alist (Athens: University of Georgia Press, 1996), 5–6, 30–31, 47–48, 60, 67, 141.

200 denounced racism in the state: Morris, Jimmy Carter, 30–32, 72, 118, 152.

201 attack Ford from both left and right: Joshua Muravchik, The Uncertain Crusade: Jimmy Carter
     and the Dilemmas of Human Rights Policy (Washington, DC: Hamilton, 1986), 2–7.

201 derailed his nomination: Smith, Morality, Reason and Power, 51; Muravchik, The Uncertain
     Crusade, 8–9; Robert A. Strong, Working in the World: Jimmy Carter and the Making of
     American Foreign Policy (Baton Rouge: Louisiana State University Press, 2000), 12–13.

202 he declared a few months after taking office: Jimmy Carter, "University of Notre Dame
     Commencement," May 22, 1977, http://millercenter.org/scripps/digitalarchive/speeches/
     spe_1977_0522_carter.

202 the B-1 bomber and the neutron bomb: Jerel A. Rosati, The Carter Administration's Quest for
     Global Community: Beliefs and Their Impact on Behavior (Columbia: University of South
     Carolina Press, 1987), 122; David Skidmore, Reversing Course: Carter's Foreign Policy, Do-
     mestic Politics, and the Failure of Reform (Nashville, TN: Vanderbilt University Press, 1996),
     45; Smith, Morality, Reason and Power, 79–81; Morris, Jimmy Carter, 272.

202 aid to a throng of anticommunist dictatorships: Rosati, Quest for Global Community, 120–22,
     129–30; Smith, Morality, Reason and Power, 54, 103–4, 148.

202 his policies marked a genuine shift: Rosati, Quest for Global Community, 154.

202 "no longer seems warranted": Rosati, Quest for Global Community, 41.

203 in Angola and a newly unified Vietnam: Smith, Morality, Reason and Power, 116–17; Rosati,
     Quest for Global Community, 119; Skidmore, Reversing Course, 30.

203 explained Undersecretary of State David Newsom: Skidmore, Reversing Course, 41–42.

203 "not totally dependent on military power": Raskin, Essays of a Citizen, 82.

Chapter 11: Fighting with Rabbits

205 should have throttled the beast: Jody Powell, The Other Side of the Story (New York: William
     Morrow, 1984), 104–8; Brooks Jackson, Associated Press, "Bunny Goes Bugs: Rabbit At-

tacks President," *Washington Post*, August 30, 1979, A1; "The Bunny Encounter: Carter Says He Splashed 'Nice Rabbit' but Denies There Was an Attack by Either," *Washington Post*, September 1, 1979, A2; Art Buchwald, "Carter Staff Pondering Harey Question: Why Not the Beast?" *Washington Post*, September 11, 1979, E1; Morris, *Jimmy Carter*, 274–75.

205 *They viewed him as a wimp:* On press coverage of Carter, see John M. Orman, *Comparing Presidential Behavior: Carter, Reagan, and the Macho Presidential Style* (New York: Greenwood, 1987), 158, 164–65.

205 *"creating the conditions for domestic reaction":* Westad, *The Global Cold War*, 195, 282–83; Leffler, *For the Soul of Mankind*, 282–85; Elizabeth Drew, "A Reporter at Large: Brzezinski," *New Yorker*, May 1, 1978, 113–16; Smith, *Morality, Reason and Power*, 153–56; Gartoff, *Détente and Confrontation*, 698–705, 713–14.

205 *Iranian crude had been an economic lifeline:* Kunz, *Butter and Guns*, 223–41.

206 *oil fields of the Persian Gulf:* Jerry W. Sanders, *Peddlers of Crisis: The Committee on the Present Danger and the Politics of Containment* (Boston: South End, 1983), 239.

206 *called America the "Great Satan":* Smith, *Morality, Reason and Power*, 181–87; Strong, *Working in the World*, 59–61.

206 *shortages of medicine, soap, toothpaste, and thread:* Leffler, *For the Soul of Mankind*, 329, 333; Westad, *The Global Cold War*, 242, 284–85, 307–8, 315, 323. For a discussion of the stagnation of the Soviet economy, beginning in the 1970s, see Philip Hanson, *The Rise and Fall of the Soviet Union: An Economic History of the USSR from 1945* (London: Pearson Education, 2003), 128–63.

207 *even if he could never quite explain why:* Irving Kristol, "An Autobiographical Memoir," in *Neoconservatism: The Autobiography of an Idea*, ed. Irving Kristol (New York: Free Press, 1995), 4.

207 *ostensibly committed to world revolution:* The more common term, *Trotskyite*, Kristol insisted, was invented by their Stalinist foes, and never used by Trotskyists themselves. Irving Kristol, "Memoirs of a Trotskyist," *New York Times*, January 23, 1977, 43.

207 *"that were cluttering up my mind":* Kristol, "An Autobiographical Memoir," 6–7, 13; Kristol, "My Cold War," in *Neoconservatism*, ed. Kristol, 483–84.

208 *they hated the New Deal, which he did not:* Murray Friedman, *The Neoconservative Revolution* (New York: Cambridge University Press, 2005), 78–79; Irving Kristol, "American Conservatism, 1945–1995," *Public Interest* 121 (1995): 82–84; Kristol, "An Autobiographical Memoir," 32.

208 *accused the McGovern Democrats of linguistic theft:* Ehrman, *The Rise of Neoconservatism*, 46.

208 *not to worry about labels:* Irving Kristol, "Confessions of a True, Self-Confessed—Perhaps the Only—Neoconservative," in *Reflections of a Neoconservative*, ed. Irving Kristol (New York: Basic Books, 1983), 74.

208 *he was ambivalent about the war:* Robert R. Tomes, *Apocalypse Then: American Intellectuals and the Vietnam War* (New York: New York University Press, 1998), 194, 219; Gary Dorrien, *The Neoconservative Mind* (Philadelphia: Temple University Press, 1993), 91–92; Irving Kristol, "American Intellectuals and Foreign Policy," in *Neoconservatism*, ed. Kristol, 76.

208 *"the main source of all evil in the world":* Irving Kristol, "The Twisted Vocabulary of Superpower Symmetry," in *Scorpions in a Bottle: Dangerous Ideas About the United States and the Soviet Union*, ed. Lissa Roche (Hillsdale, MI: Hillsdale College Press, 1986), 20.

208 *too poor to buy a sandwich:* Kristol, "Memoirs of a Trotskyist," 43.

208 *tools of cold war repression:* Mark Gerson, *The Neoconservative Vision: From the Cold War to the Culture Wars* (New York: Madison, 1996), 109.

209 *"the gun as well as the joint":* David Farber, "The Counterculture and the Antiwar Movement," in *Give Peace a Chance: Exploring the Vietnam Antiwar Movement*, ed. Melvin Small and William D. Hoover (Syracuse, NY: Syracuse University Press, 1992), 19.

209 *sent a letter to the university's president:* Matusow, *The Unraveling of America*, 332–33.

209 *"'Up against the wall, motherfucker, this is a stickup'":* Daniel Bell, "Columbia and the New Left," *Public Interest* 13 (1968): 66.

209 *"develop a certain respect for what was":* Nathan Glazer, "On Being Deradicalized," *Commentary* 50, no. 4 (1970): 74–76; James Traub, "Nathan Glazer Changes His Mind, Again," *New York Times Magazine,* June 28, 1998, 23.

209 *radically improve the estate of the urban poor:* Friedman, *The Neoconservative Revolution,* 118–19.

210 *would only make things worse:* Podhoretz, *Breaking Ranks,* 363–64.

210 *Kirkpatrick, like Kristol, admired Niebuhr:* Jeane Kirkpatrick, "Personal Virtues, Public Vices," in *The Reagan Phenomenon, and Other Speeches on Foreign Policy,* ed. Jeane Kirkpatrick (Washington, DC: American Enterprise Institute, 1983), 214.

210 *"wishing does not make it so":* Ehrman, *The Rise of Neoconservatism,* 119.

210 *"all you had to do was stand in place":* J. David Hoeveler, Jr., *Watch on the Right: Conservative Intellectuals in the Reagan Era* (Madison: University of Wisconsin Press, 1991), 85.

211 *"But Micronesia":* Sidney Blumenthal, *The Rise of the Counter-Establishment: From Conservative Ideology to Political Power* (New York: Times Books, 1986), 128.

211 *"troubling our sleep":* Norman Podhoretz, "The Culture of Appeasement," *Harper's,* October 1977, 32.

211 *Moscow had taken a decisive edge:* Sanders, *Peddlers of Crisis,* 197–204.

211 *sapped Britain's after World War I:* Podhoretz, "The Culture of Appeasement," 29–31.

212 *America's first female president:* John H. Mihalec, "Hair on the President's Chest," *Wall Street Journal,* May 11, 1984, 30.

212 *regimes in the third world—were running wild:* Kristol, "We Can't Resign as 'Policeman of the World,'" *New York Times,* May 12, 1968, 109; Irving Kristol, "Our Foreign Policy Illusions," *Wall Street Journal,* February 4, 1980, 26.

212 *"even more corrupting and demoralizing":* Irving Kristol, "Foreign Policy: The End of an Era," *Wall Street Journal,* January 18, 1979, 16.

212 *it was déjà vu:* Hoeveler, *Watch on the Right,* 153; Michael Novak, "In Praise of Jeane Kirkpatrick," December 8, 2006, http://www.aei.org/publications/filter.all,pubID.25268/pub_de tail.asp.

213 *"change with progress, optimism with virtue":* Jeane Kirkpatrick, "Dictatorships and Double Standards," *Commentary* 68, no. 5 (1979): 5, 13.

213 *"Unfortunately it does":* Sanders, *Peddlers of Crisis,* 162.

213 *tied to chairs or beds, in the ambassador's house:* Mark Bowden, *Guests of the Ayatollah: The First Battle in America's War with Militant Islam* (New York: Atlantic Monthly Press, 2006), 8–67, 127–29.

213 *number of days the Americans had been held captive:* Ellen Goodman, "And That's the Way It Is—Or Is It?" *Washington Post,* June 17, 1980.

214 *end of the post-toughness era as well:* Bowden, *Guests of the Ayatollah,* 189–91.

214 *Carter tried diplomacy and imposed sanctions:* Smith, *Morality, Reason and Power,* 199; Rosati, *The Carter Administration's Quest for Global Community,* 139.

214 *McNamara's deputy secretary of defense:* Smith, *Morality, Reason and Power,* 40.

214 *a patron to men like Gelb, Warnke, and Lake:* Rosati, *The Carter Administration's Quest for Global Community,* 111–12; I. M. Destler, Leslie Gelb, and Anthony Lake, *Our Own Worst Enemy: The Unmaking of American Foreign Policy* (New York: Simon & Schuster, 1984), 95–98; Marilyn Berger, "Cyrus R. Vance, A Confidant of Presidents, Is Dead at 84," *New York Times,* January 13, 2002.

214 *"growing American maturity in a complex world":* Frum, *How We Got Here,* 343.

214 *"Woody Woodpecker":* Drew, "A Reporter at Large: Brzezinski," 106.

214 *colleagues ran for cover:* Strong, *Working in the World,* 11.

214 *called on the Kennedy administration to bomb:* Smith, *Morality, Reason and Power,* 36.

214 *in favor of American intervention in Vietnam:* CBS News, "Dean Bundy Meets Professor Morgenthau," 198–206.

214 *America must not "chicken out":* Rosati, *The Carter Administration's Quest for Global Community,* 110; Simon Serfaty, "Brzezinski: Play It Again, Zbig," *Foreign Policy* 32 (Fall 1978): 6.

214 *from anticommunism to interdependence:* Gerry Argyris Andrianopoulos, *Kissinger and Brzezinski: The NSC and the Struggle for Control of US National Security Policy* (New York: St. Martin's, 1991), 40–42; Rosati, *The Carter Administration's Quest for Global Community,* 115, n. 9.

214 *"elements of cooperation prevail over competition":* Rosati, *The Carter Administration's Quest for Global Community,* 41, 56; Zbigniew Brzezinski, "American Policy and Global Change," address before the Trilateral Commission, Bonn, West Germany, October 30, 1977.

215 *international affairs were "not a kindergarten":* Smith, *Morality, Reason and Power,* 37, 199; Rosati, *The Carter Administration's Quest for Global Community,* 115, n. 14.

215 *Carter broke the news to the American people:* Bowden, *Guests of the Ayatollah,* 226–33, 445–68.

215 *Vance had already resigned in disgust:* Smith, *Morality, Reason and Power,* 203–4; Natasha Zaretsky, *No Direction Home: The American Family and the Fear of National Decline* (Chapel Hill: University of North Carolina Press, 2007), 229.

215 *made the Shah think he was invincible:* Gary Sick, *All Fall Down: America's Fateful Encounter with Iran* (London: I. B. Tauris, 1985), 13–14, 173.

216 *a country it had not conquered during World War II:* Ambrose and Brinkley, *Rise to Globalism,* 293–95.

216 *that the Americans kept talking about actually was:* Leffler, *For the Soul of Mankind,* 336; Westad, *The Global Cold War,* 322.

216 *ripped it to shreds:* Rosati, *The Carter Administration's Quest for Global Community,* 142–44; Smith, *Morality, Reason and Power,* 224–28, 230.

217 *imposed those restraints in the first place:* Jimmy Carter, "State of the Union Address," address to Joint Session of Congress, January 23, 1980.

217 *his administration had painstakingly negotiated:* Smith, *Morality, Reason and Power,* 243.

217 *dictatorships of Kenya, Somalia, and Oman:* Rosati, *The Carter Administration's Quest for Global Community,* 144–45.

217 *"sometimes dangerous world that really exists":* Rosati, *The Carter Administration's Quest for Global Community,* 81.

217 *"to remain the strongest nation in the world":* Rosati, *The Carter Administration's Quest for Global Community,* 88; Smith, *Morality, Reason and Power,* 230.

217 *"becomes a contagious disease":* Ambrose and Brinkley, *Rise to Globalism,* 288.

217 *"history holds its breath":* Smith, *Morality, Reason and Power,* 227–28; Steven R. Weisman, "Mondale Expects Support," *New York Times,* April 12, 1980, 5.

Chapter 12: If There Is a Bear?

218 *"Do we get to win this time?":* Susan Jeffords, *The Remasculinization of America: Gender and the Vietnam War* (Bloomington: Indiana University Press, 1989), 127–29.

218 *an earsplitting, kick-ass "Yes!":* Box Office Mojo, "Rambo First Blood: Part II," http://www.boxofficemojo.com/movies/?page-main&id=rambo2.htm.

218 *America's patriotic new president, Ronald Reagan:* Jeffords, *The Remasculinization of America,* 129, 142.

218 *a movie of his own, which followed the* Rambo *script:* Lou Cannon, *President Reagan: The Role of a Lifetime* (New York: Touchstone, 1991), 186.

218 *denied the award because of a technicality:* Initially it was believed that there had been only one witness to his heroism, not the required two. Gerry J. Gilmore, "USNS Benavidez Honors Army Medal of Honor Hero," *American Forces Press Service,* http://www.defenselink.mil/news/newsarticle.aspx?id=45442; Albin Krebs and Robert McG. Thomas, Jr., "Notes on People: A Green Beret's Bravery Gains Additional Recognition," *New York Times,* February 20, 1981, C24.

218 *"because they'd been denied permission to win":* Ronald Reagan, "Remarks on Presenting the Medal of Honor to Master Sergeant Roy P. Benavidez," *Public Papers of the President,* 155, February 24, 1981.

219 *By 1985, it was America's bestselling toy:* Gil Troy, *Morning in America: How Reagan Invented the 80s* (Princeton, NJ: Princeton University Press, 2005), 241; ICON Group International, Inc., *Hobbies: Webster's Quotations, Facts and Phrases* (San Diego: ICON Group International, 2008), 36.

219 *the ticker-tape parade they had long been denied:* Jeffords, *The Remasculinization of America*, 2.

219 *"You might say America has gone back to the gym":* Gerald J. DeGroot, *Noble Cause: America and the Vietnam War* (Harlow, England: Longman, 2000), 357.

220 *a hint of the transgressions to come:* George F. Will, " 'Fresh Starts' and Other Fictions," in *The Morning After,* ed. George F. Will (New York: Free Press, 1986), 343.

220 *had to drag him by the armpits to bed:* Cannon, *President Reagan,* 207–8.

220 *which the local butcher usually sold as pet food:* John Patrick Diggins, *Ronald Reagan: Fate, Freedom, and the Making of History* (New York: Norton, 2007), 58.

220 *"one of those rare Huck Finn–Tom Sawyer idylls":* Blumenthal, *The Rise of the Counter-Establishment,* 244; Leffler *For the Soul of Mankind,* 342.

220 *Reagan focused on the only positive one:* Cannon, *President Reagan,* 25, 179, 746; Diggins, *Ronald Reagan,* 67, 73.

221 *"inability to distinguish between fact and fantasy":* Sean Wilentz, *The Age of Reagan* (New York: HarperCollins, 2008), 130.

221 *spinoff from Unitarianism that downplayed the idea of sin:* Diggins, *Ronald Reagan,* 25.

221 *Eureka College, a Disciples' institution:* Stephen Vaughn, "The Moral Inheritance of a President: Reagan and the Dixon Disciples of Christ," *Presidential Studies Quarterly* 25, no. 1 (1995): 109–27; Beth A. Fischer, *The Reagan Reversal: Foreign Policy and the End of the Cold War* (Columbia: University of Missouri Press, 1997), 106; Paul Lettow, *Ronald Reagan and His Quest to Abolish Nuclear Weapons* (New York: Random House, 2005), 7–8; Wilentz, *The Age of Reagan,* 130.

221 *"the desires he planted in us are good":* Diggins, *Ronald Reagan,* 30.

221 *"which regularly deny the existence of evil":* Jeane Kirkpatrick, "The Reagan Reassertion of Western Values," in *The Reagan Phenomenon,* ed. Kirkpatrick, 34.

221 *fights with the communists in Hollywood in the 1940s:* Cannon, *President Reagan,* 284–87; Diggins, *Ronald Reagan,* 99–101.

221 *he didn't share its fear:* Leffler, *For the Soul of Mankind,* 340.

222 *destabilize the entire European banking system:* Dan Morgan and Robert G. Kaiser, "Group of Aides Sought Tougher Stand on Poland," *Washington Post,* January 15, 1982, A1; Dan Morgan, "West Faces Dilemma on Polish Debt," *Washington Post,* January 11, 1982, A15; George F. Will, "Reagan's Dim Candle," *Newsweek,* January 18, 1982, 100; Norman Podhoretz, "The Neo-Conservative Anguish over Reagan's Foreign Policy," *New York Times,* May 2, 1982, 30.

222 *"sensitive about being viewed as too pugnacious":* Robert Kagan, *A Twilight Struggle: American Power and Nicaragua, 1977–1990* (New York: Free Press, 1996), 270.

223 *"is Spanish for Vietnam":* Thomas Paterson, "Historical Memory and Illusive Victories: Vietnam and Central America," *Diplomatic History* 12, no. 1 (1988): 2–3.

223 *dated back to the nineteenth century:* Paterson, "Historical Memory and Illusive Victories," 3, 18.

224 *"a fucking parking lot":* Cannon, *President Reagan,* 196.

224 *declarations of war against both Cuba and Nicaragua:* William F. Buckley, "Who Was Right?" in *Right Reason,* ed. Richard Brookhiser (Boston: Little, Brown, 1985), 123–25; Buckley, "Next in Central America?" in *Right Reason,* ed. Brookhiser, 257.

224 *"reverse the totalitarian drift in Central America":* Norman Podhoretz, "Appeasement by Any Other Name," *Commentary* 76, no. 1 (1983): 38.

224 *"we will use all means to prevent this":* Irving Kristol, "The Muddle in Foreign Policy," *Wall Street Journal,* April 29, 1981, 28.

224 *a chorus of anxious references to Vietnam:* Kagan, *A Twilight Struggle,* 175.

224 *"scared the shit out of Ronald Reagan":* Cannon, *President Reagan,* 196–97.

224 *"limp-wristed, traditional cookie-pushing bullshit":* Cannon, *President Reagan,* 345, 348.

225 *opposed invading Nicaragua:* William Schneider, "'Rambo' and Reality: Having It Both Ways," in *Eagle Resurgent? The Reagan Era in American Foreign Policy,* ed. Kenneth A. Oye, Robert J. Lieber, and Donald Rothchild (Boston: Little, Brown, 1983), 60.

225 *allowing a communist takeover:* Barry Sussman, "Poll Finds a Majority Fears Entanglement in Central America," *Washington Post,* May 25, 1983, A1.

225 *members of both parties erupted in cheers:* Podhoretz, "Appeasement by Any Other Name," 27; Steven R. Weisman, "President Appeals Before Congress for Aid to Latins," *New York Times,* April 28, 1983, A1; Gerald F. Seib, "Central America Leftists Called Threat by Reagan," *Wall Street Journal,* April 28, 1983, 3.

225 *Reagan's advisers feared he might be impeached:* Cannon, *President Reagan,* 380, 718; Troy, *Morning in America,* 244; Wilentz, *The Age of Reagan,* 232, 242.

225 *World War II, Korea, and Vietnam combined:* Westad, *The Global Cold War,* 347.

226 *"and I'm not going to do it":* Cannon, *President Reagan,* 336, 337.

226 *"perhaps the saddest day of my life":* Cannon, *President Reagan,* 389–457, 442.

227 *Reagan decided to do exactly that:* Cannon, *President Reagan,* 445, 454–57, 510.

227 *declared herself "disgusted":* Bernard Weinraub, "On the Right: Long Wait for Foreign Policy Hero," *New York Times,* July 12, 1985, A12.

227 *Their boss's poll numbers soon ticked back up:* Dan Balz and Thomas B. Edsall, "The Invasion of Grenada; GOP Rallies Around Reagan," *Washington Post,* October 26, 1983, A8; David Shribman, "Poll Finds a Lack of Support for Latin Policy," *New York Times,* April 29, 1984, A1; Cannon, *President Reagan,* 510–11.

227 *since the planned runway was too small:* Cannon, *President Reagan,* 445–49.

227 *More than 70 percent of Americans backed the Grenada invasion:* Cannon, *President Reagan,* 448, 462; Schneider, "'Rambo' and Reality," 59.

227 *more than ten to one in favor:* Diggins, *Ronald Reagan,* 249.

227 *from humiliation to pride:* Cannon, *President Reagan,* 448–49.

227 *where they waved little American flags:* Jon Western, *Selling Intervention and War: The Presidency, the Media and the American Public* (Baltimore: Johns Hopkins University Press, 2005), 95; Francis X. Clines, "Medical Students Cheer Reagan at a White House Ceremony," *New York Times,* November 8, 1983, A10.

228 *more than 70 percent support at home:* Cannon, *President Reagan,* 653–54; Troy, *Morning in America,* 246.

228 *morphing into Rambo again:* Cannon, *President Reagan,* 663.

228 *declared Norman Podhoretz in 1976:* Dorrien, *Neoconservative Mind,* 178.

228 *"they do not and cannot last":* Irving Kristol, "An Automatic-Pilot Administration," *Wall Street Journal,* December 14, 1984, 26.

229 *she explained in 1981:* Jeane Kirkpatrick, "The Reagan Phenomenon and the Liberal Tradition," in *The Reagan Phenomenon,* ed. Kirkpatrick, 15.

229 *"human freedom and human dignity to its citizens":* Leffler, *For the Soul of Mankind,* 339–40.

229 *"precious little evidence to prove them wrong":* Irving Kristol, "The Succession: Understanding the Soviet Mafia," *Wall Street Journal,* November 18, 1982, 30.

230 *"oriental despotism" that stretched back to the czars:* Adam Wolfson, "The World According to Kirkpatrick: Is Ronald Reagan Listening?" *Policy Review* 31 (1985): 70.

230 *"served to reinforce these ingrained convictions":* Sanders, *Peddlers of Crisis,* 149.

230 *"people don't make wars":* Gaddis, *Strategies of Containment,* 360–61; James Mann, *The Rebellion of Ronald Reagan: A History of the End of the Cold War* (New York: Viking, 2009), 80.

230 *"another time and another era":* Diggins, *Ronald Reagan,* 3, 8, 29; Mann, *The Rebellion of Ronald Reagan,* 304.

231 *a group that sought to outlaw nuclear weapons:* Cannon, *President Reagan,* 62.

231 *didn't fit his image as a cold war hawk:* On Reagan's nuclear abolitionism, see Lettow, *Ronald Reagan and His Quest to Abolish Nuclear Weapons.*

231 *the robots will descend and kill them all:* Cannon, *President Reagan,* 61–62.

231 *"Here come the little green men again":* Cannon, *President Reagan,* 62–63.

232 *the inventor of the shield:* Thompson, *The Hawk and the Dove,* 300.

232 *until the Americans had caught up:* Beth A. Fischer, "Toeing the Hardline? The Reagan Administration and the Ending of the Cold War," *Political Science Quarterly* 112, no. 3 (1997): 483.

232 *it reflected their eternal view of world affairs:* In 1982, for instance, Podhoretz attacked Reagan for saying that once America caught up with the Soviets he would support a mutual freeze on the production of new nuclear weapons. Podhoretz, "The Neo-conservative Anguish," 30.

232 *"the gap between U.S. and Soviet military capabilities continues to grow":* Walter Pincus and Don Oberdorfer, "A-Arms Balance Holds," *Washington Post*, December 16, 1984, A1.

232 *"He had tied into NORAD!":* Cannon, *President Reagan*, 58, 85.

232 *"to see that there is never a nuclear war":* Fischer, *The Reagan Reversal*, 120.

232 *he had seen footage of the camps in a film:* Cannon, *President Reagan*, 486–90.

232 *playing out before him in real life:* "In several ways," Reagan noted, "the sequence of events described in the briefings paralleled those in the ABC movie." Barbara Farnham, "Reagan and the Gorbachev Revolution: Perceiving the End of Threat," *Political Science Quarterly* 116, no. 2 (2001): 232. See also Leffler, *For the Soul of Mankind*, 359; and Fischer, *The Reagan Reversal*, 120–21.

233 *put their military on high alert:* Fischer, *The Reagan Reversal*, 122–36; Don Oberdorfer, *From the Cold War to a New Era: The United States and the Soviet Union, 1983–1991* (Baltimore: Johns Hopkins University Press, 1998), 65–67.

233 *percentage of Americans favoring arms control had shot through the roof:* Schneider, "'Rambo' and Reality," 43; John Rielly, "American Opinion: Continuity, Not Reaganism," *Foreign Policy* 50 (1983): 96.

233 *a mutual freeze on the production of nuclear arms:* Troy, *Morning in America*, 138.

233 *a warmonger in his 1984 reelection campaign:* Cannon, *President Reagan*, 508–11, 740; Jeremi Suri, "Explaining the End of the Cold War: A New Historical Consensus?" *Journal of Cold War Studies* 4, no. 4 (2002): 92; Fischer, "Toeing the Hardline?" 496; Farnham, "Reagan and the Gorbachev Revolution," 233–34.

233 *also eager to get disarmament talks going:* Cannon, *President Reagan*, 309; Barbara Farnham, "Reagan and the Gorbachev Revolution," 230; Leffler, *For the Soul of Mankind*, 362.

233 *"nuclear weapons will be banished from the face of the earth":* Ronald Reagan, "Address to the Nation and Other Countries on United States-Soviet Relations," January 16, 1984, http://www.reagan.utexas.edu/archives/speeches/1984/11684a.htm.

233 *"thousand-fold affection returned to you":* Fischer, *The Reagan Reversal*, 40–41.

233 *"under the threat of those weapons":* Fischer, *The Reagan Reversal*, 42.

233 *"it would be the height of naivete to think otherwise":* Kristol, "An Automatic-Pilot Administration," 26.

234 *a jaw-dropping 40 percent of his government's budget:* Zubok, *Failed Empire*, 277.

234 *America looked like less of a threat:* Suri, "Explaining the End of the Cold War," 70–78.

234 *a summit meeting, without preconditions:* Fischer, "Toeing the Hardline?" 494.

234 *in Moscow to attend the funeral of Gorbache's predecessor:* Bernard Weinraub, "Bush Sent to Rites," *New York Times*, March 12, 1985, A1.

234 *did not share her "foreign policy objectives":* Bernard Weinraub, "Reagan Is Told by Kirkpatrick She Will Leave," *New York Times*, January 31, 1985, A6.

234 *"the threat or use of force in international relations":* Leffler, *For the Soul of Mankind*, 364–65.

235 *briefing papers on Soviet politics and culture:* Cannon, *President Reagan*, 57, 155, 157, 746–49.

235 *talked for almost five hours:* Fischer, *The Reagan Reversal*, 48.

235 *"bleeding when we shake hands":* Cannon, *President Reagan*, 754.

235 *"craven eagerness" to give away the nuclear store:* Dorrien, *The Neoconservative Mind*, 197–200.

235 *"like a punctured balloon":* George Will, "Reagan Botched the Daniloff Affair," *Washington Post*, September 18, 1986, A25.

235 *"a dream of a world without nuclear weapons":* Leffler, *For the Soul of Mankind*, 387–88, 392.

236 *"give a tremendous party for the whole world":* Gaddis, *Strategies of Containment*, 366; Richard

Rhodes, *Arsenals of Folly: The Making of the Nuclear Arms Race* (New York: Knopf, 2007), 230; Oberdorfer, *From the Cold War to a New Era*, 169–74, 207–8; Jack F. Matlock, Jr., *Autopsy of an Empire: The American Ambassador's Account of the Collapse of the Soviet Union* (New York: Random House, 1995), 93–98; Lettow, *Ronald Reagan and His Quest to Abolish Nuclear Weapons*, 217–26.

236 *the two sides began dismantling them:* Leffler, *For the Soul of Mankind*, 394, 401.

236 *while allowing Britain and France to keep theirs:* Leffler, *For the Soul of Mankind*, 400.

236 *"elevating wishful thinking to the status of political philosophy":* Diggins, *President Reagan*, 384.

236 *"a utopian":* "Mad Momentum," *Wall Street Journal*, April 13, 1988, 28.

236 *"is not the architect of Soviet retreat":* Jeane Kirkpatrick, *The Withering Away of the Totalitarian State and Other Surprises* (Washington, DC: AEI Press, 1990), 52; Hedrick Smith, "The Right Against Reagan," *New York Times Magazine*, January 17, section 6, 36.

236 *Dick Cheney called* glasnost *a fraud:* Diggins, *Ronald Reagan*, 384.

236 *Podhoretz proposed reconstituting it:* Dorrien, *The Neoconservative Mind*, 199; Sanders, *Peddlers of Crisis*, 152–54.

237 *Neville Chamberlain, and Adolf Hitler:* Smith, "The Right Against Reagan," 36; John Hanrahan, United Press International, "Conservatives Escalate Anti-Treaty Campaign," January 20, 1988.

237 *sounding a little like the Port Huron Statement:* Diggins, *Ronald Reagan*, 392.

237 *"a long silence yearn to break free":* Cannon, *President Reagan*, 787.

237 *"tighten our belts and spend even more on defense":* Barbara Farnham, "Reagan and the Gorbachev Revolution," 250. The historian Thomas Risse-Kappen adds that there is "not a shred of substantiation for the claim that there is any connection between the US [military] buildup and the Soviet turnabout." Diggins, *Ronald Reagan*, 405. Walter Uhler has debunked the claim that it was fear of SDI that led Gorbachev to seek to end the cold war. Walter C. Uhler, "Misreading the Soviet Threat," *Journal of Slavic Military Studies*, March 2001, www.walter-c-uhler.com/Reviews/Misreading.html. Reagan himself wrote that "I might have helped him see that the Soviet Union had less to fear from the West than he thought, and that the Soviet empire in Eastern Europe wasn't needed for the security of the Soviet Union." Gaddis, *Strategies of Containment*, 374.

238 *"a minor Dark Age":* Steel, *Walter Lippmann*, 592.

238 *"I am ashamed of our beloved nation":* Fox, *Reinhold Niebuhr*, 285.

238 *"Perhaps to save one's soul is all that is left":* Frei, *Hans J. Morgenthau*, 221–22.

238 *"succumbing feebly, day by day, to its own decadence":* Mayers, *George Kennan and the Dilemmas of US Foreign Policy*, 299.

238 *"no recovery and no return":* Hixson, *George Kennan*, 283–85.

238 *chickens were coming home to roost:* Mayers, *George Kennan and the Dilemmas of US Foreign Policy*, 314.

239 *suggested that he be "put out to pasture":* Henry Fairlie, "Mr. X: The Special Senility of the Diplomat," *New Republic*, December 24, 1977, 9–11.

239 *"springs a leak":* Paul Seabury, "George Kennan vs. Mr. 'X': The Great Container Springs a Leak," *New Republic*, December 16, 1981, 17–20.

239 *the eventual abolition of all nuclear weapons:* Hixson, *George Kennan*, 282.

239 *"is a fit occasion for satisfaction":* George Kennan, "Republicans Won the Cold War?" *New York Times*, October 28, 1992, in George Kennan, *At a Century's Ending: Reflections, 1982–1995* (New York: Norton, 1995), 187.

PART III: THE HUBRIS OF DOMINANCE

Chapter 13: Nothing Is Consummated

243 *"like a Mozart oboe concerto":* Mary McGrory, "Kennan—A Prophet Honored," *Washington Post*, April 9, 1989, B1; Don Oberdorfer, "Revolutionary Epoch Ending in Russia, Kennan Declared," *Washington Post*, April 9, 1989, A22.

244 *"Wilson was way ahead of his time":* Senate Committee on Foreign Relations, *The Future of U.S.-Soviet Relations: Hearing Before the Committee on Foreign Relations*, 101st Cong., 1st sess., 1989, 27–28; Ray Moseley, "Expert on USSR Says Communism Is Dying," *Chicago Tribune*, April 5, 1989, 1.

244 *still comparing Gorbachev to Brezhnev:* Richard Pipes, "The Russians Are Still Coming," *New York Times*, October 9, 1989, A17.

244 *still comparing him to Lenin:* Dorrien, *The Neoconservative Mind*, 199.

244 *East German communism is here to stay:* Francis Fukuyama, interviews with the author, July 15 and 21, 2008.

245 *"nuclear weapons and environmental damage":* Francis Fukuyama, *The End of History and the Last Man* (New York: Free Press, 1992), xiii, 3.

245 *Goddess of Democracy in Tiananmen Square:* Peter LaBarbera and Paul Bedard, "Walesa Thanks America for Aiding Poland," *Washington Post*, November 16, 1989, A1.

245 *"democracy is supreme":* "The Pivot of History," *New Republic*, November 16, 1918, 58.

246 *"crossed the finish line out of breath":* Robert Kagan, "A Retreat from Power?" *Commentary* 100, no. 5 (1995): 22.

246 *world's largest creditor to its largest debtor:* Samuel Huntington, "The US—Decline or Renewal?" *Foreign Affairs* 67, no.2 (1988): 78.

246 *the worst day of 1929:* E. S. Browning, "Exorcising Ghosts of Octobers Past: Despite Housing Slump, Crashes Such as in 1987 Likely to Stay Memories," *Wall Street Journal*, October 15, 2007, C1.

246 *it had sold 225,000:* James Mann, *Rise of the Vulcans: The History of Bush's War Cabinet* (New York: Viking Penguin, 2004), 161.

246 *the* New York Times' *bestseller list:* Derek Chollet and James Goldgeier, *America Between the Wars* (New York: PublicAffairs, 2008), xiii; Hawes Publications, "New York Times Bestseller List," http://www.hawes.com/1988/1988.htm.

248 *Admiral William Crowe, chairman:* Frederick Kempe, *Divorcing the Dictator: America's Bungled Affair with Noriega* (London: I. B. Tauris, 1990), 294; Dennis Hevesi, "Adm. William Crowe Dies at 82; Led Joint Chiefs," *New York Times*, October 19, 2007; Stanley Karnow, *In Our Image: America's Empire in the Philippines* (New York: Random House, 1989), 407.

248 *Norman Podhoretz's stepdaughter:* Eric Alterman, "Elliott Abrams: The Teflon Assistant Secretary," *Washington Monthly*, May 1987; Michael Crowley, "Assessment: Elliott Abrams," *Slate*, February 17, 2005, http://www.slate.com/id/2113690/; Jay Winik, *On the Brink: The Dramatic, Behind-the-Scenes Saga of the Reagan Era and the Men and Women Who Won the Cold War* (New York: Simon & Schuster, 1996), 425–32.

248 *Abrams was only thirty-two:* Winik, *On the Brink*, 433.

248 *"Northern America and Western Europe":* Irving Kristol, "What Choice Is There in Salvador?" *Wall Street Journal*, April 4, 1983, 16.

249 *the thugs were on our side:* Pamela Constable and Arturo Valenzuela, "Is Chile Next?" *Foreign Policy*, 63 (1986): 58–59, 71–73; David P. Forsythe, "Human Rights in U.S. Foreign Policy: Retrospect and Prospect," *Political Science Quarterly* 105, no. 3 (1990): 435–54; Christopher Madison, "Abrams Policy Skills to Be Put to Test in Central America Hot Seat at State," *National Journal* 17, no. 28 (1985): 1619–22; George Shultz, *Turmoil and Triumph* (New York: Scribner, 1993), 972, 974.

249 *a prominent critic of the regime:* Kempe, *Divorcing the Dictator*, 26; Emily Yoffe, "A Presidential Salary FAQ," *Slate*, January 3, 2001, http://www.slate.com/id/1006798/.

249 *Jesse Helms was appalled:* Kevin Buckley, *Panama: The Whole Story* (New York: Simon & Schuster, 1991), 27–28, 41–42; Kempe, *Divorcing the Dictator*, 176–77.

249 *he was denouncing him publicly:* Buckley, *Panama*, 90; Kempe, *Divorcing the Dictator*, 169, 294.

249 *important enough to protect:* Eytan Gilboa, "The Panama Invasion Revisited: Lessons for the Use of Force in the Post Cold War Era," *Political Science Quarterly* 110, no. 4 (1995): 541, 546–47.

250 *never smelled it up close:* Elaine Sciolino, "Washington at Work: Crowe v. Abrams: A Private Feud over Handling Panama Becomes Public," *New York Times*, October 23, 1989, A14.

250 *deferment to avoid Vietnam:* Kempe, *Divorcing the Dictator,* 299.

250 *"raised it to an art form":* William Crowe, letter to the editor, "Elliott Abrams Remains Reckless on Panama," *New York Times,* October 16, 1989.

250 *weren't fighting for it themselves:* Kempe, *Divorcing the Dictator,* 297, 303–4; Buckley, *Panama,* 138.

250 *south of the border now:* Gilboa, "The Panama Invasion Revisited," 548–49.

250 *still haunted his sleep:* Stefan Halper and Jonathan Clarke, *America Alone: The Neo-Conservatives and the Global Order* (New York: Cambridge University Press, 2004), 171.

251 *"sending those troops to Lebanon":* Marlin Fitzwater, phone interview with the author, August 7, 2008.

251 *if he gave up power:* Kempe, *Divorcing the Dictator,* 311; Buckley, *Panama,* 143–45.

251 *"will never invade Panama":* Bob Woodward, *The Commanders* (New York: Simon & Schuster, 1991), 116.

251 *"the farewell party has been indefinitely postponed":* Buckley, *Panama,* 166.

251 *there would be no deals:* Buckley, *Panama,* 147–48, 169; Kempe, *Divorcing the Dictator,* 11, 28, 313, 347.

251 *smashed him in the face with an iron bar:* Kempe, *Divorcing the Dictator,* 83, 225; Buckley, *Panama,* 152–56, 208; Woodward, *The Commanders,* 84.

252 *above her head until she collapsed:* Woodward, *The Commanders,* 84, 157–58; Buckley, *Panama,* 227.

252 *people did call him a wimp, often:* See, for instance, Margaret Warner, "Bush Battles the 'Wimp Factor,'" *Newsweek,* October 19, 1987, 28.

252 *played a mean game of horseshoes:* Richard Ben Cramer, *What It Takes* (New York: Random House, 1992), 9, 116–17; Jacob Weisberg, *The Bush Tragedy* (New York: Random House, 2008), 23, 31–32, 34.

252 *the names of his political enemies:* Woodward, *The Commanders,* 168, 191–93; Buckley, *Panama,* 247.

252 *flown to a Florida jail:* Andrew Rosenthal, "Noriega's Surrender: Overview; Noriega Gives Himself Up to U.S. Military; Is Flown to Florida to Face Drug Charges," *New York Times,* January 4, 1990, A1; Adam Pertman, "Closed Door Arguments Continue over Relocating Noriega," *Boston Globe,* January 13, 1990, A4.

252 *twenty-three U.S. soldiers died:* Woodward, *The Commanders,* 195.

252 *any U.S. president since Vietnam:* Bruce Jentleson, "The Pretty Prudent Public," *International Studies Quarterly* 36, no. 1 (1992): 55; Russell Crandall, *Gunboat Democracy: US Interventions in the Dominican Republic, Grenada and Panama* (New York: Rowman & Littlefield, 2006), 209.

253 *"alienate the Panamanian people":* Crandall, *Gunboat Democracy,* 25, 197, 204, 215; Woodward, *The Commanders,* 85, 90, 136, 138, 144, 173, 189; Karin von Hippel, *Democracy by Force: US Military Intervention in the Post–Cold War World* (Cambridge, England: Cambridge University Press, 2000), 35; Buckley, *Panama,* 179.

253 *the swelling ranks of Latin American democracies:* Crandall, *Gunboat Democracy,* 209–11, 216; Buckley, *Panama,* 179, 259.

253 *"a free country with justice and liberty":* Crandall, *Gunboat Democracy,* 210.

254 *"a trial run":* Stephen Hayes, *Cheney: The Untold Story of America's Most Powerful and Controversial Vice President* (New York: HarperCollins, 2007), 224.

254 *waving his gold watch:* Christian Alfonsi, *Circle in the Sand: Why We Went Back to Iraq* (New York: Doubleday, 2006), 21, 26, 44, 56; Thomas C. Hayes, "The Oilfield Lying Below the Iraq-Kuwait Dispute," *New York Times,* September 3, 1990.

254 *more than 45 percent:* Energy Information Administration, "International Petroleum (Oil) Reserves and Resources Data," http://www.eia.doe.gov/pub/international/iealf/crudeoilreserves.xls.

254 *dispatching troops to Saudi soil:* Alfonsi, *Circle in the Sand,* 89.

254 *and then draw its sword:* Steve A. Yetiv, *Explaining Foreign Policy: U.S. Decision-Making and the Persian Gulf War* (Baltimore: Johns Hopkins University Press, 2004), 127–29; Mann,

Rise of the Vulcans, 185–86; Richard N. Haass, *War of Necessity, War of Choice: A Memoir of Two Iraq Wars* (New York: Simon & Schuster, 2009), 94.

255 *the United States to beat Iraq to a pulp:* Tareq Y. Ismael and Andrej Kreutz, "Russian-Iraqi Relations: A Historical and Political Analysis," *Arab Studies Quarterly* 23, no. 4 (2001): 89–90.

255 *sweltering in the Saudi desert:* Woodward, *The Commanders*, 37.

256 *"been selling around here or over there":* Woodward, *The Commanders*, 35–39.

256 *dancing to calypso and revering the queen:* Karen DeYoung, *Soldier: The Life of Colin Powell* (New York: Vintage, 2007), 19, 23.

256 *in the middle of a garbage dump:* DeYoung, *Soldier*, 49, 69–70.

256 *"I perform well":* Henry Louis Gates, Jr., "Powell and the Black Elite," *New Yorker*, September 25, 1995, 73.

256 *seared into his consciousness:* Patricia Sullivan, "William Crowe, Jr.; Joint Chiefs Leader Had Diplomat's Touch," *Washington Post*, October 19, 2007, B6; DeYoung, *Soldier*, 51–68, 74–91.

256 *they began the long journey home:* DeYoung, *Soldier*, 76.

256 *had been blown to bits:* Mann, *Rise of the Vulcans*, 118.

257 *lost close to half a million men:* Woodward, *The Commanders*, 342; DeYoung, *Soldier*, 193; Alfonsi, *Circle in the Sand*, 67; Michael R. Gordon and Bernard E. Trainor, *The Generals' War: The Inside Story of the Conflict in the Gulf* (Boston: Little, Brown, 1995), 33, 130–33; James A. Bill, "Why Tehran Finally Wants a Gulf Peace; Iraq's Missile Blitz Broke Their Will to Keep Fighting," *Washington Post*, August 28, 1988, B1; Patrick E. Tyler, "Gulf War Cease-Fire Begins," *Washington Post*, August 20, 1988, A1; Lawrence Freedman and Ephraim Karsh, *The Gulf Conflict 1990–1991* (Princeton, NJ: Princeton University Press, 1995), 74, 203.

257 *Cheney's deferments during Vietnam:* Colin Powell, *My American Journey* (New York Ballantine, 1995), 393; Gary Dorrien, *Imperial Designs: Neoconservatism and the New Pax Americana* (New York: Routledge, 2004), 39.

257 *people called him "the Sphinx":* Dorrien, *Imperial Designs*, 38.

257 *a reputation for humiliating generals:* Hayes, *Cheney*, 216, 234–38; Gordon and Trainor, *The Generals' War*, 100–1.

257 *"So stick to military matters":* Powell, *My American Journey*, 451–52.

258 *almost forgot he was black—seemed pleased:* DeYoung, *Soldier*, 195; Woodward, *The Commanders*, 41–42, 299–301.

258 *itching to cut a diplomatic deal:* Yetiv, *Explaining Foreign Policy*, 117, 128; Haass, *War of Necessity, War of Choice*, 94, 107.

258 *was tentative early on:* Alfonsi, *Circle in the Sand*, 53; George [H. W.] Bush and Brent Scowcroft, *A World Transformed* (New York: Knopf, 1998), 315–18.

258 *telling them not to think:* Richard Morin and E. J. Dionne, "Vox Populi: Winds of War and Shifts of Opinion," *Washington Post*, December 23, 1990, C1; Donna Cassata, Associated Press, "Former Military Leaders Caution Against Rushing to War," November 29, 1990; Freedman and Karsh, *The Gulf Conflict*, 285.

258 *authorizing military force since 1812:* Adam Wolfson, "Humanitarian Hawks? Why Kosovo but Not Kuwait," *Policy* Review 98 (2000): 30; Jeane Kirkpatrick, *Making War to Keep Peace* (New York: Harper, 2007), 25; Max Elbaum, "The Storm at Home," *Crossroads,* April 1991, http://www.revolutionintheair.com/histstrategy/gulf1.html.

258 *"potentially devastating economic consequences":* Zbigniew Brzezinski, interview with Paul Begala and Tucker Carlson, *Crossfire,* CNN, September 21, 1990; Zbigniew Brzezinski, "Patience in the Persian Gulf, Not War," *New York Times*, October 7, 1990, 19.

258 *came out of the woodwork in opposition:* Senate Committee on Armed Services, *Crisis in the Persian Gulf Region: US Policy Options and Implications*, 101st Cong., 2nd sess., December 3, 1990; House Committee on Armed Services, *Crisis in the Persian Gulf: Sanctions, Diplomacy and War*, 101st Cong., 2nd sess., December 20, 1990.

258 *before casting their votes:* Senate Committee on Armed Services, *Crisis in the Persian Gulf Region: US Policy Options and Implications*, 101st Cong., 2nd sess., December 3, 1990.

258 *"the past that will not die":* Evan Thomas, "No Vietnam," *Newsweek,* December 10, 1990, 24.

259 *the average B-52 missed its target by 2,700 feet:* Marshall L. Michel, III, *Eleven Days of Christmas: America's Last Vietnam Battle* (New York: Encounter, 2001), 223.

259 *in the roof of the Iraqi air command:* Mark Clodfelter, "Of Demons, Storms, and Thunder," *Airpower Journal* (1991), http://www.airpower.maxwell.af.mil/airchronicles/apj/apj91/win91/clod.htm.

259 *vulnerable to attacks from the air:* Clodfelter, "Of Demons, Storms, and Thunder."

259 *"immaculate destruction":* David Halberstam, *War in the Time of Peace: Bush, Clinton and the Generals* (New York: Simon & Schuster, 2001), 56. Michael Gordon and Bernard Trainor call the Gulf War "the first war in history in which airpower not ground forces played the dominant role." Gordon and Trainor, *The Generals' War,* x.

259 *America didn't lose a single one:* Alfonsi, *Circle in the Sand,* 171.

259 *the figure was 146:* Robert W. Tucker and David C. Hendrickson, *The Imperial Temptation: The New World Order and America's Purpose* (New York: Council on Foreign Relations Press, 1992), 74; Gordon and Trainor, *The Generals' War,* 457.

259 *America actually turned a profit on the war:* Tucker and Hendrickson, *The Imperial Temptation,* 74.

259 *"the Vietnam syndrome once and for all":* Tucker and Hendrickson, *The Imperial Temptation,* 152.

259 *close to five million people cheered:* DeYoung, *Soldier,* 209.

259 *"the most powerful nation in the world":* Peter Applebome, "After the War: National Mood; War Heals Wounds at Home, but Not All," *New York Times,* March 4, 1991, A1.

260 *"We kicked ass":* Alfonsi, *Circle in the Sand,* 177.

260 *"Nothing is consummated":* Chollet and Goldgeier, *America Between the Wars,* 34.

260 *the figure had more than doubled:* Jentleson, "The Pretty Prudent Public," 67–69.

260 *independent power base:* Alfonsi, *Circle in the Sand,* 192–93.

260 *and depose Saddam itself:* Freedman and Karsh, *The Gulf Conflict,* 403.

260 *what they thought was Bush's command:* Alfonsi, *Circle in the Sand,* 215; Mann, *Rise of the Vulcans,* 193; Freedman and Karsh, *The Gulf Conflict,* 420.

261 *"caution or circumspection as to danger or risk":* "Prudence," Merriam-Webster Online Dictionary, http://www.merriam-webster.com/dictionary/prudence.

261 *to dismember America's old foe:* Bush and Scowcroft, *A World Transformed,* 44, 208, 541.

262 *to run the war through the UN, either:* Alfonsi, *Circle in the Sand,* 101, 152–53.

262 *shooting down Saddam's helicopters:* Gordon and Trainor, *The Generals' War,* 455; Lewis D. Solomon, *Paul D. Wolfowitz: Visionary Intellectual, Policymaker and Strategist* (Westport, CT: Praeger Security International, 2007), 68.

262 *the downfalls of Somoza and the Shah:* Jeane Kirkpatrick, "Our Interests in the Philippines," *Chicago Tribune,* December 15, 1985, 23.

262 *"swallowing too many happiness pills":* Irving Kristol, "Now What for US Client States?" *Wall Street Journal,* March 3, 1986, 18.

262 *Shultz in turn convinced Reagan:* Raymond Bonner, *Waltzing with a Dictator: The Marcoses and the Making of American Policy* (New York: Times Books, 1987), 432; Solomon, *Paul D. Wolfowitz,* 30–37.

262 *"the high point of my career":* Michael Dobbs, "For Wolfowitz, a Vision May Be Realized," *Washington Post,* April 7, 2003, A17.

262 *the most adamant that America come to their aid:* Gordon and Trainor, *The Generals' War,* 451, 455; Mann, *Rise of the Vulcans,* 193; Solomon, *Paul D. Wolfowitz,* 68.

263 *so high you could barely see:* Mann, *Rise of the Vulcans,* 263.

263 *inside his suit pocket:* Dorrien, *Imperial Designs,* 57; Bill Keller, "The Sunshine Warrior," *New York Times Magazine,* September 22, 2002, 48.

263 *"to create these great schemes":* Gates, "Powell and the Black Elite," 73.

263 *that the question was closed:* Gordon and Trainor, *The Generals' War,* 456.

263 *in the Gulf, had not even existed:* Mann, *Rise of the Vulcans,* 89.

264 *above much of Iraq:* Joe Stork and Martha Wenger, "The US in the Persian Gulf: From Rapid Deployment to Massive Deployment," *Middle East Report* 168 (1991): 25; Kenneth Katzman, "The Persian Gulf: Issues for U.S. Policy, 2000," *CRS Report for Congress*, November 3, 2000, 18–24; Multinational Force and Observers, "Assembling the Force," http://www.mfo.org/1/4/22/25/base2.asp, MFO.

264 *A decade later it was 72 percent:* Stockholm International Peace Research Institute, "SIPRI Arms Transfers Database," http://armstrade.sipri.org.

264 *"talks like Dirty Harry but acts like Barney Fife":* Alfonsi, *Circle in the Sand*, 208, 214.

Chapter 14: Fukuyama's Escalator

265 *a banana smeared with peanut butter:* John F. Harris, *The Survivor: Bill Clinton in the White House* (New York: Random House, 2005), 16.

265 *walked in at the stroke of 8:31:* DeYoung, *Soldier*, 186.

265 *anywhere from a few minutes to a few hours late:* Harris, *The Survivor*, 54.

265 *"the first black President":* Toni Morrison, "The Talk of the Town," *New Yorker*, October 5, 1998, 31–32.

265 *expecting his staff to work—until 2 A.M.:* Harris, *The Survivor*, 53.

265 *he asked the army not to inform his wife:* DeYoung, *Soldier*, 81.

266 *as a young boy to be fat:* Harris, *The Survivor*, 35.

266 *if their shoes were shined:* Gates, "Powell and the Black Elite," 64.

266 *whose intake increased with stress:* Harris, *The Survivor*, xxiii, 56–57.

266 *as Aspin ate thirteen hors d'oeuvres:* Bob Woodward, "The Secretary of Analysis," *Washington Post*, February 21, 1993, W8; Halberstam, *War in a Time of Peace*, 245–46.

266 *militarists like Abrams, Wolfowitz, and Cheney:* DeYoung, *Soldier*, 227.

266 *had worked in the McGovern campaign:* David Maraniss, *First in His Class: A Biography of Bill Clinton* (New York: Simon & Schuster, 1996), 264–86; Halberstam, *War in a Time of Peace*, 20.

267 *discontented young diplomats in Vietnam:* Halberstam, *War in a Time of Peace*, 180–81.

267 *Virtually none had joined SDS:* Former SDS president Todd Gitlin, e-mail message to author, June 30, 2008.

267 *"It was a draft beer so he dodged it":* Harris, *The Survivor*, 51.

267 *"pot-smoking, draft-dodging [and] womanizing":* John Lancaster, "Accused of Ridiculing Clinton, General Faces Air Force Probe," *Washington Post*, June 8, 1993, A1.

267 *he backed down:* Harris, *The Survivor*, 16–18; DeYoung, *Soldier*, 230–33.

267 *foreign policy received only 141 words:* Halberstam, *War in a Time of Peace*, 193.

267 *"focus like a laser beam on the economy":* Steven Mufson, "Clinton to Send Message with Economic Choices; Appointments Will Show Administration's Direction," *Washington Post*, November 8, 1992, A33; William Clinton, interview with Ted Koppel, *Nightline*, ABC News, November 4, 1992.

267 *not to take too much of his time:* William G. Hyland, *Clinton's World: Remaking American Foreign Policy* (New York: Praeger, 1999), 18; Halberstam, *War in a Time of Peace*, 242.

267 *Clinton's CIA director trying to get a meeting:* Halberstam, *War in a Time of Peace*, 244; "White House Has Not Been Impenetrable; Security Breached on Many Occasions," *Washington Post*, May 21, 1995, A14.

268 *"bring positive results any time soon":* Chollet and Goldgeier, *America Between the Wars*, 64. For more on the controversy regarding Tarnoff's statements, see Jim Anderson, "The Tarnoff Affair," *American Journalism Review*, March 1994, http://ajr.org/Article.asp?id=1255.

268 *"The Age of Europe has dawned":* Halberstam, *War in a Time of Peace*, 85.

268 *"We don't have a dog in this fight":* Laura Silber and Allan Little, *Yugoslavia: Death of a Nation* (New York: Penguin USA, 1996), 198–201; David C. Gompert, "The United States and Yugoslavia's Wars," in *The World and Yugoslavia's Wars*, ed. Richard H. Ullman (New York: Council on Foreign Relations, 1996), 127–28.

268 *made a fairly clean getaway:* Halberstam, *War in a Time of Peace*, 31.

268 *ultranationalist, anti-Serb bigots:* Silber and Little, *Yugoslavia*, 92–98.

268 *a small taste of things to come:* Halberstam, *War in a Time of Peace*, 96–97.

268 *it was the most vulnerable of all:* Halberstam, *War in a Time of Peace*, 121.

268 *shelling it night after night:* Silber and Little, *Yugoslavia*, 222–28.

269 *burying bodies in public gardens, even backyards:* Ian Traynor, "The Slow but Sure Sacking of Sarajevo," *Guardian*, May 11, 1992, 22; Christine Bertelson, "Escape from Sarajevo, and Death," *St. Louis Post-Dispatch*, January 19, 1993, A1; John Pomfret, "The Hidden Agony of Mostar's Muslims," *Washington Post*, August 22, 1993, A1.

269 *"In Auschwitz at least they had gas":* Halberstam, *War in a Time of Peace*, 123.

269 *to ever assert their independence again:* Halberstam, *War in a Time of Peace*, 129–30; Samantha Power, *A Problem from Hell: America and the Age of Genocide* (New York: Harper Perennial, 2007), 251, 269.

269 *as part of a UN "peacekeeping" force:* UN Department of Public Information, "Background: United Nations Protection Force," September 1996, http://www.un.org/Depts/dpko/dpko/co_mission/unprof_b.htm; Power, *A Problem from Hell*, 248–49; Silber and Little, *Yugoslavia*, 198–99.

269 *preventing the world from doing anything about it:* Halberstam, *War in a Time of Peace*, 125–27; Thomas Weiss, "Collective Spinelessness: U.N. Actions in the Former Yugoslavia," in Ullman, ed., *The World and Yugoslavia's Wars*, 59–96.

269 *striking Serb positions from the air:* Halberstam, *War in a Time of Peace*, 158, 196–97; Hyland, *Clinton's World*, 34.

270 *"the new team in turn regarded him with awe":* Harris, *The Survivor*, 49.

270 *little direct bearing on its security:* Mann, *Rise of the Vulcans*, 118.

270 *Powell replied two hundred thousand ground troops:* Halberstam, *War in a Time of Peace*, 42; Michael R. Gordon, "Powell Delivers a Resounding No on Using Limited Force in Bosnia," *New York Times*, September 28, 1992, A1; Colin L. Powell, "Why Generals Get Nervous," *New York Times*, October 8, 1992, A35; Nancy Soderberg, *The Superpower Myth: The Uses and Misuses of American Power* (New York: Wiley, 2006), 23–27.

270 *taking steps sure to enrage the Serbs:* Halberstam, *War in a Time of Peace*, 227.

270 *"enough respect not to interfere in ours":* Owen Harries, "The Collapse of the West," *Foreign Affairs* 72 (September/October 1993), http://www.foreignaffairs.org/19930901faessay8562/owen-harries/the-collapse-of-the-west.html.

270 *he no longer supported "lift and strike":* Halberstam, *War in a Time of Peace*, 228.

270 *slaughtering each other for five hundred years:* In fact, Kaplan's book was not mostly about Bosnia at all, but about Romania, Bulgaria, Greece, and other parts of the former Yugoslavia. And despite depicting Balkan hatreds as ancient, Kaplan actually supported U.S. military intervention in Bosnia. Robert D. Kaplan, *Balkan Ghosts: A Journey Through History* (New York: Vintage, 1996), ix–xii; Halberstam, *War in a Time of Peace*, 228; Elizabeth Drew, *On the Edge: The Clinton Presidency* (New York: Touchstone, 1995), 157; Paul Starobin, "The Liberal Hawk Soars," *National Journal*, May 15, 1999, 1310.

271 *atrocities committed by Bosnians against Serbs:* Halberstam, *War in a Time of Peace*, 230; Power, *A Problem from Hell*, 308.

271 *"he came back with a European one":* Halberstam, *War in a Time of Peace*, 229.

271 *the incredible shrinking United States:* Harris, *The Survivor*, 62; Michael Duffy, "The Incredible Shrinking President," *Time*, June 29, 1992, http://www.time.com/time/magazine/article/0,9171,975890,00.html.

271 *"to demonstrate that we had a heart":* Western, *Selling Intervention and War*, 133, 135–37, 155, 163, 172; Halberstam, *War in a Time of Peace*, 248–52; Maryann K. Cusimano, "Operation Restore Hope: The Bush Administration's Decision to Intervene in Somalia," Case Study No. 463, Institute for the Study of Diplomacy, 1995, 4–6; Ken Menkhaus and Louis Ortmayer, "Key Decisions in the Somalia Intervention," Case Study No. 464, Institute for the Study of Diplomacy, 1995, 2–8.

272 *inoculations that he canceled his trip:* Halberstam, *War in a Time of Peace*, 252, 254.

272 *which required sidelining Aidid:* Halberstam, *War in a Time of Peace*, 254–57; Menkhaus and

Ortmayer, "Key Decisions in the Somalia Intervention," 13–23; Boutros Boutros-Ghali, *Unvanquished: A U.S.-U.N. Saga* (New York: Random House, 1999), 93, 110.

272 *putting a bounty on Aidid's head:* Halberstam, *War in a Time of Peace,* 257–58.

272 *two meetings with the president:* Halberstam, *War in a Time of Peace,* 244.

272 *at a hotel in downtown Mogadishu:* Mark Bowden, *Black Hawk Down: A Story of Modern War* (New York: Atlantic Monthly Press, 1999), 3, 8.

272 *U.S. soldier's disfigured corpse through the streets:* Halberstam, *War in a Time of Peace,* 261–62; Bowden, *Black Hawk Down,* 108–25.

272 *"Vietmalia":* Richard Holbrooke, *To End a War* (New York: Modern Library, 1998), 217.

273 *Congress would cut off funding:* Harris, *The Survivor,* 122; Michael Kranish, "Clinton Builds Force, Sets Pullout; Says Somalia Withdrawal Can't Be Instant," *Boston Globe,* October 8, 1993, A1.

273 *he was unfit to be commander in chief:* Chollet and Goldgeier, *America Between the Wars,* 79.

273 *Critics called it Somalia Two:* Halberstam, *War in a Time of Peace,* 270–71; Harris, *The Survivor,* 123.

274 *lay down on a park bench, and tried to die:* Power, *A Problem from Hell,* 273–77, 388–89.

274 *twice already on the same trip:* Halberstam, *War in a Time of Peace,* 277; James Bennet, "Clinton in Africa: The Overview; Clinton Declares U.S., with World, Failed Rwandans," *New York Times,* March 26, 1998, A1; John F. Harris, "Clinton Tells Rwandans: World Too Slow to Act; Genocide Survivors Give Accounts of '94 Horror," *Washington Post,* March 26, 1998, A1.

274 *an intervention never carried out:* Virginia Heffernan, "Looking Back Across a Decade, with Bloody Regret," *New York Times,* April 1, 2004, E1.

275 *shelter, health care, and education:* Richard A. Falk, "Ideological Patterns in the United States Human Rights Debate: 1945–1978," in *The Dynamics of Human Rights in U.S. Foreign Policy,* ed. Natalie Kaufman Hevener (New Brunswick, NJ: Transaction, 1981), 43.

275 *sensitivity to that in the State Department:* Muravchik, *The Uncertain Crusade,* 92.

276 *"it's Spain":* Todd Gitlin, "Bosnia Isn't Vietnam, It's Spain," *Los Angeles Times,* September 14, 1993, F7.

277 *Cambodia's seven million people:* Power, *A Problem from Hell,* 143.

277 *the cause of human rights:* A surprising and honorable exception was George McGovern. Power, *A Problem from Hell,* 132–36.

277 *"with our military assistance":* Power, *A Problem from Hell,* 103.

278 *NATO should begin bombing immediately:* Leslie Gelb, "Balkan Strategy, Part II," *New York Times,* February 28, 1993, 15.

278 *including to Somalia, Bosnia, and Rwanda:* In the four decades between the UN's creation and 1991, the organization authorized eighteen peacekeeping missions. Between 1991 and 1995, it authorized twenty-two. In 1992 alone, the number of UN peacekeepers worldwide more than quadrupled, from 11,000 to 52,000. United Nations, "List of Operations, 1948–2009," http://www.un.org/Depts/dpko/dpko/list.shtml; Marrack Goulding, "The Evolution of United Nations Peacekeeping," *International Affairs* 69, no. 3 (1993): 451.

278 *about a standing UN army:* Harris, *The Survivor,* 125.

279 *rape of Bihac went on:* Halberstam, *War in a Time of Peace,* 284–85.

279 *because they were not Christian:* Halberstam, *War in a Time of Peace,* 90–91, 304–5.

279 *"distinct chapters in the history of decency":* Leon Wieseltier, "Curses," *New Republic,* October 25, 1993, 46.

280 *were leading the president astray:* Drew, *On the Edge,* 158.

280 *coverage of the Holocaust, was relentless:* Starobin, "The Liberal Hawk Soars," 1314.

280 *after reading one* Times *column:* George Stephanopoulos, *All Too Human* (Boston: Back Bay, 2000), 216.

280 *even the Quakers were for war:* Power, *A Problem from Hell,* 428, 435.

280 *broke into applause:* Harris, *The Survivor,* 42–43.

280 *the UN's Bosnia mission would collapse:* Michael Dobbs, "Embargo Vote Is Bipartisan Slap at Clinton; Dismay over Bosnia Reflects Setbacks," *Washington Post,* June 9, 1995, A22.

280 *"'the good guys' and 'the bad guys'":* Power, *A Problem from Hell,* 394, 417; David Rohde,

Endgame: The Betrayal and Fall of Srebrenica, Europe's Worst Massacre Since World War II (Boulder, CO: Westview, 1998), 326.

281 *"the president is relevant":* Harris, *The Survivor*, 178.

281 *would soon endorse Dole's presidential bid:* Halberstam, *War in a Time of Peace*, 285–86, 299.

281 *leader of the free world was "vacant":* Hyland, *Clinton's World*, 40.

281 *still writing his inaugural address:* Harris, *The Survivor*, xxiv, 11.

281 *"the faster I come back up":* Harris, *The Survivor*, 334.

282 *bomb the hell out of them:* Power, *A Problem from Hell*, 438; Halberstam, *War in a Time of Peace*, 340; Harris, *The Survivor*, 194, 196, 201.

282 *as part of the Clinton team:* Halberstam, *War in a Time of Peace*, 319, 324–28; Harris, *The Survivor*, 197.

282 *tell you how easy it was:* Halberstam, *War in a Time of Peace*, 391, 431–32.

283 *about the people Milosevic killed:* Halberstam, *War in a Time of Peace*, 395.

283 *a new willingness to discuss peace:* Halberstam, *War in a Time of Peace*, 339.

283 *their colleagues plunged to their death:* Power, *A Problem from Hell*, 294; Halberstam, *War in a Time of Peace*, 395; Holbrooke, *To End a War*, 6–13, 231–312.

283 *U.S. peacekeepers would help enforce:* Holbrooke, *To End a War*, 308–9.

284 *not a single American died:* Chollet and Goldgeier, *America Between the Wars*, 131.

284 *"The big dog barked today":* Halberstam, *War in a Time of Peace*, 340.

284 *the first secretary-general ever denied a second term:* United Nations, "Former Secretaries-General," http://www.un.org/sg/formersgs.shtml.

284 *this new institutional order:* James M. Goldgeier, *Not Whether but When: The U.S. Decision to Enlarge NATO* (Washington, DC: Brookings Institution Press, 1999), 3, 17.

285 *for NATO to stay off former Soviet soil:* Jonathan Eyal, "NATO's Enlargement: Anatomy of a Decision," *International Affairs* 73, no. 4 (1997): 709; Mark Kramer, "The Myth of a No-NATO-Enlargement Pledge to Russia," *Washington Quarterly* 32, no. 2 (2009): 39–61; Goldgeier, *Not Whether but When*, 73, 110–17, 120–21, 135–38, 140–43; Strobe Talbott, interview with author, July 10, 2008.

285 *left their bodies in the snow:* R. Jeffery Smith, "This Time, Walker Wasn't Speechless," *Washington Post*, January 22, 1999, A15.

286 *Moscow took Czechoslovakia in its grip:* Halberstam, *War in a Time of Peace*, 378–79.

286 *when she was only one year old:* Chollet and Goldgeier, *America Between the Wars*, 146.

286 *after the conference closed, the bombing began:* Ivo H. Daalder and Michael E. O'Hanlon, *Winning Ugly: NATO's War to Save Kosovo* (Washington, DC: Brookings Institution Press, 2000), 77–84; Power, *A Problem from Hell*, 447–48; Madeleine Albright, *Madam Secretary* (New York: Miramax, 2005), 505–18.

287 *fear of U.S. casualties limited its scope:* Daalder and O'Hanlon, *Winning Ugly*, 106, 122.

287 *than during the Gulf War:* Stuart Croft, "Guaranteeing Europe's Security? Enlarging NATO Again," *International Affairs* 78, no. 1 (2002): 98.

287 *"war can be won by airpower alone":* Halberstam, *War in a Time of Peace*, 471–74, 478; Power, *A Problem from Hell*, 456; Daalder and O'Hanlon, *Winning Ugly*, 200–2.

287 *two NATO planes had been shot down:* Power, *A Problem from Hell*, 459.

287 *Not a single American died in combat:* Chollet and Goldgeier, *America Between the Wars*, 214.

287 *took over Kosovo's political administration:* North Atlantic Treaty Organization, "Resolution 1244 (1999)," adopted by UN Security Council, June 10, 1999, http://www.nato.int/Kosovo/docu/u990610a.htm.

288 *the verge of financial default:* Chollet and Goldgeier, *America Between the Wars*, 166, 229.

288 *internally displaced from their villages:* Daalder and O'Hanlon, *Winning Ugly*, 40–41, 108–9.

289 *half the rate of the United States:* World Bank Development Data, http://devdata.worldbank.org/data-query.

289 *at a phenomenal 8 percent per year:* Harris, *The Survivor*, 263; President William J. Clinton, "President Clinton Addresses the Citizens of Colorado on the Global Economy," Littleton, Colorado, June 19, 1997.

289 *began to improve dramatically:* Henry J. Kaiser Family Foundation, "U.S. Teen Sexual Activ-

ity Factsheet," January 2005, http://www.kff.org/youthhivstds/upload/U-S-Teen-Sexual-Activity-Fact-Sheet.pdf; Jeffrey Grogger, Stephen J. Haider, and Jacob Klerman, "Why Did the Welfare Rolls Fall During the 1990s?" Labor and Population Program Working Paper Series 03–07, Santa Monica, California, Rand Corporation, March 2003, 1; U.S. Department of Justice, "Crime Trends," Bureau of Justice Statistics, http://bjsdata.ojp.usdoj.gov/dataonline/Search/Crime/Crime.cfm.

289 *the psychological effect was comparable:* Harris, *The Survivor,* 263.

289 *"single superpower on the cheap is astonishing":* Paul Kennedy, "The Eagle Has Landed," *Financial Times,* February 2, 2002.

290 *"The Committee to Save the World":* Joshua Cooper Ramo, "The Three Marketeers," *Time,* September 15, 1999, http://www.time.com/time/magazine/article/0,9171,990206,00.html.

291 *"sharp limits to American resources and patience":* Francis Fukuyama, "The Beginning of Foreign Policy," *New Republic,* August 17, 1992, 24–32.

291 *that demanded political reform:* Francis Fukuyama, "The Global Optimists," *New Republic,* February 6, 1995, 40–42.

292 *"so few external threats":* Chollet and Goldgeier, *America Between the Wars,* 278.

292 *predicted Christ's return to earth:* Pew Research Center for the People & the Press, "Optimism Reigns, Technology Plays a Key Role," October 24, 1999, http://people-press.org/report/51/optimism-reigns-technology-plays-key-role.

292 *the last time anything seemed possible:* Harris, *The Survivor,* 393.

292 *progress and the goodness of man:* David Brooks, "The Organization Kid," *Atlantic,* April 2001, http://www.theatlantic.com/doc/print/200104/brooks.

Chapter 15: Fathers and Sons

293 *his fourth autobiographical essay:* Kristol, "An Autobiographical Memoir." The first three were "Memoirs of a Trotskyist"; "Memoirs of a Cold Warrior," *New York Times Magazine,* February 11, 1968; and "My Cold War," *National Interest* 31 (1993): 141.

293 *publish his fourth autobiographical book:* Podhoretz's four autobiographies are *Making It* (New York: Random House, 1967), *Breaking Ranks* (New York: Harper & Row, 1979), *Ex-Friends* (New York: Free Press 1999), and *My Love Affair with America* (New York: Free Press, 2001).

293 *"as well as liberals and radicals":* Kristol, "An Autobiographical Memoir," 40.

294 *Stalinist Eastern Europe and Maoist China:* Hoeveler, *Watch on the Right,* 82–83, 153.

294 *"Sorry! It's the real world":* Jacquelyn Hardy, interview with the author, July 10, 2008.

295 *"frankly, proudly, a revolutionary power":* Jeane Kirkpatrick, "The Search for a Stable World Order," in *Legitimacy and Force: Political and Moral Dimensions,* ed. Jeane Kirkpatrick, vol. 1 (New Brunswick, NJ: Transaction, 1988), 367.

295 *"for limited, defensive purposes":* Jeane Kirkpatrick, "Moral Equivalence," in *Legitimacy and Force,* ed. Kirkpatrick, 70.

295 *"a normal country in a normal time":* Jeane Kirkpatrick, "A Normal Country in a Normal Time," *National Interest* 21 (1990): 40–45.

295 *"continue to live beyond our means":* Kirkpatrick, "A Normal Country in a Normal Time," 40–45; Jeane Kirkpatrick, "Counseling the President," *National Review,* February 10, 1989, 28.

295 *preparing for a multipolar world:* Kirkpatrick, *The Withering Away of the Totalitarian State,* 165; Kirkpatrick, "A Normal Country in a Normal Time," 40–43; Jeane Kirkpatrick, "Marriage of Convenience for the New Europe," *Financial Post,* July 10, 1990, 9; Jeane Kirkpatrick, "An 'Alice in Wonderland' Defense Budget," *Washington Post,* April 5, 1993, A21; Irving Kristol, "Defining Our National Interest," *National Interest* 21 (1990): 23.

296 *"people violently reject any such scenario":* Kristol, "Defining Our National Interest," 23.

296 *Lithuania's fledgling democracy: Does "the West" Still Exist? A Conference of the Committee for the Free World* (New York: Orwell, 1990), 39.

296 *"which Americans need to unlearn":* Kirkpatrick, "A Normal Country in a Normal Time," 43; Ralph Z. Hallow, "Neoconservatives Meet in Search of Common Ground," *Washington Times,* April 24, 1990, A3.

296 *"Something will screw it up": Does "the West" Still Exist,* 109, 117; Kristol, "In Search of Our National Interest," *Wall Street Journal,* June 7, 1990, A14.

296 *"serious economic decline":* Jeane Kirkpatrick, "Ultimate Responsibility Falls on Arab Shoulders," *Financial Post,* August 16, 1990, 11.

296 *giving sanctions time to work:* Jeane Kirkpatrick, "Gulf Crisis Proves That the World Needs a Policeman," *Financial Post,* August 28, 1990, 9.

297 *merely bomb from the air:* Jeane Kirkpatrick, "Will We Liberate Kuwait . . . ," *Washington Post,* November 12, 1990, A19.

297 *he supported an American war:* Irving Kristol, "The Gulf: Born-Again Isolationists," *Washington Post,* August 22, 1990, A21.

297 *"people who are really different from us":* Irving Kristol, "After the War, What?" *Wall Street Journal,* February 22, 1991, A10; Irving Kristol, "Tongue-Tied in Washington," *Wall Street Journal Europe,* April 15, 1991, A14.

297 *"vital American interests and lives at stake":* Jeane Kirkpatrick, "Haiti's Looking Like Another Clinton Mistake," *Post and Courier,* March 14, 1995, A11.

297 *then ignored the subject:* Kristol, "Defining our National Interest," 21; Ehrman, *The Rise of Neoconservatism,* 184; Halper and Clarke, *America Alone,* 99.

297 *putting U.S. peacekeepers on the ground:* Jeane Kirkpatrick, "U.S. Must Retaliate Strongly Against Serbs," *Post and Courier,* June 11, 1995, A21; Jeane Kirkpatrick, "U.S. Needs Clear Goals in Bosnia," *Post and Courier,* November 19, 1995, A23.

298 *and the columnist George Will:* Buckley defined his foreign policy vision as "measured internationalism" or "prudent isolationism." William F. Buckley, "American Power—For What? A Symposium," *Commentary* 109, no. 1 (2000): 23–24. In 1991, Will wrote that "concerning the question of U.S. military intervention in Iraq's civil war, President Bush has the traditional conservative's wariness about uncertain undertakings, a prudent skepticism about the promiscuous minting of abstract rights and duties, and an inclination to anchor U.S. policy in the rock of U.S. national interests. Today's imperial conservatives consider such thinking crabbed, mean-spirited and (adopting the language of liberal sensitivity-mongers) 'insensitive.' They want America to do for the world what Lyndon Johnson's Great Society was supposed to for America: fix it." George Will, "Conservative Factions Debate What's Best for Kurds," *Seattle Post-Intelligencer,* April 18, 1991, A13. Will's own writing left little doubt that he was on Bush's side.

298 *"expansive Wilsonian interventionism":* Norman Podhoretz, "Neoconservatism: A Eulogy," *Commentary* 101, no. 3 (1996): 24.

298 *"general reluctance to crusade":* Ehrman, *The Rise of Neoconservatism,* 184, 206.

298 *the younger conservative generation fired its first shot:* Charles Krauthammer, "The Unipolar Moment," *Foreign Affairs* 70, no. 1 (1990): 29.

298 *young enough to be Kirkpatrick's son:* Dorrien, *Imperial Designs,* 81; Winik, *On the Brink,* 457.

299 *"if necessary, disarm" the weapons states:* Charles Krauthammer, "The Lonely Superpower," *New Republic,* July 29, 1991, 27; Krauthammer, "The Unipolar Moment," 29–33.

300 *"is chaos":* Krauthammer, "The Unipolar Moment," 32.

300 *"Otherwise they would rather stay home":* Krauthammer, "The Lonely Superpower," 27.

302 *running the Gulf War through the UN:* Mann, *Rise of the Vulcans,* 201–2; Woodward, *The Commanders,* 106; Dorrien, *Imperial Designs,* 28; Alfonsi, *Circle in the Sand,* 101.

302 *shooting down Saddam's helicopters:* Dorrien, *Imperial Designs,* 32; Gordon and Trainor, *The Generals' War,* 455.

303 *"our role in the world":* Mann, *Rise of the Vulcans,* 208–13; Dorrien, *Imperial Designs,* 38–43; Patrick E. Tyler, "US Strategy Calls for Insuring No Rivals Develop," *New York Times,* March 8, 1992, A1; Barton Gellman, "Keeping US First: Pentagon Would Preclude a Rival Superpower," *Washington Post,* March 11, 1992, A1; Patrick E. Tyler, "Senior US Officials Assail Lone-Superpower Policy," *New York Times,* March 11, 1992, A6; Barton Gellman, "Aim of Defense Plan Supported by Bush: But President Says He Has Not Read Memo," *Washington Post,* March 12, 1992, A18.

303 *the Defense Planning Guidance:* Charles Krauthammer, "What's Wrong with the 'Pentagon Paper'?" *Washington Post,* March 13, 1992, A25.

303 *Robert Kagan and William Kristol:* Dorrien, *Imperial Designs,* 139–40; Robert Kagan, "American Power—A Guide for the Perplexed," *Commentary* 101, no. 4 (1996): 30.

303 *after America's triumph in Bosnia:* Kagan did write two essays in *Commentary* in 1994 and 1995 ("The Case for Global Activism," September 1994, 40–44; and "A Retreat from Power," 19–25) that previewed his globalist worldview, but neither got nearly as much attention as his *Foreign Affairs* essay with William Kristol in the summer of 1996, "Toward a Neo-Reaganite Foreign Policy," *Foreign Affairs* 75, no. 4 (1996), http://www.carnegie endowment.org/publications/index.cfm?fa=view&id=276.

303 *"blood and treasure on teacup wars":* Charles Krauthammer, "American Power—For What? A Symposium," *Commentary,* January 2000, 34.

303 *strategically important battles to come:* Robert Kagan, "America, Bosnia, Europe: A Compelling Interest," *Weekly Standard,* November 6, 1995, 27.

304 *"understood it was all a game":* John Podhoretz, *Hell of a Ride: Backstage at the White House Follies, 1989–1993* (New York: Simon & Schuster, 1993), 114.

304 *on foreign policy, Republicans won:* Unnamed Washington writer, interview with the author, August 7, 2009.

304 *her protégé, became his mentor:* Kagan, *A Twilight Struggle,* x.

304 *"the struggle with communism was endless":* Kagan, *A Twilight Struggle,* 55–77, 212.

305 *decades had proved that it could:* Robert Kagan, "Democracy and Double Standards," *Commentary* 104, no. 2 (1997): 19–26.

305 *in speeding history up:* Ken Jowitt, "Rage, Hubris, and Regime Change," *Policy Review* 118 (2003): 33–42.

305 *"others assumed were fixed":* Kagan and Kristol, "Toward a Neo-Reaganite Foreign Policy."

305 *bring down the Soviet empire:* Kagan, "American Power—A Guide for the Perplexed," 26.

305 *before Gorbachev even took power:* Norman Podhoretz, "The First Term: The Reagan Road to Détente," *Foreign Affairs* 63, no. 3 (1984): 447–64.

306 *without fearing U.S. aggression:* See Farnham, "Reagan and the Gorbachev Revolution," 225–52; Leffler, *For the Soul of Mankind,* 436; Suri, "Explaining the End of the Cold War," 60–92.

306 Present Danger, *which had influenced Reagan:* Norman Podhoretz, *The Present Danger: Do We Have the Will to Reverse the Decline of American Power?* (New York: Simon & Schuster, 1980); William Kristol and Robert Kagan, eds., *Present Dangers: Crisis and Opportunity in American Foreign and Defense Policy* (San Francisco: Encounter, 2000).

306 *"another Reagan":* William Kristol, "Reagan's Greatness," *Weekly Standard,* November 10, 1997, 31.

306 *the* Weekly Standard, *in 1998:* William Kristol, "Clinton's Feckless Foreign Policy," *Weekly Standard,* May 25, 1991, 11.

307 *"is reminiscent of the mid-1970's":* Kagan and Kristol, "Toward a Neo-Reaganite Foreign Policy."

307 *"does not add up to a brain tumor":* Owen Harries, "American Power—For What? A Symposium," *Commentary* 109, no. 1 (2000): 28–29.

307 *"faces now is its own weakness":* Kagan and Kristol, "Toward a Neo-Reaganite Foreign Policy."

307 *"benevolent global hegemony":* Kagan and Kristol, "Toward a Neo-Reaganite Foreign Policy."

307 *what they considered possible:* Mann, *Rise of the Vulcans,* 215.

308 Present Dangers, *as did Wolfowitz:* Elliott Abrams wrote a chapter titled "Israel and the 'Peace Process." Wolfowitz's was called "Statesmanship in the New Century." Abrams called himself a "neo-Reaganite" in his contribution to "American Power—For What? A Symposium," *Commentary* 109, no. 1 (2000): 21.

308 *Donald Rumsfeld, and Dick Cheney:* Dorrien, *Imperial Designs,* 130.

308 *"complete victory" in the Balkans:* "Letter to President Clinton," Project for the New American

Century, September 11, 1998, http://www.newamericancentury.org/kosovomilosevicsep98
.htm; William Kristol and Robert Kagan, "Victory," *Weekly Standard*, June 14, 1999.

309 *this alleged malaise:* David Brooks, "Politics and Patriotism: From Teddy Roosevelt to John
McCain," *Weekly Standard*, April 26, 1999, http://www.weeklystandard.com/check
.asp?idArticle=10411&r=onujh.

309 *"the narrower concerns of private life":* David Brooks, "A Return to National Greatness: A
Manifesto for a Lost Creed," *Weekly Standard*, March 3, 1997, 16.

309 *"holiday from history":* Schlesinger, "The New Mood in Politics," in *The Politics of Hope and
the Bitter Heritage*, 109; Charles Krauthammer, "Holiday from History," *Washington Post*,
February 14, 2003, A31.

309 *"post-greatness America":* Schlesinger, "The Decline of Greatness," in *The Politics of Hope and
the Bitter Heritage*, 37; Brooks, "A Return to National Greatness," 16.

309 *inspired JFK's fifty-mile hikes:* Roosevelt used the phrase "national greatness" in an essay titled
"National Life and Character," which was published in the collection *American Ideals* (New
York: G. P. Putnam's Sons, 1920), 268.

309 *grand public mission at all:* Brooks, "A Return to National Greatness," 16.

309 *America should journey to Mars:* Charles Krauthammer, "On to Mars: America Has Been Lost
in Space; It's Time to Find Our Nerve Again," *Weekly Standard*, January 31, 2000, 23.

309 *"except as public relations":* Jonah Goldberg, "Grading Greatness," *National Review Online*,
May 21, 2001, http://www.nationalreview.com/goldberg/goldbergprint052101.html.

310 *"the already soft edges of boomer life":* Noemie Emery, "Ask Not: John McCain Belongs to Our
Oldest Political Party—the Party of Teddy Roosevelt and FDR, JFK and Ronald Reagan—
the Patriot Party," *Weekly Standard*, February 21, 2000, http://www.weeklystandard.com/
check.asp?idArticle=10542&r=ivvnb.

310 *"sophisticated consumer demands":* Francis Fukuyama, "The End of History?" *National Interest*
16 (1989): 18.

311 *"Americans will die doing the right thing":* Jonah Goldberg, "A Continent Bleeds: Taking
Africa—and Our Responsibilities—Seriously," *National Review Online* (May 3, 2000),
http://article.nationalreview.com/?q=YmMzMDA0MWVhN2JjY2RhN2E4NmVkNTU1
ZjJjMDU0MzQ.

Chapter 16: Small Ball

312 *"in one or two sentences":* Norman Kempster, "Vietnam War Leaves Legacy of Anguish; Still
Overshadows Lives, U.S. Policies," *Los Angeles Times*, April 28, 1985, 1.

313 *be drawn into war with Iran:* Matt Bai, "The McCain Doctrines," *New York Times Magazine*, May
18, 2008, http://www.nytimes.com/2008/05/18/magazine/18mccain-t.html?ref=magazine.

313 *"trading American blood for Iraqi blood":* Jacob Weisberg, "Gulfballs," *New Republic*, March
25, 1991, 19.

313 *"another failure like Vietnam or Lebanon":* Kagan, "A Retreat from Power," 20.

313 *he grew talons and became a hawk:* John B. Judis, "Neo-McCain," *New Republic*, October 16,
2006, http://www.tnr.com/article/politics/neo-mccain.

313 *"21st century . . . Reagan Doctrine":* Brooks, "Politics and Patriotism"; David D. Kirkpatrick,
"Response to 9/11 Offers Outline of McCain Doctrine," *New York Times*, August 17, 2008,
http://www.nytimes.com/2008/08/17/us/politics/17mccain.html.

313 *a conservative magazine, JFK:* David Brooks and William Kristol, "The McCain Insurrec-
tion; The Republican Establishment and the Conservative Movement Rallied to George
W. Bush. The Voters Went for the Insurgent," *Weekly Standard*, February 14, 2000, http://
www.weeklystandard.com/check.asp?idArticle=10561&r=jyqrp; Noemie Emery, "Ask Not";
David Brooks, "The Anti-Boomer Candidate," *Weekly Standard*, February 21, 2000, http://
www.weeklystandard.com/check.asp?idArticle=10534&r=ghwet.

314 *singles complex called Chateau Dijon:* Jo Thomas, "Governor Bush's Journey; After Yale, Bush
Ambled Amiably into His Future," *New York Times*, July 22, 2000, A1; Weisberg, *The Bush*

314 *heroism of his greatest-generation dad:* David Greenberg, "Fathers and Sons: George W. Bush and His Forebears," *New Yorker,* July 12, 2004, http://www.newyorker.com/archive/2004/07/12/040712crbo_books.

314 *almost flunked out:* Nicholas Lemann, "The Redemption," *New Yorker,* January 31, 2000; Weisberg, *The Bush Tragedy,* 38–40.

314 *whether they still enjoyed sex:* Todd S. Purdum, "43+41=84," *Vanity Fair,* September 2006, http://www.vanityfair.com/politics/features/2006/09/bushes200609.

314 *an intramural stickball league:* Lou Cannon and Carl M. Cannon, *Reagan's Disciple: George W. Bush's Troubled Quest for a Presidential Legacy* (New York: PublicAffairs, 2008), 4, 5, 19; Peter Schweizer and Rochelle Schweizer, *The Bushes: Portrait of a Dynasty* (New York: Anchor, 2005), 153.

314 *his younger siblings in the car:* Weisberg, *The Bush Tragedy,* 46–47.

314 *his fiancée broke off the engagement:* Kevin Phillips, *American Dynasty: Aristocracy, Fortune and the Politics of Deceit in the House of Bush* (New York: Viking, 2004), 44.

314 *but it went belly-up:* Weisberg, *The Bush Tragedy,* xvi, 32, 34, 49; Schweizer and Schweizer, *The Bushes,* 140.

314 *his father confessed himself "disappointed":* Weisberg, *The Bush Tragedy,* 47, 49.

314 *as the family's black sheep:* Gail Sheehy, "The Accidental Candidate," *Vanity Fair,* October 2000, http://www.vanityfair.com/politics/features/2000/10/bush200010.

315 *she wouldn't get into heaven:* Weisberg, *The Bush Tragedy,* 50–52, 79–80, 83–84.

315 *in his own mind—he excelled:* Schweizer and Schweizer, *The Bushes,* 388.

315 *"the chance to watch the ball":* Schweizer and Schweizer, *The Bushes,* 133, 389.

315 *hugged the bottom of their division:* "Harold Baines," http://www.baseball-reference.com/b/baineha01.shtml; "Sammy Sosa," http://www.baseball-reference.com/s/sosasa01.shtml; "Rangers Year-by-Year Results," http://texas.rangers.mlb.com/tex/history/year_by_year_results.jsp.

315 *no rousing agenda for a second term:* Schweizer and Schweizer, *The Bushes,* 458, 504.

316 *learned from his father's defeat:* Bob Woodward, *Bush at War* (New York: Simon & Schuster, 2003), 341.

316 *was seeking Florida's top job:* Schweizer and Schweizer, *The Bushes,* 413; Weisberg, *The Bush Tragedy,* 62.

316 *state's outsize image of itself:* Sam Howe Verhovek, "Texas Governor Succeeds, Without the Flash," *New York Times,* June 14, 1995.

316 *Germany was a member of NATO:* Mann, *Rise of the Vulcans,* 255; Weisberg, *The Bush Tragedy,* 146.

316 *"Prosperity with a Purpose":* Lemann, "The Redemption."

316 *"good times for great goals":* "GOP Platform 2: Prosperity with a Purpose," ABC News, July 31, 2000, http://abcnews.go.com/Politics/story?id=123292&page=1&page=1.

316 *wasted his presidency pushing paper:* Weisberg, *The Bush Tragedy,* 65, 105.

316 *not play "small ball":* Michael Gerson, *Heroic Conservatism: Why Conservatives Need to Embrace America's Ideals (and Why They Deserve to Fail If They Don't)* (New York: Harper-Collins, 2007), 42–43.

316 *"game changer":* Ron Suskind, *The Price of Loyalty: George W. Bush, the White House, and the Education of Paul O'Neill* (New York: Simon & Schuster, 2003), 299.

317 *first foreign policy address of the campaign:* George W. Bush, "A Period of Consequence," address given at the Citadel, South Carolina, September 23, 1999, http://www.citadel.edu/r3/pao/addresses/pres_bush.html; Terry M. Neal, "Bush Outlines Defense Plan in Address at Citadel," *Washington Post,* September 24, 1999, A3.

317 *and the moment—at all:* Frank Bruni, *Ambling into History: The Unlikely Odyssey of George W. Bush* (New York: HarperCollins, 2002); Gerson, *Heroic Conservatism,* 80.

318 *"over-ruling Mr. Powell on any issue":* DeYoung, *Soldier,* 282, 287–88, 295, 297–98; Halberstam, *War in a Time of Peace,* 232, 238.

318 *regent to a dutiful boy king:* DeYoung, *Soldier,* 286–88, 301–2; Mann, *Rise of the Vulcans,* 160.

318 *except in suit and tie:* Ron Suskind, *The One Percent Doctrine: Deep Inside America's Pursuit*

of Its Enemies Since 9/11 (New York: Simon & Schuster, 2006), 151; Purdum, "43+41=84"; Schweizer and Schweizer, *The Bushes,* 501.

318 *Republicans back in charge:* DeYoung, *Soldier,* 314.

318 *when he was on the road:* Barton Gellman, *The Angler: The Cheney Vice Presidency* (New York: Penguin, 2008), 54; Bob Woodward, *Plan of Attack* (New York: Simon & Schuster, 2004), 176.

319 *in other words, to stack the deck:* Gellman, *The Angler,* 244–45; Walter Pincus, "Under Bush, the Briefing Gets Briefer," *Washington Post,* May 24, 2002, A33.

319 *called the vice president's "mole":* Gellman, *The Angler,* 32–33, 36, 38, 189. In the first Bush administration, Libby was the Pentagon's principal deputy undersecretary for strategy and resources and later deputy undersecretary of defense for policy, positions that ranked higher in the interagency process than director of the National Security Council, which was Rice's job. Gellman, *The Angler,* 43–44.

319 *a longtime Cheney associate:* Hayes, *Cheney,* 301.

319 *"most ruthless" government official he had ever met:* Roger Morris, "The Undertaker's Tally: Sharp Elbows," *Tom Dispatch,* February 13, 2007, http://www.tomdispatch.com/index print.mhtml?pid=165669; Bradley Graham, *By His Own Rules: The Ambitions, Successes, and Ultimate Failures of Donald Rumsfeld* (New York: PublicAffairs, 2009), 201.

319 *mentor to both Hadley and Libby:* Gordon and Trainor, *Cobra II,* 169; Gellman, *The Angler,* 42.

319 *thirty-six-year-old daughter, Elizabeth:* DeYoung, *Soldier,* 320; Gellman, *The Angler,* 37.

319 *was working to overthrow him:* DeYoung, *Soldier,* 314–17.

320 *"everything in the region and beyond it":* Suskind, *The Price of Loyalty,* 85, 96–97; DeYoung, *Soldier,* 345; Gordon and Trainor, *Cobra II,* 16–17.

320 *a new "Reagan doctrine":* Paul Wolfowitz, "Historical Memory: Setting History Straight," *Current,* June 2000, 22; Paul Wolfowitz, "American Power—For What? A Symposium," *Commentary* 109, no. 1 (2000): 47; Paul Wolfowitz, "Statesmanship in the New Century," in *Present Dangers,* ed. Kagan and Kristol, 323.

320 *held its first free presidential election:* Solomon, *Paul D. Wolfowitz,* 39.

321 *"history of democratic rule":* Paul Wolfowitz, "Is the Atlantic Community Obsolete?" in *The Congress of Phoenix: Rethinking Atlantic Security and Economics,* ed. Gerald Frost (Washington, DC: AEI Press, 1998), http://www.aei.org/docLib/20040217_book35.pdf.

321 *dissidents like Vaclav Havel:* Jane Mayer, "The Manipulator," *New Yorker,* June 7, 2004, http://www.newyorker.com/archive/2004/06/07/040607fa_fact1; George Packer, *The Assassin's Gate: America in Iraq* (New York: Farrar, Straus & Giroux, 2006), 12.

321 *a game changer nonetheless:* Michael Dobbs, "For Wolfowitz, a Vision May Be Realized," *Washington Post,* April 7, 2003, A17; Peter J. Boyer, "The Believer: Paul Wolfowitz Defends His War," *New Yorker,* November 1, 2004, http://www.newyorker.com/archive/2004/11/01/041101fa_fact.

322 *through the barrel of an American gun:* While the Contras may have played some role in the decision by Nicaragua's Sandinista regime to hold the free elections that led to their losing power in 1990, the much larger factor was the Soviet Union's collapse, which left the Sandinistas—like so many other third-world communist regimes—orphaned.

322 *"start at the beginning":* DeYoung, *Soldier,* 426.

322 *making incremental change:* As Douglas Feith notes, "Powell presented himself as the practical man of affairs, taking the world as he found it, focused on the here and now, intent on getting organized for the next day's set of meetings on whatever crisis was at hand." Douglas J. Feith, *War and Decision: Inside the Pentagon at the Dawn of the War on Terrorism* (New York: Harper, 2008), 60.

322 *"slowly but surely, layer by layer":* DeYoung, *Soldier,* 487.

323 *"broken, weak country":* DeYoung, *Soldier,* 305, 345; Woodward, *Plan of Attack,* 22.

323 *no threat to his neighbors:* Thomas E. Ricks, *Fiasco: The American Military Adventure in Iraq* (New York: Penguin, 2006), 27.

323 *the big game was Colin Powell:* DeYoung, *Soldier,* 323–25.

323 *he was not always in the room:* DeYoung, *Soldier,* 329.

323 *"little far forward on your skis":* Mann, *Rise of the Vulcans*, 278.

324 *to trumpet his achievement:* Woodward, *Bush at War*, 13.

324 *Cheney and Rumsfeld disapproved:* Robert Kagan and William Kristol, "A National Humiliation," *Weekly Standard*, April 16, 2001, http://www.weeklystandard.com/check .asp?idArticle=11457&r=elzif; Robert Kagan and William Kristol, "The 'Adults' Make a Mess," *Weekly Standard*, May 14, 2001, http://www.weeklystandard.com/check .asp?idArticle=583&r=oglxj.

324 *keeping the United States out of a Pacific war:* Bob Woodward, *State of Denial* (New York: Simon & Schuster, 2007), 33.

324 *might well prompt Beijing to strike:* Solomon, *Paul D. Wolfowitz*, 48; Graham, *By His Own Rules*, 679; Andrew Cockburn, *Rumsfeld: His Rise, Fall and Catastrophic Legacy* (New York: Scribner, 2007), 140.

324 *reaffirm Bush's original remark:* Kagan and Kristol, "The 'Adults' Make a Mess"; "Interview with Dick Cheney," *Fox News Sunday*, Fox News Network, April 29, 2001.

324 *largest arms sale in a decade:* Solomon, *Paul D. Wolfowitz*, 49.

324 *when Powell was around:* Dorrien, *Imperial Designs*, 38; Todd S. Purdum, "A Face Only a President Could Love," *Vanity Fair*, June 2006, http://www.vanityfair.com/politics/fea tures/2006/06/cheney200606; Hayes, *Cheney*, 6.

324 *"one might expect":* Woodward, *Plan of Attack*, 182–83.

324 *were being made someplace else:* DeYoung, *Soldier*, 335; Haass, *War of Necessity, War of Choice*, 213.

324 *sent to the vice president's staff:* Gellman, *The Angler*, 242–43, 376–77; Cullen Murphy and Todd Purdum, "Farewell to All That: An Oral History of the Bush White House," *Vanity Fair*, February 2009, http://www.vanityfair.com/politics/features/2009/02/bush-oral -history200909; Haass, *War of Necessity, War of Choice*, 220.

325 *suggested that he had lost a step:* DeYoung, *Soldier*, 335–36.

325 *September 10 cover story in* Time: Johanna McGeary, "Odd Man Out," *Time*, September 10, 2001, http://www.time.com/time/magazine/article/0,9171,1101010910–173441,00.html.

326 *"disconnected from larger purposes":* Gerson, *Heroic Conservatism*, 73. The author Robert Draper, who interviewed Bush at length, concurs with Gerson, noting that Bush "was at root a man who craved purpose—a sense of movement, of consequence. And things did not seem especially consequential in the summer of 2001." Robert Draper, *Dead Certain: The Presidency of George W. Bush* (New York: Free Press, 2008), 133.

326 *long stretches at his Texas ranch:* Fred Kaplan, *Daydream Believers: How a Few Grand Ideas Wrecked American Power* (Hoboken, NJ: Wiley, 2008), 114.

Chapter 17: The Opportunity

327 *"I will seize the opportunity":* Woodward, *Bush at War*, 32, 62, 282.

327 *traveled to the subcontinent at all:* "George W. Bush, Travelling Man?" *Washington Post*, November 28, 2007, A10.

327 *"I thought you said some band":* Mann, *Rise of the Vulcans*, 255.

327 *divided between Sunni and Shia:* Kaplan, *Daydream Believers*, 162; Weisberg, *The Bush Tragedy*, 207.

327 *"we wouldn't fight back":* Woodward, *Bush at War*, 38–39.

327 *"smash this myth":* Gerson, *Heroic Conservatism*, 82–83.

328 *a way to rally his troops:* In mid-November 2001, in a videotaped speech from Kandahar, Afghanistan, bin Laden used the "weak horse" line; http://www.npr.org/news/specials/re sponse/investigation/011213.binladen.transcript.html.

328 *"pounded sand":* Weisberg, *The Bush Tragedy*, 188; Feith, *War and Decision*, 12.

328 *"reflexive pullback":* Woodward, *Bush at War*, 20; Cockburn, *Rumsfeld*, 151.

328 *"hit a camel in the butt":* Cannon and Cannon, *Reagan's Disciple*, 184.

328 *comparing the United States to ancient Rome:* Mann, *Rise of the Vulcans*, 215.

328 *his efforts at producing WMD:* Ricks, *Fiasco*, 19, 21; Gordon and Trainor, *Cobra II*, 14.

329 *had shown in World War II:* Judy Keen, "Same President, Different Man in Oval Office," *USA Today,* October 29, 2001, http://www.usatoday.com/news/sept11/2001/10/29/bushmood -usat.htm; Woodward, *Bush at War,* 205.

329 *"our mission and our moment":* George W. Bush, "Address to a Joint Session of Congress and the American People," September 20, 2001, http://georgewbush-whitehouse.archives.gov/ news/releases/2001/09/20010920–8.html.

329 *"real calling with real heroism":* Matthew Scully, "Present at the Creation," *Atlantic,* September 2007, http://www.theatlantic.com/doc/200709/michael-gerson.

329 *the regimes that harbored them:* DeYoung, *Soldier,* 338, 346–47; Woodward, *Bush at War,* 31–32.

330 *"let Mr. Wolfowitz speak for himself":* Woodward, *Bush at War,* 43, 60–61; Mann, *Rise of the Vulcans,* 302.

330 *the 1995 bombing in Oklahoma City:* Sam Tanenhaus, "Bush's Brain Trust," *Vanity Fair,* July 2003, 114.

330 *barely anything worthwhile to bomb:* Woodward, *Bush at War,* 33, 83; Richard Clarke, *Against All Enemies: Inside America's War on Terror* (New York: Free Press, 2004), 31; Suskind, *The Price of Loyalty,* 187; Haass, *War of Necessity, War of Choice,* 196, 235.

330 *attacking Afghanistan didn't do that:* Feith, *War and Decision,* 82.

330 *Chiefs of Staff chairman Henry Shelton:* Woodward, *Bush at War,* 61.

331 *but he knew he couldn't prove it:* Woodward, *Bush at War,* 99.

331 *cronies was not big enough:* Gerson, *Heroic Conservatism,* 122.

331 *"and can move forward":* Woodward, *Bush at War,* 43.

331 *no U.S. ground forces at all:* Woodward, *Bush at War,* 79–80.

331 *"guerrilla commander's fantasy":* "Rendezvous with Afghanistan," *New York Times,* September 14, 2001, A26; "War Without Illusions," *New York Times,* September 15, 2001, A22.

331 *"create more instability":* Charles M. Sennott, "Allies Caution Bush Against Military 'Trap,' " *Boston Globe,* September 18, 2001, A21.

331 *"get the hell kicked out of you":* Woodward, *Bush at War,* 103.

331 *bin Laden, not overthrow them:* Mike Allen and Alan Sipress, "Attacks Refocus the White House on How to Fight Terrorism," *Washington Post,* September 26, 2001, A03; Feith, *War and Decision,* 78–80.

331 *getting mired in civil war:* Woodward, *Bush at War,* 123, 124.

332 *wrote Charles Krauthammer in late September:* Charles Krauthammer, "The War: A Road Map," *Washington Post,* September 28, 2001, A39.

332 *"and demand that Powell follow?":* William Kristol, "Bush vs. Powell," *Washington Post,* September 25, 2001, A23.

332 *"this is a change from the past":* Woodward, *Bush at War,* 98.

332 *defeat the Taliban on the ground:* Woodward, *Bush at War,* 314; Suskind, *The One Percent Doctrine,* 79.

333 *fighters onto Afghan soil:* Molly Moore and Kamran Khan, "Slain Rebel's Brief but Disastrous Foray," *Washington Post,* October 28, 2001, A20.

333 *news analysis in the* New York Times: R. W. Apple, Jr., "A Military Quagmire Remembered: Afghanistan as Vietnam," *New York Times,* October 31, 2001.

333 *"who have been there for 5,000 years":* Woodward, *Bush at War,* 275, 291.

333 *suggesting that they wanted him to fail:* Woodward, *Bush at War,* 262.

333 *demanded that Bush send in the army:* Fred Barnes, "Bush Only Needs to Do One Thing; Win the War," *Weekly Standard,* November 5, 2001, 12.

333 *Cheney said he had absolute faith:* Hayes, *Cheney,* 360.

333 *tried for years to hit, but never could:* Woodward, *Bush at War,* 312.

333 *the Taliban fighters were dead:* Kaplan, *Daydream Believers,* 35–36.

334 *its share of the country from 15 percent to 50 percent:* Woodward, *Bush at War,* 312.

334 *to attend school:* Ilana Ozernoy, "Liberation Day," *U.S. News & World Report,* November 26, 2001, 30; Nancy Gibbs, "Blood and Joy," *Time,* November 26, 2001, 28; Anthony Davis, "Eyewitness to a Sudden and Bloody Liberation," *Time,* November 26, 2001, 58.

334 *Baltimore, and Boston, a democrat:* Christiane Amanpour and Andrea Koppel, "Hamid Karzai No Stranger to Leadership," October 10, 2002, http://archives.cnn.com/2001/WORLD/ asiapcf/central/12/21/ret.karzai.profile/.

334 *female member of the new Afghan cabinet:* George W. Bush, "President Bush's State of the Union Address to Congress and the Nation," *New York Times,* January 30, 2002, http://www.nytimes.com/2002/01/30/us/state-union-president-bush-s-state-union-address-congress-nation.html.

334 *Special Forces troops and 110 CIA agents:* Woodward, *Bush at War,* 314.

334 *less than a single B-2 bomber:* Steven M. Kosiak, "Estimated Cost of Operation Enduring Freedom: The First Two Months," Center for Strategic and Budgetary Assessments, December 7, 2001, available at http://www.csbaonline.org; Federation of American Scientists, "B-2 Spirit," Nuclear Information Project, November 30, 1999, http://www.fas.org/nuke/guide/usa/bomber/b-2.htm.

334 *"virtually no cost in casualties":* Stanley Kurtz, "Push-Button Warriors," *National Review Online,* November 29, 2001, http://www.nationalreview.com/contributors/kurtz112901.shtml.

335 *in the former U.S.S.R.:* Andrew Bacevich, *The Limits of Power: The End of American Exceptionalism* (New York: Henry Holt, 2008), 47.

335 *"wrong about virtually everything":* Joe Loconte, "Rumsfeld's Just War; Generals Meet Theologians at the Pentagon," *Weekly Standard,* December 24, 2001, 13.

335 *"all together now—QUAGMIRE!":* Woodward, *Plan of Attack,* 37.

335 *after which Bush dubbed him "Rumstud":* Mann, *Rise of the Vulcans,* 307.

335 *87 percent of Americans:* Gallup Organization, "Presidential Job Approval in Depth," http://www.gallup.com/poll/116500/Presidential-Approval-Ratings-George-Bush.aspx.

335 *congressional and journalistic recrimination:* Mann, *Rise of the Vulcans,* 72.

336 *should not enjoy the protections of the Geneva Convention:* DeYoung, *Soldier,* 365–72; Gellman, *The Angler,* 170.

336 *"activities of the executive branch":* Gellman, *The Angler,* 106.

336 *across the border into Pakistan:* Mann, *Rise of the Vulcans,* 308.

336 *"perhaps beyond precedent":* Gerson, *Heroic Conservatism,* 122.

336 *"They're on a roll":* Seymour M. Hersh, "The Iraq Hawks: Can Their War Plan Work?" *New Yorker,* December 24, 2001, 58.

Chapter 18: The Romantic Bully

337 *motive and the means? Saddam!:* For an example of this reasoning, see Feith, *War and Decision,* 504.

337 *"likely make a different calculation":* Robert Kagan, "Power and Weakness," *Policy Review* 113 (June–July 2002), http://www.hoover.org/publications/policyreview/3460246.html.

338 *on October 9, he was rebuffed:* Woodward, *Bush at War,* 216.

338 *strategy for toppling Saddam:* Woodward, *Plan of Attack,* 1–4.

338 *Afghanistan to the Persian Gulf:* Packer, *The Assassin's Gate,* 45; Barton Gellman and Dafna Linzer, "Afghanistan, Iraq: Two Wars Collide," *Washington Post,* October 22, 2004, A1.

338 *turned to deficit:* U.S. Office of Management and Budget, *FY2008 Budget Request,* http://www.whitehouse.gov/omb/budget/fy2008/pdf/hist.pdf.

338 *"proved deficits don't matter":* Suskind, *The Price of Loyalty,* 291.

339 *and then again in 2003:* Glenn Kessler and Juliet Eilperin, "In House, a Magic Balancing Act," *Washington Post,* March 14, 2002, A7; Jonathan Weisman, "Late Deals Got Tax Cut Done," *Washington Post,* May 30, 2003, A5.

339 *he was reprimanded; then fired:* Packer, *The Assassin's Gate,* 116.

339 *less than 1 percent of that:* Kosiak, "Estimated Cost of Operation Enduring Freedom"; Murphy and Purdum, "Farewell to All That."

339 *nine months to fully deploy:* Woodward, *Bush at War,* 43; Packer, *The Assassin's Gate,* 118; Kaplan, *Daydream Believers,* 39–40.

339 *conformist, and flat-out dumb:* Cockburn, *Rumsfeld,* 153.

339 The Nutty Professor: Kaplan, *Daydream Believers*, 40–41; Gordon and Trainor, *Cobra II*, 27–28.

339 *"learned coming out of Afghanistan"*: Gordon and Trainor, *Cobra II*, 32; Woodward, *Plan of Attack*, 41.

339 *troop number down by almost a third:* Gordon and Trainor, *Cobra II*, 54–55.

340 *"has undertaken in the post–cold war era"*: Lawrence F. Kaplan and William Kristol, *The War over Iraq: Saddam's Tyranny and America's Mission* (San Francisco: Encounter, 2003), 117.

340 *relationship between a batterer and his spouse:* William Kristol, "From Truth to Deception," *Washington Post*, October 12, 2002, A31.

340 *simply buried his head in his arms:* Woodward, *State of Denial*, 19, 72; Woodward, *Bush at War*, 23, 251.

340 *"you'd find the vice president there"*: Gellman, *The Angler*, 161–62. In his memoir, Douglas Feith calls Rumsfeld "thoroughly antisentimental." Feith, *War and Decision*, 75.

341 *"truly enjoys getting people to knuckle under"*: Suskind, *The One Percent Doctrine*, 215. Robert Draper recounts a similar story of bullying, this one from Bush's time as Texas governor. In a meeting of economic advisers, Karl Rove had been speaking at great length, annoying Bush with his self-importance. "Karl," Bush suddenly yelled. "Hang up my jacket." The room fell silent, and Rove did as he was told. Draper, *Dead Certain*, 102.

341 *they did not openly weep:* Gerson, *Heroic Conservatism*, 204.

341 *But Bush did both, often:* Robert Draper, "The Prez & I," *GQ Online,* http://men.style.com/gq/features/full?id=content_7778.

341 *"complete trash, a horrible evil person"*: Draper, *Dead Certain*, 51.

341 *often broke down:* Cannon and Cannon, *Reagan's Disciple*, 195; Woodward, *State of Denial*, 270.

341 *"region of peaceful democracies"*: Scott McLellan, *What Happened: Inside the Bush White House and Washington's Culture of Deception* (New York: PublicAffairs, 2008), xii–xiii.

341 *"It's what's driving him"*: Woodward, *Plan of Attack*, 412.

341 *"he wants to talk about"*: Fred Barnes, "Bush Zeroes In," *Weekly Standard*, February 10, 2003.

342 *was the story of his life:* Gerson makes this point explicitly, writing that Bush was "convinced that societies are capable of hopeful change because individuals are capable of hopeful change, based on his own experience." Gerson, *Heroic Conservatism*, 51.

342 *untroubled by original sin:* Gerson, *Heroic Conservatism*, 51, 99.

342 *"hatred and the tactics of terror"*: George W. Bush, "President Discusses the Future of Iraq," February 26, 2003, http://georgewbush-whitehouse.archives.gov/news/releases/2003/02/print/20030226–11.html.

343 *Bush told cadets at West Point:* George W. Bush, "President Bush Delivers Graduation Speech at West Point," June 1, 2002, http://georgewbush-whitehouse.archives.gov/news/releases/2002/06/print/20020601–3.html.

343 *"That great struggle is over"*: White House, "The National Security Strategy of the United States," September 2002, 1, http://georgewbush-whitehouse.archives.gov/nsc/nss/2002/nss1.html.

344 *"it's a blueprint, a model"*: Draper, *Dead Certain*, 188.

344 *by which he meant the Iraqi National Congress:* House Budget Committee, *Hearing on FY2004 Defense Budget Request*, 108th Cong., 1st sess., 2003.

345 *soon after Saddam fell:* Gordon and Trainor, *Cobra II*, 121–22.

345 *"became an occupying power"*: Feith, *War and Decision*, 403.

345 *Bush's post-9/11 thinking than any other adviser:* Fred Barnes, *Rebel-in-Chief: Inside the Bold and Controversial Presidency of George W. Bush* (New York: Crown Forum, 2006), 48, 61.

345 *"remove the shackles on democracy"*: Mark Bowden, "Wolfowitz: The Exit Interviews," *Atlantic*, July/August 2005, http://www.theatlantic.com/doc/200507/bowden.

345 *You break it, you own it:* Cannon and Cannon, *Reagan's Disciple*, 196; DeYoung, *Soldier*, 401–2.

345 *and driving away business:* Helen Huntley, "Rule That Isn't Upsets Pottery Barn," *St. Petersburg Times*, April 20, 2004, http://www.sptimes.com/2004/04/20/Business/Rule_that_isn_t_its_r.shtml.

346 *That's where the power lay:* DeYoung, *Soldier*, 392.

346 *he was sure that Cheney wasn't:* DeYoung, *Soldier*, 417.

346 *leave America and Iraq worse off:* In January 2002, Kristol and Kagan wrote that "it is almost impossible to imagine any outcome for the world both plausible and worse than the disease of Saddam with weapons of mass destruction." Robert Kagan and William Kristol, "What to Do About Iraq," *Weekly Standard*, January 21, 2002, 23. "That things might be worse without him [Saddam] is of course a possibility," wrote Kristol and Lawrence Kaplan in their 2003 book advocating war. "But given the status quo in Iraq, it is difficult to imagine how." Kristol and Kaplan, *The War over Iraq*, 96. "It's hard to imagine how the alternative [to Saddam] could possibly be worse," wrote Max Boot in the *Standard*. Max Boot, "The False Allure of 'Stability,'" *Weekly Standard*, December 9, 2002. And in slapping down Shinseki in February 2003, Wolfowitz declared, "It is hard to conceive that it would take more forces to provide stability in post-Saddam Iraq than it would take to conduct the war itself and to secure the surrender of Saddam's security forces and his army—hard to imagine." Ricks, *Fiasco*, 97–98; House Budget Committee, *Hearing on FY2004 Defense Budget Request*, 108th Cong., 1st sess., 2003.

346 *everything that might go wrong:* Woodward, *Bush at War*, 334.

346 *"There will be civil disorder":* DeYoung, *Soldier*, 401–2.

346 *impact on America's Arab allies:* Woodward, *Plan of Attack*, 149–51; Gordon and Trainor, *Cobra II*, 81.

346 *it just wasn't in his nature:* DeYoung, *Soldier*, 509.

347 *voted yes:* Cannon and Cannon, *Reagan's Disciple*, 93.

347 *he mentioned Vietnam only once: Authorization of the Use of United States Armed Forces Against Iraq*, 107th Cong., 2nd sess., Congressional Record 148 (October 9, 2002), S10170–S10175.

347 *56 percent of older Senate Democrats:* "Final Vote Results for Roll Call 455," H.J. Res. 114, October 10, 2002, http://clerk.house.gov/evs/2002/roll455.xml; "U.S. Senate Roll Call Votes 107th Congress—2nd Session," H.J. Res. 114, October 11, 2002; *Biographical Directory of the United States Congress,* http://bioguide.congress.gov/biosearch/biosearch.asp.

348 *the very things liberals loved:* For a good example of this argument, see David Talbot, "The Making of a Hawk," Salon.com, January 3, 2002, http://www.salon.com/books/feature/2002/01/03/hawk/print.html.

348 *it was the great liberal duty of the age:* Peter Beinart, "Ism Schism," *New Republic*, September 25, 2006, 6.

348 *when Fukuyama declared that democracy had won:* Paul Berman, *Terror and Liberalism* (New York: Norton, 2004), 176.

349 *"this notion did pretty much explode":* Berman, *Terror and Liberalism*, 190.

349 *the Taliban millions might die:* Noam Chomsky, interview with Svetlana Vukovic and Svetlana Lukic, Radio B92, Belgrade, September 19, 2001, http://www.b92.net/intervju/eng/2001/0919-chomsky.phtml.

350 *another influential liberal hawk:* Michael Ignatieff, "The American Empire; The Burden," *New York Times Magazine*, January 5, 2003, http://www.nytimes.com/2003/01/05/magazine/the-american-empire-the-burden.html.

350 *"Kosovo and not do it in Iraq":* Former Clinton administration official, phone interview with the author, June 27, 2008.

350 *would bring not freedom, but chaos:* Kirkpatrick, *Making War to Keep Peace*, 279.

350 *had lost the confidence of her convictions:* Jacquelyn Hardy, phone interview with the author, July 10, 2008.

350 *under international law:* Kirkpatrick, *Making War to Keep Peace*, 281.

350 *turn George W. Bush into Napoleon:* Washington, D.C., writer, interview with the author, June 9, 2008.

351 *critical of what the* Standard *wrote:* Washington, D.C., writer (different from above), interview with the author, July 15, 2008.

351 *denounced the impending war:* Brent Scowcroft, "Don't Attack Saddam," *Wall Street Journal*,

August 15, 2002, http://www.opinionjournal.com/editorial/feature.html?id=110002133; Suskind, *The One Percent Doctrine*, 167.

351 *Bush himself stayed silent:* Haass, *War of Necessity, War of Choice*, 217.

351 *"He doesn't think he should unless he's asked":* Woodward, *State of Denial*, 114.

351 *didn't have the latest intelligence:* "Text of Bush Interview," Fox News, September 22, 2003, http://www.foxnews.com/story/0,2933,98006,00.html.

351 *America pursue regime change in Iraq:* "Letter to President Clinton on Iraq," Project for a New American Century, January 26, 1998, http://www.newamericancentury.org/iraqclintonletter.htm.

351 *would not fertilize the soil:* Francis Fukuyama, "American Conservatism: Beyond Our Shores," *Wall Street Journal*, December 24, 2002, A10.

352 *but history could not be rushed:* Robert S. Boynton, "The Neocon Who Isn't," *American Prospect Online*, October 5, 2005, http://www.prospect.org/cs/articles?articleId=10304.

352 *Fukuyama's presentation again:* Francis Fukuyama, interview with the author, July 15, 2008.

352 *"I've been put under over here":* Draper, *Dead Certain*, 186.

352 *"his permission":* Woodward, *Plan of Attack*, 269–74.

352 *"the credibility to do it":* Woodward, *Plan of Attack*, 291.

352 *"you can afford to lose a few points":* DeYoung, *Soldier*, 441, 448.

353 *Iraqi intelligence officials in Prague:* Woodward, *Plan of Attack*, 289–91; DeYoung, *Soldier*, 436–41.

353 *"leery of our own presentation":* DeYoung, *Soldier*, 444–45; Murphy and Purdum, "Farewell to All That."

353 *"conclusions based on solid intelligence":* DeYoung, *Soldier*, 449.

353 *UN weapons inspectors began refuting it almost instantly:* Ian Fisher, "Reporters on Ground Get Iraqi Rebuttal to Satellite Photos," *New York Times*, February 8, 2003, A8; Ian Fisher, "U.N.'s Chief Inspectors Will Press Iraq for Hard Data That Could Avert War," *New York Times*, February 9, 2003, A14; Peter Slevin and Colum Lynch, "U.S. Meets New Resistance at U.N.," *Washington Post*, A1; Glenn Kessler and Colum Lynch, "Blix's Report Deepens UN Rift over Iraq," *Washington Post*, March 9, 2003, A1.

354 *I don't want one, Wilkerson explained:* DeYoung, *Soldier*, 450–52.

354 *The invasion of Iraq had begun:* Woodward, *Bush at War*, 356.

354 *"lingers and plagues the heart":* Schweizer and Schweizer, *The Bushes*, 398–99.

354 *he was sleeping well:* Woodward, *State of Denial*, 155.

354 *"Not one doubt":* Woodward, *Bush at War*, 256.

354 *"Feels good":* Weisberg, *The Bush Tragedy*, 70.

354 *things were going dangerously wrong:* Gordon and Trainor, *Cobra II*, 273–74; 346–61.

354 *If this wasn't liberation, nothing was:* Gordon and Trainor, *Cobra II*, 490; Alfonsi, *Circle in the Sand*, 401.

354 *fewer than in the Gulf War:* "Iraq Coalition Casualty Count—Deaths by Year and Month," March-April 2003, http://www.icasualties.org/Iraq/ByMonth.aspx; Department of Veterans Affairs, "Fact Sheet: America's Wars," May 2008, http://www1.va.gov/opa/fact/amwars.asp.

355 *"regime's ill-gotten gains?":* Max Boot, "Good News, Operation Iraqi Freedom Went About as Well as Anyone Could Have Hoped. Why Is the Media So Glum?" *Weekly Standard*, April 15, 2003, http://www.weeklystandard.com/Content/Public/Articles/000/000/002/551wqxuo.asp.

355 *declared one article in the* Standard: Thomas Donnelly, "Lessons of a Three Week War," *Weekly Standard*, April 28, 2003, http://www.aei.org/article/17006.

355 *"without widespread loss of life":* David Brooks, "The Phony Debate," *Weekly Standard*, March 31, 2008.

355 *"Homer had given his gods":* Andrew J. Bacevich, *The New American Militarism: How Americans Are Seduced by War* (Oxford: Oxford University Press, 2005), 22.

355 *"food and water and air":* "President Bush Announces Major Combat Operations in Iraq Have Ended," Remarks by the President from the U.S.S. *Abraham Lincoln*, Released by the Office of the Press Secretary, May 1, 2003, http://georgewbush-whitehouse.archives.gov/news/releases/2003/05/20030501–15.html.

355 *"United States is on earth to achieve":* David Brooks, "Today's Progressive Spirit: The Scenes in Baghdad Flow from Understandings Realized at the American Founding," *Weekly Standard*, April 9, 2003, http://www.weeklystandard.com/Content/Public/Articles/000/000/002/519jigox.asp.

355 *and more legitimacy, without it:* Wes Vernon, "Kristol: U.N. Has Gone from 'Useless' to 'Harmful,' "March 11, 2003, http://archive.newsmax.com/archives/articles/2003/3/11/110420.shtml.

355 *France and Germany too much say:* Fred Barnes, "The Tempting of the President," *Weekly Standard*, April 21, 2003, http://www.weeklystandard.com/Content/Public/Articles/000/000/002/543ayyjy.asp.

355 *had made itself worthless:* Charles Krauthammer, "A Costly Charade at the U.N.," *Washington Post*, February 28, 2003, A23.

355 *spiked back up to 77 percent:* Ricks, *Fiasco*, 134.

356 *greatest basketball player of all time:* Linton Weeks, "23 Skiddoo; Michael Jordan Flies Down the Court for the Last Time at MCI Center," *Washington Post*, April 15, 2003, C1.

356 *million more in speaking fees:* Woodward, *Plan of Attack*, 413.

356 *"to undermine the President's policies":* Newt Gingrich, "Transforming the State Department: The Next Challenge for the Bush Administration," American Enterprise Institute, Washington, D.C., April 22, 2003, http://www.aei.org/publications/filter.all,pubID.16992/pub_detail.asp; Woodward, *State of Denial*, 251.

356 *declined to defend his secretary of state:* DeYoung, *Soldier*, 466–68.

356 *someone mentioned Powell, and they all laughed:* Woodward, *Plan of Attack*, 409–12.

Chapter 19: I'm Delighted to See Mr. Bourne

357 *"vicious dictator, and they're free!":* Department of Defense, "DoD News Briefing by Secretary Rumsfeld and General Myers," April 11, 2003, http://www.defenselink.mil/transcripts/transcript.aspx?transcriptid=2367; Woodward, *State of Denial*, 164.

358 *cost of the rampage: $12 billion:* John Kifner and John F. Burns, "As Tanks Move In, Young Iraqis Trek Out and Take Anything Not Fastened Down," *New York Times*, April 10, 2003; John F. Burns, "Looting and a Suicide Attack as Chaos Grows in Baghdad," *New York Times*, April 11, 2003; Larry Diamond, *Squandered Victory: The American Occupation and the Bungled Effort to Bring Democracy to Iraq* (New York: Times Books, 2005), 282; Woodward, *State of Denial*, 179, 183; Rajiv Chandrasekaran, *Imperial Life in the Emerald City: Inside Iraq's Green Zone* (New York: Random House, 2006), 40.

358 *"This is not my job":* Woodward, *State of Denial*, 179.

358 *picked Iraq's government clean:* James Fallows, "Blind into Baghdad," *Atlantic,* January/February 2004, http://www.theatlantic.com/doc/200401/fallows.

358 *"is an unbelievable mess":* Gordon and Trainor, *Cobra II*, 541.

358 *settled on "criminal negligence":* Diamond, *Squandered Victory*, 292.

359 *the National Guard and Reserves:* Department of Defense, "Defense Manpower Requirements Report," June 1999, http://www.defenselink.mil/prhome/docs/fy2000.pdf; Lawrence Kapp, "Reserve Component Personnel Issues," *CRS Report for Congress*, January 18, 2006, 4; Stephen Daggett and Amy Belasco, "Defense Budget for FY2003: Data Summary," *CRS Report for Congress*, March 29, 2002, 15.

359 *the entire U.S. diplomatic corps:* Cockburn, *Rumsfeld*, 106–7; Gordon and Trainor, *Cobra II*, 545–47; Linda Robinson, *Tell Me How This Ends: General David Petraeus and the Search for a Way Out of Iraq* (New York: PublicAffairs, 2008), 4; "Highlighting Public Diplomacy Needs," *American Academy of Diplomacy Newsletter* no. 59 (October 2003), http://www.academyofdiplomacy.org/publications/newsletter_archive/newsletter_issue_59.htm.

359 *responded with a blank stare:* Diamond, *Squandered Victory*, 35.

360 *United States playing only a supporting role:* Gordon and Trainor, *Cobra II*, 121–22.

360 *"the Gaullists would have been neutered":* Ricks, *Fiasco*, 124; Bradley Graham, "US Airlifts Iraqi Exile Force for Duties near Nasiriyah," *Washington Post*, April 7, 2003, A1; Linda Rob-

inson and Kevin Whitelaw, "Deploying the 'Free Iraqi Forces,'" *U.S. News & World Report*, April 7, 2003, http://www.usnews.com/usnews/news/iraq/articles/fiff030407.htm; Feith, *War and Decision*, 253.

360 *fighting force had already existed:* Military analyst Stephen Biddle made exactly that point in an analysis of America's overthrow of the Taliban. Noting the crucial role that the Northern Alliance played, he wrote that "we should be wary of suggestions that precision weapons, with or without special operations forces to direct them, have so revolutionized warfare that traditional ground forces are now superseded." Stephen Biddle, *Afghanistan and the Future of Warfare: Implications for Army and Defense Policy* (Carlisle, PA: Strategic Studies Institute, 2002), ix.

360 *force dispatched to Iraq totaled 73:* Gordon and Trainor, *Cobra II*, 122–23, 360; Bob Woodward, *The War Within: A Secret White House History* (New York: Simon & Schuster, 2008), 49; Graham, "US Airlifts Iraqi Exile Force for Duties near Nasiriyah"; Robinson and Whitelaw, "Deploying the 'Free Iraqi Forces.'"

360 *the nucleus of a post-Saddam regime:* Graham, "U.S. Airlifts Iraqi Exile Force for Duties Near Nasariyah"; Marc Santora with Patrick E. Tyler, "Pledge Made to Democracy by Exiles, Sheiks and Clerics," *New York Times*, April 16, 2003, A1.

360 *secular elite from which he hailed:* Dexter Filkins, "Where Plan A Left Ahmad Chalabi," *New York Times Magazine*, November 5, 2006, http://www.nytimes.com/2006/11/05/magazine/05CHALABI.html.

361 *mansions favored by one of Saddam's sons:* Gellman, *The Angler*, 247.

361 *sold overseas for a fat profit:* Jane Mayer, "The Manipulator," *New Yorker*, June 7, 2004, http://www.newyorker.com/archive/2004/06/07/040607fa_fact1.

361 *stealing reconstruction funds:* Mayer, "The Manipulator"; Ricks, *Fiasco*, 124.

361 *classified U.S. intelligence to Iran:* Scott Wilson, "U.S. Aids Raid on Home of Chalabi," *Washington Post*, May 21, 2004, A1.

361 *any Iraqi politician, including Saddam:* ABC News, "Poll: Iraq—Where Things Stand," March 15, 2004, http://abcnews.go.com/images/pdf/949a1IraqPoll.pdf.

361 *even more ideologically self-assured:* Cannon and Cannon, *Reagan's Disciple*, 199.

361 *pro-American democracy trapped inside:* Gordon and Trainor, *Cobra II*, 545.

361 *boldly tore off the remaining scab:* Diamond, *Squandered Victory*, 349, n. 38; Chandrasekaran, *Imperial Life in the Emerald City*, 63.

361 *holding jobs in Iraq's new government:* Sharon Otterman, "Iraq: Debaathification," *CFR Backgrounder*, April 7, 2005, http://www.cfr.org/publication/7853/iraq.html; Gordon, *Cobra II*, 549; Ricks, *Fiasco*, 158–59.

361 *interior ministry, and presidential guards:* In fact, of Iraq's 80,000 commissioned officers, only 8,000 were members of the Baath Party. Walter Slocombe, "Status of Rebuilding and Training the Iraqi Army," press conference, State Department, Washington, D.C., September 17, 2003, http://2002–2009-fpc.state.gov/24230.htm.

361 *backbone of the Iraqi economy:* Naomi Klein, "Baghdad: Year Zero," *Harper's*, September 2004, http://www.harpers.org/archive/2004/09/0080197; Roger Burbach and Jim Tarbell, *Imperial Overstretch: George W. Bush and the Hubris of Empire* (London: Zed, 2004), 186.

362 *improvised explosive devices (IEDs):* Woodward, *State of Denial*, 202, 205–6, 211; Ricks, *Fiasco*, 164.

362 *"Who's Your Baghdaddy?":* Chandrasekaran, *Imperial Life in the Emerald City*, 4, 14–15, 18, 130.

362 *fear they would poison the food:* Chandrasekaran, *Imperial Life in the Emerald City*, 9–10.

363 *seemed turned upside down:* Ricks, *Fiasco*, 206.

363 *resembles the word for "fuck":* Gordon and Trainor, *Cobra II*, 554; Feith, *War and Decision*, 434.

363 *equivalent of a raised middle finger:* Woodward, *State of Denial*, 290.

363 *"is the sound of freedom":* Chandrasekaran, *Imperial Life in the Emerald City*, 127.

363 *a more confrontational ring:* Anthony Shadid, *Night Draws Near: Iraq's People in the Shadow of America's War* (New York: Picador, 2006), 15.

363 *in an e-mail to friends:* Draper, *Dead Certain*, 204.

363 *"really understood about Iraq":* Shadid, *Night Draws Near*, 8.

363 *to roughly one thousand per month:* Hayes, *Cheney*, 426; Woodward, *State of Denial*, 261.

363 *began carrying guns to their offices:* Chandrasekaran, *Imperial Life in the Emerald City*, 181–82.

363 *more dangerous than under Saddam:* Diamond, *Squandered Victory*, 26.

363 *raids against suspected insurgents:* Ricks, *Fiasco*, 234, 236, 301–2.

363 *arrested, injured, or killed:* The International Red Cross report described U.S. raids this way: "Arresting authorities entered houses usually after dark, breaking down doors, waking up residents roughly, yelling orders, forcing family members into one room under military guard while searching the rest of the house and further breaking doors, cabinets and other property. They arrested suspects, tying their hands in the back with flexicuffs, hooding them, and taking them away. Sometimes they arrested all adult males present in a house, including elderly, handicapped and sick people. Treatment often included pushing people around, insulting, taking aim with rifles, punching and kicking and sticking with rifles. Individuals were often led away in whatever they happened to be wearing at the time of arrest—sometimes in pyjamas [*sic*] or underwear—and were denied the opportunity to gather a few essential belongings, such as clothing, hygiene items, medicine or eyeglasses." Quoted in Ricks, *Fiasco*, 235.

364 *turned the stomach of the world:* Ricks, *Fiasco*, 195–200, 238–40, 283–92.

364 *"to rob Iraq's oil":* Diamond, *Squandered Victory*, 25–26.

364 *slum known as Sadr City:* Ricks, *Fiasco*, 335–38.

364 *roughly three thousand a month:* Woodward, *State of Denial*, 368.

364 *before going out on patrol:* Ricks, *Fiasco*, 350.

365 *"this job will run you all over town":* Draper, *Dead Certain*, x, 357; Woodward, *State of Denial*, 490.

365 *as he did in late May 2003:* Woodward, *State of Denial*, 209.

365 *but as late as 2006:* Weisberg, *The Bush Tragedy*, 209.

365 *pushing to invade Panama:* Kaplan, *Daydream Believers*, 134.

365 *the last two paragraphs alone:* Woodward, *State of Denial*, 378.

365 *"ending tyranny in our world":* George W. Bush, "Second Inaugural Address," January 20, 2005, http://georgewbush-whitehouse.archives.gov/news/releases/2005/01/20050120–1.html.

366 *to elect a National Assembly:* Angus Reid Global Monitor, "Iraq: Election Tracker," Angus Reid, http://www.angus-reid.com/tracker/view/5143.

366 *brought democracy to Kyrgyzstan:* Karl Vick, "New Leadership Established in Kyrgyzstan," *Washington Post*, March 26, 2005, A8.

366 *overturn the election results by force:* David Rose, "The Gaza Bombshell," *Vanity Fair*, April 2008, http://www.vanityfair.com/politics/features/2008/04/gaza200804.

366 *in Africa, and across the globe:* Gary J. Bass, "Humanitarian Intervention in the 21st Century," *Tocqueville Review/La revue Tocqueville* 30, no. 1 (2009): 25; Fukuyama, "The End of History?" 17.

366 *"the end of the End of History":* Azar Gat, "The Return of Authoritarian Great Powers," *Foreign Affairs* 86, no. 4 (July/August 2007), available at http://www.foreignaffairs.org.

367 *without so much as a farewell party:* Joshua Muravchik, interview with the author, September 4, 2008.

367 *"institutions will be precarious at best":* Kirkpatrick, *Making War to Keep Peace*, 279.

367 *toward darkness as toward light:* Allan Gerson, "Postscript," in Kirkpatrick, *Making War to Keep Peace*, 308.

367 *"oversimplified view of human nature":* David Brooks, "The Jagged World," *New York Times*, September 3, 2006, http://select.nytimes.com/2006/09/03/opinion/03brooks.html.

367 *"war that began three years ago":* George F. Will, "Can We Make Iraq Democratic?" *City Journal*, Winter 2004, http://www.city-journal.org/html/14_1_can_we_make_iraq.html; George Will, "How Bush Can Try to Recover from Errors in Iraq War," *Chicago Sun-Times*, March 19, 2006, B7.

367 *quick to count dictatorship out:* Robert Kagan, *The Return of History and the End of Dreams* (New York: Knopf, 2008).

367 *"seemed that it has no future":* Robert Kagan, "End of Dreams, Return of History," *Policy*

Review 144 (August–September 2007), http://www.hoover.org/publications/policyre view/8552512.html.

368 *dead for more than eighty years:* Casey Blake, "Randolph Bourne's America," lecture, Columbia University, October 11, 2004, http://www.randolphbourne.columbia.edu/panel_1.pdf.

368 *founded the Randolph Bourne Institute:* See http://randolphbourne.org/.

368 *administration's wartime propaganda:* James Fallows, "We've Been Here Before," January 21, 2004, http://thebigstory.org/fallows.htm.

368 *Bourne's old professor, Charles Beard:* Andrew J. Bacevich, *American Empire: The Realities and Consequences of U.S. Diplomacy* (Cambridge, MA: Harvard University Press, 2002), 11–23.

369 *"path of organization to that end":* Robert Westbrook, "Randolph Bourne's America," lecture, Columbia University, October 11, 2004, http://www.randolphbourne.columbia.edu/panel_1.pdf.

369 *so he can seize Central Asia's oil:* Logan is actually the show's third president. Its second, Jack Keeler, decides not to seek a second term after being wounded in a terrorist attack. For a thoughtful discussion of Iraq's impact on Hollywood, see Ross Douthat, "The Return of the Paranoid Style," *Atlantic*, April 2008, http://www.theatlantic.com/doc/200804/iraq-movies.

370 *"authority was built on silly illusions":* Matt Stoller, "PDF2007: The Rise of the Netroots," *Personal Democracy Forum*, May 18, 2007, http://personaldemocracy.com/node/1432.

370 *incestuous relationship with the executive branch:* Amy Argetsinger and Roxanne Roberts, "White House Press Dinner: De-Wonked!" *Washington Post*, April 21, 2008, C3.

370 *torture "war on terror" suspects:* Eric Schmitt, "Senate Moves to Protect Military Prisoners Despite Veto Threat," *Washington Post*, October 6, 2005, A22; Gellman, *The Angler*, 357.

370 *"jump in with both feet":* Jim VandeHei and Charles Babington, "Newly Emboldened Congress Has Dogged Bush This Year," *Washington Post*, December 23, 2005, A5.

370 *at Guantánamo Bay, Cuba:* Charles Lane, "Justices Back Detainee Access to U.S. Courts; President's Powers Are Limited," *Washington Post*, June 29, 2004, A1.

370 *release information about Gitmo detainees:* Josh White, "Government Must Share All Evidence on Detainees," *Washington Post*, July 21, 2007, A2.

370 *challenge their detention in federal court:* Robert Barnes, "Justices Say Detainees Can Seek Release," *Washington Post*, June 13, 2008, A1.

370 *"extravagant, monarchial claims":* Gellman, *The Angler*, 355; Jonathan Mahler, "After the Imperial Presidency," *New York Times Magazine*, November 9, 2008, http://www.nytimes.com/2008/11/09/magazine/09power-t.html.

371 *cook agreed to prepare Rumsfeld's meal:* Thomas E. Ricks, *The Gamble: General David Petraeus and the American Military Adventure in Iraq, 2006–2008* (New York: Penguin, 2009), 77.

371 *after he was tried for murder:* Hayes, *Cheney*, 504; Richard Morin, "18%? Just How Low Is," *Washington Post*, March 5, 2006, B3.

371 *substantially reducing his influence:* Gellman, *The Angler*, 364.

371 *now less fearsome than pitiable:* Todd S. Purdum, "A Face Only a President Could Love," *Vanity Fair*, June 2006, http://www.vanityfair.com/politics/features/2006/06/cheney200606; Gellman, *The Angler*, 389.

371 *most disliked president on record:* See Kathy Frankovic, "Bush's Popularity Reaches Historic Lows," *CBS News*, January 15, 2009, http://www.cbsnews.com/stories/2009/01/15/opinion/pollpositions/main4724068.shtml; Frank Newport, "Bush Job Approval at 28%, Lowest of Administration," Gallup News, April 11, 2008, http://www.gallup.com/poll/106426/Bush-Job-Approval-28-Lowest-Administration.aspx.

371 *turned the public against them:* Cannon and Cannon, *Reagan's Disciple*, 292.

371 *pointing to their disfigured relative:* Woodward, *State of Denial*, 437–38.

371 *"This is not your daughter!":* Draper, *Dead Certain*, 350.

371 *attend his party's presidential convention:* Murphy and Purdum, "Farewell to All That."

371 *even by Iraqi standards:* Ellen Knickmeyer and K. I. Ibrahim, "Bombing Shatters Mosque in Iraq," *Washington Post*, February 23, 2006, A1.

371 *drill holes in their heads:* Robinson, *Tell Me How This Ends,* 163.

371 *a family intervention:* Weisberg, *The Bush Tragedy,* 218.

372 *Laura, and Barney, the dog:* Todd S. Purdum, "Inside Bush's Bunker," *Vanity Fair,* October 2007, http://www.vanityfair.com/politics/features/2007/10/purdum200710.

373 *then dropped out of sight:* Sabrina Tavernise, "A Shiite Militia in Baghdad Sees Its Power Wane," *New York Times,* July 27, 2008.

373 *150 a week by spring 2009:* Michael E. O'Hanlon and Jason H. Campbell, "Iraq Index: Tracking Variables of Reconstruction and Security in Post-Saddam Iraq," Brookings Institution, August 20, 2009, http://www.brookings.edu/saban/~/media/Files/Centers/Saban/Iraq%20Index/index20090820.pdf.

373 *would still be fighting there in another six:* Ricks, *The Gamble,* 325.

373 *it had cost America $3 trillion:* Joseph E. Stiglitz and Linda J. Bilmes, "The $3 Trillion War," *Vanity Fair,* April 2008, http://www.vanityfair.com/politics/features/2008/04/stiglitz200804.

373 *Iraq had taken more than 4,000:* Iraq Coalition Casualty Count, http://www.icasualties.org/Iraq/index.aspx.

374 *sizable portions of neighboring Pakistan:* Kenneth Katzman, "Afghanistan: Post-War Governance, Security, and U.S. Policy," *CRS Report for Congress,* November 26, 2008, 22–24.

374 *were suddenly everywhere:* See, for instance, Peter Baker, "Could Afghanistan Become Obama's Vietnam?" *New York Times,* August 22, 2009, WK1; and Frank Rich, "Obama at the Precipice," *New York Times,* September 26, 2009, WK12.

374 *Afghanistan in the language of despair:* For instance, see Helene Cooper, "Obama's War: Fearing Another Quagmire in Afghanistan," *New York Times,* January 25, 2009, http://www.nytimes.com/2009/01/25/weekinreview/25cooper.html; Michael Crowley, "Obama vs. Osama: Has He Picked the Right War?" *New Republic,* December 24, 2008, http://www.tnr.com/article/obama-vs-osama.

374 *Petraeus called himself a "minimalist":* Ricks, *The Gamble,* 287.

374 *"less than the vision that drove it to Baghdad":* Ricks, *The Gamble,* 156.

374 *would never be vanquished:* As Robert Gates has said, "there has to be ultimately . . . reconciliation" with the Taliban. Kristin Roberts, "Pentagon Sees Reconciliation with Taliban, Not Qaeda," Reuters, October 9, 2009, http://www.reuters.com/article/id USTRE4987PH20081009.

374 *"Maliki intends to make himself a new dictator":* Kenneth M. Pollack, "The Battle for Baghdad," *National Interest,* September/October 2009, http://www.nationalinterest.org/Article.aspx?id=22018.

375 *"Gonna burst any day now?":* Draper, *Dead Certain,* 119.

375 *share of world GDP declined every year:* Roger Altman, "The Great Crash, 2008," *Foreign Affairs* (January/February 2009): 11.

375 *"hinges on the support of its creditors":* Brad W. Setser, "Sovereign Wealth and Sovereign Power," Council on Foreign Relations, Council Special Report No. 37, September 2008, 3–4.

376 *"Le laisser-faire, c'est fini":* Altman, "The Great Crash, 2008," 11.

376 *"Non-American world":* Parag Khanna, "Waving Goodbye to Hegemony," *New York Times Magazine,* January 27, 2008, http://www.nytimes.com/2008/01/27/magazine/27world-t.html.

376 *"more like a prophet":* Paul Starobin, "Beyond Hegemony," *National Journal,* December 1, 2006, http://www.nationaljournal.com/about/njweekly/stories/2006/1201nj1.htm.

376 *"ascendancy in the world":* Paul Starobin, *After America: Narratives for the Next Global Age* (New York: Viking, 2009), 6.

376 *among Chinese, 89 percent:* Pew Global Attitudes Project, "Confidence in Obama Lifts U.S. Image Around the World," July 23, 2009, http://pewglobal.org/reports/pdf/264.pdf; Paul Taylor, Cary Funk, and Peyton Craighill, "Once Again, the Future Ain't What It Used to Be," Pew Research Center, May 2, 2006, http://pewresearch.org/pubs/311/once-again-the-future-aint-what-it-used-to-be; Starobin, *After America,* 94.

377 *he was now content to die:* John Lukacs, *George Kennan: A Study in Character* (New Haven, CT: Yale University Press, 2007), 177.

377 *destroyed, but it never was:* Lukacs, *George Kennan*, 187.

Conclusion: The Beautiful Lie

378 *"intelligent, illuminating failures":* Steffens, *The Autobiography of Lincoln Steffens*, 788.

378 *"beyond its power to accomplish":* Arthur M. Schlesinger, Sr., "What Then Is the American, This New Man?" *American Historical Review* 48, no. 2 (1943): 244. Walter Russell Mead echoes the point in his excellent book, *God and Gold*: "Optimism is the default mode of Anglo-American historical thought. How could it not be? The Whig narrative teaches us plainly that God is on our side, and centuries of victorious experience and economic progress confirm that the message is right." Walter Russell Mead, *God and Gold: Britain, America and the Making of the Modern World* (New York: Knopf, 2007), 314.

381 *Our obligations exceed our power:* Lippmann, *U.S. Foreign Policy*, 5.

382 *harm American security more than they help:* See David Kilcullen and Andrew McDonald Exum, "Death from Above, Outrage from Below," *New York Times*, May 16, 2009.

383 *if they didn't get the bomb:* For a history of Iranian foreign policy since the Islamic revolution, which makes this point, see Ray Takeyh, *Guardians of the Revolution: Iran and the World in the Age of the Ayatollahs* (New York: Oxford University Press, 2009).

384 *democratic and self-governing in its domestic affairs:* On the superpowers and Finland, see Jussi Hanhimäki, *Containing Coexistence: America, Russia, and the "Finnish Solution"* (Kent, OH: Kent State University Press, 1997); and Jussi Hanhimäki, "Self-Restraint as Containment: United States' Economic Policy, Finland, and the Soviet Union, 1945–1953," *International History Review* 17, no. 2 (May 1995): 287–305.

385 *to protect our democracy so it can thrive:* Lippmann, *U.S. Foreign Policy*, iii.

385 *health care and education combined:* Anup Shaw, "World Military Spending," *Global Issues*, http://www.globalissues.org/article/75/world-military-spending#USMilitarySpending.

385 *than of other institutions of government:* Thomas Sander, "Trust Declining in All Institutions Other Than the Military," Social Capital Blog, http://socialcapital.wordpress .com/2009/06/08/trust-declining-in-all-institutions-other-than-the-military/.

385 *primary instrument for interacting with the world:* Bureau of Public Affairs, "International Affairs—FY 2010 Budget," U.S. Department of State, http://www.state.gov/r/pa/prs/ ps/2009/05/123160.htm; Department of Defense, "A New Era of Responsibility," http:// www.whitehouse.gov/omb/assets/fy2010_new_era/Department_of_Defense.pdf.

388 *"facts which momentarily imprison them":* Fox, *Reinhold Niebuhr*, 59.

390 *"traditional myths of indomitable optimism":* C. Vann Woodward, "A Second Look at the Theme of Irony," in *The Burden of Southern History*, ed. C. Vann Woodward (Baton Rouge: Louisiana State University Press, 1993), 214, 216.

INDEX